T0348416

Hilbert Spaces with Applications

Lokenath Debnath and Piotr Mikusiński

Hilbert Spaces with Applications

Third Edition

Lokenath Debnath
University of Texas—Pan American

Piotr Mikusiński
University of Central Florida

ELSEVIER
ACADEMIC
PRESS

AMSTERDAM • BOSTON • HEIDELBERG • LONDON • NEW YORK • OXFORD
• PARIS • SAN DIEGO • SAN FRANCISCO • SINGAPORE • SYDNEY • TOKYO

Elsevier Academic Press
30 Corporate Drive, Suite 400, Burlington, MA 01803, USA
525 B Street, Suite 1900, San Diego, California 92101-4495, USA
84 Theobald's Road, London WC1X 8RR, UK

This book is printed on acid-free paper. ⊚

Library of Congress Cataloging-in-Publication Data
Application Submitted

British Library Cataloguing in Publication Data
A catalogue record for this book is available from the British Library

ISBN 13: 978-0-12-2084386
ISBN 10: 0-12-208438-1

For all information on all Elsevier Academic Press publications
visit our Web site at www.books.elsevier.com

Working together to grow
libraries in developing countries

www.elsevier.com | www.bookaid.org | www.sabre.org

ELSEVIER   BOOK AID
            International   Sabre Foundation

This book is dedicated to the memory of our fathers:
JOGESH CHANDRA DEBNATH and JAN MIKUSIŃSKI

Contents

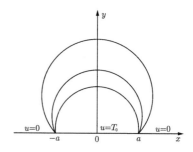

Preface
to the Third Edition

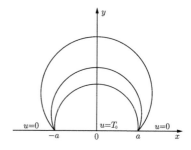

The previous two editions of our book were very well received. This new edition preserves the basic content and style of the earlier editions. It is a graduate-level text for students and a research reference for professionals in mathematics, science, and engineering. The theoretical foundations are presented in as simple a way as possible, but without sacrificing the mathematical rigor. In the part devoted to applications, we present a wide variety of topics, from classical applications to some recent developments. While the treatment of those applications is rather brief, our hope is that we present enough to stimulate interest that will encourage readers to further studies in those areas.

We have received various comments and suggestions from our colleagues, readers, and graduate students, from the United States and abroad. Those comments have been very helpful in writing this edition. We have made some additions and changes in order to modernize the contents. An effort to improve clarity of presentation and to correct a number of typographical errors was made. New examples and exercises were added. We have also taken the opportunity to entirely rewrite and reorganize several sections in an appropriate manner and to update the bibliography. Some of the major changes and additions include the following:

- Chapter 1 has been reorganized, some sections were combined, and the order of material has been modified.

- A complete characterization of finite dimensional normed spaces has been added.

- A new section on L^p spaces has been added in Chapter 2.

- The section on spectral properties of operators in Chapter 4 has been expanded.

- The presentation of the Fourier transform has been moved from Chapter 4 to Chapter 5.

- A new section on Sobolev spaces has been added to Chapter 6.

- Chapter 8 on wavelets and wavelet transforms has been revised and new material added, including a new section on orthonormal wavelets.

We would like to take this opportunity to thank all those who helped us improve the book by reading parts of the manuscript and sharing their comments with us, including Andras Balogh, Cezary Ferens, Ziad Musslimani, Zuhair Nashed, and Vladimir Varlamov. In spite of the best efforts of everyone involved, some typographical errors doubtless remain. Special thanks to June Wingler who helped us with the preparation of the LaTeX files. Finally, we wish to express our grateful thanks to Tom Singer, assistant editor, and staff of Elsevier Academic Press for their help and cooperation.

Lokenath Debnath, *University of Texas — Pan American*
Piotr Mikusiński, *University of Central Florida*
January 2005

Preface
to the Second Edition

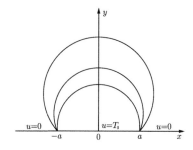

When the first edition of this book was published in 1990, it was well received, and we found the comments and criticisms of graduate students and faculty members from the United States and abroad to be helpful, beneficial, and encouraging. This second edition is the result of that input.

We have taken advantage of this new edition to update the bibliography and correct typographical errors, to include additional topics, examples, exercises, comments, and observations, and, in some cases, to entirely rewrite whole sections. The most significant difference from the first edition is the inclusion of a completely new chapter on wavelets.

We have, however, tried to preserve the character of the first edition. We intend the book to be a source of classical and modern topics dealing fully with the basic ideas and results of Hilbert space theory and functional analysis, and we also intend it to be an introduction to various methods of solution of differential and integral equations. Some of the highlights include the following:

- The book offers a detailed and clear explanation of every concept and method that is introduced, accompanied by carefully selected worked examples, with special emphasis being given to those topics in which students experience difficulty.

- A wide variety of modern examples of applications has been selected from areas of integral and ordinary differential equations, wavelets, generalized functions and partial differential equations, control theory, quantum mechanics, fluid dynamics and solid mechanics, optimization, calculus of variations, variational inequalities, approximation theory, linear and nonlinear stability analysis, and bifurcation theory.

- The book is organized with sufficient flexibility to enable instructors to select chapters appropriate to courses of differing lengths, emphases, and levels of difficulty.

- A wide spectrum of exercises has been carefully chosen and included at the end of each chapter so the reader may further develop both rigorous skills in the theory and applications of functional analysis and a deeper insight into the subject. Answers and hints to selected exercises are provided at the end of the book to provide additional help to students.

It is our pleasure to express our gratitude to those who offered their generous help at different stages of the preparation of this book. Our special thanks are due to Professor Michael Taylor, who read most of the manuscript and suggested many corrections and improvements. Professors Ahmed Zayed and Kit Chan read parts of the manuscript and offered various criticisms and suggestions that have improved the book. June Wingler, with unflagging industry and exemplary patience, typed parts of the manuscript. Finally, we wish to express our grateful thanks to Mr. Charles Glaser, executive editor, and the staff of Academic Press for their help and cooperation. Needless to say, the authors take responsibility for any remaining errors.

The final text was typeset using AmSTeX, and the figures were prepared with the aid of Adobe Illustrator 7.0.

Lokenath Debnath, Piotr Mikusiński
Orlando, January 1998

Preface
to the First Edition

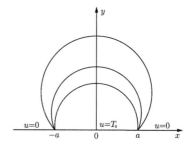

Functional analysis is one of the central areas of modern mathematics, and the theory of Hilbert spaces is the core around which functional analysis has developed. Hilbert spaces have a rich geometric structure because they are endowed with an inner product that allows the introduction of the concept of orthogonality of vectors. We believe functional analysis is best approached through a sound knowledge of Hilbert space theory. Our belief led us to prepare an earlier manuscript, which was used as class notes for courses on Hilbert space theory at the University of Central Florida and Georgia Institute of Technology. This book is essentially based on those notes.

One of the main impulses for the development of functional analysis was the study of differential and integral equations arising in applied mathematics, mathematical physics, and engineering; it was in this setting that Hilbert space methods arose and achieved their early successes. With ever greater demands for mathematical tools to provide both theory and applications for science and engineering, the utility and interest of functional analysis and Hilbert space theory seems more clearly established than ever. Keeping these things in mind, our main goal in this book has been to provide both a systematic exposition of the basic ideas and results of Hilbert space theory and functional analysis, and an introduction to various methods of solution of differential and integral equations. In addition, Hilbert space formalism is used to develop the foundations of quantum mechanics and Hilbert space methods are applied to optimization, variational and control problems, and to problems in approximation theory, nonlinear stability, and bifurcation.

One of the most important examples of a Hilbert space is the space of the Lebesgue square integrable functions. Thus, in a study of Hilbert spaces, the Lebesgue integral cannot be avoided. In several books on Hilbert spaces, the reader is asked to use the Lebesgue integral pretending that it is the Riemann integral. We prefer to include a chapter on the Lebesgue integral to give the motivated reader an opportunity to understand this beautiful and powerful extension of the Riemann integral. The presentation of the Lebesgue integral is based on a method discovered independently by H.M. MacNeille and Jan Mikusiński. The method eliminates the necessity of introducing the measure before the integral. This feature makes the approach more direct and less abstract. Since the main tool is the absolute convergence of numerical series, the theory is accessible for senior undergraduate students.

This book is appropriate for a one-semester course in functional analysis and Hilbert space theory with applications. There are two basic prerequisites for this course: linear algebra and ordinary differential equations. It is hoped that the book will prepare students for further study of advanced functional analysis and its applications. Besides, it is intended to serve as a ready reference to the reader interested in research in various areas of mathematics, physics, and engineering sciences to which the Hilbert space methods can be applied with advantage. A wide selection of examples and exercises is included, in the hope that they will serve as a testing ground for the theory and method. Finally, a special effort is made to present a large and varied number of applications to stimulate interest in the subject.

The book is divided into two parts: Part I. Theory (Chapters 1–4); Part II. Applications (Chapters 5–8). The reader should be aware that Part II is not always as rigorous as Part I.

The first chapter discusses briefly the basic algebraic concepts of linear algebra and then develops the theory of normed spaces to some extent. This chapter is by no means a replacement for a course on normed spaces. Our intent was to provide the reader who has no previous experience in the theory of normed spaces with enough background for understanding of the theory of Hilbert spaces. In this chapter, we discuss normed spaces, Banach spaces, and bounded linear mappings. A section on the contraction mapping and the fixed point theorem is also included.

In Chapter 2, we discuss the definition of the Lebesgue integral and prove the fundamental convergence theorems. The results are first stated and proved for real valued functions of a single variable, and then they are extended to complex valued functions of several real variables. A discussion of locally integrable functions, measure, and measurable functions is also included. In the last section, we prove some basic properties of convolution.

Inner product spaces, Hilbert spaces, and orthonormal systems are discussed in Chapter 3. This is followed by discussions of strong and weak convergence, orthogonal complements and projection theorems, linear functionals, and the Riesz representation theorem.

Chapter 4 is devoted to the theory of linear operators on Hilbert spaces with special emphasis on different kinds of operators and their basic properties. Bilinear functionals and quadratic forms leading to the Lax–Milgram theorem are discussed. In addition, eigenvalues and eigenvectors of linear operators are studied in some detail. These concepts play a central role in the theory of operators and their applications. The spectral theorem for self-adjoint compact operators and other related results are presented. This is followed by a brief discussion on the Fourier transforms. The last section is a short introduction to unbounded operators in a Hilbert space.

Applications of the theory of Hilbert spaces to integral and differential equations are presented in Chapter 5, and emphasis is placed on basic existence theorems and the solvability of various kinds of integral equations. Ordinary differential equations, differential operators, inverse differential operators, and Green's functions are discussed in some detail. Also included is the theory of Sturm–Liouville systems. The last section contains several examples of applications of Fourier transforms to ordinary differential equations and to integral equations.

Chapter 6 provides a short introduction to distributions and their properties. The major part of this chapter is concerned with applications of Hilbert space methods to partial differential equations. Special emphasis is given to weak solutions of elliptic boundary problems, and the use of Fourier transforms for solving partial differential equations, and, in particular, for calculating Green's functions.

In Chapter 7, the mathematical foundations of quantum mechanics are built upon the theory of Hermitian operators in a Hilbert space. This chapter includes basic concepts and equations of classical mechanics, fundamental ideas and postulates of quantum mechanics, the Heisenberg uncertainty principle, the Schrödinger and the Heisenberg pictures, and the quantum theory of the linear harmonic oscillator and of the angular momentum operators.

The final chapter is devoted to the Hilbert space methods for finding solutions of optimization problems, variational problems and variational inequalities, minimization problems of a quadratic functional, and optimal control problems for dynamical systems. Also included are brief treatments of approximation theory, linear and nonlinear stability problems, and bifurcation theory.

This book contains almost 600 examples and exercises that are either directly associated with applications or phrased in terms of the mathematical, physical, and engineering contexts in which theory arises. The exercises truly complement the text. Answers and hints to some of them are provided at the end of the book. For students and readers wishing to learn more about the subject, important references are listed in the bibliography.

In preparing this book, we have been encouraged by and have benefited from the helpful comments and criticisms of a number of graduate students and faculty members of several universities in the United States and abroad. Professors James V. Herod and Thomas D. Morley have adopted the manuscript at Georgia Institute of Technology for a graduate course on Hilbert spaces. We express our grateful thanks to them for their valuable advice and suggestions during the preparation of the book. We also wish to thank Drs. R. Ger and A. Szymański, who have carefully read parts of the manuscript and given some suggestions for improvement. It is our pleasure to acknowledge the encouragement and help of Professor P.K. Ghosh, who has provided several references and books on the subject from his personal library. We also express our grateful thanks to our friends and colleagues, including Drs. Ram N. Mohapatra, Michael D. Taylor, and Carroll A. Webber, for their interest and help during the preparation of the book. Thanks also go to Mrs. Grazyna Mikusiński for drawing all diagrams. In spite of all the best efforts of everyone involved, it is doubtless that there are still typographical errors in the book. We do hope that any remaining errors are both few and obvious and will not create undue confusion. Finally, the authors wish to express their thanks to Mrs. Alice Peters, editor, and the staff of Academic Press for their help and cooperation.

Lokenath Debnath, Piotr Mikusiński
University of Central Florida, Orlando

Normed Vector Spaces

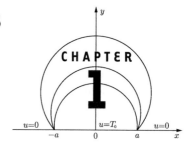

"The organic unity of mathematics is inherent in the nature of this science, for mathematics is the foundation of all exact knowledge of natural phenomena."

David Hilbert

1.1 Introduction

The basic algebraic concepts in the theory of Hilbert spaces are those of a vector space and an inner product. The inner product defines a norm, and thus every Hilbert space is a normed space. Since the norm plays a very important role in the theory, it is not possible to study Hilbert spaces without familiarity with basic concepts and properties of normed spaces. Section 1.2 provides a brief discussion of the algebraic structure of vector spaces. Section 1.3 discusses topological aspects of normed spaces. Basic properties of complete normed spaces are presented in Section 1.4. The final two sections of this chapter are devoted to linear mappings in normed spaces and the fixed point theorem.

This chapter is by no means a substitute for a course in the theory of normed spaces. We limit our discussion to those concepts which are necessary for understanding of the following chapters. Some of the definitions and theorems do not require the algebraic structure of vector spaces and can be formulated in metric spaces or general topological spaces. We choose not to make these distinctions and always assume that we are dealing with a vector space or a normed vector space.

1.2 Vector Spaces

We consider both real and complex vector spaces. The field of real numbers is denoted by $\mathbb{R}$ and the field of complex numbers by $\mathbb{C}$. Elements of $\mathbb{R}$ or $\mathbb{C}$ are called *scalars*. Sometimes it is convenient to give a definition or state a theorem without specifying the field of scalars. In such a case we use $\mathbb{F}$ to denote either $\mathbb{R}$ or $\mathbb{C}$. For instance, if $\mathbb{F}$ is used in a theorem, this means that the theorem is true for both scalar fields $\mathbb{R}$ and $\mathbb{C}$.

Definition 1.2.1.   (Vector space)

By a *vector space* we mean a nonempty set E with two operations:

$$(x, y) \mapsto x + y \text{ from } E \times E \text{ into } E \text{ called } addition,$$

$$(\lambda, x) \mapsto \lambda x \text{ from } \mathbb{F} \times E \text{ into } E \text{ called } multiplication \text{ by scalars,}$$

such that the following conditions are satisfied for all $x, y, z \in E$ and $\alpha, \beta \in \mathbb{F}$:

(a) $x + y = y + x$;

(b) $(x + y) + z = x + (y + z)$;

(c) For every $x, y \in E$ there exists a $z \in E$ such that $x + z = y$;

(d) $\alpha(\beta x) = (\alpha\beta)x$;

(e) $(\alpha + \beta)x = \alpha x + \beta x$;

(f) $\alpha(x + y) = \alpha x + \alpha y$;

(g) $1x = x$.

Elements of E are called *vectors*. If $\mathbb{F} = \mathbb{R}$, then E is called a *real vector space*, and if $\mathbb{F} = \mathbb{C}$, E is called a *complex vector space*.

From (c) it follows that for every $x \in E$ there exists a $z_x \in E$ such that $x + z_x = x$. We will show that there exists exactly one element $z \in E$ such that $x + z = x$ for all $x \in E$. That element is denoted by 0 and called the *zero vector*.

Let $x, y \in E$. By (c), there exists a $w \in E$ such that $x + w = y$. If $x + z_x = x$ for some $z_x \in E$, then, by (a) and (b),

$$y + z_x = (x + w) + z_x = (x + z_x) + w = x + w = y.$$

This shows that, if $x + z = x$ for some $x \in E$, then $y + z = y$ for any other element $y \in E$. We still need to show that such an element is unique. Indeed, if z_1 and z_2 are two such elements, then $z_1 + z_2 = z_1$ and $z_1 + z_2 = z_2$. Thus, $z_1 = z_2$.

A similar argument shows that the vector z in (c) is unique for any pair of vectors $x, y \in E$.

The unique solution z of $x + z = y$ is denoted by $y - x$. According to the definition of the zero vector, we have $x - x = 0$. The vector $0 - x$ is denoted by $-x$.

We use 0 to denote both the scalar 0 and the zero vector; this will not cause any confusion. Because of (b), the use of parentheses in expressions with more than one plus sign can be avoided.

The following properties follow easily from the definition of vector spaces:

$$\text{If } \lambda \neq 0 \text{ and } \lambda x = 0, \quad \text{then } x = 0.$$
$$\text{If } x \neq 0 \text{ and } \lambda x = 0, \quad \text{then } \lambda = 0.$$
$$0x = 0 \quad \text{and} \quad (-1)x = -x.$$

Example 1.2.2. The set $\{0\}$ is a vector space. The scalar fields $\mathbb{R}$ and $\mathbb{C}$ are the simplest nontrivial vector spaces. $\mathbb{R}$ is a real vector space; $\mathbb{C}$ can be treated as a real or a complex vector space. Here are some other simple examples of vector spaces:

$$\mathbb{R}^N = \{(x_1, \ldots, x_N): x_1, \ldots, x_N \in \mathbb{R}\},$$
$$\mathbb{C}^N = \{(z_1, \ldots, z_N): z_1, \ldots, z_N \in \mathbb{C}\},$$
$$\{(z_1, z_2, z_1 + z_2): z_1, z_2 \in \mathbb{C}\}. \qquad \square$$

Example 1.2.3. (Function spaces) Let X be an arbitrary nonempty set and let E be a vector space. Denote by F the space of all functions from X into E. Then F becomes a vector space if the addition and multiplication by scalars are defined in the following natural way:

$$(f + g)(x) = f(x) + g(x),$$
$$(\lambda f)(x) = \lambda f(x).$$

The zero vector in F is the function which assigns the zero vector of E to every element of X.

Some of the most important and interesting examples of vector spaces are function spaces. We are going to encounter some of them in the first part of this book as illustrative examples and then in the second part as the natural setting for applications.

Note that spaces $\mathbb{R}^N$ and $\mathbb{C}^N$ can be defined as function spaces: $\mathbb{R}^N$ is the space of all real valued functions defined on $\{1, \ldots, N\}$ and $\mathbb{C}^N$ is the space of all complex valued function defined on $\{1, \ldots, N\}$. $\qquad \square$

A subset E_1 of a vector space E is called a *vector subspace* (or simply a *subspace* if for every $\alpha, \beta \in \mathbb{F}$ and $x, y \in E_1$ the vector $\alpha x + \beta y$ is in E_1). Note that a subspace of a vector space is a vector space itself. According to the definition, a vector space is a subspace of itself. If we want to exclude this case, we say a proper subspace, that is, E_1 is a *proper subspace* of E if E_1 is a subspace of E and $E_1 \neq E$.

Example 1.2.4. Let Ω be an open subset of $\mathbb{R}^N$. The following are subspaces of the space of all functions from Ω into $\mathbb{C}$:

$\mathcal{C}(\Omega) =$ the space of all continuous complex valued functions defined on Ω.

$\mathcal{C}^k(\Omega) =$ the space of all complex valued functions defined on Ω with continuous partial derivatives of order k.

$\mathcal{C}^\infty(\Omega) =$ the space of infinitely differentiable functions defined on Ω.

$\mathcal{P}(\Omega) =$ the space of all polynomials of N variables (considered as functions on Ω). □

If E_1 and E_2 are subspaces of a vector space E and $E_1 \subseteq E_2$, then E_1 is a subspace of E_2. For instance, the space of all polynomials of N variables is a subspace of $\mathcal{C}^\infty(\mathbb{R}^N)$, which in turn is a subspace of $\mathcal{C}^k(\mathbb{R}^N)$ or $\mathcal{C}(\mathbb{R}^N)$. The intersection of subspaces of a vector space is always a subspace, but the same is not true for the union of subspaces.

Example 1.2.5. (Sequence spaces)   If the set X in Example 1.2.3 is the set $\mathbb{N}$ of all positive integers, then the corresponding function space is a space of sequences. The addition and multiplication by scalars are then defined as

$$(x_1, x_2, \ldots) + (y_1, y_2, \ldots) = (x_1 + y_1, x_2 + y_2, \ldots),$$

$$\lambda(x_1, x_2, \ldots) = (\lambda x_1, \lambda x_2, \ldots).$$

The space of all sequences of complex numbers is a vector space. The space of all bounded complex sequences is a proper subspace of that space. The space of all convergent sequences of complex numbers is a proper subspace of the space of all bounded sequences. □

We use the notation (x_n) or $(x_1, x_2, \ldots)$ to denote a sequence whose nth term is x_n, and $\{x_n : n \in \mathbb{N}\}$ to denote the set of all elements of the sequence. Note that $\{x_n : n \in \mathbb{N}\}$ can be a finite set even though (x_n) is an infinite sequence.

In most examples, verifying that a set is a vector space is easy or trivial. In the following example, the task is much more difficult.

Example 1.2.6. ($\mathbb{P}$-spaces)   Denote by $\mathbb{P}$, for $p \geq 1$, the space of all infinite sequences (z_n) of complex numbers such that $\sum_{n=1}^{\infty} |z_n|^p < \infty$.

We are going to show that l^p is a vector space. Since l^p is a subset of a vector space, namely the space of all sequences of complex numbers, it is enough to show that if $(x_n), (y_n) \in l^p$ and $\lambda \in \mathbb{C}$, then $(x_n + y_n) \in l^p$ and $(\lambda x_n) \in l^p$. To check the second property it suffices to note that

$$\sum_{n=1}^{\infty} |\lambda x_n|^p = |\lambda|^p \sum_{n=1}^{\infty} |x_n|^p < \infty.$$

Condition $\sum_{n=1}^{\infty} |x_n + y_n|^p < \infty$ follows immediately from Minkowski's inequality (Hermann Minkowski (1864–1909)):

$$\left(\sum_{n=1}^{\infty} |x_n + y_n|^p \right)^{1/p} \leq \left(\sum_{n=1}^{\infty} |x_n|^p \right)^{1/p} + \left(\sum_{n=1}^{\infty} |y_n|^p \right)^{1/p}.$$

The proof of Minkowski's inequality is based on Hölder's inequality (Otto Ludwig Hölder (1859–1937)). Both inequalities are proved next. $\square$

Theorem 1.2.7. (Hölder's inequality) *Let $p > 1$, $q > 1$ and $\frac{1}{p} + \frac{1}{q} = 1$. If $(x_n) \in l^p$ and $(y_n) \in l^q$, then*

$$\sum_{n=1}^{\infty} |x_n y_n| \leq \left(\sum_{n=1}^{\infty} |x_n|^p \right)^{1/p} \left(\sum_{n=1}^{\infty} |y_n|^q \right)^{1/q}.$$

Proof: Without loss of generality we can assume that $\sum_{n=1}^{\infty} |x_n| \neq 0$ and $\sum_{n=1}^{\infty} |y_n| \neq 0$. First observe that

$$x^{1/p} \leq \frac{1}{p} x + \frac{1}{q}$$

for $0 \leq x \leq 1$. Let a and b be non-negative numbers such that $a^p \leq b^q$. Then $0 \leq a^p/b^q \leq 1$ and hence we have

$$ab^{-q/p} \leq \frac{1}{p} \frac{a^p}{b^q} + \frac{1}{q}.$$

Since $-\frac{q}{p} = 1 - q$, we obtain

$$ab^{1-q} \leq \frac{1}{p} \frac{a^p}{b^q} + \frac{1}{q}.$$

Multiplying both sides by b^q, we get

$$ab \leq \frac{a^p}{p} + \frac{b^q}{q}. \tag{1.1}$$

We have proved (1.1) assuming $a^p \leq b^q$. A similar argument shows that (1.1) holds also if $b^q \leq a^p$. Therefore the inequality can be used for any $a, b \geq 0$. Using (1.1) with

$$a = \frac{|x_j|}{(\sum_{k=1}^n |x_k|^p)^{1/p}} \quad \text{and} \quad b = \frac{|y_j|}{(\sum_{k=1}^n |y_k|^q)^{1/q}},$$

where $n \in \mathbb{N}$, we get

$$\frac{|x_j|}{(\sum_{k=1}^n |x_k|^p)^{1/p}} \frac{|y_j|}{(\sum_{k=1}^n |y_k|^q)^{1/q}} \leq \frac{1}{p} \frac{|x_j|^p}{\sum_{k=1}^n |x_k|^p} + \frac{1}{q} \frac{|y_j|^q}{\sum_{k=1}^n |y_k|^q}$$

for any $1 \leq j \leq n$. By adding these inequalities for $j = 1, \ldots, n$, we obtain

$$\frac{\sum_{k=1}^n |x_j||y_j|}{(\sum_{k=1}^n |x_k|^p)^{1/p}(\sum_{k=1}^n |y_k|^q)^{1/q}} \leq \frac{1}{p} + \frac{1}{q} = 1,$$

which, by letting $n \to \infty$, gives Hölder's inequality. $\qquad\qquad\square$

Theorem 1.2.8. (Minkowski's inequality) *Let $p \geq 1$. If $(x_n), (y_n) \in l^p$, then*

$$\left(\sum_{n=1}^\infty |x_n + y_n|^p \right)^{1/p} \leq \left(\sum_{n=1}^\infty |x_n|^p \right)^{1/p} + \left(\sum_{n=1}^\infty |y_n|^p \right)^{1/p}.$$

Proof: For $p = 1$ it is enough to use the triangle inequality for the absolute value. If $p > 1$, then there exists a q such that $1/p + 1/q = 1$. Then, by Hölder's inequality, we have

$$\sum_{n=1}^\infty |x_n + y_n|^p = \sum_{n=1}^\infty |x_n + y_n||x_n + y_n|^{p-1}$$

$$\leq \sum_{n=1}^\infty |x_n||x_n + y_n|^{p-1} + \sum_{n=1}^\infty |y_n||x_n + y_n|^{p-1}$$

$$\leq \left(\sum_{n=1}^\infty |x_n|^p \right)^{1/p} \left(\sum_{n=1}^\infty |x_n + y_n|^{q(p-1)} \right)^{1/q}$$

$$+ \left(\sum_{n=1}^\infty |y_n|^p \right)^{1/p} \left(\sum_{n=1}^\infty |x_n + y_n|^{q(p-1)} \right)^{1/q}.$$

Since $q(p - 1) = p$,

$$\sum_{n=1}^{\infty} |x_n + y_n|^p \leq \left\{ \left(\sum_{n=1}^{\infty} |x_n|^p \right)^{1/p} + \left(\sum_{n=1}^{\infty} |y_n|^p \right)^{1/p} \right\} \left(\sum_{n=1}^{\infty} |x_n + y_n|^p \right)^{1/q},$$

which gives Minkowski's inequality. $\qquad \square$

Example 1.2.9. (Cartesian product of vector spaces) Let $E_1, \ldots, E_n$ be vector spaces over the same scalar field $\mathbb{F}$. Define

$$E = \left\{ (x_1, \ldots, x_n) : x_1 \in E_1, x_2 \in E_2, \ldots, x_n \in E_n \right\}$$

with operations

$$(x_1, \ldots, x_n) + (y_1, \ldots, y_n) = (x_1 + y_1, \ldots, x_n + y_n),$$

$$\lambda(x_1, \ldots, x_n) = (\lambda x_1, \ldots, \lambda x_n).$$

Then E is a vector space, which is called the *Cartesian product* or just the *product* of spaces $E_1, \ldots, E_n$. The product space is denoted by $E = E_1 \times \cdots \times E_n$.

This idea can be generalized to an arbitrary family of vector spaces. Let J be an index set, and let E_j be a vector space for every $j \in J$. The product space

$$E = \prod_{j \in J} E_j$$

is defined as the space of all functions $x : J \to \bigcup_{j \in J} E_j$ such that $x(j) \in X_j$ for every $j \in J$. The vector space operations in E are defined as in Example 1.2.3. $\square$

Let $x_1, \ldots, x_k$ be elements of a vector space E. A vector $x \in E$ is called a *linear combination* of vectors $x_1, \ldots, x_k$ if there exist scalars $\alpha_1, \ldots, \alpha_k$ such that

$$x = \alpha_1 x_1 + \cdots + \alpha_k x_k.$$

For example, any element of $\mathbb{R}^N$ is a linear combination of vectors

$$e_1 = (1, 0, 0, \ldots, 0), \quad e_2 = (0, 1, 0, \ldots, 0), \quad \ldots, \quad e_N = (0, 0, \ldots, 0, 1).$$

Similarly, any polynomial of degree k is a linear combination of monomials $1, x, x^2, \ldots, x^k$.

A finite collection of vectors $\{x_1, \ldots, x_k\}$ is called *linearly independent* if $\alpha_1 x_1 + \cdots + \alpha_k x_k = 0$ implies $\alpha_1 = \alpha_2 = \cdots = \alpha_k = 0$. An infinite collection of vectors $\mathcal{A}$ is called linearly independent if every finite subcollection of $\mathcal{A}$ is linearly independent. A collection of vectors which is not linearly independent is called *linearly dependent*. We can also say that $\mathcal{A}$ is linearly independent if no vector x in $\mathcal{A}$ is a linear combination of a finite number of vectors from $\mathcal{A}$ different from x. Vectors $e_1, \ldots, e_N$, mentioned above, are linearly independent. Also the monomials $1, x, x^2, \ldots, x^k, \ldots$ are linearly independent.

Note that linear independence may depend on the scalar field. For instance, the numbers 1 and i (i stands for the imaginary number $i = \sqrt{-1}$) represent linearly independent vectors in the space $\mathbb{C}$ over the field of real numbers. On the other hand, 1 and i are not independent in $\mathbb{C}$ over the field of complex numbers.

Let $\mathcal{A}$ be a subset of a vector space E. By span $\mathcal{A}$ we denote the set of all finite linear combinations of vectors from $\mathcal{A}$, that is,

$$\text{span } \mathcal{A} = \{\alpha_1 x_1 + \cdots + \alpha_k x_k \colon x_1, \ldots, x_k \in \mathcal{A}, \alpha_1, \ldots, \alpha_k \in \mathbb{F}, k \in \mathbb{N}\}.$$

It is easy to check that span $\mathcal{A}$ is a vector subspace of E. This subspace is called the *space spanned by* $\mathcal{A}$. It is the smallest vector subspace of E containing $\mathcal{A}$.

A set of vectors $\mathcal{B} \subset E$ is called a *basis* of E if $\mathcal{B}$ is linearly independent and span $\mathcal{B} = E$.

If there exists a finite basis in E, then E is called a *finite dimensional vector space*. Otherwise we say that E is *infinite dimensional*. It can be proved that, for a given finite dimensional space E, the number of vectors in any basis of E is the same. If, for example, E has a basis that consists of exactly n vectors, then any other basis has exactly n vectors. In such a case n is called the *dimension* of E and we write $\dim E = n$.

The following are examples of sets of vectors which are bases in $\mathbb{R}^3$:

$$\mathcal{A} = \big\{(1,0,0), (0,1,0), (0,0,1)\big\}, \tag{1.2}$$

$$\mathcal{B} = \big\{(0,1,1), (1,0,1), (1,1,0)\big\}, \tag{1.3}$$

$$\mathcal{C} = \big\{(1,2,3), (1,3,5), (3,2,3)\big\}. \tag{1.4}$$

We have $\dim \mathbb{R}^3 = 3$, and in general $\dim \mathbb{R}^N = N$. Spaces $\mathcal{C}(\Omega)$, $\mathcal{C}^k(\mathbb{R}^N)$, $\mathcal{C}^\infty(\mathbb{R}^N)$ are infinite dimensional. Note that the dimension of the real vector space $\mathbb{C}^N$ is $2N$, while the dimension of the complex vector space $\mathbb{C}^N$ is N.

1.3 Normed Spaces

In general, it does not make sense to ask what is the length or magnitude of a vector in a vector space. The concept of a norm is an abstract generalization of the length of a vector. It is defined axiomatically, that is, any real valued function satisfying certain conditions is called a norm.

Definition 1.3.1.   (Norm)
A function $x \mapsto \|x\|$ from a vector space E into $\mathbb{R}$ is called a *norm* if it satisfies the following conditions:

(a) $\|x\| = 0$ implies $x = 0$;

(b) $\|\lambda x\| = |\lambda| \|x\|$ for every $x \in E$ and $\lambda \in \mathbb{F}$;

(c) $\|x + y\| \le \|x\| + \|y\|$ for every $x, y \in E$.

Condition (c) is called the *triangle inequality*. Since

$$0 = \|0\| = \|x - x\| \le \|x\| + \|-x\| = 2\|x\|,$$

we have $\|x\| \ge 0$ for every $x \in E$. Note that (b) implies that $\|0\| = 0$. The inequality

$$\big|\|x\| - \|y\|\big| \le \|x - y\|,$$

which follows from (b) and (c), is often used.

Example 1.3.2. The function

$$\|z\| = \sqrt{|z_1|^2 + \cdots + |z_N|^2}, \quad z = (z_1, \ldots, z_N) \in \mathbb{C}^N$$

defines a norm on $\mathbb{C}^N$. This norm is called the *Euclidean norm*. The following are also norms on $\mathbb{C}^N$:

$$\|z\| = |z_1| + \cdots + |z_N|, \tag{1.5}$$

$$\|z\| = \max\{|z_1|, \ldots, |z_N|\}. \tag{1.6}$$

□

Example 1.3.3. Let Ω be a closed bounded subset of $\mathbb{R}^N$. The function $\|f\| = \max_{x \in \Omega} |f(x)|$ defines a norm on $\mathcal{C}(\Omega)$. □

Example 1.3.4. Let $z = (z_n) \in l^p$. The function defined by $\|z\| = (\sum_{n=1}^{\infty} |z_n|^p)^{1/p}$ is a norm on l^p for any $p \ge 1$. Note that Minkowski's inequality is in fact the triangle inequality for this norm. □

Definition 1.3.5. (Normed space)
A vector space with a norm is called a *normed space*.

It is possible to define different norms on the same vector space. Therefore to define a normed space we need to specify both the vector space and the norm. We say that a normed space is a pair $(E, \|\cdot\|)$, where E is a vector space and $\|\cdot\|$ is a norm defined on E. Some vector spaces are traditionally equipped with standard norms. For instance, when we say "the normed space $\mathbb{R}^N$" we mean the Euclidean norm

$$\|x\| = \sqrt{x_1^2 + \cdots + x_N^2}.$$

Similarly, the norms defined in Examples 1.3.3 and 1.3.4 are standard. If we want to consider different norms on those spaces, we have to say something like "consider the space ... with the norm defined by ...."

Note that a vector subspace of a normed space is a normed space with the same norm restricted to the subspace.

The absolute value is a norm on $\mathbb{R}$ or $\mathbb{C}$. It can be used to define convergence, because the absolute value of a difference of two numbers is the distance between those numbers and convergence is about "getting closer to the limit point." The norm plays a similar role. While $\|x\|$ can be interpreted as the magnitude of x, $\|x - y\|$ provides a measure of the distance between x and y.

Definition 1.3.6.   (Convergence in a normed space)

Let $(E, \| \cdot \|)$ be a normed space. We say that a sequence (x_n) of elements of E converges to some $x \in E$, if for every $\varepsilon > 0$ there exists a number M such that for every $n \geq M$ we have $\|x_n - x\| < \varepsilon$. In such a case we write $\lim_{n \to \infty} x_n = x$ or simply $x_n \to x$.

This definition becomes simpler if convergence of real numbers is used: $x_n \to x$ in E means $\|x_n - x\| \to 0$ in $\mathbb{R}$. The convergence in a normed space has the basic properties of convergence in $\mathbb{R}$:

A convergent sequence has a unique limit.

If $x_n \to x$ and $\lambda_n \to \lambda$ (λ_n, λ are scalars),   then $\lambda_n x_n \to \lambda x$.

If $x_n \to x$ and $y_n \to y$,   then $x_n + y_n \to x + y$.

These properties can be proved the same way as in the case of convergence in $\mathbb{R}$.

A norm on a vector space E induces a convergence in E. In other words, if we have a normed space E, then we automatically have a convergence defined in E. In practice, we often face a different situation: we have a vector space E with a given convergence and we want to know if we can find a norm on E, which would define that convergence. It is not always possible. The following two examples illustrate the problem. In the first one, the convergence can be described by a norm. In the second one, we prove that the given convergence cannot be defined by a norm.

Example 1.3.7. (Uniform convergence) Consider the space $C(\Omega)$ of all continuous functions defined on a closed bounded set $\Omega \subset \mathbb{R}^N$. We say that the sequence $f_1, f_2, \ldots \in C(\Omega)$ *converges uniformly* to f if for every $\varepsilon > 0$ there exists a constant n_0 such that $|f(x) - f_n(x)| < \varepsilon$ for all $x \in \Omega$ and all $n \geq n_0$. It is easy to see that the norm in Example 1.3.3 defines uniform convergence, that is, the sequence (f_n) converges uniformly to f if and only if

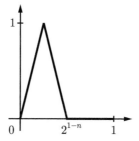

Figure 1.1   A typical function g_n in Example 1.3.8.

$\|f_n - f\| = \max_{x \in \Omega} |f_n(x) - f(x)| \to 0$ as $n \to \infty$. For this reason this norm is called the *uniform convergence norm*. $\square$

Example 1.3.8. (Pointwise convergence) Let $f, f_1, f_2, \ldots \in C([0, 1])$, the space of all continuous functions on the interval $[0, 1]$. We say that the sequence (f_n) is *pointwise convergent* to f if $|f_n(t) - f(t)| \to 0$, as $n \to \infty$, for every $t \in [0, 1]$. We will show that there is no norm on $C([0, 1])$, which defines pointwise convergence. Suppose, on the contrary, that $\| \cdot \|$ is a norm on $C([0, 1])$ such that

$$\|f_n - f\| \to 0 \quad \text{if and only if} \quad f_n(t) \to f(t) \text{ for every } t \in [0, 1]. \tag{1.7}$$

Consider the sequence of functions $g_1, g_2, \ldots$ defined by

$$g_n(t) = \begin{cases} 2^n t & \text{if } 0 \le t \le 2^{-n}, \\ 2 - 2^n t & \text{if } 2^{-n} \le t \le 2^{1-n}, \\ 0 & \text{otherwise} \end{cases}$$

(see Figure 1.1). Since $g_n \ne 0$, we have $\|g_n\| \ne 0$ for all $n \in \mathbb{N}$. If we define $f_n = g_n / \|g_n\|$, then $\|f_n\| = 1$ for all $n \in \mathbb{N}$, and thus the sequence (f_n) is not convergent to 0 (the zero function) with respect to the norm $\| \cdot \|$. On the other hand, it is easy to see that $f_n(t) \to 0$ for every $t \in [0, 1]$. This contradiction shows that a norm satisfying (1.7) cannot exist. $\square$

Definition 1.3.9.   (Equivalence of norms)
Two norms defined on the same vector space are called *equivalent* if they define the same convergence. More precisely, norms $\| \cdot \|_1$ and $\| \cdot \|_2$ on E are equivalent if for any sequence (x_n) in E and any $x \in E$,

$$\|x_n - x\|_1 \to 0 \quad \text{if and only if} \quad \|x_n - x\|_2 \to 0.$$

Example 1.3.10. The following norms on $\mathbb{R}^2$ are equivalent:

$$\left\|(x,y)\right\|_1 = \sqrt{x^2+y^2}, \qquad \left\|(x,y)\right\|_2 = |x|+|y|,$$
$$\left\|(x,y)\right\|_3 = \max\{|x|,|y|\}. \qquad \square$$

We will prove that any two norms on a finite dimensional vector space are equivalent. The proof requires some preparations. First we prove a useful criterion for equivalence of norms. The condition in the following theorem is often used as a definition of equivalence of norms.

Theorem 1.3.11. (Equivalence of norms) *Let $\|\cdot\|_1$ and $\|\cdot\|_2$ be norms on a vector space E. Then $\|\cdot\|_1$ and $\|\cdot\|_2$ are equivalent if and only if there exist positive numbers α and β such that*

$$\alpha\|x\|_1 \leq \|x\|_2 \leq \beta\|x\|_1 \quad \text{for all } x \in E.$$

Proof: Clearly, the above condition implies equivalence of norms $\|\cdot\|_1$ and $\|\cdot\|_2$. Now assume that the norms are equivalent, that is, $\|x_n - x\|_1 \to 0$ if and only if $\|x_n - x\|_2 \to 0$. Suppose there is no $\alpha > 0$ such that $\alpha\|x\|_1 \leq \|x\|_2$ for every $x \in E$. Then, for each $n \in \mathbb{N}$, there exists $x_n \in E$ such that

$$\frac{1}{n}\|x_n\|_1 > \|x_n\|_2.$$

Define

$$y_n = \frac{1}{\sqrt{n}}\frac{x_n}{\|x_n\|_2}.$$

Then $\|y_n\|_2 = \frac{1}{\sqrt{n}} \to 0$. On the other hand, $\|y_n\|_1 \geq n\|y_n\|_2 \geq \sqrt{n}$. This contradiction shows that a number α with the required property exists. Existence of the number β can be proved in a similar way. $\square$

The following technical lemma establishes a property crucial for the proof of equivalence of norms on finite dimensional spaces.

Lemma 1.3.12. *If $x_1, \ldots, x_n$ are linearly independent elements of a normed space E, then there exists a constant $c > 0$ such that*

$$\|\alpha_1 x_1 + \cdots + \alpha_n x_n\| \geq c(|\alpha_1| + \cdots + |\alpha_n|) \tag{1.8}$$

for all $\alpha_1, \ldots, \alpha_n \in \mathbb{R}$.

Proof: Since (1.8) holds for any c if $|\alpha_1| + \cdots + |\alpha_n| = 0$, we can assume that $|\alpha_1| + \cdots + |\alpha_n| > 0$. Then (1.8) is equivalent to

$$\|\beta_1 x_1 + \cdots + \beta_n x_n\| \geq c \quad \text{whenever } |\beta_1| + \cdots + |\beta_n| = 1. \tag{1.9}$$

The function $f : \mathbb{R}^n \to \mathbb{R}$ defined by

$$f(\beta_1, \ldots, \beta_n) = \|\beta_1 x_1 + \cdots + \beta_n x_n\|$$

is continuous. Since the set $B = \{(\beta_1, \ldots, \beta_n) \in \mathbb{R}^n : |\beta_1| + \cdots + |\beta_n| = 1\}$ is closed and bounded, f attains a minimum on B. Note that the minimum cannot be 0, because then we would have $\beta_1 x_1 + \cdots + \beta_n x_n = 0$ for some $|\beta_1| + \cdots + |\beta_n| = 1$, contradicting linear independence of $x_1, \ldots, x_n$. Therefore,

$$c = \min_{(\beta_1, \ldots, \beta_n) \in B} f(\beta_1, \ldots, \beta_n) = \min_{(\beta_1, \ldots, \beta_n) \in B} \|\beta_1 x_1 + \cdots + \beta_n x_n\| > 0,$$

proving (1.9). □

Theorem 1.3.13. *If E is finite dimensional, then any two norms on E are equivalent.*

Proof: Let E be a finite dimensional vector space, and let $\{e_1, \ldots, e_n\}$ be a basis in E. We will prove that an arbitrary norm $\| \cdot \|$ in E is equivalent to the norm $\| \cdot \|_0$ defined by

$$\|\alpha_1 e_1 + \cdots + \alpha_n e_n\|_0 = |\alpha_1| + \cdots + |\alpha_n|.$$

We use Theorem 1.3.11. First note that

$$\|\alpha_1 e_1 + \cdots + \alpha_n e_n\| \leq |\alpha_1| \|e_1\| + \cdots + |\alpha_n| \|e_n\|$$
$$\leq \max\{\|e_1\|, \ldots, \|e_n\|\}(|\alpha_1| + \cdots + |\alpha_n|).$$

Therefore, for any $x \in E$,

$$\|x\| \leq \beta \|x\|_0,$$

where $\beta = \max\{\|e_1\|, \ldots, \|e_n\|\}$.

Existence of a constant $\alpha > 0$ such that

$$\alpha \|x\|_0 \leq \|x\|,$$

for any $x \in E$, follows immediately from Lemma 1.3.12. □

A normed space $(E, \| \cdot \|)$ becomes a metric space if we define a metric by $d(x, y) = \|x - y\|$. The convergence defined by the norm and the convergence defined by this metric are the same. The metric in E defines a topology in E and thus all topological notions, that is, notions which can be defined in terms of open sets. It is not necessary to define the metric first and then the topology using that metric. In the remainder of this section, we define the basic topological notions in terms of the norm. We also prove some topological properties of normed spaces.

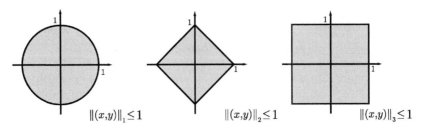

Figure 1.2   Examples of unit balls in $\mathbb{R}^2$ (Example 1.3.15).

Definition 1.3.14.   (Open ball, closed ball, sphere)
Let x be an element of a normed space E, and let r be a positive number. We use the following notation:

$$B(x, r) = \{y \in E : \|y - x\| < r\} \quad (\textit{open ball});$$
$$\overline{B}(x, r) = \{y \in E : \|y - x\| \leq r\} \quad (\textit{closed ball});$$
$$S(x, r) = \{y \in E : \|y - x\| = r\} \quad (\textit{sphere}).$$

In each case, x is called the *center* and r the *radius*.

Example 1.3.15. Figure 1.2 shows examples of balls in $\mathbb{R}^2$ with respect to the norms

$$\|(x, y)\|_1 = \sqrt{x^2 + y^2}, \qquad \|(x, y)\|_2 = |x| + |y|,$$
$$\|(x, y)\|_3 = \max\{|x|, |y|\}. \qquad\qquad\qquad\quad \square$$

Example 1.3.16. Let $E = \mathcal{C}([-\pi, \pi])$, and let $\|f\| = \max_{t \in [-\pi, \pi]} |f(t)|$. Figure 1.3 shows $\overline{B}(\{\sin t\}, 1)$. The figure should be understood as follows: $\overline{B}(\{\sin t\}, 1)$ is the set of all continuous functions on $[-\pi, \pi]$ whose graphs are in the shaded area. It is not the set of all points of the shaded area. $\quad \square$

Definition 1.3.17.   (Open and closed sets)
A subset S of a normed space E is called *open* if for every $x \in S$ there exists $\varepsilon > 0$ such that $B(x, \varepsilon) \subseteq S$. A subset S is called *closed* if its complement is open, that is, if $E \backslash S$ is open.

It is important to realize that equivalent norms define the same open sets, even though the balls are different. The same is true for closed, dense, compact sets, and other notions that can be defined in terms of open sets.

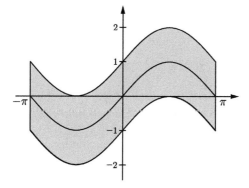

Figure 1.3 The ball $\overline{B}(\{\sin t\}, 1)$ in $\mathcal{C}([-\pi, \pi])$.

Example 1.3.18. Open balls are open sets. Closed balls and spheres are closed sets. □

Example 1.3.19. Let Ω be a closed bounded set in $\mathbb{R}^N$. Consider the space $\mathcal{C}(\Omega)$ with the norm $\|f\| = \max_{x \in \Omega} |f(x)|$. Let $f \in \mathcal{C}(\Omega)$ and $f(x) > 0$ for all $x \in \Omega$. The following sets are open in $\mathcal{C}(\Omega)$:

$$\big\{g \in \mathcal{C}(\Omega)\colon g(x) < f(x) \text{ for all } x \in \Omega\big\},$$
$$\big\{g \in \mathcal{C}(\Omega)\colon g(x) > f(x) \text{ for all } x \in \Omega\big\},$$
$$\big\{g \in \mathcal{C}(\Omega)\colon |g(x)| < f(x) \text{ for all } x \in \Omega\big\},$$
$$\big\{g \in \mathcal{C}(\Omega)\colon |g(x)| > f(x) \text{ for all } x \in \Omega\big\}.$$

The following sets are closed in $\mathcal{C}(\Omega)$:

$$\big\{g \in \mathcal{C}(\Omega)\colon g(x) \leq f(x) \text{ for all } x \in \Omega\big\},$$
$$\big\{g \in \mathcal{C}(\Omega)\colon g(x) \geq f(x) \text{ for all } x \in \Omega\big\},$$
$$\big\{g \in \mathcal{C}(\Omega)\colon |g(x)| \leq f(x) \text{ for all } x \in \Omega\big\},$$
$$\big\{g \in \mathcal{C}(\Omega)\colon |g(x)| \geq f(x) \text{ for all } x \in \Omega\big\},$$
$$\big\{g \in \mathcal{C}(\Omega)\colon g(x_0) = \lambda\big\} \quad (x_0 \text{ is a fixed point in } \Omega \text{ and } \lambda \in \mathbb{C}).$$

□

Theorem 1.3.20.

(a) *The union of any collection of open sets is open.*

(b) *The intersection of a finite number of open sets is open.*

(c) *The union of a finite number of closed sets is closed.*

(d) *The intersection of any collection of closed sets is closed.*

(e) *The empty set and the whole space are both open and closed.*

The proofs are left as exercises.

Theorem 1.3.21. *A subset S of a normed space E is closed if and only if every sequence of elements of S convergent in E has its limit in S, that is,*

$$x_1, x_2, \ldots \in S \text{ and } x_n \to x \quad implies \quad x \in S.$$

Proof: Suppose S is a closed subset of E, $x_1, x_2, \ldots \in S$, $x_n \to x$, and $x \notin S$. Since S is closed, $E \backslash S$ is open. Thus, there exists an $\varepsilon > 0$ such that $B(x, \varepsilon) \subseteq E \backslash S$. On the other hand, since $\|x - x_n\| \to 0$, we have $\|x - x_n\| < \varepsilon$ for all sufficiently large $n \in \mathbb{N}$. This contradiction shows that $x \in S$.

   Suppose now that whenever $x_1, x_2, \ldots \in S$ and $x_n \to x$, then $x \in S$. If S is not closed, then $E \backslash S$ is not open, and thus there exists $x \in E \backslash S$ such that every ball $B(x, \varepsilon)$ contains elements of S. Thus, we can find $x_1, x_2, \ldots \in S$ such that $x_n \in B(x, 1/n)$. But then $x_n \to x$ and, according to our assumption, $x \in S$. This contradicts the assumption $x \in E \backslash S$. Therefore S must be a closed set.   □

Note how the above theorem is useful in proving that the sets mentioned in Example 1.3.19 are closed.

Definition 1.3.22.   (Closure)
Let S be a subset of a normed space E. By the *closure* of S, denoted by cl S, we mean the intersection of all closed sets containing S.

In view of Theorem 1.3.20(d), the closure of a set is always a closed set. It is the smallest closed set which contains S. The following theorem gives a sequential description of the closure.

Theorem 1.3.23. *Let S be a subset of a normed space E. The closure of S is the set of limits of all convergent sequences of elements of S, that is,*

$$\text{cl}\, S = \{x \in E : \text{there exist } x_1, x_2, \ldots \in S \text{ such that } x_n \to x\}.$$

The proof is left as an exercise.

Example 1.3.24. The Weierstrass theorem says that every continuous function on an interval $[a, b]$ can be uniformly approximated by polynomials. This can also be expressed as follows: The closure of the set of all polynomials on $[a, b]$ is the whole space $\mathcal{C}([a, b])$.   □

Definition 1.3.25.   (Dense subset)
A subset S of a normed space E is called *dense* in E if $\operatorname{cl} S = E$.

Example 1.3.26. The set of all polynomials on $[a, b]$ is dense in $\mathcal{C}([a, b])$. The set of all sequences of complex numbers, which have only a finite number of nonzero terms is dense in l^p for all $p \geq 1$. $\square$

Theorem 1.3.27. *Let S be a subset of a normed space E. The following conditions are equivalent:*

(a) *S is dense in E,*

(b) *For every $x \in E$ there exist $x_1, x_2, \ldots \in S$ such that $x_n \to x$,*

(c) *Every nonempty open subset of E contains an element of S.*

The proof is left as an exercise.

Definition 1.3.28.   (Compact set)
A subset S of a normed space E is called *compact* if every sequence (x_n) in S contains a convergent subsequence whose limit belongs to S.

Example 1.3.29. In $\mathbb{R}^N$ or $\mathbb{C}^N$, a set is compact if and only if it is bounded and closed. $\square$

This simple characterization of compact sets is valid for any finite dimensional normed spaces. In infinite dimensional normed spaces the situation is more complicated.

Definition 1.3.30.   (Bounded subset)
A subset S of a normed space E is called *bounded* if $S \subseteq B(0, r)$ for some $r > 0$.

It is easy to show that S is bounded if and only if $\|\lambda_n x_n\| \to 0$ for any $x_n \in S$ and any scalars $\lambda_n \to 0$.

Theorem 1.3.31. *Compact sets are closed and bounded.*

Proof: Let S be a compact subset of a normed space E. Suppose $x_1, x_2, \ldots \in S$ and $x_n \to x$. Then (x_n) contains a subsequence (x_{p_n}), which converges to some $y \in S$. On the other hand, we have $x_{p_n} \to x$. Thus, $x = y$ and $x \in S$. This shows that S is closed.

Suppose now that S is not bounded. Then there exists a sequence $x_1, x_2, \ldots \in S$ such that $\|x_n\| \geq n$ for all $n \in \mathbb{N}$. Clearly, (x_n) does not contain a convergent subsequence, and hence S is not compact. $\qquad \square$

The next example shows that closed and bounded sets in infinite dimensional spaces need not be compact.

Example 1.3.32. Consider the space $C([0, 1])$. The closed unit ball $\overline{B}(0, 1)$ is a closed and bounded set, but it is not compact. To see this consider the sequence of functions defined by $x_n(t) = t^n$. Then $x_n \in \overline{B}(0, 1)$ for all $n \in \mathbb{N}$. Clearly, no subsequence of (x_n) converges in $C([0, 1])$. $\qquad \square$

It turns out that finite dimensional normed spaces can be characterized by the property that the unit ball is compact. The proof of this fact is based on a theorem that is usually referred to as Riesz's lemma.

Theorem 1.3.33. (Riesz's lemma)   *Let X be a closed proper subspace of a normed space E. For every $\varepsilon \in (0, 1)$ there exists an $x_\varepsilon \in E$ such that $\|x_\varepsilon\| = 1$ and $\|x_\varepsilon - x\| \geq \varepsilon$ for all $x \in X$.*

Proof: Let $z \in E \backslash X$ and let $d = \inf_{x \in X} \|z - x\|$. Since X is closed, $d > 0$. If $\varepsilon \in (0, 1)$, then

$$d \leq \|z - x_0\| \leq \frac{d}{\varepsilon}$$

for some $x_0 \in X$. We will show that

$$x_\varepsilon = \frac{z - x_0}{\|z - x_0\|}$$

has the desired property. Indeed, for any $x \in X$ we have

$$\|x_\varepsilon - x\| = \left\| \frac{z - x_0}{\|z - x_0\|} - x \right\|$$

$$= \frac{1}{\|z - x_0\|} \|z - x_0 - \|z - x_0\|x\|$$

$$\geq \frac{1}{\|z - x_0\|} d \geq \frac{\varepsilon}{d} d = \varepsilon,$$

where the first inequality follows from the definition of d since $x_0 + \|z - x_0\|x \in X$. $\qquad \square$

Theorem 1.3.34. *A normed space E is finite dimensional if and only if the closed unit ball in E is compact.*

Proof: Let E be finite dimensional and let $\{e_1, \ldots, e_N\}$ be a basis in E. By Theorem 1.3.13, we can assume, without loss of generality, that the norm on E is defined by

$$\|\alpha_1 e_1 + \cdots + \alpha_N e_N\| = |\alpha_1| + \cdots + |\alpha_N|.$$

It is a simple exercise to show that the unit ball with respect to this norm is compact.

Now assume that E is infinite dimensional. Using Riesz's lemma, we construct a sequence $x_n \in E$ such that $\|x_n\| = 1$ for all $n \in \mathbb{N}$ and $\|x_m - x_n\| \geq \frac{1}{2}$ for all $m, n \in \mathbb{N}$, $m \neq n$. Let $x_1 \in E$ be any element such that $\|x_1\| = 1$. Since span$\{x_1\}$ is a closed and proper subspace of E, there exists an $x_2 \in E$ such that $\|x_2\| = 1$ and $\|x_1 - x_2\| \geq \frac{1}{2}$. Now suppose we have already constructed $x_1, \ldots, x_n$ with the desired properties. Since span$\{x_1, \ldots, x_n\}$ is a closed and proper subspace of E, there exists an $x_{n+1} \in E$ such that $\|x_{n+1}\| = 1$ and $\|x_k - x_{n+1}\| \geq \frac{1}{2}$ for all $k = 1, 2, \ldots, n$.

Suppose (x_n) has a convergent subsequence (x_{p_n}) such that $x_{p_n} \to x$, for some $x \in E$. Then

$$\frac{1}{2} \leq \|x_{p_n} - x_{p_{n+1}}\| \leq \|x_{p_n} - x\| + \|x_{p_{n+1}} - x\|$$

for all $n \in \mathbb{N}$, which contradicts convergence of (x_{p_n}). $\qquad\square$

1.4 Banach Spaces

Every Cauchy sequence of numbers converges. Every absolutely convergent series of numbers converges. These are very important properties of real and complex numbers. Many crucial arguments concerning numbers rely on them. One expects that similar properties of a normed space would be of great importance. This is true. However, not all normed spaces have these properties. Those that do are called Banach spaces after the name of Stafan Banach (1892–1945).

Definition 1.4.1.　(Cauchy sequence)
A sequence of vectors (x_n) in a normed space is called a *Cauchy sequence* if for every $\varepsilon > 0$ there exists a number M such that $\|x_m - x_n\| < \varepsilon$ for all $m, n > M$.

Cauchy sequences are named after Augustin Louis Cauchy (1789–1857).

Theorem 1.4.2. *The following conditions are equivalent:*

(a) *(x_n) is a Cauchy sequence;*

(b) *$\|x_{p_n} - x_{q_n}\| \to 0$ as $n \to \infty$, for every pair of increasing sequences of positive integers (p_n) and (q_n);*

(c) $\|x_{p_{n+1}} - x_{p_n}\| \to 0$ as $n \to \infty$, for every increasing sequence of positive integers (p_n).

The proof is left as an exercise.

Observe that *every convergent sequence is a Cauchy sequence*. In fact, if $\|x_n - x\| \to 0$, then

$$\|x_{p_n} - x_{q_n}\| \le \|x_{p_n} - x\| + \|x_{q_n} - x\| \to 0,$$

for every pair of increasing sequences of indices (p_n) and (q_n). The converse is not true, in general.

Example 1.4.3. Let $\mathcal{P}([0,1])$ be the space of polynomials on $[0,1]$ with the norm of uniform convergence $\|P\| = \max_{[0,1]} |P(x)|$. Define

$$P_n(x) = 1 + x + \frac{x^2}{2!} + \cdots + \frac{x^n}{n!}$$

for $n = 1, 2, \ldots$. Then (P_n) is a Cauchy sequence, but it does not converge in $\mathcal{P}([0,1])$ because its limit is not a polynomial. $\qquad\square$

Lemma 1.4.4. *If (x_n) is a Cauchy sequence in a normed space, then the sequence of norms $(\|x_n\|)$ converges.*

Proof: Since $|\|x\| - \|y\|| \le \|x - y\|$, we have $|\|x_m\| - \|x_n\|| \le \|x_m - x_n\| \to 0$ as $m, n \to \infty$. This shows that the sequence of norms is a Cauchy sequence of real numbers, hence it is convergent. $\qquad\square$

Note that this lemma implies that *every Cauchy sequence is bounded*, that is, if (x_n) is a Cauchy sequence, then there is a number M such that $\|x_n\| \le M$ for all n.

Definition 1.4.5.   (Banach space)
A normed space E is called *complete* if every Cauchy sequence in E converges to an element of E. A complete normed space is called a *Banach space*.

Example 1.4.6. We will show that the space l^2 is complete. Let (a_n) be a Cauchy sequence in l^2. If

$$a_n = (\alpha_{n,1}, \alpha_{n,2}, \ldots),$$

then given any $\varepsilon > 0$, there exists a number n_0 such that

$$\sum_{k=1}^{\infty} |\alpha_{m,k} - \alpha_{n,k}|^2 < \varepsilon^2 \tag{1.10}$$

for all $m, n \geq n_0$. Note that this implies that for every fixed $k \in \mathbb{N}$ and for every $\varepsilon > 0$ there exists a number n_0 such that

$$|\alpha_{m,k} - \alpha_{n,k}| < \varepsilon$$

for all $m, n \geq n_0$. But this means that, for every k, the sequence $(\alpha_{n,k})$ is a Cauchy sequence in $\mathbb{C}$ and thus convergent. Denote

$$\alpha_k = \lim_{n \to \infty} \alpha_{n,k}, \quad k = 1, 2, \ldots \quad \text{and} \quad a = (\alpha_n).$$

We are going to prove that a is an element of l^2 and that the sequence (a_n) converges to a. Indeed, from (1.10), by letting $m \to \infty$, we obtain

$$\sum_{k=1}^{\infty} |\alpha_k - \alpha_{n,k}|^2 \leq \varepsilon^2 \tag{1.11}$$

for every $n \geq n_0$. Since

$$\sum_{k=1}^{\infty} |\alpha_{n_0,k}|^2 < \infty,$$

by Minkowski's inequality, we have

$$\sqrt{\sum_{k=1}^{\infty} |\alpha_k|^2} = \sqrt{\sum_{k=1}^{\infty} \left(|\alpha_k| - |\alpha_{n_0,k}| + |\alpha_{n_0,k}| \right)^2}$$

$$\leq \sqrt{\sum_{k=1}^{\infty} \left(|\alpha_k| - |\alpha_{n_0,k}| \right)^2} + \sqrt{\sum_{k=1}^{\infty} |\alpha_{n_0,k}|^2}$$

$$\leq \sqrt{\sum_{k=1}^{\infty} |\alpha_k - \alpha_{n_0,k}|^2} + \sqrt{\sum_{k=1}^{\infty} |\alpha_{n_0,k}|^2} < \infty.$$

This proves that the sequence $a = (\alpha_n)$ is an element of l^2. Moreover, since ε is arbitrarily small, (1.11) implies

$$\lim_{n \to \infty} \|a - a_n\| = \lim_{n \to \infty} \sqrt{\sum_{k=1}^{\infty} \left(|\alpha_k - \alpha_{n,k}| \right)^2} = 0,$$

which means that the sequence (a_n) is convergent to a in l^2. □

Example 1.4.7. Another important example of a Banach space is the space $C([a, b])$ of continuous (real or complex valued) functions on an interval $[a, b]$. Recall that the norm on $C([a, b])$ is defined by $\|f\| = \max_{[a,b]} |f(x)|$.

Let (f_n) be a Cauchy sequence in $C([a, b])$. For an arbitrary $\varepsilon > 0$ there exists an $n_0 \in \mathbb{N}$ such that

$$\|f_n - f_m\| < \varepsilon \quad \text{for all } n, m \geq n_0,$$

and thus also

$$\left| f_n(x) - f_m(x) \right| < \varepsilon \quad \text{for all } n, m \geq n_0 \text{ and all } x \in [a, b]. \tag{1.12}$$

Note that this implies that $(f_n(x))$ is a Cauchy sequence for every $x \in [a, b]$. Completeness of $\mathbb{R}$ (or $\mathbb{C}$) allows us to define

$$f(x) = \lim_{n \to \infty} f_n(x), \quad x \in [a, b].$$

Now, by letting $m \to \infty$ in (1.12), we get

$$\left| f_n(x) - f(x) \right| \leq \varepsilon \quad \text{for all } n \geq n_0 \text{ and all } x \in [a, b]. \tag{1.13}$$

Let $x_0 \in [a, b]$. Since f_{n_0} is continuous on $[a, b]$, there exists a $\delta > 0$ such that $\left| f_{n_0}(x_0) - f_{n_0}(y) \right| < \varepsilon$ for every $y \in [a, b]$ such that $|x_0 - y| < \delta$. Then

$$\left| f(x_0) - f(y) \right| \leq \left| f(x_0) - f_{n_0}(x_0) \right| + \left| f_{n_0}(x_0) - f_{n_0}(y) \right| + \left| f_{n_0}(y) - f(y) \right|$$
$$< \varepsilon + \varepsilon + \varepsilon = 3\varepsilon,$$

whenever $|x_0 - y| < \delta$. Since x_0 and ε are arbitrary, the above proves continuity of f. Finally, since (1.13) implies

$$\|f_n - f\| \leq \varepsilon \quad \text{for all } n \geq n_0,$$

the sequence (f_n) converges to f uniformly. $\qquad \square$

Definition 1.4.8.   (Convergent and absolutely convergent series)
A series $\sum_{n=1}^{\infty} x_n$ in a normed space E is called *convergent* if the sequence of partial sums converges in E, that is, there exists $x \in E$ such that $\|x_1 + x_2 + \cdots + x_n - x\| \to 0$ as $n \to \infty$. In that case we write $\sum_{n=1}^{\infty} x_n = x$. If $\sum_{n=1}^{\infty} \|x_n\| < \infty$, then the series is called *absolutely convergent*.

In general, an absolutely convergent series need not converge, as can be seen from Example 1.4.3.

Theorem 1.4.9. *A normed space is complete if and only if every absolutely convergent series converges.*

Proof: Let E be a Banach space. Suppose $x_n \in E$ and $\sum_{n=1}^{\infty} \|x_n\| < \infty$. Define

$$s_n = x_1 + \cdots + x_n, \quad n = 1, 2, \ldots$$

We will show that (s_n) is a Cauchy sequence. Let $\varepsilon > 0$ and let k be a positive integer such that

$$\sum_{n=k+1}^{\infty} \|x_n\| < \varepsilon.$$

Then, for every $m > n > k$, we have

$$\|s_m - s_n\| = \|x_{n+1} + \cdots + x_m\| \leq \sum_{r=n+1}^{\infty} \|x_r\| < \varepsilon.$$

This proves that the sequence (s_n) is a Cauchy sequence in E. Since E is complete, (s_n) converges in E, which means that the series $\sum_{n=1}^{\infty} x_n$ converges.

Assume now, that E is a normed space in which every absolutely convergent series converges. We need to prove that E is complete. Let (x_n) be a Cauchy sequence in E. Then, for every $k \in \mathbb{N}$, there exists a $p_k \in \mathbb{N}$ such that

$$\|x_m - x_n\| < 2^{-k} \quad \text{for all } m, n \geq p_k.$$

Without loss of generality, we can assume that the sequence (p_n) is strictly increasing. Since the series $\sum_{k=1}^{\infty} (x_{p_{k+1}} - x_{p_k})$ is absolutely convergent, it is convergent, and thus the sequence

$$x_{p_k} = x_{p_1} + (x_{p_2} - x_{p_1}) + \cdots + (x_{p_k} - x_{p_{k-1}})$$

converges to an element $x \in E$. Consequently,

$$\|x_n - x\| \leq \|x_n - x_{p_n}\| + \|x_{p_n} - x\| \to 0,$$

because (x_n) is a Cauchy sequence. □

Theorem 1.4.10. *A closed vector subspace of a Banach space is a Banach space itself.*

Proof: Let $(E, \|\cdot\|)$ be a Banach space and let F be a closed vector subspace of E. If (x_n) is a Cauchy sequence in F, then it is a Cauchy sequence in E, and therefore there exists an $x \in E$ such that $x_n \to x$. Since F is a closed subset of E, we have $x \in F$. Thus every Cauchy sequence in F converges to an element of F. □

Some spaces arising naturally in applications are not complete. It turns out that it is always possible to enlarge such a space to a complete space. The main ideas of the construction are described next.

Let $(E, \| \cdot \|)$ be a normed space. A normed space $(\tilde{E}, \| \cdot \|_1)$ is called a *completion* of $(E, \| \cdot \|)$ if

(a) There exists a one-to-one mapping $\Phi : E \to \tilde{E}$ such that

$$\Phi(\alpha x + \beta y) = \alpha \Phi(x) + \beta \Phi(y) \quad \text{for all } x, y \in E \text{ and } \alpha, \beta \in \mathbb{F},$$

(b) $\|x\| = \|\Phi(x)\|_1$ for every $x \in E$,

(c) $\Phi(E)$ is dense in $\tilde{E}$,

(d) $\tilde{E}$ is complete.

A space $\tilde{E}$, satisfying these conditions, can be formally defined as the space of equivalence classes of Cauchy sequences of elements of E. Two Cauchy sequences (x_n) and (y_n) of elements of E are called *equivalent* if $\lim_{n \to \infty} \|x_n - y_n\| = 0$. The set of all Cauchy sequences equivalent to a given Cauchy sequence (x_n) is denoted by $[(x_n)]$ and called the *equivalence class* of (x_n). The set of all equivalence classes of Cauchy sequences of elements of E, denoted by $\tilde{E}$, becomes a vector space when the addition and multiplication by scalars are defined as follows:

$$[(x_n)] + [(y_n)] = [(x_n + y_n)] \quad \text{and} \quad \lambda[(x_n)] = [(\lambda x_n)].$$

The norm on $\tilde{E}$ is defined by

$$\big\| [x_n] \big\|_1 = \lim_{n \to \infty} \|x_n\|.$$

By Lemma 1.4.4, $\lim_{n \to \infty} \|x_n\|$ exists for every Cauchy sequence (x_n). It easy to show that if $[(x_n)]$ and $[(y_n)]$ are equivalent, then $\lim_{n \to \infty} \|x_n\| = \lim_{n \to \infty} \|y_n\|$.

Now define $\Phi : E \to \tilde{E}$ by $\Phi(x) = [(x, x, \ldots)]$. Clearly, Φ satisfies (a) and (b). To show that $\Phi(E)$ is dense in $\tilde{E}$ note that every element $[(x_n)]$ of $\tilde{E}$ is the limit of the sequence $(\Phi(x_n))$.

Now we will prove that $\tilde{E}$ is complete. Let (X_n) be a Cauchy sequence in $\tilde{E}$. Since $\Phi(E)$ is dense in $\tilde{E}$, for every $n \in \mathbb{N}$ there exists $x_n \in E$ such that

$$\big\| \Phi(x_n) - X_n \big\|_1 < \frac{1}{n}.$$

From the inequalities

$$\|x_n - x_m\| = \big\| \Phi(x_n) - \Phi(x_m) \big\|_1$$

$$\leq \big\| \Phi(x_n) - X_n \big\|_1 + \|X_n - X_m\|_1 + \big\| X_m - \Phi(x_m) \big\|_1$$

$$\leq \|X_n - X_m\|_1 + \frac{1}{n} + \frac{1}{m},$$

we see that (x_n) is a Cauchy sequence in E. Define $X = [(x_n)]$. It remains to show that

$$\lim_{n \to \infty} \|X_n - X\|_1 = 0.$$

Indeed, we have

$$\|X_n - X\|_1 \leq \left\|X_n - \Phi(x_n)\right\|_1 + \left\|\Phi(x_n) - X\right\|_1 < \left\|\Phi(x_n) - X\right\|_1 + \frac{1}{n} \to 0,$$

because $\lim_{n \to \infty} \|\Phi(x_n) - X\|_1 = 0$.

1.5 Linear Mappings

First, we introduce some notation that will be used in the remainder of the book. Let E_1 and E_2 be vector spaces, and let L be a mapping from E_1 into E_2. If $y = L(x)$, then y is called the *image* of x.

If A is a subset of E_1, then $L(A)$ denotes the *image* of the set A, that is, $L(A)$ is the set of all elements in E_2, which are images of elements of A. If B is a subset of E_2, then $L^{-1}(B)$ denotes the *inverse image* of B, that is, $L^{-1}(B)$ is the set of all elements in E_1 whose images are elements of B. In symbols:

$$L(A) = \{L(x): x \in A\}, \qquad L^{-1}(B) = \{x \in E_1: L(x) \in B\}.$$

Note that the use of notation L^{-1} does not imply that L is invertible.

We often consider mappings, which are defined on proper subsets of vector spaces. Then it is important to specify the *domain* of L, denoted by $\mathcal{D}(L)$. The set $L(\mathcal{D}(L))$ is called the *range* of L and denoted by $\mathcal{R}(L)$, that is,

$$\mathcal{R}(L) = \{y \in E_2: L(x) = y \text{ for some } x \in \mathcal{D}(L)\}.$$

By the *null space* of L, denoted by $\mathcal{N}(L)$, we mean the set of all elements $x \in \mathcal{D}(L)$ such that $L(x) = 0$. Finally, by the *graph* of L, denoted by $\mathcal{G}(L)$, we mean the subset of $E_1 \times E_2$ defined as follows:

$$\mathcal{G}(L) = \{(x, y): x \in \mathcal{D}(L) \text{ and } y = L(x)\}.$$

Definition 1.5.1.   (Linear mapping)
A mapping $L: E_1 \to E_2$ is called a *linear mapping* if $L(\alpha x + \beta y) = \alpha L(x) + \beta L(y)$ for all $x, y \in E_1$ and all scalars α, β.

In the context of mappings between vector spaces we often write Lx instead of $L(x)$.

In Definition 1.5.1 we are assuming that the domain of L is a vector space. If the domain of L is a subset E_1 which is not a vector space, then we have to be more careful when defining linearity of L. We can say that a mapping L from a subset $S = \mathcal{D}(L)$ of a vector space E_1 into a vector space E_2 is linear if $L(\alpha x + \beta y) = \alpha Lx + \beta Ly$ for all $x, y \in S$ and all scalars α, β such that $\alpha x + \beta y \in S$. On the other hand, such a mapping L has a unique extension to a linear mapping from the vector space span $\mathcal{D}(L)$ into E_2. Indeed, if $y \in$ span $\mathcal{D}(L)$, then $y = \lambda_1 x_1 + \cdots + \lambda_n x_n$, for some $x_1, \ldots, x_n \in \mathcal{D}(L)$ and some scalars $\lambda_1, \ldots, \lambda_n$, and we can define $Ly = \lambda_1 Lx_1 + \cdots + \lambda_n Lx_n$. For this reason, we can always assume, without loss of generality, that the domain of a linear mapping is a vector space. Under this assumption, it can be easily shown that $\mathcal{R}(L)$, $\mathcal{N}(L)$, and $\mathcal{G}(L)$ are vector spaces.

In the remaining part of this section we will assume that both spaces E_1 and E_2 are normed spaces. We use the same symbol $\|\cdot\|$ to denote the norm on E_1 as well as on E_2. This will not lead to any misunderstanding.

Definition 1.5.2.   (Continuous mapping)
Let E_1 and E_2 be normed spaces. A mapping F from E_1 into E_2 is called *continuous at* $x_0 \in E_1$ if, for any sequence (x_n) of elements of E_1 convergent to x_0, the sequence $(F(x_n))$ converges to $F(x_0)$, that is, F is continuous at x_0 if $\|x_n - x_0\| \to 0$ implies $\|F(x_n) - F(x_0)\| \to 0$. If F is continuous at every $x \in E_1$, then we simply say that F is *continuous*.

Several examples of continuous mappings are discussed in Chapter 4 (Section 4.2). Here we only make the following simple but useful observation.

Example 1.5.3. The norm on a normed space E is a continuous mapping from E into $\mathbb{R}$. Indeed, if $\|x_n - x\| \to 0$, then $|\|x_n\| - \|x\|| \leq \|x_n - x\| \to 0$.   □

Continuity can be described in many different ways. The conditions in the following theorem characterize continuity in terms of open and closed sets. The proof of the theorem is left as an exercise.

Theorem 1.5.4. *Let* $F : E_1 \to E_2$. *The following conditions are equivalent:*

(a) *F is continuous;*

(b) *The inverse image $F^{-1}(U)$ of any open subset U of E_2 is open in E_1;*

(c) *The inverse image $F^{-1}(S)$ of any closed subset S of E_2 is closed in E_1.*

From now on we are going to limit our discussion to linear mappings.

Theorem 1.5.5. *If a linear mapping $L : E_1 \to E_2$ is continuous at some $x_0 \in E_1$, then it is continuous.*

Proof: Assume L is continuous at $x_0 \in E_1$. Let x be an arbitrary element of E_1 and let (x_n) be a sequence convergent to x. Then the sequence $(x_n - x + x_0)$ converges to x_0 and thus we have $\|Lx_n - Lx\| = \|L(x_n - x + x_0) - Lx_0\| \to 0$, which completes the proof. $\qquad\square$

Definition 1.5.6. (Bounded linear mapping)
A linear mapping $L : E_1 \to E_2$ is called *bounded* if there exists a number $\alpha > 0$ such that $\|Lx\| \le \alpha \|x\|$ for all $x \in E_1$.

Note that the condition in this definition is equivalent to saying that L is bounded by α on the unit sphere in E_1. More precisely, $\|Lx\| \le \alpha$ for all $x \in E_1$ such that $\|x\| = 1$.

Theorem 1.5.7. *A linear mapping is continuous if and only if it is bounded.*

Proof: If L is bounded and $x_n \to 0$, then $\|Lx_n\| \le \alpha \|x_n\| \to 0$. Thus, L is continuous at 0 and hence, by Theorem 1.5.5, L is continuous.

If L is not bounded, then for every $n \in \mathbb{N}$ there exists an $x_n \in E_1$ such that $\|Lx_n\| > n\|x_n\|$. Define

$$y_n = \frac{x_n}{n\|x_n\|}, \qquad n = 1, 2, \ldots .$$

Then $y_n \to 0$. Since $\|Ly_n\| > 1$ for all $n \in \mathbb{N}$, L is not continuous. $\qquad\square$

Since every linear mapping between finite dimensional spaces is bounded, it is continuous. The above theorem implies also that, for linear mappings, continuity and uniform continuity are equivalent.

The space of all linear mappings from a vector space E_1 into a vector space E_2 becomes a vector space if the addition and multiplication by scalars are defined as follows:

$$(L_1 + L_2)x = L_1 x + L_2 x \quad \text{and} \quad (\lambda L)x = \lambda(Lx).$$

If E_1 and E_2 are normed spaces, then the set of all bounded linear mappings from E_1 into E_2, denoted by $\mathcal{B}(E_1, E_2)$, is a vector subspace of the space defined above.

Theorem 1.5.8. *If E_1 and E_2 are normed spaces, then $\mathcal{B}(E_1, E_2)$ is a normed space with the norm defined by*

$$\|L\| = \sup_{\|x\|=1} \|Lx\|. \tag{1.14}$$

Proof: We will only show that the norm (1.14) satisfies the triangle inequality. If $L_1, L_2 \in \mathcal{B}(E_1, E_2)$ and $x \in E_1$ is such that $\|x\| = 1$, then

$$\|L_1 x + L_2 x\| \leq \|L_1 x\| + \|L_2 x\|.$$

This implies

$$\|L_1 x + L_2 x\| \leq \sup_{\|x\|=1} \|L_1 x\| + \sup_{\|x\|=1} \|L_2 x\| = \|L_1\| + \|L_2\|,$$

and hence

$$\|L_1 + L_2\| = \sup_{\|x\|=1} \|L_1 x + L_2 x\| \leq \|L_1\| + \|L_2\|. \qquad \square$$

It follows from (1.14) that $\|Lx\| \leq \|L\| \|x\|$ for all $x \in E_1$. In fact, $\|L\|$ is the least number α such that $\|Lx\| \leq \alpha \|x\|$ for all $x \in E_1$.

The norm defined in (1.14) is the standard norm on $\mathcal{B}(E_1, E_2)$ and this is the norm we mean when we say "the normed space $\mathcal{B}(E_1, E_2)$." Convergence with respect to this norm is called *uniform convergence*.

Theorem 1.5.9. *If E_1 is a normed space and E_2 is a Banach space, then $\mathcal{B}(E_1, E_2)$ is a Banach space.*

Proof: We only need to show that $\mathcal{B}(E_1, E_2)$ is complete. Let (L_n) be a Cauchy sequence in $\mathcal{B}(E_1, E_2)$ and let x be an arbitrary element of E_1. Then

$$\|L_m x - L_n x\| \leq \|L_m - L_n\| \|x\| \to 0 \quad \text{as } m, n \to \infty,$$

which shows that $(L_n x)$ is a Cauchy sequence in E_2. By completeness of E_2, there is a unique element $y \in E_2$ such that $L_n x \to y$. Since x is an arbitrary element of E_1, this defines a mapping L from E_1 into E_2:

$$Lx = \lim_{n \to \infty} L_n x.$$

We will show that $L \in \mathcal{B}(E_1, E_2)$ and $\|L_n - L\| \to 0$.

Clearly, L is a linear mapping. Since Cauchy sequences are bounded, there exists a constant α such that $\|L_n\| \leq \alpha$ for all $n \in \mathbb{N}$. Consequently,

$$\|Lx\| = \left\| \lim_{n \to \infty} L_n x \right\| = \lim_{n \to \infty} \|L_n x\| \leq \alpha \|x\|.$$

Therefore L is bounded and thus $L \in \mathcal{B}(E_1, E_2)$. It remains to prove that $\|L_n - L\| \to 0$. Let $\varepsilon > 0$, and let k be such that $\|L_m - L_n\| < \varepsilon$ for every $m, n \geq k$. If $\|x\| = 1$ and $m, n \geq k$, then

$$\|L_m x - L_n x\| \leq \|L_m - L_n\| < \varepsilon.$$

By letting $n \to \infty$, (m remains fixed), we obtain $\|L_m x - Lx\| \le \varepsilon$ for every $m \ge k$ and every $x \in E_1$ with $\|x\| = 1$. This means that $\|L_m - L\| \le \varepsilon$ for all $m > k$, which completes the proof. $\qquad\square$

Theorem 1.5.10. *If L is a continuous linear mapping from a subspace of a normed space E_1 into a Banach space E_2, then L has a unique extension to a continuous linear mapping defined on the closure of the domain $\mathcal{D}(L)$. In particular, if $\mathcal{D}(L)$ is dense in E_1, then L has a unique extension to a continuous linear mapping defined on the whole space E_1.*

Proof: If $x \in \mathrm{cl}\, \mathcal{D}(L)$, then there exists a sequence (x_n) in $\mathcal{D}(L)$ convergent to x. Since (x_n) is a Cauchy sequence,

$$\|L x_m - L x_n\| = \big\| L(x_m - x_n) \big\| \le \|L\| \, \|x_m - x_n\| \to 0, \qquad \text{as } m, n \to \infty.$$

Thus, (Lx_n) is a Cauchy sequence in E_2. Since E_2 is complete, there is a $z \in E_2$ such that $Lx_n \to z$. We want to define the value of the extension $\tilde{L}$ at x as $\tilde{L}x = z$, that is,

$$\tilde{L}x = \lim_{n\to\infty} L x_n, \quad x_n \in \mathcal{D}(L) \text{ and } x_n \to x.$$

This definition will be correct only if we can show that the limit z is the same for all sequences in $\mathcal{D}(L)$ convergent to x. Indeed, if $y_n \in \mathcal{D}(L)$ and $y_n \to x$, then

$$L y_n = L y_n - L x_n + L x_n = L(y_n - x_n) + L x_n \to z,$$

because $y_n - x_n \to 0$, and hence also $L(y_n - x_n) \to 0$. Clearly, $\tilde{L}$ is a linear mapping and $\tilde{L}x = Lx$ whenever $x \in \mathcal{D}(L)$. It remains to show that $\tilde{L}$ is continuous. Let $x \in \mathrm{cl}\, \mathcal{D}(L)$, $\|x\| = 1$. There exist $x_1, x_2, \ldots \in \mathcal{D}(L)$ such that $x_n \to x$. Then $\|x_n\| \to \|x\| = 1$ and

$$\big\| \tilde{L}x \big\| = \lim_{n\to\infty} \|L x_n\| \le \|L\|.$$

Thus, $\tilde{L}$ is bounded, hence continuous, and $\|\tilde{L}\| = \|L\|$. $\qquad\square$

Theorem 1.5.11. *If $L : E_1 \to E_2$ is a continuous linear mapping, then the null space $\mathcal{N}(L)$ is a closed subspace of E. Moreover, if the domain $\mathcal{D}(L)$ is closed, then the graph $\mathcal{G}(L)$ is a closed subspace of $E_1 \times E_2$.*

The proof is left as an exercise.

Spaces $\mathcal{B}(E, \mathbb{F})$ of bounded linear mappings from a normed space E into the scalar field $\mathbb{F}$ are of special interest. Elements of $\mathcal{B}(E, \mathbb{F})$ are called *functionals*.

The space $\mathcal{B}(E, \mathbb{F})$ is usually denoted by E' and called the *dual space* of E. Theorems proved in this section apply to dual spaces of normed spaces. Note that since the scalar field is a complete space, the dual space of a normed space is always a Banach space.

The last theorem we prove in this section is the Banach–Steinhaus theorem, also known as the uniform boundedness principle (Hugo Dyonizy Steinhaus (1887–1972)). It is one of the most important theorems in the theory of normed spaces. The standard proof of this theorem is based on a topological argument known as the Baire category theorem (see, for instance, E. Kreyszig (1978)) due to René-Louis Baire (1874–1932). We present a proof based on the diagonal theorem. This method was first introduced by Jan Mikusiński (1970) and then extensively used by Piotr Antosik and Charles Swartz (see Antosik and Swartz (1985) and Swartz (1992)). The advantage of this approach is that it is simple and no new concepts are needed.

Theorem 1.5.12. (Diagonal theorem)   *Let E be a normed space and let (x_{ij}), $i, j \in \mathbb{N}$, be an infinite matrix of elements of E. If*

(a) $\lim_{i \to \infty} x_{ij} = 0$ *for every $j \in \mathbb{N}$ and*

(b) *every increasing sequence of indices (p_j) has a subsequence (q_j) such that*

$$\lim_{i \to \infty} \sum_{j=1}^{\infty} x_{q_i q_j} = 0,$$

then $\lim_{i \to \infty} x_{ii} = 0$.

Proof: Suppose $\lim_{i \to \infty} x_{ii} \neq 0$. Then there exists an increasing sequence of indices (p_i) and some $\varepsilon > 0$ such that $\|x_{p_i p_i}\| \geq \varepsilon$ for all $i \in \mathbb{N}$. By (b), the sequence (p_i) has a subsequence (q_i) such that $\lim_{i \to \infty} \sum_{j=1}^{\infty} x_{q_i q_j} = 0$. Note that each row and each column in the matrix $(x_{q_i q_j})$ converges to 0. Set $r_1 = q_1$. Now let r_2 be the first index such that $r_2 > r_1$, $\|x_{q_i r_1}\| < \varepsilon/4$ for all $q_i \geq r_2$, and $\|x_{r_1 r_2}\| < \varepsilon/8$. Next, let r_3 be the first index such that $r_3 > r_2$, $\|x_{q_i r_2}\| < \varepsilon/8$ for all $q_i \geq r_3$, and $\|x_{r_j r_3}\| < \varepsilon/16$ for $j = 1, 2$. In the nth step we let r_n be the first index such that $r_n > r_{n-1}$, $\|x_{q_i r_{n-1}}\| < \varepsilon/2^n$ for all $q_i \geq r_n$, and $\|x_{r_j r_n}\| < \varepsilon/2^{n+1}$ for $j = 1, \ldots, n - 1$. Continuing this process, we construct an infinite matrix $(x_{r_i r_j})$ such that

$$\|x_{r_i r_j}\| < \varepsilon/2^{j+1} \quad \text{for all } i \text{ such that } i \neq j.$$

In view of (b), (r_j) has a subsequence (s_j) such that $\lim_{i \to \infty} \sum_{j=1}^{\infty} x_{s_i s_j} = 0$. Consider the matrix $(x_{s_i s_j})$. For every $i \in \mathbb{N}$, we have

$$\left\| \sum_{j=1}^{\infty} x_{s_i s_j} \right\| = \left\| x_{s_i s_i} + \sum_{i \neq j} x_{s_i s_j} \right\|$$

$$\geq \left| \|x_{s_i s_i}\| - \left\| \sum_{i \neq j} x_{s_i s_j} \right\| \right|$$

$$\geq \left| \|x_{s_i s_i}\| - \sum_{i \neq j} \|x_{s_i s_j}\| \right| \geq \frac{\varepsilon}{2}.$$

This, however, is impossible since $\lim_{i \to \infty} \sum_{j=1}^{\infty} x_{s_i s_j} = 0$. This contradiction proves the theorem. □

Theorem 1.5.13. (Banach–Steinhaus theorem) *Let $\mathcal{T}$ be a family of bounded linear mappings from a Banach space X into a normed space Y. If for every $x \in X$ there exists a constant M_x such that $\|Tx\| \leq M_x$ for all $T \in \mathcal{T}$, then there exists a constant $M > 0$ such that $\|T\| \leq M$ for all $T \in \mathcal{T}$.*

Proof: Suppose there is no such M. Then there exist $T_1, T_2, \ldots \in \mathcal{T}$ and $x_1, x_2, \ldots \in X$, such that $\|x_n\| \leq 1$, for all $n \in \mathbb{N}$, and $\|T_n x_n\| \to \infty$ as $n \to \infty$. For some increasing sequence of indices (p_n), we must have $\|T_{p_n} x_{p_n}\| \geq n2^n$ or

$$\left\| \frac{1}{n} T_{p_n} \frac{x_{p_n}}{2^n} \right\| \geq 1. \tag{1.15}$$

Consider the matrix (y_{ij}) defined by $y_{ij} = \frac{1}{i} T_{p_i} \frac{x_{p_j}}{2^j}$, $i, j \in \mathbb{N}$. Since the series $\sum_{j=1}^{\infty} \frac{x_{p_j}}{2^j}$ is absolutely convergent and X is complete, there is a $z \in X$ such that $z = \sum_{j=1}^{\infty} \frac{x_{p_j}}{2^j}$. Then

$$\left\| \sum_{j=1}^{\infty} y_{ij} \right\| = \left\| \sum_{j=1}^{\infty} \frac{1}{i} T_{p_i} \frac{x_{p_j}}{2^j} \right\| = \frac{1}{i} \left\| T_{p_i} \left(\sum_{j=1}^{\infty} \frac{x_{p_j}}{2^j} \right) \right\| = \frac{1}{i} \| T_{p_i}(z) \| \leq \frac{C}{i},$$

where C is a constant that depends on z. Consequently $\lim_{i \to \infty} \sum_{j=1}^{\infty} y_{ij} = 0$. Note that the same argument can be repeated for the matrix $(y_{q_i q_i})$ where (q_i) is an arbitrary increasing sequence of indices. Since also $\lim_{i \to \infty} y_{ij} = 0$ for all $j \in \mathbb{N}$, the assumptions of the diagonal theorem are satisfied and we must have

$$\lim_{i \to \infty} \frac{1}{i} T_{p_i} \frac{x_{p_i}}{2^i} = \lim_{i \to \infty} y_{ii} = 0.$$

But this contradicts (1.15). □

1.6 Contraction Mappings and the Banach Fixed Point Theorem

The name *fixed point theorem* is usually given to a result which says that, if a mapping f satisfies certain conditions, then there is a point z such that $f(z) = z$. Such a point z is called a *fixed point* of f. Theorems of this kind have numerous important applications. Some of them, in the theory of differential and integral equations, will be discussed in Chapter 5.

Example 1.6.1. Let $E = \mathcal{C}([0, 1])$ be the space of complex-valued continuous functions defined on the closed interval $[0, 1]$. Let T be defined by

$$(Tx)(t) = x(0) + \int_0^t x(\tau)d\tau.$$

Clearly, for any $a \in \mathbb{C}$, the function $x(t) = ae^t$ is a fixed point of T.   □

Theorem 1.6.4 proved in this section is a version of a theorem called the contraction theorem or the Banach fixed point theorem. The theorem is usually formulated for metric spaces because the algebraic structure of the space is not essential.

Definition 1.6.2.   (Contraction mapping)
A mapping f from a subset A of a normed space E into E is called a *contraction mapping* (or simply a *contraction*) if there exists a positive number $\alpha < 1$ such that

$$\|f(x) - f(y)\| \le \alpha \|x - y\| \quad \text{for all } x, y \in A. \tag{1.16}$$

Note that *contraction mappings are continuous.*

Example 1.6.3. Consider the nonlinear algebraic equation $x^3 - x - 1 = 0$. This equation has three roots. There are several ways of putting the equation in the form $Tx = x$, for example

$$Tx = (1 + x)^{1/3}, \quad Tx = x^3 - 1, \quad Tx = \frac{1}{x^2 - 1}.$$

The original equation has a root in $[1, 2]$. The mapping T defined by $T(x) = (1 + x)^{1/3}$ is a contraction on $[1, 2]$. Indeed, by the mean value theorem, we have

$$|Tx - Ty| = \left|(1 + x)^{1/3} - (1 + y)^{1/3}\right| \le \frac{2^{1/3}}{6}|x - y|.$$

Note that the other two mappings are not contractions. $\quad\square$

A number of important examples of contraction mappings will be discussed in Chapter 5.

Theorem 1.6.4. (Banach fixed point theorem) *Let F be a closed subset of a Banach space E and let f be a contraction mapping from F into F. Then there exists a unique $z \in F$ such that $f(z) = z$.*

Proof: Let $0 < \alpha < 1$ be such that

$$\|f(x) - f(y)\| \le \alpha \|x - y\|$$

for all $x, y \in F$. Let x_0 be an arbitrary point in F and let $x_n = f(x_{n-1})$ for $n = 1, 2, \ldots$. We will show that (x_n) is a Cauchy sequence. First observe that, for any $n \in \mathbb{N}$,

$$\|x_{n+1} - x_n\| \le \alpha \|x_n - x_{n-1}\| \le \alpha^2 \|x_{n-1} - x_{n-2}\| \le \cdots \le \alpha^n \|x_1 - x_0\|.$$

Hence, for any $m, n \in \mathbb{N}$ such that $m < n$, we have

$$\|x_n - x_m\| \le \|x_n - x_{n-1}\| + \|x_{n-1} - x_{n-2}\| + \cdots + \|x_{m+1} - x_m\|$$
$$\le \left(\alpha^{n-1} + \alpha^{n-2} + \cdots + \alpha^m\right)\|x_1 - x_0\|$$
$$\le \frac{\|x_1 - x_0\|}{1 - \alpha}\alpha^m \to 0 \quad \text{as } m \to \infty.$$

Thus, (x_n) is a Cauchy sequence. Since F is a closed subset of a complete space, there exists a $z \in F$ such that $x_n \to z$ as $n \to \infty$. We are going to show that z is the unique point such that $f(z) = z$. Indeed, since

$$\|f(z) - z\| \le \|f(z) - x_n\| + \|x_n - z\|$$
$$= \|f(z) - f(x_{n-1})\| + \|x_n - z\|$$
$$\le \alpha \|z - x_{n-1}\| + \|x_n - z\| \to 0 \quad \text{as } n \to \infty,$$

we have $\|f(z) - z\| = 0$, and thus $f(z) = z$. Suppose now $f(w) = w$ for some $w \in F$. Then

$$\|z - w\| = \|f(z) - f(w)\| \le \alpha \|z - w\|.$$

Since $0 < \alpha < 1$, we must have $\|z - w\| = 0$, which implies $z = w$. $\quad\square$

Note that the proof provides a practical way of finding or approximating the fixed point. The use of this method will be discussed in Chapter 5.

Example 1.6.5. We will apply the method of successive approximation to find the solution of $x^3 - x - 1 = 0$ in $[1, 2]$. In Example 1.6.3 we have shown that

$Tx = (1 + x)^{1/3}$ is a contraction mapping on $[1, 2]$ and its fixed point is the solution of the equation. We set the initial guess at $x_0 = 1$. Then we find

$$x_1 = Tx_0 \approx 1.2599,$$

$$x_2 = Tx_1 \approx 1.3123,$$

$$x_3 = Tx_2 \approx 1.3224,$$

$$x_4 = Tx_3 \approx 1.3243,$$

$$x_5 = Tx_4 \approx 1.3246,$$

$$x_6 = Tx_5 \approx 1.3247,$$

$$x_7 = Tx_6 \approx 1.3247.$$

Thus, the sixth iteration gives the root to four decimal places.   □

The number α in the definition of contraction mapping is assumed to be strictly less than 1. If this assumption is replaced by $\alpha \leq 1$ then the fixed point theorem in the above version is no longer true. Even if the condition (1.16) is replaced by strict inequality

$$\|Tx - Ty\| < \|x - y\| \quad \text{for all } x, y \in A,$$

T need not have a fixed point.

Example 1.6.6. Consider the function $f(x) = x + e^{-x}$ as a mapping from $\mathbb{R}^+$ into $\mathbb{R}^+$, where $\mathbb{R}^+$ denotes the set of all non-negative real numbers. For any $x, y \in \mathbb{R}^+$ we have, by the mean value theorem,

$$\left| f(x) - f(y) \right| < |x - y|,$$

since $|f'(\xi)| < 1$ for all $\xi \in \mathbb{R}^+$. However, f is not a contraction, because there is no $\alpha < 1$ such that $|f(x) - f(y)| \leq \alpha|x - y|$ for all $x, y \in \mathbb{R}^+$. Note that f does not have a fixed point.   □

1.7 Exercises

1. Prove that for every $x, y \in E$ there exists a unique $z \in E$ such that $x + z = y$.

2. Prove that for any vectors x, y, and z we have the following:

   (a) $x + (y - x) = y$.

   (b) $x - (y - z) = x - y + z$.

3. Let x be an element of a vector space and let λ be a scalar. Prove the following:

(a) If $\lambda \neq 0$ and $\lambda x = 0$, then $x = 0$.

(b) If $x \neq 0$ and $\lambda x = 0$, then $\lambda = 0$.

(c) $0x = 0$ and $(-1)x = -x$.

4. Check that the following are vector spaces:

(a) $\mathbb{R}^N$.

(b) $\mathbb{C}^N$.

(c) $C(\Omega)$, where Ω is a subset of $\mathbb{R}^N$.

(d) $C^k(\mathbb{R}^N)$.

(e) $C^\infty(\mathbb{R}^N)$.

5. Show that the family of all solutions of the ordinary differential equation $y'' + y = 0$ is a vector space.

6. Show that the family of all solutions of the integral equation $y(x) = \int_a^b K(x, t)y(t)dt$ is a vector space.

7. Show that the family of all solutions of the nonlinear ordinary differential equation $y'' = y^2$ is not a vector space.

8. Let $p, q > 1$ and $\frac{1}{p} + \frac{1}{q} = 1$. Prove that $x^{\frac{1}{p}} \leq \frac{1}{p}x + \frac{1}{q}$ for every $0 \leq x \leq 1$.

9. Prove that l^p is a proper vector subspace of l^q whenever $1 \leq p < q$.

10. Show that any vector of $\mathbb{R}^3$ is a linear combination of vectors $(1, 0, 0)$, $(1, 1, 0)$, and $(1, 1, 1)$.

11. Prove that every four vectors in $\mathbb{R}^3$ are linearly dependent.

12. Show that the functions $f_n(x) = x^n$, $n = 0, 1, 2, \ldots$, are linearly independent.

13. Show that the functions $f_n(x) = e^{nx}$, $n = 0, 1, 2, \ldots$, are linearly independent.

14. Prove that spaces $C(\Omega), C^k(\mathbb{R}^N), C^\infty(\mathbb{R}^N)$ are infinite dimensional.

15. Denote by l_0 the space of all infinite sequences of complex numbers (z_n) such that $z_n = 0$ for all but a finite number of indices n. Find a basis of l_0.

16. According to the definition of a basis $\mathcal{B}$ of a vector space E, any vector $x \in E$ can be represented in the form $x = \lambda_1 e_1 + \cdots + \lambda_n e_n$, where $e_1, \ldots, e_n \in \mathcal{B}$ and $\lambda_1, \ldots, \lambda_n$ are nonzero scalars. Prove that such a representation is unique.

17. Let $E_1, \ldots, E_n$ be vector spaces over the same scalar field $\mathbb{F}$. If $\dim E_k = d_k$, for $k = 1, \ldots, n$, what is the dimension of the Cartesian product $E_1 \times \cdots \times E_n$?

18. Show that $|\,\|x\| - \|y\|\,| \leq \|x - y\|$ for any pair of vectors in a normed space.

19. Show that

    (a) $\|(z_1, \ldots, z_N)\| = \sum_{n=1}^{N} |z|$,

    (b) $\|(z_1, \ldots, z_N)\| = \max\{|z_1|, \ldots, |z_N|\}$

    are norms on $\mathbb{C}^N$.

20. Let Ω be a closed bounded subset of $\mathbb{R}^N$. Show that $\|f\| = \max_{x \in \Omega} |f(x)|$ is a norm on $\mathcal{C}(\Omega)$.

21. Show that $\|z\| = \left(\sum_{n=1}^{\infty} |z_n|^p\right)^{1/p}$ is a norm on l^p for any $p \geq 1$.

22. Prove that in any normed space:

    (a) A convergent sequence has a unique limit.

    (b) If $x_n \to x$ and $\lambda_n \to \lambda$ (λ_n, λ are scalars), then $\lambda_n x_n \to \lambda x$.

    (c) If $x_n \to x$ and $y_n \to y$, then $x_n + y_n \to x + y$.

23. Show that two norms are equivalent if and only if they define the same bounded sets.

24. Using Theorem 1.3.11 prove that the following norms on $\mathbb{R}^2$ are equivalent:
$$\|(x, y)\|_1 = \sqrt{x^2 + y^2}, \qquad \|(x, y)\|_2 = |x| + |y|,$$
$$\|(x, y)\|_3 = \max\{|x|, |y|\}.$$

25. Prove Theorem 1.3.20.

26. Consider the convergence in $\mathcal{C}(\mathbb{R})$ defined as follows: $f_n \to f$ if the sequence (f_n) converges to f uniformly on every bounded interval $[a, b]$. Is it possible to define this convergence by a norm on $\mathcal{C}(\mathbb{R})$?

27. Show that the only dense linear subspace of $\mathbb{C}^N$ is $\mathbb{C}^N$ itself.

28. Find a sequence of elements of l^1 that is convergent in l^2 but divergent in l^1. What about the converse? Can you give similar example for any pair of spaces l^p and l^q with $p \neq q$?

29. Show, without using Theorem 1.3.13, that the following norms on $\mathbb{C}^N$ are equivalent:
$$\|(z_1, \ldots, z_N)\| = \sqrt{|z_1|^2 + \cdots + |z_N|^2},$$

$$\|(z_1, \ldots, z_N)\| = \sum_{n=1}^{N} |z_n|,$$

$$\|(z_1, \ldots, z_N)\| = \max\{|z_1|, \ldots, |z_N|\}.$$

30. Find two norms on $\mathcal{C}([0, 1])$ which are not equivalent.

31. Prove that a subset S of a normed space is bounded if and only if $\|\lambda_n x_n\| \to 0$ for any $x_n \in S$ and any scalars $\lambda_n \to 0$.

32. Show that, in a normed space, $x_n \to x$ implies $\frac{1}{n}(x_1 + \cdots + x_n) \to x$.

33. Show that, in a normed space, if every subsequence of a sequence (x_n) contains a subsequence convergent to x, then (x_n) itself converges to x.

34. Prove Theorem 1.3.27.

35. Let E_1 and E_2 be normed spaces. Show that a function $f : E_1 \to E_2$ is continuous if and only if for every $x \in E_1$ and $\varepsilon > 0$ there exists a $\delta > 0$ such that $\|f(x) - f(y)\| < \varepsilon$ whenever $\|x - y\| < \delta$.

36. Let E_1 and E_2 be normed spaces and let $f : E_1 \to E_2$ be a continuous function. Prove that, if $S \subset E_1$ is compact, then $f(S)$ is compact in E_2.

37. Let $f : E \to \mathbb{R}$ be a continuous function and let $S \subset E$ be compact. Prove that f attains the minimum and the maximum values on S.

38. Is the sequence of functions $f_n(x) = \frac{nx}{1+nx^2}$ a Cauchy sequence in $\mathcal{C}([0, 1])$?

39. Prove Theorem 1.4.2.

40. Show that the sum of two Cauchy sequences is a Cauchy sequence.

41. Prove completeness of l^p for arbitrary $p \geq 1$.

42. Give an example of an incomplete normed space.

43. Show that, in a Banach space, every sequence (x_n) convergent to zero contains a subsequence (x_{p_n}) such that the series $\sum_{n=1}^{\infty} x_{p_n}$ converges.

44. Consider the space $\mathcal{C}([a, b])$ with the norm defined as $\|f\| = \int_a^b |f(t)| dt$. Is this a Banach space?

45. Show that $L(f)(x) = \int_0^x f(t) dt$ defines a continuous linear mapping from $\mathcal{C}([0, 1])$ into itself.

46. Give an example of a linear mapping from a normed space into a normed space which is not continuous.

47. Prove that a linear mapping from a normed space into a normed space is continuous if and only if it maps bounded sets to bounded sets.

48. Prove Theorem 1.5.4.

49. Prove Theorem 1.5.11.

50. Show that $\mathcal{B}(E_1, E_2)$ is a vector space.

51. A sequence of mappings $L_n \in \mathcal{B}(E_1, E_2)$ *converges strongly* to $L \in \mathcal{B}(E_1, E_2)$ if for every $x \in E_1$ we have $\|L_n x - Lx\| \to 0$ as $n \to \infty$. Show that uniform convergence implies strong convergence, but not conversely.

52. Let $E = \mathcal{C}^{\infty}([a, b])$ be the space of all infinitely differentiable functions on the interval $[a, b]$ with $\|f\| = \max_{[a,b]} |f(x)|$. Is the differential operator $D = \frac{d}{dx}$ a contraction mapping?

53. Define $T : \mathbb{R}^2 \to \mathbb{R}^2$ by $T(x, y) = (y^{1/3}, x^{1/3})$. What are the fixed points of T? In which quadrants of the xy-plane is T a contraction?

54. Is the mapping $T : \mathbb{R} \to \mathbb{R}$ defined by

$$Tx = \begin{cases} x - \frac{1}{2}e^x & \text{for } x \le 0, \\ -\frac{1}{2} + \frac{1}{2}x & \text{for } x > 0 \end{cases}$$

a contraction?

55. Consider the space $\mathbb{R}^N$. Elements of $\mathbb{R}^N$ are denoted by $x = (x_1, \ldots, x_N)$, $y = (y_1, \ldots, y_N)$, and so on. The norm on $\mathbb{R}^N$ is defined by $\|x - y\| = \max_{1 \le n \le N} |x_n - y_n|$. Let $T : \mathbb{R}^N \to \mathbb{R}^N$ defined by $Tx = y$ where $y_k = \sum_{n=0}^{N} a_{kn}x_n + b_k$, $k = 1, \ldots, N$. Under what conditions is T a contraction mapping?

The Lebesgue Integral

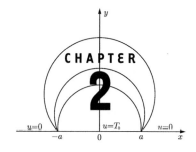

"In contrast to the classical Riemann integral, my new concept of the integral provides a better answer to the question of the connection between differentiation and integration (main theorem of calculus)."

Henri Leon Lebesgue

2.1 Introduction

The main purpose of this book is to present basic methods and applications of Hilbert spaces. One of the most important examples of Hilbert spaces, from the point of view of both theory and applications, is the space of Lebesgue square integrable functions on $\mathbb{R}^N$. Thus, the Lebesgue integral is essential for understanding some of the most important aspects of Hilbert space theory.

Since Henri Lebesgue (1875–1941), in a series of five short notes published between 1899 and 1901, introduced a new integral, now universally called the Lebesgue integral, numerous approaches to the theory have been developed. In the most common approach, one studies first the concept of measure and then defines the integral. This is not a very efficient way if what is needed is the integral, not measure. Moreover, the theory of measure is often considered too difficult to be taught in an undergraduate course.

In this book, we introduce the Lebesgue integral following the idea of H.M. MacNeille (1907–1973) and J. Mikusiński (1913–1987). This approach is fast and direct (integral is introduced without auxiliary concepts like measure) and can be easily taught at an undergraduate level. Since the case of real-valued functions on $\mathbb{R}$ exhibits the essential ideas, we first present the construction in great detail in that case. Then we generalize it to complex-valued functions on $\mathbb{R}$. Finally, we describe the extension of the construction to complex-valued functions on $\mathbb{R}^N$.

2.2 Step Functions

By a *step function* on the real line $\mathbb{R}$ we mean a finite linear combination of characteristic functions of semiopen intervals $[a, b) \subseteq \mathbb{R}$. Thus, for every step function f, there are intervals $[a_1, b_1), \ldots, [a_n, b_n)$ and numbers $\lambda_1, \ldots, \lambda_n \in \mathbb{R}$ such that

$$f = \lambda_1 f_1 + \cdots + \lambda_n f_n, \tag{2.1}$$

where f_k is the characteristic function of $[a_k, b_k)$, that is, $f_k(x) = 1$ if $x \in [a_k, b_k)$, and $f(x) = 0$ otherwise. Clearly, representation of a step function in the form (2.1) is not unique. On the other hand, if we assume that intervals $[a_k, b_k)$ are disjoint and the minimal number of intervals is used, then the representation is unique. Such a representation can be obtained in the following way:

Let f be a step function and let $a_0, a_1, \ldots, a_n$ be all points of discontinuity of f. In other words, $a_0, a_1, \ldots, a_n$ are the points where the graph of f has a jump. We can always assume that the points are indexed so that $a_0 < a_1 < \cdots < a_n$. Denote by g_k $(k = 1, \ldots, n)$ the characteristic function of the interval $[a_{k-1}, a_k)$. Then

$$f = \alpha_1 g_1 + \cdots + \alpha_n g_n,$$

where $\alpha_k = f(a_{k-1}), k = 1, \ldots, n$. This representation satisfies the required conditions. It is called the *basic representation* of f. Since this definition does not make much sense if $f = 0$, we call $f = 0$ the basic representation of the zero function.

The collection of all step functions on $\mathbb{R}$ is a vector space. The absolute value of a step function is again a step function. If $f = \alpha_1 f_1 + \cdots + \alpha_n f_n$ is the basic representation of a step function f, then $|f| = |\alpha_1| f_1 + \cdots + |\alpha_n| f_n$.

For any real-valued functions f and g, we have

$$\min\{f, g\} = \frac{1}{2}(f + g - |f - g|) \quad \text{and} \quad \max\{f, g\} = \frac{1}{2}(f + g + |f - g|).$$

Figure 2.1   A "typical" step function.

Thus, if f and g are step functions, then $\min\{f, g\}$ and $\max\{f, g\}$ are also step functions. Finally, it is easy to see that the collection of all step functions is closed under translations, that is, if f is a step function and $\tau_z f$ is the function defined by $\tau_z f(x) = f(x - z)$ for some $z \in \mathbb{R}$, then $\tau_z f$ is a step function. Note that if f is the characteristic function of $[a, b)$, then $\tau_z f$ is the characteristic function of $[a + z, b + z)$. Hence, if $f = \lambda_1 f_1 + \cdots + \lambda_n f_n$ is the basic representation of f, then $\tau_z f = \lambda_1 \tau_z f_1 + \cdots + \lambda_n \tau_z f_n$ is the basic representation of $\tau_z f$.

By the *support* of a nonzero function f, denoted by $\operatorname{supp} f$, we mean the set of all points $x \in \mathbb{R}$ for which $f(x) \neq 0$. The support of a nonzero step function is always a finite union of semiopen intervals. If $f = 0$, $\operatorname{supp} f = \emptyset$.

Definition 2.2.1. (Integral of a step function)
The *integral* $\int f$ of a step function $f(x) = \lambda_1 f_1(x) + \cdots + \lambda_n f_n(x)$, where f_k is the characteristic function of $[a_k, b_k)$, $k = 1, \ldots, n$, is defined by

$$\int f = \lambda_1(b_1 - a_1) + \cdots + \lambda_n(b_n - a_n).$$

Clearly, the value $\int f$ is equal to the Riemann integral of f. From the properties of the Riemann integral it follows that the defined integral does not depend on a particular representation. This fact is of importance for construction of the Lebesgue integral. The independence can be proved without using properties of the Riemann integral. The reader is asked to provide an elementary proof as an exercise.

Theorem 2.2.2. *For any step functions f and g we have*

(a) $\int (f + g) = \int f + \int g$;

(b) $\int \lambda f = \lambda \int f$, $\lambda \in \mathbb{R}$;

(c) $f \leq g$ *implies* $\int f \leq \int g$;

(d) $|\int f| \leq \int |f|$;

(e) $\int \tau_z f = \int f$, $z \in \mathbb{R}$.

Proof: Properties (a) and (b) follow directly from Definition 2.2.1. To prove (c) we first show that $f \geq 0$ implies $\int f \geq 0$. Indeed, if $f = 0$, then $\int f = 0$ by (b). If $f \geq 0$ and f does not vanish identically on $\mathbb{R}$, then all the coefficients in the basic representation of f are positive, and thus $\int f > 0$. Now, if $f \leq g$, then $g - f \geq 0$ and hence $\int (g - f) \geq 0$, which gives us $\int f \leq \int g$, by (a) and (b). Since $f \leq |f|$ and $-f \leq |f|$, we have $\int f \leq \int |f|$

and $\int (-f) \leq \int |f|$, by (c), which implies $|\int f| \leq \int |f|$, by (b). Property (e) is obvious in view of the earlier remarks.   □

A rather obvious property of the integral of step functions is formulated in the following lemma. It will be used in the proof of Theorem 2.2.6. The easy proof is left as an exercise.

Lemma 2.2.3. *Let f be a step function whose support is contained in the union of semiopen intervals $[a_1, b_1), \ldots, [a_n, b_n)$. If $|f| < M$, for some constant M, then*

$$\int |f| \leq M \sum_{k=1}^{n} (b_k - a_k).$$

Lemma 2.2.4. *Let $[a_1, b_1), [a_2, b_2), \ldots$ be a partition of an interval $[a, b)$, that is, the intervals $[a_1, b_1), [a_2, b_2), \ldots$ are disjoint and*

$$\bigcup_{n=1}^{\infty} [a_n, b_n) = [a, b). \tag{2.2}$$

Then

$$\sum_{n=1}^{\infty} (b_n - a_n) = b - a. \tag{2.3}$$

Proof: If $c \in (a, b]$, then $[a_1, b_1) \cap [a, c), [a_2, b_2) \cap [a, c), \ldots$ is a partition of $[a, c)$. Denote by S the set of all points $c \in (a, b]$ such that

$$\sum_{a_n < b_{c,n}} (b_{c,n} - a_n) = c - a, \tag{2.4}$$

where $b_{c,n} = \min\{b_n, c\}$ and the summation is over all those n for which $a_n < b_{c,n}$. Note that S is not empty because $a = a_{n_0}$ for some $n_0 \in \mathbb{N}$ and then $b_{n_0} \in S$. It suffices to prove that $b \in S$.

We first prove that $\mathrm{lub}\, S \in S$ ($\mathrm{lub}\, S$ denotes the least upper bound of S). Indeed, if $s = \mathrm{lub}\, S$ and $\{s_n\}$ is a nondecreasing sequence of elements of S convergent to s, then

$$s_n - a = \sum_{a_m < b_{s_n,m}} (b_{s_n,m} - a_m) \leq \sum_{a_m < b_{s,m}} (b_{s,m} - a_m) \leq s - a. \tag{2.5}$$

Since $s_n - a \to s - a$, (2.5) implies $\sum_{a_m < b_{s,m}} (b_{s,m} - a_m) = s - a$, and consequently $s \in S$.

Now we show that $s = b$. Suppose $s < b$. Then $s \in [a_k, b_k)$ for some $k \in \mathbb{N}$, and thus $b_k \in S$. But this means that $b_k > s$, contradicting the definition of s.   □

This lemma may sound like something obvious, something that should not require a proof. On the other hand, the proof is unexpectedly difficult. The following example shows that the same property formulated for rational numbers is false, although it sounds equally obvious. This indicates that some special properties of the reals are essential here.

Example 2.2.5. We use the following notation: if $a, b \in \mathbb{Q}$, then $[a, b)_\mathbb{Q} = \{x \in \mathbb{Q}: a \leq x < b\}$. We will show that for any $\varepsilon > 0$, $c \subset \mathbb{Q}$, there are intervals $[a_n, b_n)_\mathbb{Q}$ such that $[0, 1)_\mathbb{Q} \subseteq \bigcup_{n=1}^\infty [a_n, b_n)_\mathbb{Q}$ and $\sum_{n=1}^\infty (b_n - a_n) < \varepsilon$.

Indeed, since the set $[0, 1)_\mathbb{Q}$ is countable, we can write $[0, 1)_\mathbb{Q} = \{q_1, q_2, \ldots\}$. For $n = 1, 2, \ldots$, define $a_n = q_n$ and $b_n = q_n + \varepsilon/2^{n+1}$. The intervals $[a_n, b_n)_\mathbb{Q}$ will not be disjoint and need not be contained in $[0, 1)$, but we can easily define new intervals $[a_n, b_n)_\mathbb{Q}$ that have those properties. The details of such a construction are left as an exercise. □

The next theorem describes an important property of the integral of step functions.

Theorem 2.2.6. *Let (f_n) be a nonincreasing sequence of non-negative step functions such that $\lim_{n \to \infty} f_n(x) = 0$ for every $x \in \mathbb{R}$. Then $\lim_{n \to \infty} \int f_n = 0$.*

Proof: Since the sequence $(\int f_n)$ is nonincreasing and bounded from below (by 0), it converges. Let

$$\lim_{n \to \infty} \int f_n = \varepsilon. \tag{2.6}$$

Suppose $\varepsilon > 0$. Let $[a, b)$ be an interval containing the support of f_1 (and thus the support of every f_n, $n = 1, 2, \ldots$). Let $\alpha = \frac{\varepsilon}{2(b-a)}$. For $n = 1, 2, \ldots$ define

$$A_n = \{x \in [a, b): f_n(x) < \alpha\} \quad \text{and} \quad B_1 = A_1, \quad B_n = A_n \setminus A_{n-1} \text{ for } n \geq 2.$$

Note that B_n's are disjoint (because $A_{n-1} \subseteq A_n$) and $\bigcup_{n=1}^\infty B_n = [a, b)$ (because $\bigcup_{n=1}^\infty A_n = [a, b)$). Since f_n's are step functions, B_n's are finite unions of disjoint semiopen intervals, say

$$B_n = [a_{n,1}, b_{n,1}) \cup \cdots \cup [a_{n,k_n}, b_{n,k_n}).$$

The intervals

$$[a_{1,1}, b_{1,1}), \ldots, [a_{1,k_1}, b_{1,k_1}), \ldots, [a_{n,1}, b_{n,1}), \ldots, [a_{n,k_n}, b_{n,k_n}), \ldots$$

satisfy the assumptions of Lemma 2.2.4. Therefore

$$\sum_{n=1}^\infty \sum_{k=1}^{k_n} (b_{n,k} - a_{n,k}) = b - a.$$

Let $n_0 \in \mathbb{N}$ be such that

$$\sum_{n=n_0+1}^{\infty} \sum_{k=1}^{k_n} (b_{n,k} - a_{n,k}) < \delta, \tag{2.7}$$

where $\delta = \frac{\varepsilon}{2 \max |f_1|}$. Set $B = B_1 \cup \cdots \cup B_{n_0}$ and define two auxiliary functions g and h:

$$g(x) = \begin{cases} f_{n_0}(x) & \text{for } x \in B, \\ 0 & \text{otherwise,} \end{cases} \quad \text{and} \quad h(x) = \begin{cases} 0 & \text{for } x \in B, \\ f_{n_0}(x) & \text{otherwise.} \end{cases}$$

Since $B \subset A_{n_0}$, we have $f_{n_0}(x) < \alpha$ for $x \in B$. Consequently, $g(x) < \alpha$ for all $x \in \mathbb{R}$, which gives

$$\int g < \alpha(b - a) = \frac{\varepsilon}{2}, \tag{2.8}$$

by Lemma 2.2.3. Using the same lemma for h, we obtain

$$\int h < \delta \max |f_{n_0}| \le \delta \max |f_1| = \frac{\varepsilon}{2}, \tag{2.9}$$

because of (2.7). As $f_{n_0} = g + h$, we have

$$\int f_{n_0} = \int g + \int h < \varepsilon,$$

by (2.8) and (2.9). Since the sequence $(\int f_n)$ is nonincreasing, we conclude

$$\lim_{n \to \infty} \int f_n \le \int f_{n_0} < \varepsilon,$$

which contradicts (2.6). Therefore $\varepsilon = 0$, which completes the proof. $\square$

In the following corollary infinity is allowed as the limit.

Corollary 2.2.7. *Let (f_n) be nondecreasing sequences of step functions. If $\lim_{n \to \infty} f_n(x) \ge 0$ for every $x \in \mathbb{R}$, then $\lim_{n \to \infty} \int f_n \ge 0$.*

Proof: The step functions $\max\{0, -f_n\}$, $n = 1, 2, \ldots$, form a nonincreasing sequence, which converges to zero for every $x \in \mathbb{R}$. Hence, by Theorem 2.2.6, we have $\lim_{n \to \infty} \int \max\{0, -f_n\} = 0$. Since, $f_n = \max\{0, f_n\} - \max\{0, -f_n\}$ for all $n \in \mathbb{N}$, we have

$$\int f_n = \int \max\{0, f_n\} - \int \max\{0, -f_n\}.$$

Now, by letting $n \to \infty$, we obtain the desired inequality. $\square$

2.3 Lebesgue Integrable Functions

The technical preparations of the previous section allow us now to give a simple definition of Lebesgue integrable functions.

Definition 2.3.1.   (Lebesgue integrable function)
A real valued function f defined on $\mathbb{R}$ is called *Lebesgue integrable* if there exists a sequence of step functions (f_n) such that the following two conditions are satisfied:

(a) $\displaystyle\sum_{n=1}^{\infty} \int |f_n| < \infty$;

(b) $\displaystyle f(x) = \sum_{n=1}^{\infty} f_n(x)$ for every $x \in \mathbb{R}$ such that $\displaystyle\sum_{n=1}^{\infty} |f_n(x)| < \infty$.

The integral of f is then defined by

$$\int f = \sum_{n=1}^{\infty} \int f_n. \tag{2.10}$$

If a function f and a sequence of step functions (f_n) satisfy (a) and (b), then we write

$$f \simeq \sum_{n=1}^{\infty} f_n \quad \text{or} \quad f \simeq f_1 + f_2 + \cdots.$$

Note that because of (a), the series $\sum_{n=1}^{\infty} \int f_n$ is convergent. However, to prove that the Lebesgue integral (2.10) is well defined, we need to show that the number $\int f$ is independent of a particular representation $f \simeq f_1 + f_2 + \cdots$. This follows easily from the next lemma.

Lemma 2.3.2. *If $f \simeq f_1 + f_2 + \cdots$ and $f \geq 0$, then $\int f_1 + \int f_2 + \cdots \geq 0$.*

Proof: Let $\varepsilon > 0$ and let $n_0 \in \mathbb{N}$ be such that

$$\sum_{n=n_0+1}^{\infty} \int |f_n| < \varepsilon, \tag{2.11}$$

(n_0 exists by (a) in Definition 2.3.1). Define

$$g_n = f_1 + \cdots + f_{n_0} + |f_{n_0+1}| + \cdots + |f_{n_0+n}|,$$

for $n = 1, 2, \ldots$. Since (g_n) is a nondecreasing sequence of step functions such that $\lim_{n\to\infty} g_n(x) \geq 0$ for all $x \in \mathbb{R}$, we have $\lim_{n\to\infty} \int g_n \geq 0$, by Corollary 2.2.7. Thus,

$$0 \leq \int f_1 + \cdots + \int f_{n_0} + \sum_{n=n_0+1}^{\infty} \int |f_n|$$

$$\leq \int f_1 + \cdots + \int f_{n_0} + \sum_{n=n_0+1}^{\infty} \left(\int f_n + \int |f_n| \right) + \sum_{n=n_0+1}^{\infty} \int |f_n|$$

$$= \sum_{n=1}^{\infty} \int f_n + 2 \sum_{n=n_0+1}^{\infty} \int |f_n|$$

$$\leq \sum_{n=1}^{\infty} \int f_n + 2\varepsilon,$$

proving the lemma, since ε is arbitrary.   □

Corollary 2.3.3. *If $f \simeq \sum_{n=1}^{\infty} f_n$ and $f \simeq \sum_{n=1}^{\infty} g_n$, then $\sum_{n=1}^{\infty} \int f_n = \sum_{n=1}^{\infty} \int g_n$.*

Proof: Since $0 \simeq f_1 - g_1 + f_2 - g_2 + \cdots$, we have, by Lemma 2.3.2,

$$\int f_1 - \int g_1 + \int f_2 - \int g_2 + \cdots \geq 0$$

and hence

$$\sum_{n=1}^{\infty} \int f_n - \sum_{n=1}^{\infty} \int g_n \geq 0,$$

because both series are absolutely convergent. Similarly, we obtain

$$\sum_{n=1}^{\infty} \int g_n - \sum_{n=1}^{\infty} \int f_n \geq 0,$$

and therefore the sums of both series are equal.   □

Corollary 2.3.3 implies that step functions are integrable and that in this case integrals in Definition 2.2.1 and Definition 2.3.1 are equal (which justifies the use of the same symbol in both cases). Indeed, if f is a step function, then $f \simeq f + 0 + 0 + \cdots$. It can be proved (see Section 2.10) that every Riemann integrable function is Lebesgue integrable and both integrals are equal.

The space of all Lebesgue integrable functions defined on $\mathbb{R}$ is denoted by $L^1(\mathbb{R})$. In the remainder of this book, Lebesgue integrable functions are called simply *integrable*.

Theorem 2.3.4. $L^1(\mathbb{R})$ *is a vector space and $\int$ is a linear functional on $L^1(\mathbb{R})$. Moreover, if $f, g \in L^1(\mathbb{R})$ and $f \leq g$, then $\int f \leq \int g$.*

Proof: If $f \simeq \sum_{n=1}^{\infty} f_n$, $g \simeq \sum_{n=1}^{\infty} g_n$, and $\lambda \in \mathbb{R}$, then

$$f + g \simeq f_1 + g_1 + f_2 + g_2 + \cdots$$

and

$$\lambda f \simeq \lambda f_1 + \lambda f_2 + \cdots,$$

and consequently $f + g \in L^1(\mathbb{R})$ and $\lambda f \in L^1(\mathbb{R})$. Moreover,

$$\int (f + g) = \int f + \int g \quad \text{and} \quad \int \lambda f = \lambda \int f.$$

If $f \leq g$, then $g - f \geq 0$, and hence $\int (g - f) \geq 0$, by Lemma 2.3.2. This implies $\int g - \int f \geq 0$, proving the theorem. $\qquad\square$

Definition 2.3.1 is fairly simple, although it may be a little disturbing. Condition (a) does not involve f in any way. Condition (b) says that f equals the sum of series $\sum_{n=1}^{\infty} f_n$ at those points where the series converges absolutely. How big is the set? If we could find a sequence of step functions f_n satisfying (a) such that $\sum_{n=1}^{\infty} |f_n|$ diverges at every point of an open interval (a, b), then we could write $f \simeq f_1 + f_2 + \cdots$ for any function f that vanishes outside of (a, b). Since we have proved that the integral is well defined, we know that such a sequence (f_n) cannot exists. In fact, the series cannot diverge at every point of a nonempty open interval. Actually, we can prove more: If $f \simeq f_1 + f_2 + \cdots$ and Z denotes the set of all points where the series $\sum_{n=1}^{\infty} |f_n(x)|$ diverges, then the integral of the characteristic function of Z equals 0. Indeed, if g is the characteristic function of Z, then

$$f + g \simeq f_1 + f_2 + \cdots,$$

and thus

$$\int (f + g) = \int f.$$

This, in view of Theorem 2.3.4, implies $\int g = 0$.

If $\int g = 0$, then the set Z must be, in some sense, a small set. Such a set is called a null set. We can say that series $\sum_{n=1}^{\infty} f_n(x)$ converges to $f(x)$ for all x except a null set. This type of convergence is called convergence almost everywhere. Null sets and convergence almost everywhere will be discussed in Section 2.7.

At this point the reader may expect some examples of Lebesgue integrable functions. We know that every step function is Lebesgue integrable, but this is definitely not a satisfactory example. For instance, we would like to see examples of functions which are Lebesgue integrable but not Riemann integrable.

Unfortunately all we have at this time is the definition of the integral, and it would be awkward to try to prove that a function is integrable by expanding the function into a series of step functions. It would be like trying to evaluate $\int_0^1 e^x dx$ using Riemann sums. The reader has to be patient and let the theory develop a little bit further, and then we will easily construct nontrivial examples of Lebesgue integrable functions.

2.4 The Absolute Value of an Integrable Function

In the previous section, we mention that all Riemann integrable functions are Lebesgue integrable. (Recall that, by definition, a Riemann integrable function vanishes outside a bounded interval.) For improper integrals it is not so. For instance, the integral

$$\int_{-\infty}^{\infty} \frac{\sin x}{x} dx$$

converges, although the function is not Lebesgue integrable, as we can see from the next theorem.

Theorem 2.4.1. *If $f \in L^1(\mathbb{R})$ and $f \simeq f_1 + f_2 + \cdots$, then $|f| \in L^1(\mathbb{R})$ and we have*

$$\left| \int f \right| \le \int |f| \le \sum_{n=1}^{\infty} \int |f_n|.$$

Proof: Let $f \simeq f_1 + f_2 + \cdots$. Define

$$Z = \left\{ x \in \mathbb{R}: \sum_{n=1}^{\infty} |f_n(x)| < \infty \right\}$$

and

$$s_n = f_1 + \cdots + f_n, \quad \text{for } n = 1, 2, \ldots.$$

Then $f(x) = \lim_{n \to \infty} s_n(x)$ for every $x \in Z$. Hence also

$$|f(x)| = \lim_{n \to \infty} |s_n(x)|,$$

or, equivalently,

$$|f(x)| = |s_1(x)| + (|s_2(x)| - |s_1(x)|) + (|s_3(x)| - |s_2(x)|) + \cdots$$

for every $x \in Z$. If we let $g_1 = |s_1|$ and $g_n = |s_n| - |s_{n-1}|$ for $n \ge 2$, then

$$|f(x)| = \sum_{n=1}^{\infty} g_n(x)$$

for every $x \in Z$. Moreover, since

$$|g_n| = \big||s_n| - |s_{n-1}|\big| \leq |s_n - s_{n-1}| = |f_n|, \qquad (2.12)$$

we have $\int |g_n| \leq \int |f_n|$, and hence $\sum_{n=1}^{\infty} \int |g_n| < \infty$. It may seem that $|f| \simeq g_1 + g_2 + \cdots$, but there may exist points not in Z for which the series $\sum_{n=1}^{\infty} |g_n(x)|$ converges and we cannot guarantee that $\sum_{n=1}^{\infty} g_n(x) = |f(x)|$ at those points. We can get rid of those points by adding and subtracting terms of the series $\sum_{n=1}^{\infty} f_n$:

$$|f| \simeq g_1 + f_1 - f_1 + g_2 + f_2 - f_2 + \cdots. \qquad (2.13)$$

This does not change convergence at points of Z, but makes the series diverge at other points. Since

$$\int |g_1| + \int |f_1| + \int |f_1| + \int |g_2| + \int |f_2| + \int |f_2| + \cdots < \infty,$$

the expansion (2.13) is valid and thus $|f|$ is integrable. Moreover,

$$\int |f| = \int g_1 + \int f_1 - \int f_1 + \int g_2 + \int f_2 - \int f_2 + \cdots$$

$$= \int g_1 + \int g_2 + \cdots \leq \int |g_1| + \int |g_2| + \cdots$$

$$\leq \int |f_1| + \int |f_2| + \cdots.$$

Finally, since $f \leq |f|$ and $-f \leq |f|$, we have $\int f \leq \int |f|$ and $-\int f \leq \int |f|$, and hence $\left|\int f\right| \leq \int |f|$. □

Corollary 2.4.2. *If $f, g \in L^1(\mathbb{R})$, then $\min\{f, g\}, \max\{f, g\} \in L^1(\mathbb{R})$.*

Proof: Since

$$\min\{f, g\} = \frac{1}{2}(f + g - |f - g|) \quad \text{and} \quad \max\{f, g\} = \frac{1}{2}(f + g + |f - g|),$$

the assertion follows by Theorems 2.3.4 and 2.4.1. □

Integrability of $|f|$ does not imply integrability of f. However, it is not possible to present an example of such a function. We can only prove that such a function exists. The proof of existence requires mathematical tools that will not be introduced in this book. The interested reader should refer to books on real analysis or measure theory (see, for example, Friedman (1982) or Halmos (1974)).

The following theorem establishes a useful property of translations of integrable functions (see, for example, the proof of Theorem 2.15.3). Note that if

f is integrable, then the translated function $\tau_z f$ is integrable for any $z \in \mathbb{R}$ and $\int \tau_z f = \int f$.

Theorem 2.4.3. *If* $f \in L^1(\mathbb{R})$, *then* $\lim_{z \to 0} \int |\tau_z f - f| = 0$.

Proof: Note that the proof is easy if f is a step function. Now let f be an arbitrary integrable function and let $\varepsilon > 0$. If $f \simeq f_1 + f_2 + \cdots$, then there exists an $n_0 \in \mathbb{N}$ such that

$$\sum_{n=n_0+1}^{\infty} \int |f_n| < \frac{\varepsilon}{3}.$$

We have

$$\int |\tau_z f - f| \leq \int \left| \sum_{n=1}^{n_0} \tau_z f_n - \sum_{n=1}^{n_0} f_n \right| + \sum_{n=n_0+1}^{\infty} \int |\tau_z f_n| + \sum_{n=n_0+1}^{\infty} \int |f_n|$$

$$= \int \left| \sum_{n=1}^{n_0} \tau_z f_n - \sum_{n=1}^{n_0} f_n \right| + 2 \sum_{n=n_0+1}^{\infty} \int |f_n|$$

$$< \int \left| \sum_{n=1}^{n_0} \tau_z f_n - \sum_{n=1}^{n_0} f_n \right| + \frac{2\varepsilon}{3}.$$

Since $\sum_{n=1}^{n_0} f_n$ is a step function, we have

$$\lim_{z \to 0} \int \left| \sum_{n=1}^{n_0} \tau_z f_n - \sum_{n=1}^{n_0} f_n \right| = 0.$$

Consequently, $\int |\tau_z f - f| < \varepsilon$ for sufficiently small z. □

2.5 Series of Integrable Functions

In the construction described in the preceding sections we start with an integral defined for step functions and extend it to a larger class of functions by expending them in a series of step functions $f \simeq f_1 + f_2 + \cdots$. Are we going to obtain new functions if we repeat the construction and use integrable functions instead of step functions? It turns out that we do not get anything new. This result is a form of completeness of the space of Lebesgue integrable functions. It is of fundamental importance for the theory of the Lebesgue integral.

First we extend the use of the notation $f \simeq f_1 + f_2 + \cdots$ to a series of integrable functions.

Definition 2.5.1.

Let f be a real-valued function and let (f_n) be a sequence of integrable functions. If

(a) $\displaystyle\sum_{n=1}^{\infty} \int |f_n| < \infty,$

(b) $\displaystyle f(x) = \sum_{n=1}^{\infty} f_n(x)$ for every $x \in \mathbb{R}$ such that $\displaystyle\sum_{n=1}^{\infty} |f_n(x)| < \infty,$

then we write $f \simeq f_1 + f_2 + \cdots$ or $f \simeq \sum_{n=1}^{\infty} f_n.$

The following technical lemma will be used in the proof of the important Theorem 2.5.3.

Lemma 2.5.2. *If $f \in L^1(\mathbb{R})$, then for every $\varepsilon > 0$ there exists a sequence of step functions (f_n) such that $f \simeq f_1 + f_2 + \cdots$ and $\sum_{n=1}^{\infty} \int |f_n| \leq \int |f| + \varepsilon$.*

Proof: Let $f \simeq g_1 + g_2 + \cdots$ be an arbitrary expansion of f in a series of step functions. Then there exists an $n_0 \in \mathbb{N}$ such that $\sum_{n=n_0+1}^{\infty} \int |g_n| < \frac{\varepsilon}{2}$. Define

$$f_1 = g_1 + \cdots + g_{n_0} \quad \text{and} \quad f_n = g_{n_0+n-1} \quad \text{for } n \geq 2.$$

Then obviously $f \simeq f_1 + f_2 + \cdots$. Since $\int |f_1| - \int |f| \leq \int |f_1 - f|$ and $f - f_1 \simeq f_2 + f_3 + \cdots$, we get

$$\int |f_1| - \int |f| \leq \sum_{n=2}^{\infty} \int |f_n|$$

and hence

$$\int |f_1| - \sum_{n=2}^{\infty} \int |f_n| \leq \int |f|.$$

Consequently,

$$\sum_{n=1}^{\infty} \int |f_n| = \int |f_1| + \sum_{n=2}^{\infty} \int |f_n|$$

$$= \int |f_1| - \sum_{n=2}^{\infty} \int |f_n| + 2\sum_{n=2}^{\infty} \int |f_n|$$

$$\leq \int |f| + 2\sum_{n=2}^{\infty} \int |f_n|$$

$$= \int |f| + 2 \sum_{n=n_0+1}^{\infty} \int |g_n|$$

$$< \int |f| + \varepsilon. \qquad \square$$

Theorem 2.5.3. *If $f \simeq f_1 + f_2 + \cdots$, where $f_1, f_2, \ldots$ are integrable functions, then f is integrable and $\int f = \int f_1 + \int f_2 + \cdots$.*

Proof: By Lemma 2.5.2, there exist step functions $f_{n,k}$ $(n, k \in \mathbb{N})$ such that

$$f_n \simeq f_{n,1} + f_{n,2} + \cdots$$

and

$$\int |f_{n,1}| + \int |f_{n,2}| + \cdots \leq \int |f_n| + 2^{-n}$$

for all $n \in \mathbb{N}$. Let (h_n) be a sequence arranged from all the functions $g_{n,k}$. Then $f \simeq h_1 + h_2 + \cdots$. Consequently, $f \in L^1(\mathbb{R})$ and

$$\int f = \int h_1 + \int h_2 + \cdots = \int f_1 + \int f_2 + \cdots,$$

because all the involved series converge absolutely. $\qquad \square$

Corollary 2.5.4. *Let $f_1, f_2, \ldots \in L^1(\mathbb{R})$. If $\sum_{n=1}^{\infty} \int |f_n| < \infty$, then there exists an integrable function f such that $f \simeq f_1 + f_2 + \cdots$.*

Proof: The function f can be defined as follows:

$$f(x) = \begin{cases} \displaystyle\sum_{n=1}^{\infty} f_n(x) & \text{whenever } \displaystyle\sum_{n=1}^{\infty} |f_n(x)| < \infty, \\ 0 & \text{otherwise.} \end{cases} \qquad \square$$

2.6 Norm in $L^1(\mathbb{R})$

One of the important features of the Lebesgue integral is that techniques of normed spaces can be used. We will see that, with some precautions, $L^1(\mathbb{R})$ can be treated as a Banach space.

Definition 2.6.1.   (L^1-norm)
The functional $\| \cdot \| : L^1(\mathbb{R}) \to \mathbb{R}$ defined by $\|f\| = \int |f|$ is called the *norm in $L^1(\mathbb{R})$* or the *L^1-norm*.

The functional $\|\cdot\|$ is well-defined in view of Theorem 2.4.1. By Theorem 2.3.4, we have

$$\|\lambda f\| = \int |\lambda f| = \int |\lambda||f| = |\lambda| \int |f| = |\lambda|\|f\|.$$

Since $|f + g| \le |f| + |g|$, we have

$$\|f + g\| = \int |f + g| \le \int |f| + \int |g| = \|f\| + \|g\|,$$

by the same theorem. However, we cannot say that the pair $(L^1(\mathbb{R}), \|\cdot\|)$ is a normed space. There are nonzero functions f for which $\int |f| = 0$. Consider, for example, the function $f(x) = 0$ for every x except 0 and $f(0) = 1$. This is a nonzero function, but $\int |f| = 0$. This difficulty can be resolved as described next.

Definition 2.6.2.   (Null function)
A function f is called a *null function* if $\int |f| = 0$.

In this definition, we are implicitly assuming that $|f|$ is integrable, but we are not assuming that f is integrable. As we can see from the following theorem, in this case, integrability of f follows from integrability of $|f|$.

Theorem 2.6.3.   *If f is a null function and $|g| \le |f|$, then g is a null function.*

Proof: Observe that the only difficulty here is integrability of g. To prove that g is integrable note that

$$g \simeq |f| + |f| + \cdots. \tag{2.14}$$

In fact, since f is a null function, we have $\int |f| + \int |f| + \cdots = 0 + 0 + \cdots < \infty$. Moreover, if the series $f(x) + f(x) + \cdots$ is absolutely convergent at some $x \in \mathbb{R}$, then $f(x) = 0$. But then $g(x) = 0$ and we have $g(x) = |f(x)| + |f(x)| + \cdots$. This proves that (2.14) holds, and thus g is integrable. $\qquad\square$

Two functions f and g are called *equivalent* if $f - g$ is a null function. It is easy to check that the defined relation is an equivalence relation. Now we define the space $\mathcal{L}^1(\mathbb{R})$ as the space of equivalence classes of Lebesgue integrable functions. The equivalence class of $f \in L^1(\mathbb{R})$ is denoted by $[f]$, that is,

$$[f] = \left\{ g \in L^1(\mathbb{R}): \int |f - g| = 0 \right\}.$$

With the usual definitions

$$[f] + [g] = [f + g], \qquad \lambda[f] = [\lambda f], \qquad \|[f]\| = \int |f|,$$

$(\mathcal{L}^1(\mathbb{R}), \|\cdot\|)$ becomes a normed space. In Section 2.8 we will prove that $\mathcal{L}^1(\mathbb{R})$ is complete.

The whole construction may seem artificial, but it is necessary if we want $\|\cdot\|$ to be a norm. In practice, we often do not distinguish between $\mathcal{L}^1(\mathbb{R})$ and $L^1(\mathbb{R})$ and formulate everything in terms of $L^1(\mathbb{R})$. This does not cause any real difficulties, as long as one is aware of the problem. Actually, when dealing with $\mathcal{L}^1(\mathbb{R})$ new difficulties arise. For instance, note that for $F \in \mathcal{L}^1(\mathbb{R})$ we cannot specify the value of F at a point.

Definition 2.6.4.   (Convergence in norm)
We say that a sequence of functions $f_1, f_2, \ldots \in L^1(\mathbb{R})$ *converges to a function* $f \in L^1(\mathbb{R})$ *in norm*, denoted by $f_n \to f$ i.n., if $\int |f_n - f| \to 0$.

Notice that this is the usual convergence in a normed space. We use a special name for this convergence because another type of convergence will be also used. As a convergence defined by a norm, it has the following properties:

> *If $f_n \to f$ i.n. and $\lambda \in \mathbb{R}$, then $\lambda f_n \to \lambda f$ i.n.*

> *If $f_n \to f$ i.n. and $g_n \to g$ i.n., then $f_n + g_n \to f + g$ i.n.*

Moreover

> *If $f_n \to f$ i.n., then $|f_n| \to |f|$ i.n.,*

which follows immediately from the inequality $\|f_n\| - \|f\| \le |f_n - f|$.

The following two theorems are simple but important.

Theorem 2.6.5. *If $f_n \to f$ i.n., then $\int f_n \to \int f$.*

Proof: $|\int f_n - \int f| = |\int (f_n - f)| \le \int |f_n - f| \to 0$. $\square$

Theorem 2.6.6. *If $f \simeq f_1 + f_2 + \cdots$, then the series $f_1 + f_2 + \cdots$ converges to f in norm.*

Proof: Let $\varepsilon > 0$. There exists an integer n_0 such that $\sum_{n=n_0}^{\infty} \int |f_n| < \varepsilon$. Since

$$f - f_1 - \cdots - f_n \simeq f_{n+1} + f_{n+2} + \cdots,$$

for every $n > n_0$, we have

$$\int |f - f_1 - \cdots - f_n| \le \int |f_{n+1}| + \int |f_{n+2}| + \cdots < \varepsilon,$$

by Theorem 2.4.1. $\square$

2.7 Convergence Almost Everywhere

If $f \simeq f_1 + f_2 + \cdots$, then the series $f_1(x) + f_2(x) + \cdots$ need not converge at every $x \in \mathbb{R}$. On the other hand, the set of those points where we do not have convergence is a "small set." This suggests a new type of convergence, which is very natural when the Lebesgue integral is considered.

Definition 2.7.1. (Null set)
A set $X \subseteq \mathbb{R}$ is called a *null set* or a *set of measure zero* if its characteristic function is a null function.

Every countable set is a null set. A countable union of null sets is a null set. There are null sets which are not countable, for example, the Cantor set (see Exercise 20). The following theorem is an immediate consequence of Theorem 2.6.3.

Theorem 2.7.2. *Every subset of a null set is a null set.*

Definition 2.7.3. (Equality almost everywhere)
Let f and g be functions defined on $\mathbb{R}$. If the set of all $x \in \mathbb{R}$ for which $f(x) \neq g(x)$ is a null set, then we say that f *equals* g *almost everywhere* and write $f = g$ a.e.

Theorem 2.7.4. $f = g$ a.e. if and only if $\int |f - g| = 0$.

Proof: Let h be the characteristic function of the set Z of all $x \in \mathbb{R}$ for which $f(x) \neq g(x)$.
If $f = g$ a.e., then $\int |h| = \int h = 0$. Therefore

$$|f - g| \simeq h + h + \cdots,$$

which implies $\int |f - g| = 0$.
Conversely, if $\int |f - g| = 0$, then

$$h \simeq |f - g| + |f - g| + \cdots,$$

and hence $\int h = 0$. This shows that Z is a null set, that is, $f = g$ a.e. □

Note that Theorem 2.7.4 implies that f is a null function if and only if $f = 0$ a.e. Similarly, the equivalence of functions defined in Section 2.6 and equality almost everywhere are the same notion. This important property is used often in arguments concerning the Lebesgue integral. Another useful consequence of Theorem 2.7.4 is that we do not need to know the value of a function at every point in order to find its integral. It is sufficient to know the values almost everywhere, that is, everywhere except a null set. The function need not even be defined at every point.

Theorem 2.7.5. *Suppose* $f_n \to f$ *i.n. Then* $f_n \to g$ *i.n. if and only if* $f = g$ *a.e.*

Proof: If $f_n \to f$ i.n. and $f = g$ a.e., then

$$\|f_n - g\| = \int |f_n - g| \leq \int |f_n - f| + \int |f - g| = \int |f_n - f| = \|f_n - f\| \to 0.$$

If $f_n \to f$ i.n. and $f_n \to g$ i.n., then

$$\int |f - g| \leq \int |f - f_n| + \int |f_n - g| \to 0,$$

completing the proof. □

Definition 2.7.6.   (Convergence almost everywhere)
We say that a sequence of functions $f_1, f_2, \ldots$ defined on $\mathbb{R}$ *converges to* f *almost everywhere*, denoted by $f_n \to f$ a.e., if $f_n(x) \to f(x)$ for every x except a null set.

Convergence almost everywhere has properties similar to convergence in norm.

If $f_n \to f$ a.e. and $\lambda \in \mathbb{R}$, then $\lambda f_n \to \lambda f$ a.e.

If $f_n \to f$ a.e. and $g_n \to g$ a.e., then $f_n + g_n \to f + g$ a.e.

If $f_n \to f$ a.e., then $|f_n| \to |f|$ a.e.

Theorem 2.7.7. *Suppose* $f_n \to f$ *a.e. Then* $f_n \to g$ *a.e. if and only if* $f = g$ *a.e.*

Proof: If $f_n \to f$ a.e. and $f_n \to g$ a.e., then $f_n - f_n \to f - g$ a.e., which means that $f - g = 0$ a.e.

Now, assume $f_n \to f$ a.e. and $f = g$ a.e. Denote by A the set of all $x \in \mathbb{R}$ such that the sequence $(f_n(x))$ does not converge to $f(x)$ and by B the set of all $x \in \mathbb{R}$ such that $f(x) \neq g(x)$. Then A and B are null sets and so is $A \cup B$. Since $f_n(x) \to g(x)$ for every x not in $A \cup B$, we have $f_n \to g$ a.e. □

The following two examples show that convergence in norm and convergence almost everywhere are essentially different.

Example 2.7.8. For $n = 1, 2, \ldots$, define

$$f_n(x) = \begin{cases} \dfrac{1}{\sqrt{n}} & \text{for } x \in [-n, n], \\ 0 & \text{otherwise.} \end{cases}$$

Then $f_n(x) \to 0$ for every $x \in \mathbb{R}$, and thus $f_n \to 0$ a.e. On the other hand, since

$$\int |f_n| = 2\sqrt{n} \to \infty,$$

the sequence is not convergent in norm.   □

Example 2.7.9. Consider the sequence (f_n) of functions defined as follows:

$$f_n(x) = \begin{cases} 1 & \text{if } 2^{k-1} \le n < 2^k \text{ and } x \in \left[\frac{n}{2^{k-1}} - 1, \frac{n+1}{2^{k-1}} - 1\right), \\ 0 & \text{otherwise,} \end{cases}$$

$n = 1, 2, \ldots$ (see Figure 2.2).
  Since

$$\int f_n = \frac{1}{2^{k-1}} \quad \text{for } 2^{k-1} \le n < 2^k,$$

$\int |f_n| = \int f_n \to 0$, and thus the sequence (f_n) converges to 0 in norm. On the other hand, for every $x \in [0, 1), f_n(x) = 0$ for infinitely many $n \in \mathbb{N}$ and $f_n(x) = 1$ for infinitely many $n \in \mathbb{N}$. Therefore, the sequence (f_n) does not converge almost everywhere.   □

  Example 2.7.8 can be used to show that, in general, even uniform convergence does not imply convergence in norm. However, if we assume that $f, f_1, f_2, \ldots$ are integrable functions that vanish outside a bounded interval $[a, b]$ and the sequence (f_n) converges to f uniformly, then $f_n \to f$ i.n. Indeed,

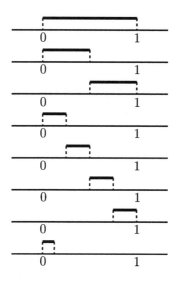

Figure 2.2 Functions $f_1, \ldots, f_8$ in Example 2.7.9.

in this case we have

$$\int |f_n - f| \le (b - a) \sup_{[a,b]} |f_n(x) - f(x)| \to 0.$$

Theorem 2.7.10. *Let $f_1, f_2, \ldots \in L^1(\mathbb{R})$, and $\int |f_1| + \int |f_2| + \cdots < \infty$. Then the series $f_1 + f_2 + \cdots$ converges almost everywhere.*

Proof: By Corollary 2.5.4, there exists a function $f \in L^1(\mathbb{R})$ such that $f \simeq f_1 + f_2 + \cdots$. Since $f(x) = \sum_{n=1}^{\infty} f_n(x)$ for every x such that $\sum_{n=1}^{\infty} |f_n(x)| < \infty$, it suffices to show that the set of all points $x \in \mathbb{R}$ for which the series $\sum_{n=1}^{\infty} |f_n(x)|$ is not absolutely convergent is a null set. Let g be the characteristic function of that set. Then $g \simeq f_1 - f_1 + f_2 - f_2 + \cdots$, and consequently

$$\int |g| = \int g = \int f_1 - \int f_1 + \int f_2 - \int f_2 + \cdots = 0. \qquad \square$$

Theorem 2.7.10 leads us to an important corollary.

Corollary 2.7.11. *If $f \simeq f_1 + f_2 + \cdots$, then $f = f_1 + f_2 + \cdots$ a.e.*

Theorem 2.7.12. *Let $f_1, f_2, \ldots \in L^1(\mathbb{R})$, and $\int |f_1| + \int |f_2| + \cdots < \infty$. Then $f = f_1 + f_2 + \cdots$ a.e. if and only if $f = f_1 + f_2 + \cdots$ i.n.*

Proof: By Corollary 2.5.4, there exists a function $g \in L^1(\mathbb{R})$ such that $g \simeq f_1 + f_2 + \cdots$. Then, by Theorem 2.6.6 we have $g = f_1 + f_2 + \cdots$ i.n., and by Corollary 2.7.11 we have $g = f_1 + f_2 + \cdots$ a.e.

Now, if $f = f_1 + f_2 + \cdots$ a.e., then $f = g$ a.e., by Theorem 2.7.7. Hence $f = f_1 + f_2 + \cdots$ i.n., by Theorem 2.7.5.

Conversely, if $f = f_1 + f_2 + \cdots$ i.n., then $f = g$ a.e., by Theorem 2.7.5. Hence $f = f_1 + f_2 + \cdots$ a.e., by Theorem 2.7.7. $\qquad \square$

2.8 Fundamental Convergence Theorems

Now we are ready to prove the main theorems of this chapter. The first theorem implies that $L^1(\mathbb{R})$ (or, more precisely, $\mathcal{L}^1(\mathbb{R})$, defined in Section 2.6) is a Banach space.

Theorem 2.8.1. *The space $L^1(\mathbb{R})$ is complete.*

Proof: In view of Theorem 1.4.9, it suffices to prove that every absolutely convergent series converges in norm. Let $f_n \in L^1(\mathbb{R})$, $n = 1, 2, \ldots$, and $\sum_{n=1}^{\infty} \int |f_n| < \infty$. Then, by Corollary 2.5.4, there exists an $f \in L^1(\mathbb{R})$ such that $f \simeq \sum_{n=1}^{\infty} f_n$. This in turn implies, by Theorem 2.6.6, that the series $\sum_{n=1}^{\infty} f_n$ converges to f in norm, proving the theorem. $\qquad \square$

In view of the previous theorem, the space $L^1(\mathbb{R})$ can be defined as the completion of the space of step functions with respect to convergence in norm. This looks like a very simple definition of the Lebesgue integrable functions. However, in such an approach elements of $L^1(\mathbb{R})$ are not functions and it is not at all clear if and how we can associate a function with an element of $L^1(\mathbb{R})$. Moreover, the important relations between convergence in norm and convergence almost everywhere do not follow easily from such a definition.

We know that convergence in norm and convergence almost everywhere are essentially different. The next theorem connects both convergences in an elegant way.

Theorem 2.8.2. *If $f_n \to f$ i.n., then there exists a subsequence (f_{p_n}) of (f_n) such that $f_{p_n} \to f$ a.e.*

Proof: Since $\int |f_n - f| \to 0$, there exists an increasing sequence of positive integers (p_n) such that $\int |f_{p_n} - f| < 2^{-n}$. Then

$$\int |f_{p_{n+1}} - f_{p_n}| \le \int |f_{p_{n+1}} - f| + \int |f - f_{p_n}| < \frac{3}{2^{n+1}},$$

and consequently,

$$\int |f_{p_1}| + \int |f_{p_2} - f_{p_1}| + \int |f_{p_3} - f_{p_2}| + \cdots < \infty.$$

Thus, there exists a $g \in L^1(\mathbb{R})$ such that

$$g \simeq f_{p_1} + (f_{p_2} - f_{p_1}) + (f_{p_3} - f_{p_2}) + \cdots,$$

and, by Corollary 2.7.11,

$$g = f_{p_1} + (f_{p_2} - f_{p_1}) + (f_{p_3} - f_{p_2}) + \cdots \quad \text{a.e.}$$

This means $f_{p_n} \to g$ a.e. Since also $f_{p_n} \to g$ i.n. and $f_{p_n} \to f$ i.n., we conclude $f = g$ a.e., by Theorem 2.7.5. Therefore $f_{p_n} \to f$ a.e., by Theorem 2.7.7.   □

Theorems 2.8.1 and 2.8.2 are attributed to Frigyes Riesz (1880–1956). Convergence in norm has the following important property:

$$f_n \to f \text{ i.n.} \quad \text{implies} \quad \int f_n \to \int f.$$

In other words, the limit in norm can be interchanged with integration:

$$\int \lim_{n \to \infty} f_n = \lim_{n \to \infty} \int f_n.$$

Convergence almost everywhere does not share this property. On the other hand, in practice it is often much easier to show that a sequence of functions converges almost everywhere than to show convergence in norm. Theorems 2.8.3 and 2.8.4 give conditions which are usually easy to check and at the same time imply convergence in norm.

A sequence of functions (f_n) is called *nonincreasing almost everywhere* if the set of points $x \in \mathbb{R}$ where $(f_n(x))$ is increasing is a null set. It is called *nondecreasing almost everywhere* if the set of points $x \in \mathbb{R}$ where $(f_n(x))$ is decreasing is a null set. A sequence that is either nonincreasing almost everywhere or nondecreasing almost everywhere is called *monotone almost everywhere*.

The following theorem is due to Beppo Levi (1876–1961). Note a certain similarity between Theorem 2.8.3 and Theorem 2.2.6.

Theorem 2.8.3. (Monotone convergence theorem)   *If (f_n) is a sequence of integrable functions that is monotone almost everywhere and $|\int f_n| \le M$ for some constant M and all $n \in \mathbb{N}$, then there exists an integrable function f such that $f_n \to f$ i.n. and $f_n \to f$ a.e. Moreover, $|\int f| \le M$.*

Proof: Without loss of generality, we can assume that the sequence is nondecreasing almost everywhere and the functions are non-negative. In such a case

$$\int |f_1| + \int |f_2 - f_1| + \cdots + \int |f_n - f_{n-1}| = \int |f_n| \le M,$$

for every $n \in \mathbb{N}$. By letting $n \to \infty$, we obtain

$$\int |f_1| + \int |f_2 - f_1| + \cdots \le M.$$

By Corollary 2.5.4, there exists an $f \in L^1(\mathbb{R})$ such that $f \simeq f_1 + (f_2 - f_1) + \cdots$. Hence, $f_n \to f$ i.n., by Theorem 2.6.6, and $f_n \to f$ a.e., by Corollary 2.7.11. Finally,

$$\left| \int f \right| = \left| \int f_1 + \int (f_2 - f_1) + \int (f_3 - f_2) - \cdots \right|$$

$$\le \int |f_1| + \int |f_2 - f_1| + \int |f_3 - f_2| + \cdots \le M. \qquad \square$$

Theorem 2.8.4. (The Lebesgue dominated convergence theorem)   *If a sequence of integrable functions (f_n) converges almost everywhere to a function f and there exists an integrable function h such that $|f_n| \le h$ for every $n \in \mathbb{N}$, then f is integrable and $f_n \to f$ i.n.*

Proof: For $m, n = 1, 2, \ldots$, define

$$g_{m,n} = \max \{ |f_m|, \ldots, |f_{m+n}| \}.$$

Then, for every fixed $m \in \mathbb{N}$, the sequence $(g_{m,1}, g_{m,2}, \ldots)$ is nondecreasing and, since

$$\left| \int g_{m,n} \right| = \int g_{m,n} \leq \int h < \infty,$$

there is an integrable function g_m such that $g_{m,n} \to g_m$ a.e. as $n \to \infty$, by the monotone convergence theorem.

Note that the sequence (g_n) is nonincreasing almost everywhere and $\int |g_n| \leq \int |f_1|$ for all $n \in \mathbb{N}$. Again, by the monotone convergence theorem, there exists an integrable function g such that $g_n \to g$ a.e. and $g_n \to g$ i.n. Now we consider two cases.

Case 1: Suppose $f = 0$. Then $f_n \to 0$ a.e., and therefore $g_n \to 0$ a.e. Since the sequence (f_n) converges in norm, we obtain $g_n \to 0$ i.n. Hence,

$$\int |f_n| \leq \int g_n \to 0,$$

which proves the theorem in the first case.

Case 2: When f is an arbitrary function, then for every increasing sequence of positive integers (p_n), we have

$$h_n = f_{p_{n+1}} - f_{p_n} \to 0 \text{ a.e.}$$

and $|h_n| \leq 2h$ for every $n \in \mathbb{N}$. By Case 1, we must have $h_n \to 0$ i.n. This shows that the sequence (f_n) is a Cauchy sequence in $L^1(\mathbb{R})$ and therefore it converges in norm to some $\tilde{f} \in L^1(\mathbb{R})$, by Theorem 2.8.1. On the other hand, by Theorem 2.8.2, there exists an increasing sequence of positive integers q_n such that $f_{q_n} \to \tilde{f}$ a.e. But $f_{q_n} \to f$ a.e., and thus $\tilde{f} = f$ a.e. This, in view of Theorem 2.7.5, implies that $f_n \to f$ i.n. □

The following useful theorem follows rather easily from the monotone convergence theorem. It is a version of the so-called Fatou's lemma (Pierre Joseph Louis Fatou (1878–1929)).

Theorem 2.8.5. (Fatou's lemma) *Let (f_n) be a sequence of non-negative integrable functions such that $\int f_n \leq M$ for some M and every $n \in \mathbb{N}$. If $f_n \to f$ a.e., then f is integrable and $\int f \leq M$.*

Proof: Let $\varphi_{n,k} = \min\{f_n, f_{n+1}, \ldots, f_{n+k}\}$, for $n, k \in \mathbb{N}$. For a fixed $n \in \mathbb{N}$, the sequence $(\varphi_{n,1}, \varphi_{n,2}, \ldots)$ is a decreasing sequence of integrable functions such that $| \int \varphi_{n,k} | \leq \int \varphi_{n,1} < \infty$. Thus, by the monotone convergence theorem, it converges almost everywhere to an integrable function φ_n, that is,

$$\varphi_n = \inf\{f_n, f_{n+1}, f_{n+2}, \ldots\} \text{ a.e.}$$

Since $\int \varphi_n \leq \int f_n \leq M$ and the sequence (φ_n) is nondecreasing almost everywhere, the sequence (φ_n) converges almost everywhere to an integrable function g and we have $\int g \leq M$, again by the monotone convergence theorem. But $\varphi_n \to f$ a.e. Thus $f = g$ a.e. and $\int f \leq M$.   □

2.9 Locally Integrable Functions

The integral $\int f$ corresponds to integration over the entire real line, so the symbols $\int_{-\infty}^{\infty} f$ or $\int_{\mathbb{R}} f$ could be used instead. In applications we often need to integrate functions over bounded intervals. This concept can be easily defined using the integral $\int f$.

Definition 2.9.1.   (Integral over an interval)

By the integral of a function f over an interval $[a, b]$, denoted by $\int_a^b f$, we mean the value of the integral $\int f \chi_{[a,b]}$, where $\chi_{[a,b]}$ denotes the characteristic function of $[a, b]$ (and $f \chi_{[a,b]}$ is the product of f and $\chi_{[a,b]}$).

In other words, $\int_a^b f$ is the integral of the function equal to f on $[a, b]$ and zero otherwise. The choice of the closed interval $[a, b]$ in the definition has no significance. We could use (a, b), or the semiopen intervals, and we would have a completely equivalent definition. In proofs it is often more convenient to use $\chi_{[a,b)}$ instead of $\chi_{[a,b]}$, because of the definition of step functions.

Theorem 2.9.2. *If $f \in L^1(\mathbb{R})$, then the integral $\int_a^b f$ exists for every interval $[a, b]$.*

   Proof: Let $f \simeq f_1 + f_2 + \cdots$. Define, for $n = 1, 2, \ldots$,

$$g_n(x) = \begin{cases} f_n(x) & \text{if } x \in [a, b], \\ 0 & \text{otherwise.} \end{cases}$$

It is easy to check that $f \chi_{[a,b]} \simeq g_1 + g_2 + \cdots$.   □

The converse of the above theorem is not true. For instance, for the constant function $f = 1$, the integral $\int_a^b f$ exists for every $-\infty < a < b < \infty$, although $f \notin L^1(\mathbb{R})$. This suggests the following definition:

Definition 2.9.3.   (Locally integrable functions)

A function f defined on $\mathbb{R}$ is called *locally integrable*, if the integral $\int_a^b f$ exists for every $-\infty < a < b < \infty$.

Although in the definition we require integrability of f over every bounded interval, it is sufficient to check that the integral $\int_{-n}^{n} f$ exists for every $n \in \mathbb{N}$. The proof of this simple fact is left as an exercise.

The locally integrable functions form a vector space. Note that Theorem 2.9.2 implies that $L^1(\mathbb{R})$ is a subspace of the space of locally integrable functions. The absolute value of a locally integrable function is locally integrable. However, the product of locally integrable functions need not be locally integrable (see Exercise 39).

Theorem 2.9.4. *Let f and g be locally integrable functions. If g is bounded on $[a, b]$ for every $-\infty < a < b < \infty$, then the product fg is a locally integrable function.*

Proof: Let f and g be as described in the theorem. For $-\infty < a < b < \infty$ define $F = f \chi_{[a,b)}$ and $G = g \chi_{[a,b)}$. Let $F \simeq f_1 + f_2 + \cdots$, where each f_n's are step functions with supports contained in $[a, b)$. We will show that $FG \simeq f_1 g + f_2 g + \cdots$.

First note that $f_1 g, f_2 g, \ldots$ are integrable functions. If $g < M$ on $[a, b)$ for some $M > 0$, then $\int |f_n g| < M \int |f_n|$ for all $n \in \mathbb{N}$. Thus,

$$\sum_{n=1}^{\infty} \int |f_n g| < M \sum_{n=1}^{\infty} \int |f_n| < \infty.$$

Moreover, if $\sum_{n=1}^{\infty} |g(x) f_n(x)| < \infty$ for some $x \in \mathbb{R}$, then $\sum_{n=1}^{\infty} |f_n(x)| < \infty$ or $g(x) = 0$. In either case, $F(x)G(x) = \sum_{n=1}^{\infty} f_n(x)g(x)$ for that x. $\qquad\square$

Theorem 2.9.5. *If f is a locally integrable function such that $|f| \leq g$ for some $g \in L^1(\mathbb{R})$, then $f \in L^1(\mathbb{R})$.*

Proof: Let $f_n = f \chi_{[-n,n]}$ for $n = 1, 2, \ldots$. Then $f_1, f_2, \ldots$ are integrable functions, the sequence (f_n) converges to f everywhere and $|f_n| \leq g$ for every $n \in \mathbb{N}$. Thus, by the Lebesgue dominated convergence theorem, $f \in L^1(\mathbb{R})$. $\qquad\square$

In applications it often convenient to use the symbol $\int_a^b f$ even if $b \leq a$. We adopt the usual convention:

$$\int_a^b f = -\int_b^a f \quad \text{and} \quad \int_a^a f = 0.$$

The subspace of $L^1(\mathbb{R})$ of all functions vanishing outside of some interval $[a, b]$ is a vector space. It is denoted by $L^1([a, b])$. As a subspace of $L^1(\mathbb{R})$, it is a normed space (with functions equal almost everywhere identified). The norm in $L^1([a, b])$ is then defined by

$$\|f\| = \int_a^b |f|.$$

Completeness of $L^1([a, b])$ follows easily from completeness of $L^1(\mathbb{R})$. Clearly, all properties of the Lebesgue integral apply to functions vanishing outside of some interval. Instead of intervals other sets can be used as well. This will be discussed further in Section 2.11.

2.10 The Lebesgue Integral and the Riemann Integral

In this section we are going to prove that Riemann integrable functions (Georg Friedrich Bernhard Riemann (1826–1866)) are Lebesgue integrable. Then we will prove some theorems, which show that useful properties of the Riemann integral are not lost when the more general Lebesgue integral is considered.

Theorem 2.10.1. *A Riemann integrable function is Lebesgue integrable and the integrals are equal.*

Proof: Let f be a Riemann integrable function vanishing outside an interval $[a, b)$ and let $|f| < M$ for some constant M. For every $n \in \mathbb{N}$, we partition the interval $[a, b)$ into n intervals of equal length,

$$[a, a + c), [a + c, a + 2c), \ldots, [a + (n - 1)c, a + nc),$$

where $c = (b - a)/n$. Then we define functions

$$g_n(x) = \begin{cases} \inf\{f(t): t \in [a + (k - 1)c, a + kc)\} & \text{if } x \in [a + (k - 1)c, a + kc), \\ 0 & \text{otherwise,} \end{cases}$$

and

$$h_n(x) = \begin{cases} \sup\{f(t): t \in [a + (k - 1)c, a + kc)\} & \text{if } x \in [a + (k - 1)c, a + kc), \\ 0 & \text{otherwise.} \end{cases}$$

Since (g_n) is a bounded nondecreasing sequence of functions, it converges at every $x \in \mathbb{R}$. Define $g(x) = \lim_{n \to \infty} g_n(x)$. Functions g_n are integrable, being step functions, and thus, by the Lebesgue dominated convergence theorem, g is integrable and

$$\int g = \lim_{n \to \infty} \int g_n.$$

Similarly, the function defined by $h(x) = \lim_{n \to \infty} h_n(x)$ is integrable and

$$\int h = \lim_{n \to \infty} \int h_n.$$

On the other hand, since f is Riemann integrable, we have

$$\lim_{n \to \infty} \int g_n = \lim_{n \to \infty} \int h_n,$$

and thus

$$\int |h - g| = \int h - g = \lim_{n \to \infty} \int (h_n - g_n) = 0. \tag{2.15}$$

Hence, $g = h$ a.e., and since $g \le f \le h$, we have $g = f$ a.e. (as well as $h = f$ a.e.). This proves that f is Lebesgue integrable. Clearly, the Riemann integral of f and the Lebesgue integral of f are equal, both being equal to $\lim_{n \to \infty} \int g_n$. $\qquad \square$

Theorem 2.10.2. *Let f be a Lebesgue integrable function on a bounded or unbounded interval (a, b), and let $a \le c \le b$. Then the function $F(x) = \int_c^x f$ is continuous in (a, b) and the limits*

$$F(a) = \lim_{x \to a+} F(x) \quad \text{and} \quad F(b) = \lim_{x \to b-} F(x)$$

exist and are finite.

Proof: Let x be a point such that $a \le x < b$, and let (x_n) be a decreasing sequence in (a, b) convergent to x, that is, $x < x_{n+1} < x_n < b$ and $x_n \to x$. Let $g_n = \chi_{(x, x_n)}$, $n = 1, 2, \dots$. Then the sequence of products $f g_n$ is convergent to zero at every point of (a, b). Since $|f g_n| \le |f|$, we have

$$F(x_n) - F(x) = \int_x^{x_n} f = \int_a^b f g_n \to 0,$$

by the Lebesgue dominated convergence theorem. Thus, F is right-hand continuous at $x \in (a, b)$ and has a limit at a. Similarly, we can prove left-hand continuity at every x, such that $a < x \le b$, by considering the characteristic functions of intervals (x_n, x) where $a < x_n < x_{n+1} < x$ and $x_n \to x$. Since F is right-hand and left-hand continuous, it is continuous in (a, b) and the limits $F(a)$ and $F(b)$ exist. $\qquad \square$

Theorem 2.10.3. *Let f be a function integrable over (a, b) and let $F(x) = \int_c^x f$ for some $c \in (a, b)$. If f is continuous at a point $x_0 \in (a, b)$, then F is differentiable at x_0 and we have $F'(x_0) = f(x_0)$.*

Proof: For $h > 0$, we have

$$\frac{F(x_0 + h) - F(x_0)}{h} = \frac{\int_c^{x_0 + h} f - \int_c^{x_0} f}{h} = \frac{1}{h} \int_{x_0}^{x_0 + h} f. \tag{2.16}$$

Since f is continuous at x_0, for an arbitrary $\varepsilon > 0$ there is a $\delta > 0$ such that

$$f(x_0) - \varepsilon < f(t) < f(x_0) + \varepsilon \quad \text{for all } t \in (x_0, x_0 + \delta), \tag{2.17}$$

Integrating (2.17) from x_0 to $x_0 + h$, we get

$$h\big(f(x_0) - \varepsilon\big) < \int_{x_0}^{x_0 + h} f < h\big(f(x_0) + \varepsilon\big), \quad \text{for } 0 < h < \delta.$$

Hence, by (2.16), we have

$$f(x_0) - \varepsilon < \frac{F(x_0 + h) - F(x_0)}{h} < f(x_0) + \varepsilon, \quad \text{for } 0 < h < \varepsilon.$$

Since ε is arbitrary, this shows that

$$\lim_{h \to 0+} \frac{F(x_0 + h) - F(x_0)}{h} = f(x_0).$$

Using a similar argument for $h < 0$, we obtain

$$\lim_{h \to 0-} \frac{F(x_0 + h) - F(x_0)}{h} = f(x_0).$$

This gives the desired result. $\qquad\qquad\square$

The following theorem establishes a change of variables formula for the Lebesgue integral. The theorem is a special case of a more general theorem on the change of variables for the Lebesgue integral, but it is sufficient for most applications.

Theorem 2.10.4. (Change of variables) *Let g be a nondecreasing differentiable function defined on a bounded or unbounded interval (a, b) such that g' is integrable over (a, b). Denote by $g(a)$ and $g(b)$ the limits (finite or infinite) of g at a and b, respectively, that is, $g(a) = \lim_{x \to a+} g(x)$ and $g(b) = \lim_{x \to b-} g(x)$. If f is an integrable function over $(g(a), g(b))$, then the product $f(g(t))g'(t)$ is integrable over (a, b) and*

$$\int_{g(a)}^{g(b)} f(t)dt = \int_a^b f\big(g(t)\big)g'(t)dt.$$

Proof: First we prove that the product $f(g(t))g'(t)$ is integrable over (a, b). Let

$$f \simeq \lambda_1 f_1 + \lambda_2 f_2 + \cdots, \tag{2.18}$$

where $f_n = \chi_{[a_n, b_n)}$, for some intervals $[a_n, b_n) \subseteq (g(a), g(b))$. We will show that the function $\Phi(t) = f(g(t))g'(t)$ expands into

$$\Phi \simeq \lambda_1 \Phi_1 + \lambda_2 \Phi_2 + \cdots, \tag{2.19}$$

where $\Phi_n(t) = f_n(g(t))g'(t)$.

Since $g(a) \le a_n < b_n \le g(b)$ and g is continuous (being differentiable), for every $n \in \mathbb{N}$ there are numbers $\alpha_n < \beta_n$ such that $g(\alpha_n) = a_n$ and $g(\beta_n) = b_n$. Then

$$\int |\Phi_n| = \int \Phi_n = \int_{\alpha_n}^{\beta_n} g'(t)dt = g(\beta_n) - g(\alpha_n) = b_n - a_n = \int f_n. \quad (2.20)$$

Since $|\lambda_1| \int f_1 + |\lambda_2| \int f_2 + \cdots < \infty$, we also have $\int |\lambda_1 \Phi_1| + \int |\lambda_2 \Phi_2| + \cdots < \infty$. Moreover, if the series $\lambda_1 \Phi_1(t) + \lambda_2 \Phi_2(t) + \cdots$ converges absolutely for some t, then it converges to $\Phi(t)$. This is trivially true if $g'(t) = 0$. If $g'(t) = k \ne 0$, then letting $g(t) = x$, we have

$$\lambda_1 \Phi_1(t) + \lambda_2 \Phi_2(t) + \cdots = \big(\lambda_1 f_1(x) + \lambda_2 f_2(x) + \cdots\big)k = f(x)k = \Phi(t),$$

with the series being convergent absolutely. Thus, (2.19) holds, which proves integrability of $f(g(t))g'(t)$.

Now, by (2.19), (2.20), and (2.18), we have

$$\int_a^b f\big(g(t)\big)g'(t)dt = \lambda_1 \int \Phi_1 + \lambda_2 \int \Phi_2 + \cdots$$

$$= \lambda_1 \int f_1 + \lambda_2 \int f_2 + \cdots$$

$$= \int_{g(a)}^{g(b)} f(t)dt. \qquad \square$$

2.11 Lebesgue Measure on $\mathbb{R}$

Our presentation of the Lebesgue integral does not require the concept of measure. On the other hand, once the integral is defined, we can easily define measurable sets and Lebesgue measure. In this section, we briefly discuss these notions.

Lebesgue measure is an extension of the concept of length of an interval onto a large class of sets. Not all sets are measurable. The proof of existence of nonmeasurable sets is not effective. This means that we cannot construct a nonmeasurable set, we can only prove that such a construction is possible. We are not going to discuss details of this problem. The interested reader is referred to Friedman (1982).

Definition 2.11.1.   (Measurable set)

A set S is called *measurable* if the characteristic function of S is a locally integrable function.

To every measurable set we assign a non-negative number or infinity, which is called the measure of that set.

Definition 2.11.2.   (Lebesgue measure)

Let S be a measurable set. If the characteristic function χ_S is an integrable function, then by the *measure* of S, denoted by $\mu(S)$, we mean the value of the integral $\mu(S) = \int \chi_S$. If χ_S is not integrable, then we define $\mu(S) = \infty$.

Null sets are sets whose measure is 0. The following theorem gives a useful tool for proving that a set is a null set.

Theorem 2.11.3. *A set $S \subseteq \mathbb{R}$ is a null set if and only if for every $\varepsilon > 0$ there exist intervals $[a_1, b_1), [a_2, b_2), \ldots$ such that*

$$S \subseteq \bigcup_{n=1}^{\infty} [a_n, b_n) \quad and \quad \sum_{n=1}^{\infty} (b_n - a_n) < \varepsilon. \tag{2.21}$$

Proof: Let $S \subseteq \mathbb{R}$ be a null set, and let f be the characteristic function of S. Fix an $\varepsilon > 0$. By Lemma 2.5.2, there exists a sequence of step functions (f_n) such that $f \simeq f_1 + f_2 + \cdots$ and $\sum_{n=1}^{\infty} \int |f_n| < \varepsilon/3$. For all $n \in \mathbb{N}$ define

$$A_n = \left\{ x \in \mathbb{R} : \left| f_1(x) \right| + \cdots + \left| f_n(x) \right| \geq \frac{1}{2} \right\}.$$

Note that these sets are finite unions of intervals because $|f_1| + \cdots + |f_n|$ is a step function for every $n \in \mathbb{N}$. Let

$$A_n = [a_{n,1}, b_{n,1}) \cup \cdots \cup [a_{n,k_n}, b_{n,k_n}).$$

Then $\sum_{k=1}^{k_n} (b_{n,k} - a_{n,k}) < 2\varepsilon/3$ for every $n \in \mathbb{N}$. Since $A_1 \subseteq A_2 \subseteq \cdots$ and $S \subseteq \bigcup_{n=1}^{\infty} A_n$, we can find disjoint intervals $[a_1, b_1), [a_2, b_2), \ldots$ satisfying (2.21).

Now assume that, for every $\varepsilon > 0$, there exist intervals $[a_1, b_1), [a_2, b_2), \ldots$ satisfying (2.21). Without loss of generality, we can assume that the intervals are disjoint. Let, for every $n \in \mathbb{N}$,

$$[a_{n,1}, b_{n,1}), [a_{n,2}, b_{n,2}), \ldots$$

be a sequence of disjoint intervals such that

$$S \subseteq \bigcup_{k=1}^{\infty} [a_{n,k}, b_{n,k}) \tag{2.22}$$

and

$$\sum_{k=1}^{\infty} (b_{n,k} - a_{n,k}) < \frac{1}{n}. \tag{2.23}$$

Denote by $g_{n,k}$ the characteristic function of the interval $[a_{n,k}, b_{n,k})$, and define

$$f_n(x) = \sum_{k=1}^{\infty} g_{n,k}(x).$$

Note that for every $n \in \mathbb{N}$, f_n is a well-defined integrable function and $\int f_n < \frac{1}{n}$. Consider the sequence of functions $h_n = \min\{f_1, f_2, \ldots, f_n\}$. Since this is a nonincreasing sequence of non-negative functions, the limit $h(x) = \lim_{n \to \infty} h_n(x)$ exists for every $x \in \mathbb{R}$ and h is integrable, by the monotone convergence theorem. Moreover, $\int h_n \leq \int f_n < \frac{1}{n}$ implies that h is a null function.

Let g denote the characteristic function of the set S. By (2.22), $g \leq f_n$, and hence $g \leq h_n$, for all $n \in \mathbb{N}$. Thus, $g \leq h$ and, by Theorem 2.6.3, g is a null function. This means that S is a null set. □

The property of measure proved in the next theorem is called σ-*additivity*. It is one of the fundamental properties of the Lebesgue measure.

Theorem 2.11.4. *Let* $S_1, S_2, \ldots$ *be a sequence of disjoint measurable sets. Then the union* $S = \bigcup_{n=1}^{\infty} S_n$ *is measurable and*

$$\mu(S) = \sum_{n=1}^{\infty} \mu(S_n). \tag{2.24}$$

Proof: Suppose first that there exists a bounded interval $[a, b]$ such that $S_n \subseteq [a, b]$ for every $n \in \mathbb{N}$. Then

$$\chi_S \simeq \chi_{S_1} + \chi_{S_2} + \cdots.$$

Consequently, S is measurable and we have

$$\mu(S) = \int \chi_S = \sum_{n=1}^{\infty} \int \chi_{S_n} = \sum_{n=1}^{\infty} \mu(S_n). \tag{2.25}$$

Now let $S, S_1, S_2, \ldots$ be arbitrary sets satisfying the assumptions of the theorem. Note that, by the first part of the proof, S is measurable. To prove (2.24), we consider two cases.

Case 1: χ_S is integrable. Then χ_{S_n} is integrable for every $n \in \mathbb{N}$, because $\chi_{S_n} \leq \chi_S$, and we have

$$\chi_S \simeq \chi_{S_1} + \chi_{S_2} + \cdots.$$

Then (2.24) can be proved as in (2.25).

Case 2: $\mu(S) = \infty$. Suppose that $\sum_{n=1}^{\infty} \mu(S_n) < \infty$. Then $\sum_{n=1}^{\infty} \int \chi_{S_n} < \infty$ and thus there exists an integrable function f such that $\sum_{n=1}^{\infty} \chi_{S_n} = f$ a.e. But $\sum_{n=1}^{\infty} \chi_{S_n} = \chi_S$, and hence $f = \chi_S$ a.e. This means that χ_S is integrable, contrary to the assumption. Therefore $\sum_{n=1}^{\infty} \mu(S_n) = \infty$. $\quad\square$

The notion of the integral over an interval $[a, b]$, introduced in Section 2.9, can be extended to the integral over any measurable set Ω:

$$\int_{\Omega} f = \int f \chi_{\Omega}.$$

Integrability of the product $f\chi_{\Omega}$ follows the form of the definition of measurable sets and Theorem 2.9.5. For any measurable set $\Omega \subseteq \mathbb{R}$, we can define the space $L^1(\Omega)$ of all integrable functions vanishing outside of Ω. With the norm inherited from $L^1(\mathbb{R})$, $L^1(\Omega)$ is a Banach space.

In Section 2.4 the question whether integrability of $|f|$ implies integrability of f is considered. The problem is closely related to the problem of existence of nonmeasurable sets, which in turn is related to existence of nonmeasurable functions. Using Theorem 2.11.7, one can prove the following: *If f is a measurable function, then $|f| \in L^1(\mathbb{R})$ implies $f \in L^1(\mathbb{R})$.*

Definition 2.11.5.   (Measurable functions)

A function f is called *measurable* if there exists a sequence of step functions $f_1, f_2, \ldots$ such that $f_n \to f$ a.e.

Obviously, every integrable function is measurable. An easy argument shows that every locally integrable function is measurable. There are measurable functions which are not locally integrable. Indeed, consider the function defined by

$$f(x) = \begin{cases} 1/x & \text{for } x > 0, \\ 0 & \text{for } x \leq 0. \end{cases}$$

Theorem 2.11.6. *The measurable functions form a vector space. The absolute value of a measurable function is a measurable function. The product of measurable functions is a measurable function.*

Proof: Let f and g be measurable functions, and let (f_n) and (g_n) be sequences of step functions such that $f_n \to f$ a.e. and $g_n \to g$ a.e. Then

$$f_n + g_n \to f + g \text{ a.e.,}$$

$$\lambda f_n \to \lambda f \text{ a.e.,}$$

$$|f_n| \to |f| \text{ a.e.,}$$

$$f_n g_n \to fg \text{ a.e.}$$

Since $(f_n + g_n)$, (λf_n), $(f_n g_n)$, and $(|f_n|)$ are sequences of step functions, the assertions follow. $\square$

Observe that Theorem 2.11.6 implies that if f and g are measurable functions, then $\max\{f, g\}$ and $\min\{f, g\}$ are measurable functions.

Theorem 2.11.7. *If f is a measurable function and $|f| \le g$ for some locally integrable function g, then f is locally integrable.*

Proof: Let $[a, b)$ be an arbitrary bounded interval in $\mathbb{R}$. We need to show that $f \chi_{[a,b)}$ is an integrable function. Suppose first that $f \ge 0$. Let (f_n) be a sequence of step functions such that $f_n \to f$ a.e. Define $g_n = \chi_{[a,b)} \min\{f_n, g\}$. Then (g_n) is a sequence of integrable functions convergent to $f \chi_{[a,b)}$ almost everywhere. Thus, by the Lebesgue dominated convergence theorem, $f \chi_{[a,b)}$ is an integrable function.

For an arbitrary measurable function f, we first define $f^+ = \frac{1}{2}(|f| + f)$ and $f^- = \frac{1}{2}(|f| - f)$. Then we have $f^+, f^- \ge 0$ and $f = f^+ - f^-$. Therefore, the theorem follows from the first part of the proof and Theorem 2.11.6. $\square$

Other properties of the Lebesgue measure, measurable sets, and measurable functions can be found in the exercises.

2.12 Complex-Valued Lebesgue Integrable Functions

In this section we extend the definition of Lebesgue integrable functions to include functions with complex values. First we define a complex-valued step function: f is a *complex-valued step function* if there exist complex numbers $\lambda_1, \ldots, \lambda_n$ and intervals $[a_1, b_1), \ldots, [a_n, b_n)$ such that

$$f(x) = \lambda_1 \chi_{[a_1, b_1)} + \cdots + \lambda_n \chi_{[a_n, b_n]}. \tag{2.26}$$

The integral $\int f$ of the step function (2.26) is defined by

$$\int f = \lambda_1(b_1 - a_1) + \cdots + \lambda_n(b_n - a_n).$$

The defined integral has the same properties as the integral of real step functions. Note that, if f is a step function, then its real part $\operatorname{Re} f$ and imaginary part $\operatorname{Im} f$ are step functions and we have $\int f = \int \operatorname{Re} f + i \int \operatorname{Im} f$.

Definition 2.12.1.   (Lebesgue integral for complex-valued functions)
A complex-valued function f is *Lebesgue integrable* if there exists a sequence of step functions (f_n) such that the following two conditions are satisfied:

(a) $\displaystyle\sum_{n=1}^{\infty} \int |f_n| < \infty;$

(b) $\displaystyle f(x) = \sum_{n=1}^{\infty} f_n(x)$ for every $x \in \mathbb{R}$ such that $\displaystyle\sum_{n=1}^{\infty} |f_n(x)| < \infty.$

The integral of f is then defined by

$$\int f = \sum_{n=1}^{\infty} \int f_n.$$

If a function f and a sequence of step functions (f_n) satisfy (a) and (b), then we write as before

$$f \simeq \sum_{n=1}^{\infty} f_n \quad \text{or} \quad f \simeq f_1 + f_2 + \cdots.$$

We do not have to repeat the argument presented for real functions to prove that the defined integral does not depend on a particular representation by a series of step functions and thus is well defined. This follows immediately from the next theorem.

Theorem 2.12.2. *A complex-valued function f is integrable if and only if its real part $\operatorname{Re} f$ and its imaginary part $\operatorname{Im} f$ are integrable. Moreover, if f is integrable, then*

$$\int f = \int \operatorname{Re} f + i \int \operatorname{Im} f. \tag{2.27}$$

Proof: Let $f \simeq f_1 + f_2 + \cdots$, where $f_1, f_2, \ldots$ are complex-valued step functions. Then

$$\int |\operatorname{Re} f_n| \leq \int |f_n| \quad \text{and} \quad \int |\operatorname{Im} f_n| \leq \int |f_n|,$$

and hence

$$\sum_{n=1}^{\infty} \int |\operatorname{Re} f_n| < \infty \quad \text{and} \quad \sum_{n=1}^{\infty} \int |\operatorname{Im} f_n| < \infty.$$

We can thus write

$$\operatorname{Re} f \simeq \operatorname{Re} f_1 + \operatorname{Im} f_1 - \operatorname{Im} f_1 + \operatorname{Re} f_2 + \operatorname{Im} f_2 - \operatorname{Im} f_2 + \cdots$$

and

$$\operatorname{Im} f \simeq \operatorname{Im} f_1 + \operatorname{Re} f_1 - \operatorname{Re} f_1 + \operatorname{Im} f_2 + \operatorname{Re} f_2 - \operatorname{Re} f_2 + \cdots.$$

This proves that $\operatorname{Re} f$ and $\operatorname{Im} f$ are integrable functions and (2.27) holds.

If $\operatorname{Im} f \simeq f_1 + f_2 + \cdots$ and $\operatorname{Re} f \simeq g_1 + g_2 + \cdots$, then $f \simeq f_1 + ig_1 + f_2 + ig_2 + \cdots$, which proves integrability of f. $\quad\square$

Theorem 2.12.3. *If a complex-valued function f is integrable, then the real-valued function $|f|$ is integrable and $|\int f| \leq \int |f|$.*

Proof: Let $f \simeq f_1 + f_2 + \cdots$ and $s_n = f_1 + \cdots + f_n$. Then, in view of Corollary 2.7.11, $s_n \to f$ a.e., and hence also $|s_n| \to |f|$ a.e. Since $\sum_{n=1}^{\infty} \int |f_n| < \infty$, the function $g = \sum_{n=1}^{\infty} |f_n|$ is integrable, and we have

$$|s_n| \to |f| \text{ a.e.} \quad \text{and} \quad |s_n| < g.$$

Therefore, by the Lebesgue dominated convergence theorem, $|f|$ is integrable.

For every $n \in \mathbb{N}$, we have

$$\left| \int f_1 + \cdots + \int f_n \right| = \left| \int s_n \right| \leq \int |s_n|.$$

Hence, by letting $n \to \infty$, we obtain $|\int f| \leq \int |f|$. $\quad\square$

Definition 2.12.4. (Locally integrable complex-valued functions) A complex-valued function defined on $\mathbb{R}$ is called *locally integrable* if its real part and imaginary part are locally integrable.

The next theorem follows immediately from Theorem 2.12.3.

Theorem 2.12.5. *The absolute value* $|f|$ *of a locally integrable complex-valued function f is locally integrable.*

At this point the reader should look back at the properties of real-valued integrable functions to see how they generalize to complex-valued functions. For example, the important Theorems 2.8.1, 2.8.2, and 2.8.4 remain true for complex-valued functions. Most of those results can be easily obtained from the real-valued cases by considering the real and imaginary parts of functions and using the theorems of this section.

2.13 The Spaces $L^p(\mathbb{R})$

In this section, we introduce a family of spaces that are among the most important examples of Banach spaces.

Definition 2.13.1. $(L^p(\mathbb{R}))$
For a real $p > 1$, by $L^p(\mathbb{R})$ we denote the space of all complex-valued locally integrable functions f such that $|f|^p \in L^1(\mathbb{R})$.

We will prove that, for every $p > 1$, $L^p(\mathbb{R})$ with the norm defined by

$$\|f\|_p = \left(\int |f|^p \right)^{1/p} \tag{2.28}$$

is a Banach space. First we need to show that $L^p(\mathbb{R})$ is a vector space, then that (2.28) is a norm in $L^p(\mathbb{R})$, and finally that the space is complete. Even the fact that $L^p(\mathbb{R})$ is a vector space is not obvious. The approach we take is similar to the one presented in Chapter 1 and concerning the space l^p (see Example 1.2.6).

Theorem 2.13.2. (Hölder's inequality)   *Let* $1 < p < \infty$, $1 < q < \infty$, *and* $\frac{1}{p} + \frac{1}{q} = 1$. *If* $f \in L^p(\mathbb{R})$ *and* $g \in L^q(\mathbb{R})$, *then* $fg \in L^1(\mathbb{R})$ *and*

$$\|fg\|_1 \leq \|f\|_p \|g\|_q. \tag{2.29}$$

Proof: Since the inequality is obvious if $\|f\|_p = 0$ or $\|g\|_q = 0$, we assume that $\|f\|_p \neq 0$ and $\|g\|_q \neq 0$. It was established in Chapter 1 (see (1.1)) that for any non-negative numbers a and b, we have

$$ab \leq \frac{a^p}{p} + \frac{b^q}{q}.$$

For any $x \in \mathbb{R}$, by taking

$$a = \frac{|f(x)|}{\|f\|_p} \quad \text{and} \quad b = \frac{|g(x)|}{\|g\|_q},$$

we obtain

$$\frac{|f(x)g(x)|}{\|f\|_p\|g\|_q} \le \frac{|f(x)|^p}{p\|f\|_p^p} + \frac{|g(x)|^q}{q\|g\|_q^q}.$$

Integrating both sides of this inequality gives

$$\frac{\int |fg|}{\|f\|_p\|g\|_q} \le \frac{1}{p} + \frac{1}{q} = 1.$$

$\square$

Theorem 2.13.3. (Minkowski's inequality) *Let $1 \le p < \infty$. If $f, g \in L^p(\mathbb{R})$, then $f + g \in L^p(\mathbb{R})$ and*

$$\|f + g\|_p \le \|f\|_p + \|g\|_p. \tag{2.30}$$

Proof: For any non-negative numbers a and b and $p \ge 1$, we have

$$(a + b)^p \le 2^{p-1}(a^p + b^p).$$

(We leave the proof of this elementary inequality as an exercise.) Hence,

$$|f + g|^p \le (|f| + |g|)^p \le 2^{p-1}(|f|^p + |g|^p),$$

proving $f + g \in L^p(\mathbb{R})$.

We already know that (2.30) holds for $p = 1$, so we can assume that $p > 1$. Let $q > 1$ be such that $\frac{1}{p} + \frac{1}{q} = 1$. Since

$$|f + g|^p = |f + g||f + g|^{p-1} \le |f||f + g|^{p-1} + |g||f + g|^{p-1} \tag{2.31}$$

and $|f + g|^{p-1} \in L^q(\mathbb{R})$, Hölder's inequality implies $|f||f + g|^{p-1}$, $|g||f + g|^{p-1} \in L^1(\mathbb{R})$ and

$$\int |f||f + g|^{p-1} \le \|f\|_p \big\||f + g|^{p-1}\big\|_q$$
$$= \|f\|_p \left(\int |f + g|^{(p-1)q}\right)^{1/q} = \|f\|_p\|f + g\|_p^{p/q},$$

since $(p - 1)q = p$. Similarly,

$$\int |g||f + g|^{p-1} \le \|g\|_p\|f + g\|_p^{p/q}.$$

Now, from (2.31), we obtain

$$\|f+g\|_p^p \le \int |f||f+g|^{p-1} + \int |g||f+g|^{p-1}$$
$$\le \|f\|_p \|f+g\|_p^{p/q} + \|g\|_p \|f+g\|_p^{p/q}$$
$$= (\|f\|_p + \|g\|_p)\|f+g\|_p^{p/q}.$$

Since $p - \frac{p}{q} = 1$, in equality (2.30) follows.   □

Now it is easy to check that $L^p(\mathbb{R})$ is a vector space for every $p \ge 1$ and that (2.28) is a norm in $L^p(\mathbb{R})$. To be exact, we should say that (2.28) defines a norm in the space $\mathcal{L}^p(\mathbb{R})$ of equivalence classes of functions in $L^p(\mathbb{R})$ that are equal almost everywhere, but we will not make such a distinction. (See the discussion in Section 2.6.)

Theorem 2.13.4. *The space $L^p(\mathbb{R})$ is complete for all $1 \le p < \infty$.*

Proof: Let (f_n) be a Cauchy sequence in $L^p(\mathbb{R})$ and let $M > 0$ be a fixed number. Then, by Hölder's inequality,

$$\int_{-M}^{M} |f_m - f_n| \le \left(\int_{-M}^{M} |f_m - f_n|^p\right)^{1/p} \left(\int_{-M}^{M} 1\right)^{1/q}$$
$$= (2M)^{1/q}\|f_m - f_n\|_p \to 0 \quad \text{as } m, n \to \infty,$$

where $\frac{1}{p} + \frac{1}{q} = 1$. Thus, (f_n) is a Cauchy sequence in $L^1([-M, M])$ and hence it converges to a function $f \in L^1([-M, M])$, that is,

$$\int_{-M}^{M} |f - f_n| \to 0 \quad \text{as } n \to \infty.$$

By Theorem 2.8.2, the sequence (f_n) has a subsequence convergent to f almost everywhere in $[-M, M]$.

Now we apply the described method for $M = 1, 2, \ldots$. First we find a subsequence $(f_{1,n})$ of (f_n) that converges almost everywhere in $[-1, 1]$. Next, since $(f_{1,n})$ is a Cauchy sequence in $L^p([-2, 2])$, it has a subsequence $(f_{2,n})$ that converges almost everywhere in $[-2, 2]$. Continuing in this fashion, we construct a collection of sequences $(f_{m,n})$, $m = 1, 2, \ldots$, such that

$(f_{m,n})$ is a subsequence of (f_n) for every $m \in \mathbb{N}$,

$(f_{m+1,n})$ is a subsequence of $(f_{m,n})$ for every $m \in \mathbb{N}$,

$(f_{m,n})$ converges almost everywhere in $[-m, m]$, as $n \to \infty$, for every $m \in \mathbb{N}$.

Since $(f_{n,n})$ is a subsequence of (f_n), there exists a sequence of indices $p_n \to \infty$ such that $f_{n,n} = f_{p_n}$ for all $n \in \mathbb{N}$. Clearly, (f_{p_n}) is a subsequence of (f_n) that converges almost everywhere in $\mathbb{R}$ to some function f. Since (f_{p_n}) is a Cauchy sequence in $L^p(\mathbb{R})$, given any $\varepsilon > 0$, we have

$$\int |f_{p_m} - f_{p_n}|^p < \varepsilon$$

for sufficiently large m and n. By letting $m \to \infty$, we obtain

$$\int |f - f_{p_n}|^p = \|f - f_{p_n}\|_p^p \leq \varepsilon,$$

by Fatou's lemma (Theorem 2.8.5). This proves that $f \in L^p(\mathbb{R})$. Moreover,

$$\|f - f_n\|_p \leq \|f - f_{p_n}\|_p + \|f_{p_n} - f_n\|_p < 2\varepsilon^{1/p}$$

for all sufficiently large n. Since ε is an arbitrary positive number, the sequence (f_n) converges to f in $L^p(\mathbb{R})$.   □

If Ω is a measurable subset of $\mathbb{R}$, then $L^p(\Omega)$ can be defined as the space of all complex-valued measurable functions f on Ω such that $\int_\Omega |f|^p < \infty$. The norm in $L^p(\Omega)$ is defined by

$$\|f\|_p = \left(\int_\Omega |f|^p \right)^{1/p}.$$

It is easy to see that the theorems in this section remain true if $L^p(\mathbb{R})$ is replaced by $L^p(\Omega)$. In fact, $L^p(\Omega)$ can be identified with the subspace of $L^p(\mathbb{R})$ of all functions vanishing outside of Ω.

The space $L^\infty(\mathbb{R})$ is also used, but it is defined differently. First we introduce the *essential supremum* of a measurable function f, denoted by ess sup f, as the smallest number M such that $f \leq M$ almost everywhere. If no such number exists, then we write ess sup $f = \infty$. Then $L^\infty(\mathbb{R})$ is defined as the space of all measurable functions f such that ess sup $|f| < \infty$ and the norm in $L^\infty(\mathbb{R})$ is defined by

$$\|f\|_\infty = \text{ess sup } |f|. \tag{2.32}$$

It is easy to see that Hölder's and Minkowski's inequalities hold when $p = \infty$ (and $q = 1$), and thus $L^\infty(\mathbb{R})$ is a vector space and (2.32) is a norm. It can also be shown that $L^\infty(\mathbb{R})$ is complete.

2.14 Lebesgue Integrable Functions on $\mathbb{R}^N$

The theory presented in the first 13 sections of this chapter generalizes to functions defined on $\mathbb{R}^N$ without any essential difficulties. Some changes are necessary in Section 2.2, but the remaining sections (with the exception of Section 2.10) are written in such a way that they can be read with $\mathbb{R}^N$ in mind. We include in this section some basic definitions to indicate what changes are necessary in Section 2.2.

By a *semiopen interval* in $\mathbb{R}^N$, we mean a set I, which can be represented as

$$I = [a_1, b_1) \times \cdots \times [a_N, b_N). \qquad (2.33)$$

In other words, $x = (x_1, \ldots, x_N) \in I$ if $a_k \le x_k < b_k$ for $k = 1, 2, \ldots, N$. For the interval (2.33), we define $m(I) = (b_1 - a_1) \cdots (b_N - a_N)$. If the dimension $N = 1$ then $m(I)$ is just the length of I; if $N = 2$ then $m(I)$ is the area of I, if $N = 3$ then $m(I)$ is the volume of I.

By a *step function* we mean a finite linear combination of characteristic functions of semiopen intervals:

$$f = \lambda_1 \chi_{I_1} + \cdots + \lambda_n \chi_{I_n}. \qquad (2.34)$$

For the step function f in (2.34), we define

$$\int f = \lambda_1 m(I_1) + \cdots + \lambda_n m(I_n).$$

Definition 2.14.1.   (Lebesgue integrable function on $\mathbb{R}^N$)
A real or complex-valued function f defined on $\mathbb{R}^N$ is called *Lebesgue integrable* if there exists a sequence of step functions (f_n) such that the following two conditions are satisfied:

(a) $\displaystyle\sum_{n=1}^{\infty} \int |f_n| < \infty;$

(b) $\displaystyle f(x) = \sum_{n=1}^{\infty} f_n(x)$ for every $x \in \mathbb{R}^N$ such that $\displaystyle\sum_{n=1}^{\infty} |f_n(x)| < \infty.$

The integral of f is then defined by

$$\int f = \sum_{n=1}^{\infty} \int f_n.$$

The space of all Lebesgue integrable functions on $\mathbb{R}^N$ is denoted by $L^1(\mathbb{R}^N)$.

Definition 2.14.2.   (Locally integrable function on $\mathbb{R}^N$)

A function f defined on $\mathbb{R}^N$ is called *locally integrable* if for every bounded interval I, the product $f \chi_I$ is an integrable function.

Lebesgue measure on $\mathbb{R}^N$ can be defined as in Section 2.11: A set $S \subseteq \mathbb{R}^N$ is called *measurable* if the characteristic function of S is a locally integrable function. If the characteristic function χ_S is an integrable function, then by the measure of S, denoted by $\mu(S)$, we mean the value of the integral $\mu(S) = \int \chi_S$. If χ_S is locally integrable but not integrable, then we define $\mu(S) = \infty$. A function f is called *measurable* if there exists a sequence of step functions $f_1, f_2, \ldots$ such that $f_n \to f$ a.e.

For a measurable set $\Omega \subseteq \mathbb{R}^N$, by $\int_\Omega f$ we mean the integral of the function equal to f on Ω and 0 everywhere else, that is, $\int_\Omega f = \int f \chi_\Omega$.

Definition 2.14.3.   ($L^p(\mathbb{R}^N)$)

For a real $p > 1$, by $L^p(\mathbb{R}^N)$, we denote the space of all complex-valued locally integrable functions f on $\mathbb{R}^N$ such that $|f|^p \in L^1(\mathbb{R}^N)$.

For every $p > 1$, $L^p(\mathbb{R}^N)$ with the norm defined by

$$\|f\|_p = \left(\int |f|^p \right)^{1/p} \tag{2.35}$$

is a Banach space and Hölder's inequality holds: If $1 < p < \infty$, $1 < q < \infty$, $\frac{1}{p} + \frac{1}{q} = 1$, $f \in L^p(\mathbb{R}^N)$, and $g \in L^q(\mathbb{R}^N)$, then $fg \in L^1(\mathbb{R}^N)$ and

$$\|fg\|_1 \leq \|f\|_p \|g\|_q. \tag{2.36}$$

The space of all functions $f \in L^p(\mathbb{R}^N)$ vanishing outside a measurable set $\Omega \subseteq \mathbb{R}^N$ is denoted by $L^p(\Omega)$. It is a vector subspace of $L^p(\mathbb{R}^N)$. If Ω is of finite measure, then $L^p(\Omega) \subset L^r(\Omega)$ whenever $p \geq r$.

The essential supremum of a measurable function f, denoted by $\mathrm{ess\,sup} f$, is the smallest number M such that $f \leq M$ almost everywhere. If no such number exists, then we write $\mathrm{ess\,sup} f = \infty$. Then $L^\infty(\mathbb{R}^N)$ is defined as the space of all measurable functions f such that $\mathrm{ess\,sup} |f| < \infty$ and the norm in $L^\infty(\mathbb{R}^N)$ is defined by

$$\|f\|_\infty = \mathrm{ess\,sup} |f|.$$

Applications frequently involve multiple integrals. Changing the order of integration is crucial in many arguments. For example, given (a, b) and (c, d), intervals in $\mathbb{R}$, and f, a function on $(a, b) \times (c, d)$, we want to be able to use the equality

$$\int_a^b \int_c^d f(x, y)dydx = \int_c^d \int_a^b f(x, y)dxdy.$$

For the Riemann integral, the equality is usually proved under the assumption of continuity of f, or at least piecewise continuity. The Lebesgue integral allows an essential relaxation of these assumptions. In the following theorem, we show that integrability of f over $(a, b) \times (c, d)$ is sufficient. The theorem is a special case of Fubini's theorem (Guido Fubini (1879–1943)). The formulation of the theorem requires clarification of the notation.

Let (a, b) and (c, d) be bounded or unbounded intervals in $\mathbb{R}$ and let $J = (a, b) \times (c, d)$. By the integral $\int_J f$, we mean the integral of f as defined in this section (Definition 2.14.1). For a fixed $x \in (a, b)$ the integral

$$\int_c^d f(x, y)dy \tag{2.37}$$

denotes the Lebesgue integral of the function $f_x(y) = f(x, y)$, that is, $\int_c^d f(x, y)dy = \int_c^d f_x$. If the function f_x is integrable over (c, d) for every $x \in (a, b)$, then (2.37) defines a function on (a, b):

$$F(x) = \int_c^d f(x, y)dy.$$

If $F \in L^1((a, b))$, then F can be integrated over (a, b). The resulting integral will be denoted as a double integral of f:

$$\int_a^b F = \int_a^b \int_c^d f(x, y)dydx.$$

Notice that the question whether F is integrable or not makes sense even if F is defined almost everywhere, that is, if the integral (2.37) exists almost everywhere in (a, b).

In a similar fashion, only changing the order, we define the integral

$$\int_c^d \int_a^b f(x, y)dxdy.$$

Theorem 2.14.4. (Fubini's theorem)   *Let (a, b) and (c, d) be bounded or unbounded intervals in $\mathbb{R}$ and let $J = (a, b) \times (c, d)$. If f is an integrable function on J, then the function*

$$F(x) = \int_c^d f(x, y)dy$$

is defined almost everywhere in (a, b), *F is integrable in* (a, b), *and we have*

$$\int_J f = \int_c^d F = \int_a^b \int_c^d f(x, y)dydx. \tag{2.38}$$

Proof: Since f is integrable over J, there are intervals $I_n = [a_n, b_n) \times [c_n, d_n) \subseteq J$ and scalars $\lambda_1, \lambda_2, \ldots$ such that

$$f \simeq \lambda_1 \chi_{I_1} + \lambda_2 \chi_{I_2} + \cdots.$$

If we denote

$$g_n = \chi_{[a_n, b_n)}, \quad h_n = \chi_{[c_n, d_n)}, \quad \text{and} \quad f_n(x, y) = g_n(x)h_n(y),$$

then we have

$$f \simeq \lambda_1 f_1 + \lambda_2 f_2 + \cdots. \tag{2.39}$$

Note that

$$\int f_n = \int g_n \int h_n.$$

(Remember that $f_n \in L^1(\mathbb{R}^2)$ and $g_n, h_n \in L^1(\mathbb{R})$, so that the symbol $\int$ has different meanings on the left-hand side and the right-hand side of the above equality.) Since

$$|\lambda_1| \int g_1 \int h_1 + |\lambda_2| \int g_2 \int h_2 + \cdots = |\lambda_1| \int f_1 + |\lambda_2| \int f_2 + \cdots < \infty,$$

there is an integrable function F (a function of x) such that

$$F \simeq \left(\lambda_1 \int h_1\right)g_1 + \left(\lambda_2 \int h_2\right)g_2 + \cdots. \tag{2.40}$$

Let $x_0 \in (a, b)$ be a point where the series (2.40) converges absolutely. Then there exists an integrable function G (a function of y) such that

$$G \simeq \left(\lambda_1 g_1(x_0)\right)h_1 + \left(\lambda_2 g_2(x_0)\right)h_2 + \cdots. \tag{2.41}$$

If the series (2.41) converges absolutely at some $y \in (c, d)$, then

$$\lambda_1 g_1(x_0)h_1(y) + \lambda_2 g_2(x_0)h_2(y) + \cdots = f(x_0, y),$$

by (2.39). Thus, the function $f(x_0, y)$ is integrable over (c, d), and we have

$$\int_c^d f(x_0, y)dy = \lambda_1 g_1(x_0) \int h_1 + \lambda_2 g_2(x_0) \int h_2 + \cdots \tag{2.42}$$

for every x_0, where the series (2.40) converges absolutely. Consequently, the function

$$F(x) = \int_c^d f(x, y)dy$$

is defined almost everywhere in (a, b), and the first part of the theorem is proved. Now, in view of (2.40), F is integrable and we have

$$\int_a^b F = \int_a^b \int_c^d f(x, y)dydx = \lambda_1 \int g_1 \int h_1 + \lambda_2 \int g_2 \int h_2 + \cdots.$$

On the other hand,

$$\lambda_1 \int g_1 \int h_1 + \lambda_2 \int g_2 \int h_2 + \cdots = \int_J f.$$

Thus, (2.38) is proved.   □

2.15 Convolution

If the integral $\int f(x - y)g(y)dy$ exists for all $x \in \mathbb{R}$, or at least almost everywhere, then it defines a function which is called the *convolution* of f and g, denoted by $f * g$. In this section, we discuss some basic properties of convolution.

Theorem 2.15.1. *If $f, g \in L^1(\mathbb{R})$, then the function $f(x - y)g(y)$ is integrable for almost all $x \in \mathbb{R}$. Moreover, the convolution*

$$(f * g)(x) = \int f(x - y)g(y)dy$$

is an integrable function, and we have

$$\int |f * g| \le \int |f| \int |g|. \tag{2.43}$$

Proof: First we show that the function $H(x, y) = f(x - y)g(y)$ is integrable in $\mathbb{R}^2$. Since $f, g \in L^1(\mathbb{R})$, there are step functions $f_1, f_2, \ldots$ and $g_1, g_2, \ldots$ such that

$$f \simeq f_1 + f_2 + \cdots \quad \text{and} \quad g \simeq g_1 + g_1 + \cdots.$$

Now define

$$F_n(x, y) = f_n(x - y) \quad \text{and} \quad G_n(x, y) = g_n(y).$$

We will show that

$$H \simeq \sum_{n=1}^{\infty} \sum_{k=1}^{n} F_k G_{n-k+1}. \tag{2.44}$$

In fact, since

$$\int \left| \sum_{k=1}^{n} F_k G_{n-k+1} \right| \le \sum_{k=1}^{n} \int |F_k G_{n-k+1}| \le \sum_{k=1}^{n} \int |f_k| \int |g_{n-k+1}| \tag{2.45}$$

and the series

$$\sum_{n=1}^{\infty} \int |f_n| \quad \text{and} \quad \sum_{n=1}^{\infty} \int |g_n|$$

are absolutely convergent, we have

$$\sum_{n=1}^{\infty} \int \left| \sum_{k=1}^{n} F_k G_{n-k+1} \right| < \infty.$$

Moreover, if for some $(x, y) \in \mathbb{R}^N$ the series

$$\sum_{n=1}^{\infty} \sum_{k=1}^{n} F_k(x, y) G_{n-k+1}(x, y)$$

converges absolutely and, for some $n_0, m_0 \in \mathbb{N}$, we have $f_{n_0}(x - y) \ne 0$ and $g_{m_0}(y) \ne 0$, then both series

$$\sum_{n=1}^{\infty} f_n(x - y) \quad \text{and} \quad \sum_{n=1}^{\infty} g_n(y)$$

converge absolutely. If, on the other hand, $f_n(x - y) = 0$ for all $n \in \mathbb{N}$ or $g_n(y) = 0$ for all $n \in \mathbb{N}$, then $H(x, y) = 0$. In either case we have

$$H(x, y) = \sum_{n=1}^{\infty} \sum_{k=1}^{n} F_k(x, y) G_{n-k+1}(x, y),$$

at that point (x, y). Therefore, (2.44) holds, which means that $H \in L^1(\mathbb{R}^2)$. By Fubini's Theorem 2.14.4, the integral $\int f(x - y) g(y) dy$ is defined almost everywhere in $\mathbb{R}$ (with respect to x) and defines an integrable function, that is, $f * g \in L^1(\mathbb{R})$.

To show that (2.43) holds, note that since H is integrable, so is $|H|$ and thus

$$\int |f * g| = \int \left| \int f(x - y)g(y)dy \right| dx$$

$$\leq \int\int |f(x - y)||g(y)|dydx$$

$$= \int\int |f(x - y)||g(y)|dxdy \quad \text{(by Fubini's theorem)}$$

$$= \int |f(x - y)|dx \int |g(y)|dy$$

$$\left(\int |f(x - y)|dx \text{ is independent of } y \right)$$

$$= \int |f| \int |g|. \qquad \square$$

Theorem 2.15.2. *If $f, g \in L^1(\mathbb{R})$, then $f * g = g * f$.*

Proof follows easily by the change of variables (Theorem 2.10.4).

Theorem 2.15.3. *If f is an integrable function and g is a bounded, locally integrable function, then the convolution $f * g$ is a continuous function.*

Proof: First observe that since $|f(x - y)g(y)| \leq M|f(x - y)|$ for some constant M and every $x \in \mathbb{R}$ the integral $\int f(x - y)g(y)dy$ is defined at every $x \in \mathbb{R}$, by Theorem 2.9.5. Now we will show that the convolution $f * g$ is a continuous function.

For any $x, t \in \mathbb{R}$, we have

$$|(f * g)(x + t) - (f * g)(x)| = \left| \int f(x + t - y)g(y)dy - \int f(x - y)g(y)dy \right|$$

$$= \left| \int (f(x + t - y) - f(x - y))g(y)dy \right|$$

$$\leq \int |f(x + t - y) - f(x - y)||g(y)|dy$$

$$\leq M \int |f(t - y) - f(-y)|dy.$$

Now Theorem 2.4.3 implies $\lim_{t \to \infty} \int |f(t - y) - f(-y)|dy = 0$, proving continuity of $f * g$. $\qquad \square$

2.16 Exercises

1. Denote by $\mathcal{A}$ the family consisting of all finite unions of semiopen intervals $[a, b)$ and the empty set. Prove the following properties of $\mathcal{A}$:

(a) If $A_1, \ldots, A_n \in \mathcal{A}$, then $A_1 \cup \cdots \cup A_n \in \mathcal{A}$.

(b) If $A_1, \ldots, A_n \in \mathcal{A}$, then $A_1 \cap \cdots \cap A_n \in \mathcal{A}$.

(c) If $A, B \in \mathcal{A}$, then $A \setminus B \in \mathcal{A}$.

2. Show that step functions form a vector space.

3. Prove that the integral of a step function is independent of a particular representation (2.1).

4. Show that for any step functions f and g we have the following:

(a) $\operatorname{supp}(f + g) \subseteq \operatorname{supp} f \cup \operatorname{supp} g$.

(b) $\operatorname{supp} fg = \operatorname{supp} f \cap \operatorname{supp} g$.

(c) $\operatorname{supp} |f| = \operatorname{supp} f$.

(d) $\operatorname{supp} \lambda f = \operatorname{supp} f, \lambda \in \mathbb{R}, \lambda \neq 0$.

(e) If $|f| \leq |g|$, then $\operatorname{supp} f \subseteq \operatorname{supp} g$.

5. Prove Lemma 2.2.3.

6. Complete the argument in Example 2.2.5.

7. Show that, if f is integrable, then the translated function $\tau_z f$ is integrable for any $z \in \mathbb{R}$ and $\int \tau_z f = \int f$.

8. Prove Theorem 2.4.3 for step functions.

9. Expand the following functions into a series of step functions, that is, find step functions $f_1, f_2, \ldots$ such that $f \simeq f_1 + f_2 + \cdots$:

(a) $f(x) = \begin{cases} 1 & \text{if } x = 0, \\ 0 & \text{if } x \neq 0. \end{cases}$

(b) $f(x) = \begin{cases} 1 & \text{if } x \in [a, b], \\ 0 & \text{if } x \notin [a, b], \end{cases} \quad a < b.$

(c) $f(x) = \max\{0, 1 - |x|\}$.

(d) f is a piecewise continuous function with bounded support.

10. Show that $f \in L^1(\mathbb{R})$ if and only if there are intervals $[a_1, b_1), [a_2, b_2), \ldots$ and numbers $\lambda_1, \lambda_2, \ldots$ such that $f \simeq \lambda_1 \chi_{[a_1, b_1)} + \lambda_2 \chi_{[a_2, b_2)} + \cdots$.

11. Show that if $f \simeq f_1 + f_2 + \cdots$ then $f + g \simeq g + f_1 + f_2 + \cdots$ for any step function g. In particular, if $f \simeq f_1 + f_2 + \cdots$ then $f - f_1 - \cdots - f_n \simeq f_{n+1} + f_{n+2} + \cdots$.

12. If $f \in L^1(\mathbb{R})$ and f vanishes outside of a bounded interval $[a, b)$, then there are step functions $f_1, f_2, \ldots$ vanishing outside of $[a, b)$ such that $f \simeq f_1 + f_2 + \cdots$.

13. If $f \in L^1(\mathbb{R})$ and $f \geq 0$, is it always possible to find non-negative step functions $f_1, f_2, \ldots$ such that $f \simeq f_1 + f_2 + \cdots$?

14. Show that the characteristic function of the set of all rational numbers is Lebesgue integrable but not Riemann integrable.

15. Define $f^+ = \max\{0, f\}$ and $f^- = \max\{0, -f\}$. Prove that $f \in L^1(\mathbb{R})$ if and only if $f^+ \in L^1(\mathbb{R})$ and $f^- \in L^1(\mathbb{R})$.

16. Show that, if f is a continuous, integrable function, then there are step functions $f_1, f_2, \ldots$ such that $f \simeq f_1 + f_2 + \cdots$ and $|f| \simeq |f_1| + |f_2| + \cdots$.

17. By a *tent* function we mean a function of the form

$$
f(x) = \begin{cases}
2(x - a)/(b - a) & \text{if } a \leq x \leq (a + b)/2, \\
2(b - x)/(b - a) & \text{if } (a + b)/2 \leq x \leq b, \\
0 & \text{otherwise,}
\end{cases}
$$

where $a < b$ (see Figure 2.3). Show that $f \in L^1(\mathbb{R})$ if and only if there exist tent functions $f_1, f_2, \ldots$ and numbers $\lambda_1, \lambda_2, \ldots$ such that $f \simeq \lambda_1 f_1 + \lambda_2 f_2 + \cdots$.

18. Show that the relation $f \sim g$ if $\int |f - g| = 0$ is an equivalence in $L^1(\mathbb{R})$.

19. Prove the following:

    (a) Every countable subset of $\mathbb{R}$ is a null set.

    (b) A countable union of null sets is a null set.

20. Consider the following sequence of subsets of the interval $[0, 1]$:

$$
S_0 = [0, 1],
$$

$$
S_1 = [0, 1] \setminus \left(\frac{1}{3}, \frac{2}{3} \right),
$$

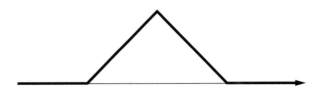

Figure 2.3   A tent function.

$$S_2 = [0, 1] \setminus \left(\left(\frac{1}{9}, \frac{2}{9} \right) \cup \left(\frac{4}{9}, \frac{5}{9} \right) \cup \left(\frac{7}{9}, \frac{8}{9} \right) \right),$$

$$\vdots$$

$$S_n = [0, 1] \setminus \bigcup_{k=0}^{3^{n-1}-1} \left(\frac{1+3k}{3^n}, \frac{2+3k}{3^n} \right),$$

$$\vdots$$

Define

$$C = \bigcap_{n=0}^{\infty} S_n.$$

This set is called the *Cantor set* (Georg Ferdinand Ludwig Philipp Cantor (1845–1918)).

(a) Show that C is a null set.

(b) Show that C is not countable.

21. Prove the following properties of convergence almost everywhere:

   (a) If $f_n \to f$ a.e. and $\lambda \in \mathbb{R}$, then $\lambda f_n \to \lambda f$ a.e.

   (b) If $f_n \to f$ a.e. and $g_n \to g$ a.e., then $f_n + g_n \to f + g$ a.e.

   (c) If $f_n \to f$ a.e., then $|f_n| \to |f|$ a.e.

22. Show that every Lebesgue integrable function can be approximated in norm and almost everywhere by a sequence of continuous functions.

23. Let $f \in L^1(\mathbb{R})$. Define

$$f_n(x) = \begin{cases} f(x) & \text{if } |x| \le n, \\ 0 & \text{otherwise.} \end{cases}$$

Show that $f_n \to f$ i.n.

24. Show that there exists an unbounded continuous function $f \in L^1(\mathbb{R})$.

25. Show that if f is a uniformly continuous function on $\mathbb{R}$ and $f \in L^1(\mathbb{R})$, then f is bounded and $\lim_{|x| \to \infty} f(x) = 0$.

26. Show that locally integrable functions form a vector space.

27. Let $f \in L^1(\mathbb{R})$ and let g be a bounded, locally integrable function. Show that $fg \in L^1(\mathbb{R})$ and $\int |fg| \le \sup_{x \in \mathbb{R}} |g(x)| \int |f|$.

28. Show that the space $L^1(J)$ is complete for any interval $J \subseteq \mathbb{R}$.

29. Prove: If a sequence of locally integrable functions (f_n) converges almost everywhere to a function f and $|f_n| \leq h$ for every $n \in \mathbb{N}$, where h is a locally integrable function, then f is locally integrable.

30. In Example 2.7.9 we define a sequence of functions (f_n) convergent to 0 in norm but divergent at every point of $[0, 1]$. Find a subsequence of (f_n) convergent to 0 almost everywhere.

31. Prove: If a sequence of integrable functions (f_n) converges almost everywhere to a function f and $|f_n(x)| \leq h(x)$ for almost all $x \in \mathbb{R}$, all $n \in \mathbb{N}$, and some integrable function h, then f is integrable and $f_n \to f$ i.n.

32. Show that the function

$$f(x) = \begin{cases} \dfrac{\sin x}{x} & \text{if } x \neq 0, \\ 1 & \text{if } x = 0 \end{cases}$$

is not a Lebesgue integrable, although the improper Riemann integral $\int_{-\infty}^{\infty} f(x)\,dx$ converges.

33. Let $\mathcal{M}$ denote the collection of all measurable subsets of $\mathbb{R}$. Prove the following:

    (a) $\emptyset, \mathbb{R} \in \mathcal{M}$.

    (b) If $A_1, A_2, \ldots \in \mathcal{M}$, then $\bigcup_{n=1}^{\infty} A_n \in \mathcal{M}$.

    (c) If $A_1, A_2, \ldots \in \mathcal{M}$, then $\bigcap_{n=1}^{\infty} A_n \in \mathcal{M}$.

    (d) If $A, B \in \mathcal{M}$, then $A \backslash B \in \mathcal{M}$.

    (e) Intervals are measurable sets.

    (f) Open subsets of $\mathbb{R}$ are measurable.

    (g) Closed subsets of $\mathbb{R}$ are measurable.

34. Let $\mathcal{M}$ be the collection of all measurable subsets of $\mathbb{R}$ and let μ be the Lebesgue measure on $\mathbb{R}$. Prove the following:

    (a) If $A_1, A_2, \ldots \in \mathcal{M}$, then

    $$\mu\left(\bigcup_{n=1}^{\infty} A_n\right) \leq \sum_{n=1}^{\infty} \mu(A_n).$$

    (b) If $A, B \in \mathcal{M}$ and $A \subseteq B$, then $\mu(B \backslash A) = \mu(B) - \mu(A)$.

    (c) If $A_1, A_2, \ldots \in \mathcal{M}$ and $A_1 \subseteq A_2 \subseteq A_3 \subseteq \cdots$, then

    $$\mu\left(\bigcup_{n=1}^{\infty} A_n\right) = \lim_{n \to \infty} \mu(A_n).$$

(d) If $A_1, A_2, \ldots \in \mathcal{M}$ and $A_1 \supseteq A_2 \supseteq A_3 \supseteq \cdots$, then

$$\mu\left(\bigcap_{n=1}^{\infty} A_n\right) = \lim_{n\to\infty} \mu(A_n).$$

35. Let f be a real-valued function on $\mathbb{R}$. Show that the following conditions are equivalent:

(a) f is measurable.

(b) $\{x \in \mathbb{R}: f(x) \le \alpha\}$ is a measurable set for all $\alpha \in \mathbb{R}$.

(c) $\{x \in \mathbb{R}: f(x) < \alpha\}$ is a measurable set for all $\alpha \in \mathbb{R}$.

(d) $\{x \in \mathbb{R}: f(x) \ge \alpha\}$ is a measurable set for all $\alpha \in \mathbb{R}$.

(e) $\{x \in \mathbb{R}: f(x) > \alpha\}$ is a measurable set for all $\alpha \in \mathbb{R}$.

36. Prove: Let $A_1, A_2, \ldots$ be measurable sets such that $\lim_{n\to\infty} \mu(A_n) = 0$. Then, for every $f \in L^1(\mathbb{R})$, we have $\lim_{n\to\infty} \int_{A_n} f = 0$.

37. Prove that $L^1(\Omega)$ is a Banach space for any measurable set $\Omega \subseteq \mathbb{R}^N$. (The norm in $L^1(\Omega)$ is defined by $\|f\| = \int_\Omega |f|$ and functions equal almost everywhere are identified.)

38. Show that for any non-negative numbers a and b and $p \ge 1$, we have $(a+b)^p \le 2^{p-1}(a^p + b^p)$.

39. Let

$$g(x) = \begin{cases} 1/\sqrt{x} & \text{for } 0 < |x| < 1, \\ 0 & \text{otherwise.} \end{cases}$$

Show that $g \in L^1(\mathbb{R})$ but $g^2 \notin L^1(\mathbb{R})$.

40. Let $f(x) = \min\{1, 1/|x|\}$. Show that $f \notin L^1(\mathbb{R})$ but $f \in L^2(\mathbb{R})$.

41. Show that $L^p([a, b]) \subseteq L^r([a, b])$ for any $1 \le r < p \le \infty$.

42. Show that Hölder's and Minkowski's inequalities hold when $p = \infty$ and $q = 1$.

43. Let $f_n, f \in L^\infty(\mathbb{R})$. Show that $\|f_n - f\|_\infty \to 0$ if and only if there exists a set of measure zero $S \subset \mathbb{R}$ such that $f_n \to f$ uniformly on $\mathbb{R}\backslash S$.

44. Let $f, g, h \in L^1(\mathbb{R})$. Prove the following properties of the convolution:

(a) $(f * g) * h = f * (g * h)$ (Associative),

(b) $(\alpha f + \beta g) * h = \alpha(f * h) + \beta(g * h)$ (Distributive)

where α and β are constants.

45. Calculate $(f * g)(x)$ for the following functions:

   (a) $f(x) = \sin x$, $g(x) = e^{-|x|}$.

   (b) $f(x) = x\chi_{[-2,1]}(x)$, $g(x) = x^2 + 2$.

   (c) $f(x) = e^x$, $g(x) = \chi_{[0,\infty]}(x)$.

   (d) $f(x) = \chi_{[-4,0]}(x)$, $g(x) = \chi_{[-1,1]}(x)$.

46. (a) If $(f, g) \in L^1(\mathbb{R})$, show that $(f * g)(x)$ exists for almost all x and $(f * g) \in L^1(\mathbb{R})$.

   (b) Prove that $\int_{-\infty}^{\infty}(f * g)(x)dx = \int_{-\infty}^{\infty} f(t)dt \int_{-\infty}^{\infty} g(s)ds$ and $\|f * g\|_1 \leq \|f\|_1\|g\|_1$.

47. If $g(x) = \frac{1}{2a}H(a - |x|)$, show that $(f * g)(x)$ is the average value of f in $[x - a, x + a]$.

48. Let f be the characteristic function of the interval $[-1, 1]$. Calculate the convolutions $f * f$ and $f * f * f$.

49. Let $f \in L^1(\mathbb{R})$ and let g be a bounded differentiable function on $\mathbb{R}$. Show that $f * g$ is differentiable. If, in addition, g' is bounded show that $(f * g)' = f * g'$.

50. Let f be a locally integrable function on $\mathbb{R}$ and let g be a differentiable function with bounded support in $\mathbb{R}$. Show that $f * g$ is differentiable and $(f * g)' = f * g'$.

51. Prove the following properties of differentiability of the convolution:

   (a) $(f * g)'' = f' * g'$.

   (b) $(f * g)^{(n+m)}(x) = f^{(n)}(x) * g^{(m)}(x)$.

52. **(Young's inequality)** Suppose $1 \leq p, q, r \leq \infty$ satisfy the equation

$$\frac{1}{p} + \frac{1}{q} = 1 + \frac{1}{r}.$$

   If $f \in L^p(\mathbb{R})$ and $g \in L^q(\mathbb{R})$, show that $f * g \in L^r(\mathbb{R})$ and we have

$$\|f * g\|_r \leq \|f\|_p\|g\|_q$$

   (William Henry Young (1863–1942)).

53. Prove that $L^\infty(\mathbb{R}^N)$ is complete.

54. Show that $L^p(\mathbb{R}^N)$ is separable for $1 \leq p < \infty$.

55. Show that $L^\infty([0, 1])$ is not separable.

56. Show that the space $\mathcal{C}_c(\mathbb{R}^N)$ of continuous functions with compact support is dense in $L^p(\mathbb{R}^N)$ for $1 \leq p < \infty$.

57. Show that $\mathcal{C}([0, 1])$ is not dense in $L^\infty([0, 1])$.

58. Show that the space $\mathcal{D}(\mathbb{R}^N)$ of smooth functions with compact support is dense in $L^p(\mathbb{R}^N)$ for $1 \leq p < \infty$.

Hilbert Spaces and Orthonormal Systems

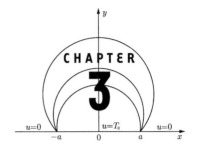

"As long as a branch of knowledge offers an abundance of problems, it is full of vitality."

David Hilbert

"Hilbert spaces constitute at present the most important examples of Banach spaces, not only because they are the most natural and closest generalization in the realm of 'infinite dimensions,' of our classical Euclidean geometry, but chiefly for the fact they have been, up to now, the most useful spaces in the applications to functional analysis."

Jean Dieudonné

3.1 Introduction

The theory of Hilbert spaces was initiated by David Hilbert (1862–1943) in his 1912 work on quadratic forms in infinitely many variables, which he applied to the theory of integral equations. He published a series of papers on the subject during 1904–1910 and formulated the theory of the space l^2 (square summable sequences) in connection with integral equations. The idea of the l^2-space and the spectral theory of bounded quadratic forms were published in 1906. All these papers were reproduced in the form of a book entitled *Grundzüge einer allgemeinen Theorie der linearen Integralgleichungen* in 1912. The book had a tremendous influence on mathematical analysis and its applications.

Years later, John von Neumann (1903–1957) first formulated an axiomatic theory of Hilbert spaces and developed the modern theory of operators on Hilbert spaces. His remarkable contribution to this area has provided the mathematical foundation of quantum mechanics. Von Neumann's work has also provided a physical interpretation of quantum mechanics in terms of abstract relations in an infinite dimensional Hilbert space.

This chapter is concerned with inner product spaces (called also *pre-Hilbert spaces* or *unitary* spaces) and Hilbert spaces. The basic ideas and properties will be discussed with special attention given to orthonormal systems. The theory is illustrated by numerous examples.

3.2 Inner Product Spaces

Definition 3.2.1.   (Inner product space)

Let E be a complex vector space. A mapping $\langle \cdot, \cdot \rangle : E \times E \to \mathbb{C}$ is called an *inner product* in E if for any $x, y, z \in E$ and $\alpha, \beta \in \mathbb{C}$ the following conditions are satisfied:

(a) $\langle x, y \rangle = \overline{\langle y, x \rangle}$ (the bar denotes the complex conjugate);

(b) $\langle \alpha x + \beta y, z \rangle = \alpha \langle x, z \rangle + \beta \langle y, z \rangle$;

(c) $\langle x, x \rangle \geq 0$;

(d) $\langle x, x \rangle = 0$ implies $x = 0$.

A vector space with an inner product is called an *inner product space*.

According to the definition, the inner product of two vectors is a complex number. By (a), $\langle x, x \rangle = \overline{\langle x, x \rangle}$, which means that $\langle x, x \rangle$ is a real number for every $x \in E$. It follows from (b) that

$$\langle x, \alpha y + \beta z \rangle = \overline{\langle \alpha y + \beta z, x \rangle} = \overline{\alpha \langle y, x \rangle + \beta \langle z, x \rangle} = \overline{\alpha} \langle x, y \rangle + \overline{\beta} \langle x, z \rangle.$$

In particular

$$\langle \alpha x, y \rangle = \alpha \langle x, y \rangle \quad \text{and} \quad \langle x, \alpha y \rangle = \overline{\alpha} \langle x, y \rangle.$$

Hence, if $\alpha = 0$, we have $\langle 0, y \rangle = \langle x, 0 \rangle = 0$.

Example 3.2.2. The simplest, although important, example of an inner product space is the space of complex numbers $\mathbb{C}$. The inner product is defined by $\langle x, y \rangle = x\overline{y}$. □

Example 3.2.3. The space $\mathbb{C}^N$ of ordered N-tuples $x = (x_1, \ldots, x_N)$ of complex numbers, with the inner product defined by

$$\langle x, y \rangle = \sum_{k=1}^{N} x_k \overline{y}_k, \quad x = (x_1, \ldots, x_N), \ y = (y_1, \ldots, y_N),$$

is an inner product space. ☐

Example 3.2.4. The space l^2 of all sequences $(x_1, x_2, x_3, \ldots)$ of complex numbers such that $\sum_{k=1}^{\infty} |x_k|^2 < \infty$ (see Section 1.2), with the inner product defined by

$$\langle x, y \rangle = \sum_{k=1}^{\infty} x_k \bar{y}_k, \quad x = (x_1, x_2, x_3, \ldots), \ y = (y_1, y_2, y_3, \ldots),$$

is an infinite dimensional inner product space. As we will see later, this space is, in a sense, the most important example of an inner product space (see Theorem 3.4.27). ☐

Example 3.2.5. Consider the space of sequences $(x_1, x_2, x_3, \ldots)$ of complex numbers with only a finite number of nonzero terms. This is an inner product space with the inner product defined as in Example 3.2.4. ☐

Example 3.2.6. The space $\mathcal{C}([a, b])$ of all continuous complex valued functions on the interval $[a, b]$, with the inner product

$$\langle f, g \rangle = \int_a^b f(x) \overline{g(x)} dx,$$

is an inner product space. ☐

Example 3.2.7. The space $L^2(\mathbb{R})$ (see Section 2.13) with the inner product defined by

$$\int_{-\infty}^{\infty} f(x) \overline{g(x)} dx,$$

and, more generally, the space $L^2(\mathbb{R}^N)$ (see Section 2.14) with the inner product defined by

$$\int_{\mathbb{R}^N} f(x) \overline{g(x)} dx,$$

are very important inner product spaces.

In applications we often use a subset Ω of $\mathbb{R}^N$ and the space $L^2(\Omega)$ with the inner product defined by the integral over Ω. For example, the space $L^2([a, b])$ is the right setting in many cases. ☐

Example 3.2.8. Let E be the Cartesian product of inner product spaces E_1 and E_2, that is, $E = E_1 \times E_2 = \{(x, y): x \in E_1, y \in E_2\}$. The space E is an inner product space with the inner product defined by

$$\langle (x_1, y_1), (x_2, y_2) \rangle = \langle x_1, x_2 \rangle + \langle y_1, y_2 \rangle.$$

Note that E_1 and E_2 can be identified with the subspaces $E_1 \times \{0\}$ and $\{0\} \times E_2$, respectively.

Similarly, we can define the inner product on $E_1 \times \cdots \times E_n$. This method can be used to construct new examples of inner product spaces.   □

An inner product space is a vector space with an inner product. It turns out that every inner product space is also a normed space with the norm defined by

$$\|x\| = \sqrt{\langle x, x \rangle}.$$

First notice that this functional is well defined because $\langle x, x \rangle$ is always a non-negative (real) number and $\|x\| = 0$ if and only if $x = 0$. Moreover,

$$\|\lambda x\| = \sqrt{\langle \lambda x, \lambda x \rangle} = \sqrt{\lambda \overline{\lambda} \langle x, x \rangle} = |\lambda| \|x\|.$$

It thus remains to prove the triangle inequality. This is not as simple as the first two conditions. We first prove the so-called Schwarz's inequality (Hermann Amandus Schwarz (1843–1921)), which will be used in the proof of the triangle inequality.

Theorem 3.2.9. (Schwarz's inequality)   *For any two elements x and y of an inner product space, we have*

$$\left| \langle x, y \rangle \right| \le \|x\| \|y\|. \tag{3.1}$$

The equality $|\langle x, y \rangle| = \|x\| \|y\|$ holds if and only if x and y are linearly dependent.

Proof: If $y = 0$, then (3.1) is satisfied because both sides are equal to zero. Assume then $y \ne 0$. We have

$$0 \le \langle x + \alpha y, x + \alpha y \rangle = \langle x, x \rangle + \overline{\alpha} \langle x, y \rangle + \alpha \langle y, x \rangle + |\alpha|^2 \langle y, y \rangle. \tag{3.2}$$

Now put $\alpha = -\langle x, y \rangle / \langle y, y \rangle$ in (3.2), and then multiply by $\langle y, y \rangle$ to obtain

$$0 \le \langle x, x \rangle \langle y, y \rangle - \left| \langle x, y \rangle \right|^2.$$

This gives Schwarz's inequality.

If x and y are linearly dependent, then $x = \alpha y$ for some $\alpha \in \mathbb{C}$. Hence,

$$\left| \langle x, y \rangle \right| = \left| \langle x, \alpha x \rangle \right| = |\overline{\alpha}| \langle x, x \rangle = |\alpha| \|x\| \|x\| = \|x\| \|\alpha x\| = \|x\| \|y\|.$$

Now, let x and y be vectors such that $|\langle x, y \rangle| = \|x\| \|y\|$, or, equivalently,

$$\langle x, y \rangle \langle y, x \rangle = \langle x, x \rangle \langle y, y \rangle. \tag{3.3}$$

We will show that $\langle y, y \rangle x - \langle x, y \rangle y = 0$, which proves that x and y are linearly dependent. Indeed, by (3.3) we have

$$\langle\langle y, y\rangle x - \langle x, y\rangle y, \langle y, y\rangle x - \langle x, y\rangle y\rangle$$
$$= \langle y, y\rangle^2\langle x, x\rangle - \langle y, y\rangle\langle y, x\rangle\langle x, y\rangle - \langle x, y\rangle\langle y, y\rangle\langle y, x\rangle + \langle x, y\rangle\langle y, x\rangle\langle y, y\rangle$$
$$= 0,$$

completing the proof. □

Corollary 3.2.10. (Triangle inequality) *For any two elements x and y of an inner product space we have*

$$\|x + y\| \leq \|x\| + \|y\|. \tag{3.4}$$

Proof: When $\alpha = 1$, Equation (3.2) can be written as

$$\|x + y\|^2 = \langle x + y, x + y\rangle = \langle x, x\rangle + 2\operatorname{Re}\langle x, y\rangle + \langle y, y\rangle$$
$$\leq \langle x, x\rangle + 2|\langle x, y\rangle| + \langle y, y\rangle$$
$$\leq \|x\|^2 + 2\|x\|\|y\| + \|y\|^2 \quad \text{(by Schwarz's inequality)}$$
$$= (\|x\| + \|y\|)^2,$$

where $\operatorname{Re} z$ denotes the real part of $z \in \mathbb{C}$. □

The preceding discussion justifies the following definition:

Definition 3.2.11. (Norm in an inner product space)
By the *norm in an inner product space E* we mean the functional defined by $\|x\| = \sqrt{\langle x, x\rangle}$.

We have proved that every inner product space is a normed space. It is only natural to ask whether every normed space is an inner product space. More precisely: is it possible to define in a normed space $(E, \|\cdot\|)$ an inner product $\langle\cdot, \cdot\rangle$ such that $\|x\| = \sqrt{\langle x, x\rangle}$ for every $x \in E$? In general, the answer is negative. In the following theorem, we prove a property of the norm in an inner product space, which is a necessary and sufficient condition for a normed space to be an inner product space; see Exercise 12.

Theorem 3.2.12. (Parallelogram law) *For any two elements x and y of an inner product space, we have*

$$\|x + y\|^2 + \|x - y\|^2 = 2(\|x\|^2 + \|y\|^2). \tag{3.5}$$

Proof: We have

$$\|x + y\|^2 = \langle x + y, x + y\rangle = \langle x, x\rangle + \langle x, y\rangle + \langle y, x\rangle + \langle y, y\rangle$$

and hence

$$\|x + y\|^2 = \|x\|^2 + \langle x, y \rangle + \langle y, x \rangle + \|y\|^2. \tag{3.6}$$

Now replace y by $-y$ in (3.6) to obtain

$$\|x - y\|^2 = \|x\|^2 - \langle x, y \rangle - \langle y, x \rangle + \|y\|^2. \tag{3.7}$$

By adding (3.6) and (3.7), we obtain the parallelogram law. $\square$

One of the most important consequences of having the inner product is the possibility of defining orthogonality of vectors. This makes the theory of Hilbert spaces very different from the general theory of Banach spaces.

Definition 3.2.13.   (Orthogonal vectors)
Two vectors x and y in an inner product space are called *orthogonal*, denoted by $x \perp y$, if $\langle x, y \rangle = 0$.

If $x \perp y$, then $\langle y, x \rangle = \overline{\langle x, y \rangle} = 0$, and thus $y \perp x$. In other words, the relation $\perp$ is symmetric.

The next theorem is another example of the geometric character of the norm defined by an inner product.

Theorem 3.2.14. (Pythagorean formula)   *For any pair of orthogonal vectors, we have*

$$\|x + y\|^2 = \|x\|^2 + \|y\|^2. \tag{3.8}$$

Proof: If $x \perp y$, then $\langle x, y \rangle = \langle y, x \rangle = 0$, and thus the equality follows immediately from (3.6). $\square$

In the definition of the inner product space we assume that E is a complex vector space. It is possible to define a real inner product space with the inner product of any two vectors being a real number. Then condition (b) in Definition 3.2.1 becomes $\langle x, y \rangle = \langle y, x \rangle$. All the preceding theorems hold in the real case. If, in Examples 3.2.2–3.2.7, the word *complex* is replaced by *real* and $\mathbb{C}$ by $\mathbb{R}$, we obtain a number of examples of real inner product spaces. A finite dimensional real inner product space is called a *Euclidean space*.

If $x = (x_1, \ldots, x_N)$ and $y = (y_1, \ldots, y_N)$ are vectors in $\mathbb{R}^N$, then the inner product $\langle x, y \rangle = \sum_{k=1}^{N} x_k y_k$ can be defined equivalently as $\langle x, y \rangle = \|x\| \|y\| \cos \theta$, where θ is the angle between vectors x and y. In this case, Schwarz's inequality follows from

$$\frac{|\langle x, y \rangle|}{\|x\| \|y\|} = |\cos \theta| \leq 1, \quad x \neq 0, \ y \neq 0.$$

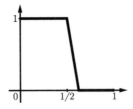

Figure 3.1 A "typical" function f_n in Example 3.3.5.

3.3 Hilbert Spaces

Definition 3.3.1. (Hilbert space)
A complete inner product space is called a *Hilbert space*.

By completeness of an inner product space E, we mean completeness of E as a normed space with the norm defined by the inner product.

Now we are going to discuss completeness of the inner product spaces mentioned in Section 3.2 and also give some new examples of inner product spaces and Hilbert spaces.

Example 3.3.2. Since $\mathbb{C}$ is complete, it is a Hilbert space, and so is $\mathbb{C}^N$. □

Example 3.3.3. l^2 is a Hilbert space. The completeness was proved in Section 1.4 (see Example 1.4.6). □

Example 3.3.4. The space E described in Example 3.2.5 is an inner product space, which is not a Hilbert space. It is not complete. The sequence

$$x_n = \left(1, \frac{1}{2}, \frac{1}{3}, \ldots, \frac{1}{n}, 0, 0, \ldots\right)$$

is a Cauchy sequence:

$$\lim_{n,m\to\infty} \|x_n - x_m\| = \lim_{n,m\to\infty} \left[\sum_{k=\min\{m,n\}+1}^{\max\{m,n\}} \frac{1}{k^2}\right]^{1/2} = 0.$$

However, the sequence does not converge in E, because its limit $(1, \frac{1}{2}, \frac{1}{3}, \ldots)$ is not in E. (The sequence (x_n) converges in l^2.) □

Example 3.3.5. The space discussed in Example 3.2.6 is another example of an incomplete inner product space. In fact, consider the following sequence of functions in $\mathcal{C}([0, 1])$, (see Figure 3.1):

$$f_n(x) = \begin{cases} 1 & \text{if } 0 \le x \le \frac{1}{2}, \\ 1 - 2n\left(x - \frac{1}{2}\right) & \text{if } \frac{1}{2} \le x \le \frac{1}{2n} + \frac{1}{2}, \\ 0 & \text{if } \frac{1}{2n} + \frac{1}{2} \le x \le 1. \end{cases}$$

Evidently, f_n's are continuous. Moreover,

$$\|f_n - f_m\| \le \left(\frac{1}{n} + \frac{1}{m}\right)^{1/2} \to 0, \quad \text{as } m, n \to \infty.$$

Thus, (f_n) is a Cauchy sequence. It is easy to see that the sequence is pointwise convergent to the function

$$f(x) = \begin{cases} 1 & \text{if } 0 \le x \le \frac{1}{2}, \\ 0 & \text{if } \frac{1}{2} < x \le 1. \end{cases}$$

The limit function is not continuous and thus not an element of $\mathcal{C}([0, 1])$. Therefore the sequence (f_n) is not convergent in $\mathcal{C}([0, 1])$. Consequently, $\mathcal{C}([0, 1])$ is not a Hilbert space. $\qquad\square$

Example 3.3.6. The spaces $L^2(\mathbb{R})$ and $L^2([a, b])$ are Hilbert spaces (see Theorem 2.13.4). $\qquad\square$

Example 3.3.7. Let ρ be a measurable function defined on the interval $[a, b]$ such that $\rho(x) > 0$ almost everywhere in $[a, b]$. Denote by $L^{2,\rho}([a, b])$ the space of all complex-valued measurable functions on $[a, b]$ such that

$$\int_a^b |f(x)|^2 \rho(x) dx < \infty.$$

This is a Hilbert space with the inner product

$$\langle f, g \rangle = \int_a^b f(x)\overline{g(x)}\rho(x)dx.$$

To prove completeness, consider a Cauchy sequence (f_n) in $L^{2,\rho}([a, b])$. Then

$$\|f_m - f_n\|_{L^{2,\rho}([a,b])}^2 = \int_a^b |f_m(x) - f_n(x)|^2 \rho(x)dx \to 0 \quad \text{as } n \to \infty.$$

Define

$$F_n = f_n\sqrt{\rho}, \quad n \in \mathbb{N}.$$

Since

$$\|F_m - F_n\|^2_{L^2([a,b])} = \int_a^b |F_m(x) - F_n(x)|^2 dx$$

$$= \int_a^b |f_m(x)\sqrt{\rho(x)} - f_n(x)\sqrt{\rho(x)}|^2 dx$$

$$= \int_a^b |f_m(x) - f_m(x)|^2 \rho(x) dx$$

$$= \|f_m - f_n\|^2_{L^{2,\rho}([a,b])},$$

(F_n) is a Cauchy sequence in $L^2([a,b])$. Thus, there exists $F \in L^2([a,b])$ such that

$$\|F_n - F\|^2_{L^2([a,b])} = \int_a^b |F_n(x) - F(x)|^2 dx \to 0.$$

We can show that $\frac{F}{\sqrt{\rho}} \in L^{2,\rho}([a,b])$ and $f_n \to \frac{F}{\sqrt{\rho}}$ in $L^{2,\rho}([a,b])$, proving completeness of $L^{2,\rho}([a,b])$. $\qquad\square$

Example 3.3.8. (Sobolev spaces)   Let Ω be an open set in $\mathbb{R}^N$. Denote by $\tilde{H}^m(\Omega)$, $m = 1, 2, \ldots$, the space of all complex-valued functions $f \in C^m(\Omega)$ such that $D^\alpha f \in L^2(\Omega)$ for all $|\alpha| \leq m$, where $\alpha = (\alpha_1, \ldots, \alpha_N)$, $\alpha_1, \ldots, \alpha_N$ are non-negative integers, $|\alpha| = \alpha_1 + \cdots + \alpha_N$, and

$$D^\alpha f = \frac{\partial^{|\alpha|} f}{\partial x_1^{\alpha_1} \partial x_2^{\alpha_2} \ldots \partial x_N^{\alpha_N}}.$$

For example, if $N = 2$, $\alpha = (2, 1)$, we have

$$D^\alpha f = \frac{\partial^3 f}{\partial x_1^2 \partial x_2}.$$

For $f \in \tilde{H}^m(\Omega)$, we thus have

$$\int_\Omega \left| \frac{\partial^{|\alpha|} f}{\partial x_1^{\alpha_1} \partial x_2^{\alpha_2} \ldots \partial x_N^{\alpha_N}} \right|^2 < \infty$$

for every multi-index $\alpha = (\alpha_1, \alpha_2, \ldots, \alpha_N)$ such that $|\alpha| \leq m$. The inner product in $\tilde{H}^m(\Omega)$ is defined by

$$\langle f, g \rangle = \int_\Omega \sum_{|\alpha| \leq m} D^\alpha f \overline{D^\alpha g}.$$

If $\Omega \subseteq \mathbb{R}^2$, then the inner product in $\tilde{H}^2(\Omega)$ is

$$\langle f, g \rangle = \int_\Omega (f\overline{g} + f_x\overline{g_x} + f_y\overline{g_y} + f_{xx}\overline{g_{xx}} + f_{yy}\overline{g_{yy}} + f_{xy}\overline{g_{xy}}).$$

If $\Omega = (a, b) \subset \mathbb{R}$, the inner product in $\tilde{H}^m(a, b)$ is

$$\langle f, g \rangle = \int_a^b \sum_{n=0}^m \frac{d^n f}{dx^n} \overline{\frac{d^n g}{dx^n}}.$$

$\tilde{H}^m(\Omega)$ is an inner product space, but it is not a Hilbert space because it is not complete. The completion of $\tilde{H}^m(\Omega)$, denoted by $H^m(\Omega)$, is a Hilbert space. The space $H^m(\Omega)$ is a particular case of a general class of spaces denoted by $W^{m,p}(\Omega)$ (see Section 6.3), introduced by Sergei Lvovich Sobolev (1908–1989). We have $H^m(\Omega) = W^{m,2}(\Omega)$. Because of the applications to partial differential equations, spaces $H^m(\Omega)$ belong to the most important examples of Hilbert spaces.   □

Since every inner product space is a normed space, it is equipped with a convergence, namely the convergence defined by the norm. This convergence will be called *strong convergence*.

Definition 3.3.9.   (Strong convergence)
A sequence (x_n) of vectors in an inner product space E is called *strongly convergent* to a vector x in E if $\|x_n - x\| \to 0$ as $n \to \infty$.

The word *strong* is added in order to distinguish *strong convergence* from *weak convergence*.

Definition 3.3.10.   (Weak convergence)
A sequence (x_n) of vectors in an inner product space E is called *weakly convergent* to a vector x in E if $\langle x_n, y \rangle \to \langle x, y \rangle$ as $n \to \infty$, for every $y \in E$.

The condition in the above definition can be also stated as $\langle x_n - x, y \rangle \to 0$ as $n \to \infty$, for every $y \in E$. It will be convenient to reserve the notation "$x_n \to x$" for the strong convergence and use "$x_n \overset{w}{\to} x$" to denote weak convergence.

Theorem 3.3.11. *A strongly convergent sequence is weakly convergent (to the same limit), that is, $x_n \to x$ implies $x_n \overset{w}{\to} x$.*

Proof: Suppose the sequence (x_n) converges strongly to x. This means $\|x_n - x\| \to 0$ as $n \to \infty$. By Schwarz's inequality, we have

$$|\langle x_n - x, y \rangle| \leq \|x_n - x\| \|y\| \to 0 \quad \text{as } n \to \infty,$$

and thus $\langle x_n - x, y \rangle \to 0$ as $n \to \infty$ for every $y \in E$. □

In general, the converse of Theorem 3.3.11 is not true. A suitable example will be given in Section 3.4.

For any fixed y in an inner product space E, the mapping $\langle \cdot, y \rangle : E \to \mathbb{C}$ is a linear functional on E. Theorem 3.3.11 says that such a functional is continuous for every $y \in E$. Obviously, the mapping $\langle x, \cdot \rangle : E \to \mathbb{C}$ is also continuous.

Theorem 3.3.12. *If $x_n \to x$ and $y_n \to y$, then $\langle x_n, y_n \rangle \to \langle x, y \rangle$.*

Proof: If $x_n \to x$ and $y_n \to y$, then

$$\begin{aligned}
|\langle x_n, y_n \rangle - \langle x, y \rangle| &\leq |\langle x_n, y_n \rangle - \langle x, y_n \rangle| + |\langle x, y_n \rangle - \langle x, y \rangle| \\
&= |\langle x_n - x, y_n \rangle| + |\langle x, y_n - y \rangle| \\
&\leq \|x_n - x\| \|y_n\| + \|x\| \|y_n - y\| \to 0,
\end{aligned}$$

since the sequence (y_n) is bounded. □

As an immediate consequence of this theorem, we obtain the following important property of strong convergence:

$$x_n \to x \quad \text{implies} \quad \|x_n\| \to \|x\|. \tag{3.9}$$

In the next section it will become clear that Theorem 3.3.12 and property (3.9) need not hold for weak convergence. On the other hand, we have the following theorem.

Theorem 3.3.13. *If $x_n \xrightarrow{w} x$ and $\|x_n\| \to \|x\|$, then $x_n \to x$.*

Proof: If $x_n \xrightarrow{w} x$, then for all y we have

$$\langle x_n, y \rangle \to \langle x, y \rangle \quad \text{as } n \to \infty.$$

Hence,

$$\langle x_n, x \rangle \to \langle x, x \rangle = \|x\|^2.$$

Now

$$\begin{aligned}
\|x_n - x\|^2 &= \langle x_n - x, x_n - x \rangle \\
&= \langle x_n, x_n \rangle - \langle x_n, x \rangle - \langle x, x_n \rangle + \langle x, x \rangle \\
&= \|x_n\|^2 - 2\operatorname{Re}\langle x_n, x \rangle + \|x\|^2 \to \|x\|^2 - 2\|x\|^2 + \|x\|^2 = 0
\end{aligned}$$

as $n \to \infty$. Thus, the sequence (x_n) is strongly convergent to x.   □

The next theorem is often useful in proving weak convergence.

Theorem 3.3.14. *Let S be a subset of an inner product space E such that* span S *is dense in E. If (x_n) is a bounded sequence in E and*

$$\langle x_n, y \rangle \; \to \; \langle x, y \rangle \quad \text{for every } y \subset S,$$

then $x_n \overset{w}{\to} x$.

Proof: Clearly, if $\langle x_n, y \rangle \to \langle x, y \rangle$ for every $y \in S$, then $\langle x_n, y \rangle \to \langle x, y \rangle$ for every $y \in$ span S. Let $z \in E$ and let ε be an arbitrary positive number. Since span S is dense in E, there exists $y_0 \in$ span S such that

$$\|z - y_0\| < \frac{\varepsilon}{3M},$$

where M is a positive constant such that $\|x\| \le M$ and $\|x_n\| \le M$ for all $n \in \mathbb{N}$. Since $\langle x_n, y \rangle \to \langle x, y \rangle$ for every $y \in$ span S, there exists $n_0 \in \mathbb{N}$ such that

$$\left| \langle x_n, y_0 \rangle - \langle x, y_0 \rangle \right| < \frac{\varepsilon}{3} \quad \text{for all } n > n_0.$$

Now, for any $n > n_0$, we have

$$\left| \langle x_n, z \rangle - \langle x, z \rangle \right| \le \left| \langle x_n, z \rangle - \langle x_n, y_0 \rangle \right| + \left| \langle x_n, y_0 \rangle - \langle x, y_0 \rangle \right|$$
$$+ \left| \langle x, y_0 \rangle - \langle x, z \rangle \right|$$
$$< \|x_n\| \|z - y_0\| + \frac{\varepsilon}{3} + \|x\| \|y_0 - z\|$$
$$< M \frac{\varepsilon}{3M} + \frac{\varepsilon}{3} + M \frac{\varepsilon}{3M} = \varepsilon.$$

Since z and ε are arbitrary, we conclude that $x_n \overset{w}{\to} x$.   □

The following theorem describes an important property of weakly convergent sequences in Hilbert spaces. The proof is not elementary; it is based on the Banach–Steinhaus theorem.

Theorem 3.3.15. *Weakly convergent sequences in a Hilbert space are bounded, that is, if (x_n) is a weakly convergent sequence, then there exists a number M such that $\|x_n\| \le M$ for all $n \in \mathbb{N}$.*

Proof: Let (x_n) be a weakly convergent sequence in a Hilbert space H. Define

$$f_n(x) = \langle x, x_n \rangle, \quad n \in \mathbb{N}.$$

Then $f_n : H \to \mathbb{C}$ is a bounded linear functional for every $n \in \mathbb{N}$, by Theorem 3.3.11. Since, for every $x \in H$, the sequence $(\langle x, x_n \rangle)$ converges, it is bounded, that is, there exists a constant M_x such that

$$\left| f_n(x) \right| = \left| \langle x, x_n \rangle \right| \le M_x \quad \text{for all } n \in \mathbb{N}.$$

By the Banach–Steinhaus theorem (Theorem 1.5.13), there exists a constant M such that

$$\| f_n \| \le M \quad \text{for all } n \in \mathbb{N}.$$

Since

$$\left| f_n(x) \right| = \left| \langle x, x_n \rangle \right| \le \| x \| \| x_n \|$$

for all $x \in H$, we have $\| f_n \| \le \| x_n \|$. On the other hand,

$$\left| f_n(x_n) \right| = \left| \langle x_n, x_n \rangle \right| = \| x_n \|^2.$$

Consequently, $\| f_n \| = \| x_n \|$, and thus

$$\| x_n \| \le M \quad \text{for all } n \in \mathbb{N}. \qquad \square$$

3.4 Orthogonal and Orthonormal Systems

By a basis of a vector space E we mean a linearly independent family $\mathcal{B}$ of vectors from E such that any vector $x \in E$ can be written as $x = \sum_{n=1}^{m} \lambda_n x_n$ where $x_n \in \mathcal{B}$ and the λ_n's are scalars. In inner product spaces, orthonormal bases are of much greater importance. Instead of finite combinations $\sum_{n=1}^{m} \lambda_n x_n$, infinite sums are allowed and the condition of linear independence is replaced by orthogonality. One of the immediate advantages of these changes is that, in all important examples, it is possible to describe orthonormal bases. For example, $L^2([-\pi, \pi])$ has countable orthonormal bases consisting of simple functions (see Example 3.4.16), while every basis of $L^2([-\pi, \pi])$ is uncountable and we can only prove that such a basis exists without being able to describe its elements. In this and the next section, we give all necessary definitions and discuss basic properties of orthonormal bases.

Definition 3.4.1. (Orthogonal and orthonormal systems)
A family S of nonzero vectors in an inner product space E is called an *orthogonal system* if $x \perp y$ for any two distinct elements of S. If, in addition, $\| x \| = 1$ for all $x \in S$, then S is called an *orthonormal system*.

Every orthogonal set of nonzero vectors can be normalized: If S is an orthogonal system, then $S_1 = \{ \frac{x}{\| x \|} : x \in S \}$ is an orthonormal system.

Note that if x is orthogonal to each of $y_1, \ldots, y_n$, then x is orthogonal to every linear combination of vectors $y_1, \ldots, y_n$. In fact, if $y = \sum_{k=1}^n \lambda_k y_k$, then we have

$$\langle x, y \rangle = \left\langle x, \sum_{k=1}^n \lambda_k y_k \right\rangle = \sum_{k=1}^n \overline{\lambda_k} \langle x, y_k \rangle = 0.$$

Theorem 3.4.2. *Orthogonal systems are linearly independent.*

Proof: Let S be an orthogonal system. Suppose $\sum_{k=1}^n \alpha_k x_k = 0$, for some $x_1, \ldots, x_n \in S$ and $\alpha_1, \ldots, \alpha_n \in \mathbb{C}$. Then

$$0 = \sum_{m=1}^n \langle 0, \alpha_m x_m \rangle = \sum_{m=1}^n \left\langle \sum_{k=1}^n \alpha_k x_k, \alpha_m x_m \right\rangle = \sum_{m=1}^n |\alpha_m|^2 \|x_m\|^2.$$

This implies that $\alpha_m = 0$ for each $m \in \mathbb{N}$. Thus, $x_1, \ldots, x_n$ are linearly independent. $\qquad\square$

Definition 3.4.3.   (Orthonormal sequence)
A sequence of vectors which constitutes an orthonormal system is called an *orthonormal sequence*.

In applications, it is often convenient to use sequences indexed by the set of all integers, $\mathbb{Z}$. The condition of orthonormality of a sequence (x_n) can be expressed in terms of the *Kronecker delta* symbol (Leopold Kronecker (1823–1891)):

$$\langle x_m, x_n \rangle = \delta_{mn} = \begin{cases} 0 & \text{if } m \neq n, \\ 1 & \text{if } m = n. \end{cases} \tag{3.10}$$

Example 3.4.4. For $e_n = (0, \ldots, 0, 1, 0, \ldots)$ with 1 in the nth position, the set $S = \{e_1, e_2, \ldots\}$ is an orthonormal system in l^2. $\qquad\square$

Example 3.4.5. Let $\varphi_n(x) = \frac{e^{inx}}{\sqrt{2\pi}}$, $n \in \mathbb{Z}$. The set $\{\varphi_n: n \in \mathbb{Z}\}$ is an orthonormal system in $L^2([-\pi, \pi])$. Indeed, for $m \neq n$, we have

$$\langle \varphi_m, \varphi_n \rangle = \frac{1}{2\pi} \int_{-\pi}^\pi e^{i(m-n)x} dx = \frac{e^{\pi i(m-n)} - e^{-\pi i(m-n)}}{2\pi i(m-n)} = 0.$$

On the other hand,

$$\langle \varphi_n, \varphi_n \rangle = \frac{1}{2\pi} \int_{-\pi}^\pi e^{i(n-n)x} dx = 1.$$

Thus, $\langle \varphi_m, \varphi_n \rangle = \delta_{mn}$ for every pair of integers m and n. $\qquad\square$

Example 3.4.6. The *Legendre polynomials* (Adrien-Marie Legendre (1752–1833)) defined by

$$P_0(x) = 1,$$

$$P_n(x) = \frac{1}{2^n n!} \frac{d^n}{dx^n} (x^2 - 1)^n, \quad n = 1, 2, 3, \ldots \qquad (3.11)$$

form an orthogonal system in $L^2([-1, 1])$. For convenience, we write $(x^2 - 1)^n = p_n(x)$ so that

$$\int_{-1}^{1} P_n(x) x^m dx = \frac{1}{2^n n!} \int_{-1}^{1} p_n^{(n)}(x) x^m dx. \qquad (3.12)$$

We evaluate this integral for $m < n$ by recursion. First we note that

$$p_n^{(k)}(x) = 0 \quad \text{for } x = \pm 1 \text{ and } k = 0, 1, 2, \ldots, n-1.$$

Hence, by integration by parts in (3.12), we get

$$\int_{-1}^{1} p_n^{(n)}(x) x^m dx = -m \int_{-1}^{1} p_n^{(n-1)}(x) x^{m-1} dx.$$

Repeated application of this operation ultimately leads to

$$(-1)^m m! \int_{-1}^{1} p_n^{(n-m)}(x) dx = (-1)^m m! \left[p_n^{(n-m-1)}(x) \right]_{-1}^{1} = 0 \quad (m < n).$$

Consequently,

$$\int_{-1}^{1} P_n(x) x^m dx = 0, \quad \text{for } m < n. \qquad (3.13)$$

Since P_m is a polynomial of degree m, it follows that

$$\langle P_n, P_m \rangle = \int_{-1}^{1} P_n(x) P_m(x) dx = 0, \quad \text{for } n \neq m. \qquad (3.14)$$

This proves orthogonality of the Legendre polynomials. To obtain an orthonormal system from the Legendre polynomials, we need to evaluate the norm of P_n in $L^2([-1, 1])$:

$$\|P_n\| = \sqrt{\int_{-1}^{1} (P_n(x))^2 dx}.$$

By repeated integration by parts, we first obtain

$$\int_{-1}^{1} (1 - x^2)^n dx = \int_{-1}^{1} (1 - x)^n (1 + x)^n dx$$

$$= \frac{n}{n+1} \int_{-1}^{1} (1-x)^{n-1}(1+x)^{n+1} dx = \cdots$$

$$= \frac{n(n-1)\cdots 2 \cdot 1}{(n+1)(n+2)\cdots 2n} \int_{-1}^{1} (1+x)^{2n} dx$$

$$= \frac{(n!)^2 2^{2n+1}}{(2n)!(2n+1)}. \tag{3.15}$$

A similar procedure gives

$$\int_{-1}^{1} \left(p_n^{(n)}(x)\right)^2 dx = 0 - \int_{-1}^{1} p_n^{(n-1)}(x) p_n^{(n+1)}(x) dx$$

$$= \cdots$$

$$= (-1)^n \int_{-1}^{1} p_n(x) p_n^{(2n)}(x) dx$$

$$= (2n)! \int_{-1}^{1} (1-x)^n (1+x)^n dx, \tag{3.16}$$

where we have used the fact that the $2n$th derivative of $p_n(x) = (x^2 - 1)^n$ is the same as the derivative of the term of exponent $2n$. The $2n$th derivatives of all the other terms of the sum are zero. Now from (3.11), (3.15), and (3.16) we obtain

$$\int_{-1}^{1} \left(P_n(x)\right)^2 dx = \frac{1}{(2^n n!)^2}(2n)! \frac{(n!)^2 2^{2n+1}}{(2n)!(2n+1)} = \frac{2}{2n+1}. \tag{3.17}$$

Thus, the polynomials $\sqrt{n + \frac{1}{2}} P_n(x)$ form an orthonormal system in $L^2([-1, 1])$.   □

Example 3.4.7. Denote by H_n the *Hermite polynomial* (Charles Hermite (1822–1901)) of degree n, that is,

$$H_n(x) = (-1)^n e^{x^2} \frac{d^n}{dx^n} e^{-x^2}. \tag{3.18}$$

The functions $\varphi_n(x) = e^{-x^2/2} H_n(x)$ form an orthogonal system in $L^2(\mathbb{R})$. The inner product

$$\langle \varphi_n, \varphi_m \rangle = (-1)^{n+m} \int_{-\infty}^{\infty} e^{x^2} \frac{d^n}{dx^n} e^{-x^2} \frac{d^m}{dx^m} e^{-x^2} dx$$

can be evaluated by integration by parts, which gives

$$(-1)^{n+m} \langle \varphi_n, \varphi_m \rangle = \left[e^{x^2} \frac{d^n}{dx^n} e^{-x^2} \frac{d^{m-1}}{dx^{m-1}} e^{-x^2} \right]_{-\infty}^{\infty}$$

$$- \int_{-\infty}^{\infty} \frac{d}{dx} \left[e^{x^2} \frac{d^n}{dx^n} e^{-x^2} \right] \frac{d^{m-1}}{dx^{m-1}} e^{-x^2} dx, \quad (3.19)$$

and hence all terms under the differential sign contain the factor e^{-x^2}. Since for any $k \in \mathbb{N}$, we have

$$x^k e^{-x^2} \to 0 \quad \text{as } x \to \infty,$$

the first term in (3.19) vanishes. Therefore, repeated integration by parts leads to

$$\langle \varphi_n, \varphi_m \rangle = 0 \quad \text{for } n \neq m. \quad (3.20)$$

To obtain an orthonormal system, we evaluate the norm:

$$\|\varphi_n\|^2 = \int_{-\infty}^{\infty} e^{-x^2} \left(H_n(x) \right)^2 dx = \int_{-\infty}^{\infty} e^{-x^2} \left[e^{x^2} \frac{d^n}{dx^n} e^{-x^2} \right]^2 dx.$$

Integration by parts n times yields

$$\|\varphi_n\|^2 = (-1)^n \int_{-\infty}^{\infty} e^{-x^2} \frac{d^n}{dx^n} \left[e^{x^2} \frac{d^n}{dx^n} e^{-x^2} \right] dx.$$

Since H_n is a polynomial of degree n, direct differentiation gives

$$e^{x^2} \frac{d^n}{dx^n} e^{-x^2} = (-2x)^n + \cdots$$

and

$$\frac{d^n}{dx^n} \left[e^{x^2} \frac{d^n}{dx^n} e^{-x^2} \right] = \frac{d^n}{dx^n} \left((-2x)^n + \cdots \right) = (-1)^n 2^n n!.$$

Consequently,

$$\|\varphi_n\|^2 = 2^n n! \int_{-\infty}^{\infty} e^{-x^2} dx = 2^n n! \sqrt{\pi}. \quad (3.21)$$

Thus, the functions

$$\psi_n(x) = \frac{1}{\sqrt{2^n n! \sqrt{\pi}}} e^{-x^2/2} H_n(x)$$

form an orthonormal system in $L^2(\mathbb{R})$. $\qquad \square$

In the preceding examples, the original sequence of functions is orthogonal but not orthonormal. Although the calculations involved might be complicated, it is always possible to normalize the functions and obtain an orthonormal sequence. It turns out that the same is possible if the original sequence of functions (or, in general, a sequence of vectors in an inner product space) is linearly independent, not necessarily orthogonal. The method of transforming such a sequence into an orthonormal sequence is called the *Gram–Schmidt orthonormalization process* (Jorgen Pedersen Gram (1850–1916), Erhard Schmidt (1876–1959)). The process can be described as follows:

Given a sequence (y_n) of linearly independent vectors in an inner product space, define sequences (w_n) and (x_n) inductively by

$$w_1 = y_1, \qquad\qquad x_1 = \frac{w_1}{\|w_1\|},$$

$$w_k = y_k - \sum_{n=1}^{k-1} \langle y_k, x_n \rangle x_n, \qquad x_k = \frac{w_k}{\|w_k\|}, \qquad \text{for } k = 2, 3, \ldots.$$

The sequence (w_n) is orthogonal. We show that by induction. First note that

$$\langle w_2, w_1 \rangle = \langle y_2 - \langle y_2, x_1 \rangle x_1, y_1 \rangle = \langle y_2, y_1 \rangle - \langle y_2, x_1 \rangle \langle x_1, y_1 \rangle$$

$$= \langle y_2, y_1 \rangle - \frac{\langle y_2, y_1 \rangle \langle y_1, y_1 \rangle}{\|y_1\|^2} = 0.$$

Assume now that $w_1, \ldots, w_{k-1}$ are orthogonal. Then, for any $m < k$,

$$\langle w_k, w_m \rangle = \langle y_k, w_m \rangle - \frac{\sum_{n=1}^{k-1} \langle y_k, w_n \rangle \langle w_n, w_m \rangle}{\|w_m\|^2}$$

$$= \langle y_k, w_m \rangle - \frac{\langle y_k, w_m \rangle \langle w_m, w_m \rangle}{\|w_m\|^2} = 0.$$

Therefore, vectors $w_1, \ldots, w_k$ are orthogonal. It follows, by induction, that the sequence (w_n) is orthogonal, and thus (x_n) is orthonormal. It is easy to check that any linear combination of vectors $x_1, \ldots, x_n$ is also a linear combination of $y_1, \ldots, y_n$ and vice versa. In other words, $\text{span}\{x_1, \ldots, x_n\} = \text{span}\{y_1, \ldots, y_n\}$ for every $n \in \mathbb{N}$.

In Section 3.3 we proved that the Pythagorean formula holds for any pair of orthogonal vectors in an inner product space. It turns out that it can be generalized to any finite number of orthogonal vectors.

Theorem 3.4.8. (Pythagorean formula)   *If $x_1, \ldots, x_n$ are orthogonal vectors in an inner product space, then*

$$\left\| \sum_{k=1}^{n} x_k \right\|^2 = \sum_{k=1}^{n} \|x_k\|^2. \qquad (3.22)$$

Proof: If $x_1 \perp x_2$, then $\|x_1 + x_2\|^2 = \|x_1\|^2 + \|x_2\|^2$, by (3.8). Thus the theorem is true for $n = 2$. Assume now that the (3.22) holds for $n - 1$, that is,

$$\left\| \sum_{k=1}^{n-1} x_k \right\|^2 = \sum_{k=1}^{n-1} \|x_k\|^2.$$

Set $x = \sum_{k=1}^{n-1} x_k$ and $y = x_n$. Since $x \perp y$, we have

$$\left\| \sum_{k=1}^{n} x_k \right\|^2 = \|x + y\|^2 = \|x\|^2 + \|y\|^2 = \sum_{k=1}^{n-1} \|x_k\|^2 + \|x_n\|^2 = \sum_{k=1}^{n} \|x_k\|^2.$$

This proves the theorem. □

The Pythagorean formula will help us prove the following important property of orthonormal sets attributed to Friedrich Wilhelm Bessel (1784–1846).

Theorem 3.4.9. (Bessel's equality and inequality) *Let $x_1, \dots, x_n$ be an orthonormal set of vectors in an inner product space E. Then, for every $x \in E$, we have*

$$\left\| x - \sum_{k=1}^{n} \langle x, x_k \rangle x_k \right\|^2 = \|x\|^2 - \sum_{k=1}^{n} |\langle x, x_k \rangle|^2 \tag{3.23}$$

and

$$\sum_{k=1}^{n} |\langle x, x_k \rangle|^2 \le \|x\|^2. \tag{3.24}$$

Proof: In view of the Pythagorean formula (3.22), we have

$$\left\| \sum_{k=1}^{n} \alpha_k x_k \right\|^2 = \sum_{k=1}^{n} \|\alpha_k x_k\|^2 = \sum_{k=1}^{n} |\alpha_k|^2$$

for arbitrary complex numbers $\alpha_1, \dots, \alpha_n$. Hence

$$\left\| x - \sum_{k=1}^{n} \alpha_k x_k \right\|^2 = \left\langle x - \sum_{k=1}^{n} \alpha_k x_k, x - \sum_{k=1}^{n} \alpha_k x_k \right\rangle$$

$$= \|x\|^2 - \left\langle x, \sum_{k=1}^{n} \alpha_k x_k \right\rangle - \left\langle \sum_{k=1}^{n} \alpha_k x_k, x \right\rangle + \sum_{k=1}^{n} |\alpha_k|^2 \|x_k\|^2$$

$$= \|x\|^2 - \sum_{k=1}^{n} \overline{\alpha_k} \langle x, x_k \rangle - \sum_{k=1}^{n} \alpha_k \overline{\langle x, x_k \rangle} + \sum_{k=1}^{n} \alpha_k \overline{\alpha_k}$$

$$= \|x\|^2 - \sum_{k=1}^{n} |\langle x, x_k \rangle|^2 + \sum_{k=1}^{n} |\langle x, x_k \rangle - \alpha_k|^2. \tag{3.25}$$

In particular, if $\alpha_k = \langle x, x_k \rangle$, this results yields (3.23). From (3.23) it follows that

$$0 \le \|x\|^2 - \sum_{k=1}^{n} |\langle x, x_k \rangle|^2,$$

which gives (3.24). □

Note that expression (3.25) is minimized by taking $\alpha_k = \langle x, x_k \rangle$. This choice of α_k's minimizes $\|x - \sum_{k=1}^{n} \alpha_k x_k\|$, and thus it provides the best approximation of x by a linear combination of vectors $x_1, \ldots, x_n$. This property of orthonormal systems is of fundamental importance for many approximation techniques.

If (x_n) is an orthonormal sequence, then by letting $n \to \infty$ in (3.24), we obtain

$$\sum_{k=1}^{\infty} |\langle x, x_k \rangle|^2 \le \|x\|^2 \tag{3.26}$$

and consequently,

$$\lim_{n \to \infty} \langle x, x_n \rangle = 0.$$

Therefore, *orthonormal sequences are weakly convergent to zero*. On the other hand, since $\|x_n\| = 1$ for all $n \in \mathbb{N}$, orthonormal sequences are not strongly convergent.

Inequality (3.26) implies that the series $\sum_{k=1}^{\infty} |\langle x, x_k \rangle|^2$ converges for every $x \in E$. In other words, the sequence $(\langle x, x_n \rangle)$ is an element of l^2. We can say that an orthonormal sequence in E induces a mapping from E into l^2. The expansion

$$x \sim \sum_{n=1}^{\infty} \langle x, x_n \rangle x_n \tag{3.27}$$

is called a *generalized Fourier series* of x (Jean Baptiste Joseph Fourier (1768–1830)). The scalars $\alpha_n = \langle x, x_n \rangle$ are called the *generalized Fourier coefficients* of x with respect to the orthonormal sequence (x_n). As mentioned earlier, this set of coefficients gives the best approximation for any finite set of x_n's. In general, we do not know whether the series in (3.27) is convergent. However, as the next theorem shows, completeness of the space ensures the convergence.

Theorem 3.4.10. *Let (x_n) be an orthonormal sequence in a Hilbert space H, and let (α_n) be a sequence of complex numbers. Then the series $\sum_{n=1}^{\infty} \alpha_n x_n$ converges if and only if $\sum_{n=1}^{\infty} |\alpha_n|^2 < \infty$ and in that case*

$$\left\| \sum_{n=1}^{\infty} \alpha_n x_n \right\|^2 = \sum_{n=1}^{\infty} |\alpha_n|^2. \tag{3.28}$$

Proof: For every $m > k > 0$, we have

$$\left\| \sum_{n=k}^{m} \alpha_n x_n \right\|^2 = \sum_{n=k}^{m} |\alpha_n|^2, \tag{3.29}$$

by the Pythagorean formula (3.22). If $\sum_{n=1}^{\infty} |\alpha_n|^2 < \infty$, then from (3.29) we obtain that the sequence $s_m = \sum_{n=1}^{m} \alpha_n x_n$ is a Cauchy sequence. This implies convergence of the series $\sum_{n=1}^{\infty} \alpha_n x_n$, because of completeness of H.

Conversely, if the series $\sum_{n=1}^{\infty} \alpha_n x_n$ converges, then (3.29) implies the convergence of $\sum_{n=1}^{\infty} |\alpha_n|^2$, because the sequence of numbers $\sigma_m = \sum_{n=1}^{m} |\alpha_n|^2$ is a Cauchy sequence in $\mathbb{R}$.

To obtain (3.28), it is enough to take $k = 1$ and let $m \to \infty$ in (3.29). $\square$

This theorem and (3.26) imply that in a Hilbert space H the series $\sum_{n=1}^{\infty} \langle x, x_n \rangle x_n$ converges for every $x \in H$. However, it can happen that it converges to an element different from x.

Example 3.4.11. Let $H = L^2([-\pi, \pi])$, and let $x_n(t) = \frac{1}{\sqrt{\pi}} \sin nt$ for $n = 1, 2, \ldots$. The sequence (x_n) is an orthonormal set in H. On the other hand, for $x(t) = \cos t$, we have

$$\sum_{n=1}^{\infty} \langle x, x_n \rangle x_n(t) = \sum_{n=1}^{\infty} \left[\frac{1}{\sqrt{\pi}} \int_{-\pi}^{\pi} \cos t \sin nt\, dt \right] \frac{\sin nt}{\sqrt{\pi}}$$

$$= \sum_{n=1}^{\infty} 0 \cdot \sin nt = 0 \neq \cos t. \qquad \square$$

Definition 3.4.12. (Complete orthonormal sequence)
An orthonormal sequence (x_n) in an inner product space E is said to be *complete* if for every $x \in E$ we have

$$x = \sum_{n=1}^{\infty} \langle x, x_n \rangle x_n. \tag{3.30}$$

It is important to remember that since the right-hand side of (3.30) is an infinite series, the equality means

$$\lim_{n \to \infty} \left\| x - \sum_{k=1}^{n} \langle x, x_k \rangle x_k \right\| = 0,$$

where $\| \cdot \|$ is the norm in E. For example, if $E = L^2([-\pi, \pi])$ and (f_n) is an orthonormal sequence in E, then by

$$f = \sum_{n=1}^{\infty} \langle f, f_n \rangle f_n,$$

we mean

$$\lim_{n \to \infty} \int_{-\pi}^{\pi} \left| f(t) - \sum_{k=1}^{n} \alpha_k f_k(t) \right|^2 dt = 0, \quad \text{where } \alpha_k = \int_{-\pi}^{\pi} f(t) \overline{f_k(t)} dt.$$

This, in general, does not imply pointwise convergence, so we cannot claim that $f(x) = \sum_{n=1}^{\infty} \alpha_n f_n(x)$.

Definition 3.4.13.   (Orthonormal basis)

An orthonormal system B in an inner product space E is called an *orthonormal basis* if every $x \in E$ has a unique representation

$$x = \sum_{n=1}^{\infty} \alpha_n x_n,$$

where $\alpha_n \in \mathbb{C}$ and x_n's are distinct elements of B.

Note that a complete orthonormal sequence (x_n) is an orthonormal basis. It suffices to show the uniqueness. Indeed, if

$$x = \sum_{n=1}^{\infty} \alpha_n x_n \quad \text{and} \quad x = \sum_{n=1}^{\infty} \beta_n x_n,$$

then

$$0 = \|x - x\|^2 = \left\| \sum_{n=1}^{\infty} \alpha_n x_n - \sum_{n=1}^{\infty} \beta_n x_n \right\|^2 = \left\| \sum_{n=1}^{\infty} (\alpha_n - \beta_n) x_n \right\|^2 = \sum_{n=1}^{\infty} |\alpha_n - \beta_n|^2$$

by Theorem 3.4.10. This means that $\alpha_n = \beta_n$ for all $n \in \mathbb{N}$, proving the uniqueness.

If (x_n) is a complete orthonormal sequence in an inner product space E, then the set

$$\text{span}\{x_1, x_2, \ldots\} = \left\{ \sum_{k=1}^{n} \alpha_k x_k \colon n \in \mathbb{N}, \ \alpha_1, \ldots, \alpha_k \in \mathbb{C} \right\}$$

is dense in E.

The following two theorems give important characterizations of complete orthonormal sequences in Hilbert spaces.

Theorem 3.4.14. *An orthonormal sequence (x_n) in a Hilbert space H is complete if and only if $\langle x, x_n \rangle = 0$ for all $n \in \mathbb{N}$ implies $x = 0$.*

Proof: Suppose (x_n) is a complete orthonormal sequence in H. Then for every $x \in H$, we have

$$x = \sum_{n=1}^{\infty} \langle x, x_n \rangle x_n.$$

Thus, if $\langle x, x_n \rangle = 0$ for every $n \in \mathbb{N}$, then $x = 0$.

Conversely, suppose $\langle x, x_n \rangle = 0$ for all $n \in \mathbb{N}$ implies $x = 0$. Let x be an element of H. Define

$$y = \sum_{n=1}^{\infty} \langle x, x_n \rangle x_n.$$

The sum y exists in H by (3.26) and Theorem 3.4.10. Since, for every $n \in \mathbb{N}$,

$$\langle x - y, x_n \rangle = \langle x, x_n \rangle - \left\langle \sum_{k=1}^{\infty} \langle x, x_k \rangle x_k, x_n \right\rangle$$

$$= \langle x, x_n \rangle - \sum_{k=1}^{\infty} \langle x, x_k \rangle \langle x_k, x_n \rangle$$

$$= \langle x, x_n \rangle - \langle x, x_n \rangle = 0,$$

we have $x - y = 0$, and hence $x = \sum_{n=1}^{\infty} \langle x, x_n \rangle x_n$.   □

The formula in the following theorem, known as Parseval's formula (Marc-Antoine Parseval des Chênes (1755–1836)), can be interpreted as an extension of the Pythagorean formula to infinite sums.

Theorem 3.4.15. (Parseval's formula) *An orthonormal sequence (x_n) in a Hilbert space H is complete if and only if*

$$\|x\|^2 = \sum_{n=1}^{\infty} |\langle x, x_n \rangle|^2 \tag{3.31}$$

for every $x \in H$.

Proof: Let $x \in H$. By (3.23), for every $n \in \mathbb{N}$, we have

$$\left\| x - \sum_{k=1}^{n} \langle x, x_k \rangle x_k \right\|^2 = \|x\|^2 - \sum_{k=1}^{n} |\langle x, x_k \rangle|^2. \tag{3.32}$$

If (x_n) is a complete sequence, then the expression on the left in (3.32) converges to zero as $n \to \infty$. Hence,

$$\lim_{n\to\infty} \left[\|x\|^2 - \sum_{k=1}^{n} |\langle x, x_k \rangle|^2 \right] = 0.$$

Therefore, (3.31) holds.

Conversely, if (3.31) holds, then the expression on the right in (3.32) converges to zero as $n \to \infty$, and thus

$$\lim_{n\to\infty} \left\| x - \sum_{k=1}^{n} \langle x, x_k \rangle x_k \right\|^2 = 0.$$

This proves that the sequence (x_n) is complete. $\square$

Example 3.4.16. The orthonormal system

$$\varphi_n(x) = \frac{e^{inx}}{\sqrt{2\pi}}, \quad n = 0, \pm 1, \pm 2, \ldots,$$

given in Example 3.4.5, is complete in $L^2([-\pi, \pi])$. The proof of completeness is not simple. It will be discussed in Section 3.5.

A simple change of scale allows us to represent a function $f \in L^2([0, a])$ in the form

$$f(x) = \sum_{n=-\infty}^{\infty} \beta_n e^{2n\pi ix/a}$$

where

$$\beta_n = \frac{1}{a} \int_0^a f(t) e^{-2n\pi it/a} dt. \qquad \square$$

Example 3.4.17. The sequence of functions

$$\frac{1}{\sqrt{2\pi}}, \frac{\cos x}{\sqrt{\pi}}, \frac{\sin x}{\sqrt{\pi}}, \frac{\cos 2x}{\sqrt{\pi}}, \frac{\sin 2x}{\sqrt{\pi}}, \ldots$$

is a complete orthonormal system in $L^2([-\pi, \pi])$. The orthogonality follows from the following identities by simple integration:

$$2\cos nx \cos mx = \cos(n+m)x + \cos(n-m)x,$$

$$2\sin nx \sin mx = \cos(n-m)x - \cos(n+m)x,$$

$$2\cos nx \sin mx = \sin(n+m)x - \sin(n-m)x.$$

Since

$$\int_{-\pi}^{\pi} \cos^2 nx \, dx = \int_{-\pi}^{\pi} \sin^2 mx \, dx = \pi,$$

the sequence is also orthonormal. Completeness follows from completeness of the sequence in Example 3.4.16, in view of the following identities:

$$e^0 = 1 \quad \text{and} \quad e^{inx} = (\cos nx + i \sin nx). \qquad \square$$

Example 3.4.18. Each of the following two sequences of functions is a complete orthonormal system in $L^2([0, \pi])$:

$$\frac{1}{\sqrt{\pi}}, \sqrt{\frac{2}{\pi}} \cos x, \sqrt{\frac{2}{\pi}} \cos 2x, \sqrt{\frac{2}{\pi}} \cos 3x, \ldots,$$

$$\sqrt{\frac{2}{\pi}} \sin x, \sqrt{\frac{2}{\pi}} \sin 2x, \sqrt{\frac{2}{\pi}} \sin 3x, \ldots. \qquad \square$$

Example 3.4.19. (Rademacher functions and Walsh functions) *Rademacher functions* $R(m, x)$ (Hans Rademacher (1892–1969)) can be introduced in many different ways. We will use the definition based on the sine function:

$$R(m, x) = \text{sgn}\big(\sin(2^m \pi x)\big), \quad m = 0, 1, 2, \ldots, \; x \in [0, 1],$$

where sgn denotes the signum function defined by

$$\text{sgn}(x) = \begin{cases} 1 & \text{if } x > 0, \\ 0 & \text{if } x = 0, \\ -1 & \text{if } x < 0. \end{cases}$$

Rademacher functions form an orthonormal system in $L^2([0, 1])$. Obviously,

$$\int_0^1 |R(m, x)|^2 \, dx = 1 \quad \text{for all } m.$$

To show that for $m \neq n$, we have

$$\int_0^1 R(m, x) \overline{R(n, x)} \, dx = 0,$$

first notice that $\int_a^b R(m, x) dx = 0$ whenever $2^m(b - a)$ is an even number. Thus, for $m > n \geq 0$, we have

$$\int_0^1 R(m,x)\overline{R(n,x)}dx = \int_0^1 R(m,x)R(n,x)dx$$

$$= \sum_{k=1}^{2^n} \int_{(k-1)/2^n}^{k/2^n} R(m,x)R(n,x)dx$$

$$= \sum_{k=1}^{2^n} \operatorname{sgn}\left(R\left(n,\frac{2k-1}{2}\right)\right) \int_{(k-1)/2^n}^{k/2^n} R(m,x)dx = 0,$$

because all the integrals equal 0.

The sequence of Rademacher functions is not complete. Indeed, consider the function

$$f(x) = \begin{cases} 0 & \text{if } 0 \le x < \frac{1}{4}, \\ 1 & \text{if } \frac{1}{4} \le x \le \frac{3}{4}, \\ 0 & \text{if } \frac{3}{4} < x \le 1. \end{cases}$$

Then

$$\int_0^1 R(0,x)f(x)dx = \frac{1}{2} \quad \text{and} \quad \int_0^1 R(m,x)f(x)dx = 0 \quad \text{for } m \ge 1,$$

but $f(x) \ne \frac{1}{2}R(0,x)$.

Rademacher functions can be used to construct *Walsh functions* (John Walsh (1786–1847)), which form a complete orthonormal system. Walsh functions are denoted by $W(m,x)$, $m = 0, 1, 2, \ldots$. For $m = 0$, we set $W(0,x) = 1$. For other values of m, we first represent m as a binary number, that is,

$$m = \sum_{k=1}^{n} 2^{k-1}a_k = a_1 + 2^1 a_2 + 2^2 a_3 + \cdots + 2^{n-1}a_n,$$

where $a_1, a_2, \ldots, a_n = 0$ or 1. Then we define

$$W(m,x) = \prod_{k=1}^{n} (R(k,x))^{a_k} = (R(1,x))^{a_1}(R(2,x))^{a_2} \cdots (R(n,x))^{a_n}$$

(where $(R(m,x))^0 \equiv 1$). For instance, since 53 is written as 110101 in binary form, we have

$$W(53,x) = R(1,x)R(3,x)R(5,x)R(6,x).$$

Clearly, we have

$$R(n,x) = W(2^{n-1},x), \quad n \in \mathbb{N}.$$

Figure 3.2 shows several Walsh functions.   □

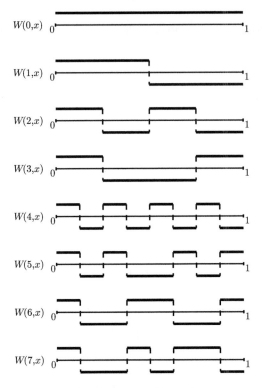

$W(0,x)$
$W(1,x)$
$W(2,x)$
$W(3,x)$
$W(4,x)$
$W(5,x)$
$W(6,x)$
$W(7,x)$

Figure 3.2 Walsh functions.

Definition 3.4.20. (Separable spaces)
A Hilbert space is called *separable* if it contains a complete orthonormal sequence. Finite dimensional Hilbert spaces are considered separable.

Example 3.4.21. Space $L^2([-\pi, \pi])$ is separable. Example 3.4.16 shows a complete orthonormal sequence in $L^2([-\pi, \pi])$. □

Example 3.4.22. Space l^2 is separable. □

Example 3.4.23. (Nonseparable Hilbert space) Let H be the space of all complex-valued functions defined on $\mathbb{R}$, which vanish everywhere except a countable number of points in $\mathbb{R}$ and such that

$$\sum_{f(x)\neq 0} |f(x)|^2 < \infty.$$

The inner product in H can be defined as

$$\langle f, g \rangle = \sum_{f(x)g(x) \neq 0} f(x)\overline{g(x)}.$$

This space is not separable because for any sequence of functions $f_n \in H$, there are nonzero functions f such that $\langle f, f_n \rangle = 0$ for all $n \in \mathbb{N}$.   □

Throughout this book, we will consider only separable Hilbert spaces.

Recall that a set S in a Banach space E is called dense in E if every element of E can be approximated by a sequence of elements of S. More precisely, for every $x \in E$ there exist $x_n \in S$ such that $\|x - x_n\| \to 0$ as $n \to \infty$.

Theorem 3.4.24. *Every separable Hilbert space contains a countable dense subset.*

Proof: Let (x_n) be a complete orthonormal sequence in a Hilbert space H. The set

$$S = \left\{ (\alpha_1 + i\beta_1)x_1 + \cdots + (\alpha_n + i\beta_n)x_n : \alpha_1, \ldots, \alpha_n, \beta_1, \ldots, \beta_n \in \mathbb{Q}, n \in \mathbb{N} \right\}$$

is obviously countable. Since, for every $x \in H$,

$$\left\| \sum_{k=1}^{n} \langle x, x_k \rangle x_k - x \right\| \to 0 \quad \text{as } n \to \infty,$$

S is dense in H.   □

The statement in Theorem 3.4.24 is often used as a definition of separability.

Theorem 3.4.25. *Every orthogonal set in a separable Hilbert space is countable.*

Proof: Let S be an orthogonal set in a separable Hilbert space H, and let S_1 be the set of normalized vectors from S, that is, $S_1 = \{x/\|x\|: x \in S\}$. For any distinct $x, y \in S_1$ we have

$$\|x - y\|^2 = \langle x - y, x - y \rangle$$
$$= \langle x, x \rangle - \langle x, y \rangle - \langle y, x \rangle + \langle y, y \rangle$$
$$= 1 - 0 - 0 + 1 \quad \text{(by the orthogonality)}$$
$$= 2.$$

This means that the distance between any two distinct elements of S_1 is $\sqrt{2}$.

Now consider the collection of $1/\sqrt{2}$-neighborhoods about every element of S_1. Clearly, no two of these neighborhoods can have a common point. Since every dense subset of H must have at least one point in every neighborhood and H has a countable dense subset, S_1 has to be countable. Thus, S is countable, proving the theorem. $\qquad\square$

Definition 3.4.26. (Hilbert space isomorphism)
A Hilbert space H_1 is said to be *isomorphic* to a Hilbert space H_2 if there exists a one-to-one linear mapping T from H_1 onto H_2 such that

$$\langle T(x), T(y)\rangle = \langle x, y\rangle \tag{3.33}$$

for every $x, y \in H_1$. Such a mapping T is called a *Hilbert space isomorphism* of H_1 onto H_2.

Note that (3.33) implies $\|T\| = 1$, because $\|T(x)\| = \|x\|$ for every $x \in H_1$.

Theorem 3.4.27. *Let H be a separable Hilbert space.*

(a) *If H is infinite dimensional, then it is isomorphic to l^2;*

(b) *If $\dim H = N$, then it is isomorphic to $\mathbb{C}^N$.*

Proof: Let (x_n) be a complete orthonormal sequence in H. If H is infinite dimensional, then (x_n) is an infinite sequence. Let x be an element of H. Define $T(x) = (\alpha_n)$, where $\alpha_n = \langle x, x_n\rangle$, $n = 1, 2, \ldots$. By Theorem 3.4.10, T is a one-to-one mapping from H onto l^2. It is clearly a linear mapping. Moreover, for $\alpha_n = \langle x, x_n\rangle$ and $\beta_n = \langle y, x_n\rangle$, $x, y \in H$, $n \in \mathbb{N}$, we have

$$\langle T(x), T(y)\rangle = \langle (\alpha_n), (\beta_n)\rangle$$

$$= \sum_{n=1}^{\infty} \alpha_n \overline{\beta_n} = \sum_{n=1}^{\infty} \langle x, x_n\rangle \overline{\langle y, x_n\rangle}$$

$$= \sum_{n=1}^{\infty} \langle x, \langle y, x_n\rangle x_n\rangle = \left\langle x, \sum_{n=1}^{\infty} \langle y, x_n\rangle x_n\right\rangle = \langle x, y\rangle.$$

Thus, T is an isomorphism from H onto l^2.

The proof of (b) is left as an exercise. $\qquad\square$

It is easy to check that isomorphism of Hilbert spaces is an equivalence relation.

Since any infinite dimensional separable Hilbert space is isomorphic to l^2, it follows that any two such spaces are isomorphic. The same is true for real Hilbert spaces; any real infinite dimensional separable Hilbert space is isomorphic to the real space l^2. In some sense, there is only one real and one complex infinite dimensional separable Hilbert space.

3.5 Trigonometric Fourier Series

In this section we prove that the sequence

$$\varphi_n(x) = \frac{e^{inx}}{\sqrt{2\pi}}, \quad n = 0, \pm 1, \pm 2, \ldots,$$

is a complete orthonormal sequence in $L^2([-\pi, \pi])$. The orthogonality has been established in Example 3.4.5. The proof of completeness is much more complicated. For the purpose of this proof it will be convenient to identify elements of $L^1([-\pi, \pi])$ with 2π-periodic locally integrable functions on $\mathbb{R}$, because then we have

$$\int_{-\pi}^{\pi} f(t)dt = \int_{-\pi-x}^{\pi-x} f(t)dt = \int_{-\pi}^{\pi} f(t-x)dt$$

for any $f \in L^1([-\pi, \pi])$ and any $x \in \mathbb{R}$.

Let $f \in L^1([-\pi, \pi])$ and

$$f_n = \sum_{k=-n}^{n} \langle f, \varphi_k \rangle \varphi_k, \quad n = 0, 1, 2, \ldots.$$

Then

$$f_n(x) = \sum_{k=-n}^{n} \frac{1}{2\pi} \int_{-\pi}^{\pi} f(t)e^{-ikt}dt\, e^{ikx} = \sum_{k=-n}^{n} \frac{1}{2\pi} \int_{-\pi}^{\pi} f(t)e^{ik(x-t)}dt.$$

We intend to show that for every $f \in L^1([-\pi, \pi])$, we have

$$\lim_{n\to\infty} \frac{f_0 + f_1 + \cdots + f_n}{n+1} = f$$

in the $L^1([-\pi, \pi])$ norm. First observe that

$$\frac{f_0(x) + f_1(x) + \cdots + f_n(x)}{n+1} = \sum_{k=-n}^{n} \left(1 - \frac{|k|}{n+1}\right) \langle f, \varphi_k \rangle \varphi_k(x)$$

$$= \sum_{k=-n}^{n} \frac{1}{2\pi} \left(1 - \frac{|k|}{n+1}\right) \int_{-\pi}^{\pi} f(t)e^{-ikt}dt\, e^{ikx}$$

$$= \frac{1}{2\pi} \int_{-\pi}^{\pi} f(t) \left(\sum_{k=-n}^{n} \left(1 - \frac{|k|}{n+1}\right)e^{ik(x-t)}\right)dt.$$

$$(3.34)$$

Lemma 3.5.1. *For every $n \in \mathbb{N}$ and $x \in \mathbb{R}$ we have*

$$\sum_{k=-n}^{n} \left(1 - \frac{|k|}{n+1}\right) e^{ikx} = \frac{1}{n+1} \frac{\sin^2 \frac{(n+1)x}{2}}{\sin^2 \frac{x}{2}}.$$

Proof: First note that

$$\sin^2 \frac{x}{2} = \frac{1}{2}(1 - \cos x) = -\frac{1}{4}e^{-ix} + \frac{1}{2} - \frac{1}{4}e^{ix}.$$

Then routine calculations give

$$\left(-\frac{1}{4}e^{-ix} + \frac{1}{2} - \frac{1}{4}e^{ix}\right) \sum_{k=-n}^{n} \left(1 - \frac{|k|}{n+1}\right) e^{ikx}$$

$$= \frac{1}{n+1} \left(-\frac{1}{4}e^{-i(n+1)x} + \frac{1}{2} - \frac{1}{4}e^{i(n+1)x}\right),$$

proving the lemma. □

Definition 3.5.2. (Summability kernel)
By a *summability kernel*, we mean a sequence (κ_n) of 2π-periodic continuous functions satisfying:

$$\int_{-\pi}^{\pi} \kappa_n(t)dt = 2\pi \quad \text{for all } n \in \mathbb{N}, \tag{3.35}$$

$$\int_{-\pi}^{\pi} |\kappa_n(t)|dt \leq M \quad \text{for some } M \text{ and all } n \in \mathbb{N}, \tag{3.36}$$

$$\lim_{n \to \infty} \int_{\delta}^{2\pi - \delta} |\kappa_n(t)|dt = 0 \quad \text{for all } \delta \in (0, \pi). \tag{3.37}$$

Lemma 3.5.3. *The sequence of functions*

$$K_n(t) = \sum_{k=-n}^{n} \left(1 - \frac{|k|}{n+1}\right) e^{ikt}$$

is a summability kernel.

Proof: Since $\int_{-\pi}^{\pi} e^{ikt}dt = 2\pi$ if $k = 0$ and $\int_{-\pi}^{\pi} e^{ikt}dt = 0$ for any other integer k, we obtain

$$\int_{-\pi}^{\pi} K_n(t)dt = \sum_{k=-n}^{n} \left(1 - \frac{|k|}{n+1}\right) \int_{-\pi}^{\pi} e^{ikt}dt = 2\pi.$$

From Lemma 3.5.1, it follows that $K_n \geq 0$ and thus

$$\int_{-\pi}^{\pi} |K_n(t)| dt = \int_{-\pi}^{\pi} K_n(t) dt = 2\pi.$$

Finally, let $\delta \in (0, \pi)$. For $t \in (\delta, 2\pi - \delta)$, we have $\sin \frac{t}{2} \geq \sin \frac{\delta}{2}$ and hence

$$K_n(t) = \frac{1}{n+1} \frac{\sin^2 \frac{(n+1)x}{2}}{\sin^2 \frac{x}{2}} \leq \frac{1}{(n+1) \sin^2 \frac{\delta}{2}}.$$

Thus,

$$\int_{\delta}^{2\pi-\delta} K_n(t) dt \leq \frac{2\pi}{(n+1) \sin^2 \frac{\delta}{2}}.$$

For a fixed δ the right-hand side tends to 0 as $n \to \infty$. □

The kernel in Lemma 3.5.3 is called the *Fejér's kernel* (Lipót Fejér (1880–1959)). The reason for introducing here the notion of a summability kernel is the following important theorem.

Theorem 3.5.4. *Let (κ_n) be a summability kernel and let $f \in L^1([-\pi, \pi])$. Then*

$$\lim_{n \to \infty} \frac{1}{2\pi} \int_{-\pi}^{\pi} \kappa_n(t) f(x-t) dt = f(x)$$

in the $L^1([-\pi, \pi])$ norm, that is,

$$\lim_{n \to \infty} \int_{-\pi}^{\pi} \left| \frac{1}{2\pi} \int_{-\pi}^{\pi} \kappa_n(t) f(x-t) dt - f(x) \right| dx = 0. \tag{3.38}$$

Proof: By (3.35), we have

$$f(x) = \frac{1}{2\pi} \int_{-\pi}^{\pi} \kappa_n(t) f(x) dt$$

and hence

$$\frac{1}{2\pi} \int_{-\pi}^{\pi} \kappa_n(t) f(x-t) dt - f(x) = \frac{1}{2\pi} \int_{-\pi}^{\pi} \kappa_n(t) \big(f(x-t) - f(x) \big) dt.$$

Consequently,

$$\int_{-\pi}^{\pi} \left| \frac{1}{2\pi} \int_{-\pi}^{\pi} \kappa_n(t) f(x-t) dt - f(x) \right| dx$$

$$= \int_{-\pi}^{\pi} \left| \frac{1}{2\pi} \int_{-\pi}^{\pi} \kappa_n(t) \big(f(x-t) - f(x) \big) dt \right| dx$$

$$= \int_{-\pi}^{\pi} \left| \frac{1}{2\pi} \int_{-\delta}^{\delta} \kappa_n(t) \big(f(x-t) - f(x) \big) dt \right| dx$$

$$+ \int_{-\pi}^{\pi} \left| \frac{1}{2\pi} \int_{\delta}^{2\pi-\delta} \kappa_n(t) \big(f(x-t) - f(x) \big) dt \right| dx$$

$$\leq \frac{1}{2\pi} \int_{-\pi}^{\pi} \int_{-\delta}^{\delta} \left| \kappa_n(t) \big(f(x-t) - f(x) \big) \right| dt \, dx$$

$$+ \frac{1}{2\pi} \int_{-\pi}^{\pi} \int_{\delta}^{2\pi-\delta} \left| \kappa_n(t) \big(f(x-t) - f(x) \big) \right| dt \, dx.$$

Since

$$\frac{1}{2\pi} \int_{-\pi}^{\pi} \int_{-\delta}^{\delta} \left| \kappa_n(t) \big(f(x-t) - f(x) \big) \right| dt \, dx$$

$$\leq \frac{1}{2\pi} \left(\max_{|t| \leq \delta} \int_{-\pi}^{\pi} \left| f(x-t) - f(x) \right| dx \right) \int_{-\pi}^{\pi} \left| \kappa_n(t) \right| dt,$$

and since, by Theorem 2.4.3, we have

$$\lim_{t \to 0} \int_{-\pi}^{\pi} \left| f(x-t) - f(x) \right| dx = 0,$$

for an arbitrary $\varepsilon > 0$ we can find $\delta > 0$ small enough to ensure that

$$\frac{1}{2\pi} \int_{-\pi}^{\pi} \int_{-\delta}^{\delta} \left| \kappa_n(t) \big(f(x-t) - f(x) \big) \right| dt \, dx < \varepsilon \qquad (3.39)$$

for all $n \in \mathbb{N}$, since, by (3.36), $\int_{-\pi}^{\pi} |\kappa_n(t)| dt$ is bounded.

On the other hand, we have

$$\int_{-\pi}^{\pi} \int_{\delta}^{2\pi-\delta} \left| \kappa_n(t) \big(f(x-t) - f(x) \big) \right| dt \, dx$$

$$\leq \left(\max_{|t| \leq \pi} \int_{-\pi}^{\pi} \left| f(x-t) - f(x) \right| dx \right) \int_{\delta}^{2\pi-\delta} \left| \kappa_n(t) \right| dt$$

$$\leq 2 \int_{-\pi}^{\pi} \left| f(x) \right| dx \int_{\delta}^{2\pi-\delta} \left| \kappa_n(t) \right| dt \to 0 \quad \text{as } n \to \infty, \qquad (3.40)$$

by (3.37). Combining (3.39) and (3.40), we obtain the proof of (3.38). $\qquad \square$

Theorem 3.5.5. *If $f \in L^1([-\pi, \pi])$ and $\langle f, \varphi_n \rangle = 0$ for all integers n, then $f = 0$ a.e.*

Proof: If

$$\int_{-\pi}^{\pi} f(t)e^{-int}\,dt = 0,$$

for all integers n, then

$$f_n(x) = \sum_{k=-n}^{n} \frac{1}{2\pi} \int_{-\pi}^{\pi} f(t)e^{ik(x-t)}\,dt = 0,$$

and consequently,

$$\frac{f_0(x) + f_1(x) + \cdots + f_n(x)}{n+1} = \frac{1}{2\pi} \int_{-\pi}^{\pi} f(t) \left(\sum_{k=-n}^{n} \left(1 - \frac{|k|}{n+1} \right) e^{ik(x-t)} \right) dt$$

$$= 0.$$

On the other hand, since f and all the functions e^{ikx} are 2π-periodic, we have

$$\frac{1}{2\pi} \int_{-\pi}^{\pi} f(t) \left(\sum_{k=-n}^{n} \left(1 - \frac{|k|}{n+1} \right) e^{ik(x-t)} \right) dt$$

$$= \frac{1}{2\pi} \int_{-\pi}^{\pi} f(x-t) \left(\sum_{k=-n}^{n} \left(1 - \frac{|k|}{n+1} \right) e^{ikt} \right) dt$$

and hence, by Theorem 3.5.4 and Lemma 3.5.3,

$$\lim_{n\to\infty} \frac{f_0 + f_1 + \cdots + f_n}{n+1} = f$$

in the $L^1([-\pi, \pi])$ norm. Therefore, $f = 0$ a.e., proving the theorem. □

Theorem 3.5.6. *The sequence*

$$\varphi_n(x) = \frac{e^{inx}}{\sqrt{2\pi}}, \quad n = 0, \pm 1, \pm 2, \ldots,$$

is complete.

Proof: If $f \in L^2([-\pi, \pi])$, then $f \in L^1([-\pi, \pi])$. Thus, by Theorem 3.5.5, if $\langle f, \varphi_n \rangle = 0$ for all integers n, then $f = 0$ a.e., that is, $f = 0$ in $L^2([-\pi, \pi])$. This proves completeness of the sequence by Theorem 3.4.14. □

Theorem 3.5.6 implies that, for every $f \in L^2([-\pi, \pi])$, we have

$$f = \sum_{k=-\infty}^{\infty} \alpha_n \varphi_n \tag{3.41}$$

where

$$\varphi_n(x) = \frac{e^{inx}}{\sqrt{2\pi}} \quad \text{and} \quad \alpha_n = \frac{1}{\sqrt{2\pi}} \int_{-\pi}^{\pi} f(t)e^{-ikt} dt.$$

In this case Parseval's formula yields

$$\| f \|^2 = \int_{-\pi}^{\pi} |f(x)|^2 dx = \sum_{n=-\infty}^{\infty} |\alpha_n|^2.$$

The series (3.41) is called the *Fourier series* of f and the numbers α_n are called the *Fourier coefficients* of f. It is important to remember that, in general, (3.41) does not imply pointwise convergence. The problem of pointwise convergence of Fourier series is much more difficult. In 1966, L. Carleson proved that Fourier series of functions in $L^2([-\pi, \pi])$ converge almost everywhere.

3.6 Orthogonal Complements and Projections

By a subspace of an inner product space E, we mean a vector subspace of E. A subspace of an inner product space is an inner product space itself. A closed subspace of a Hilbert space is a Hilbert space, because a closed subspace of a complete normed space is complete.

Definition 3.6.1. (Orthogonal complement)
Let S be a nonempty subset of an inner product space E. An element $x \in E$ is said to be orthogonal to S, denoted by $x \perp S$, if $\langle x, y \rangle = 0$ for every $y \in S$. The set of all elements of E orthogonal to S, denoted by $S^\perp$, is called the *orthogonal complement* of S. In symbols:

$$S^\perp = \{x \in E, \ x \perp S\}.$$

The orthogonal complement of $S^\perp$ is denoted by $S^{\perp\perp} = (S^\perp)^\perp$.

The notation $S^\perp$ is imprecise, because the complement depends on E. If E_1 and E_2 are two different inner product spaces and $S \subset E_1 \cap E_2$, then obviously the complement of S in E_1 is not the same as the complement in E_2. The notation $S^\perp$ is used when it is clear what the inner product space is. For instance, in Definition 3.6.1 the underlying space is E and thus all complements are with respect to E.

If $x \perp y$ for every $y \in E$, then $x = 0$. Thus $E^\perp = \{0\}$. Similarly, $\{0\}^\perp = H$.

Two subsets A and B of an inner product space are said to be *orthogonal* if $x \perp y$ for every $x \in A$ and $y \in B$. This is denoted by $A \perp B$. Note that if $A \perp B$, then $A \cap B = \{0\}$ or $\emptyset$.

Theorem 3.6.2. *For any subset S of an inner product space E, the set $S^\perp$ is a closed subspace of E.*

Proof: If $\alpha, \beta \in \mathbb{C}$ and $x, y \in S^\perp$, then

$$\langle \alpha x + \beta y, z \rangle = \alpha \langle x, z \rangle + \beta \langle y, z \rangle = 0$$

for every $z \in S$. Thus, $S^\perp$ is a vector subspace of E. We next prove that $S^\perp$ is closed.

Let $(x_n) \in S^\perp$ and $x_n \to x$ for some $x \in E$. From continuity of the inner product, we have

$$\langle x, y \rangle = \left\langle \lim_{n \to \infty} x_n, y \right\rangle = \lim_{n \to \infty} \langle x_n, y \rangle = 0,$$

for every $y \in S$. This shows that $x \in S^\perp$, and thus, $S^\perp$ is closed.   □

This theorem implies that $S^\perp$ is a Hilbert space for any subset S of a Hilbert space H. Note that S does not have to be a vector space. Since $S \perp S^\perp$, we have $S \cap S^\perp = \{0\}$ or $S \cap S^\perp = \emptyset$.

Definition 3.6.3.   (Convex set)
A set U in a vector space is called *convex* if for any $x, y \in U$ and $\alpha \in (0, 1)$ we have $\alpha x + (1 - \alpha)y \in U$.

Note that a vector subspace is a convex set.

The following theorem, concerning the minimization of the norm, is of fundamental importance in approximation theory.

Theorem 3.6.4. (The closest point property) *Let S be a closed convex subset of a Hilbert space H. For every point $x \in H$ there exists a unique point $y \in S$ such that*

$$\|x - y\| = \inf_{z \in S} \|x - z\|. \tag{3.42}$$

Proof: Let (y_n) be a sequence in S such that

$$\lim_{n \to \infty} \|x - y_n\| = \inf_{z \in S} \|x - z\|.$$

Denote $d = \inf_{z \in S} \|x - z\|$. Since $\frac{1}{2}(y_m + y_n) \in S$, we have

$$\left\| x - \frac{1}{2}(y_m + y_n) \right\| \geq d, \quad \text{for all } m, n \in \mathbb{N}.$$

Moreover, by the parallelogram law (3.5),

$$\|y_m - y_n\|^2 = 4\left\|x - \frac{1}{2}(y_m + y_n)\right\|^2 + \|y_m - y_n\|^2 - 4\left\|x - \frac{1}{2}(y_m + y_n)\right\|^2$$

$$= \left\|(x - y_m) + (x - y_n)\right\|^2 + \left\|(x - y_m) - (x - y_n)\right\|^2$$

$$\quad - 4\left\|x - \frac{1}{2}(y_m + y_n)\right\|^2$$

$$= 2\big(\|x - y_m\|^2 + \|x - y_n\|^2\big) - 4\left\|x - \frac{1}{2}(y_m + y_n)\right\|^2.$$

Since

$$2\big(\|x - y_m\|^2 + \|x - y_n\|^2\big) \to 4d^2, \quad \text{as } m, n \to \infty,$$

and

$$\left\|x - \frac{1}{2}(y_m + y_n)\right\|^2 \ge d^2,$$

we have $\|y_m - y_n\|^2 \to 0$, as $m, n \to \infty$. Thus, (y_n) is a Cauchy sequence. Since H is complete and S is closed, the limit $\lim_{n\to\infty} y_n = y$ exists and $y \in S$. From the continuity of the norm, we obtain

$$\|x - y\| = \left\|x - \lim_{n\to\infty} y_n\right\| = \lim_{n\to\infty} \|x - y_n\| = d.$$

We have proved that there exists a point in S satisfying (3.42). It remains to prove the uniqueness. Suppose there is another point y_1 in S satisfying (3.42). Then, since $\frac{1}{2}(y + y_1) \in S$, we have

$$\|y - y_1\|^2 = 4d^2 - 4\left\|x - \frac{y + y_1}{2}\right\|^2 \le 0.$$

This can only happen if $y = y_1$. □

Theorem 3.6.4 gives an existence and uniqueness result, which is crucial for optimization problems. However, it does not tell us how to find that optimal point. The characterization of the optimal point in the case of a real inner product space, stated in the following theorem, is often useful in such problems.

Theorem 3.6.5. *Let S be a convex subset of a real inner product space E, $y \in S$ and let $x \in E$. Then the following conditions are equivalent:*

(a) $\|x - y\| = \inf_{z \in S} \|x - z\|$,

(b) $\langle x - y, z - y \rangle \le 0$ *for all $z \in S$.*

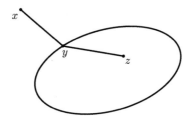

Figure 3.3   Geometric interpretation of Theorem 3.6.5 in $\mathbb{R}^2$.

Proof: Let $z \in S$. Since S is convex, $\lambda z + (1 - \lambda)y \in S$ for every $\lambda \in (0, 1)$. Then, by (a), we have

$$\|x - y\| \leq \|x - \lambda z - (1 - \lambda)y\| = \|(x - y) - \lambda(z - y)\|.$$

Hence, as E is a real inner product space, we get

$$\|x - y\|^2 \leq \|x - y\|^2 - 2\lambda\langle x - y, z - y\rangle + \lambda^2\|z - y\|^2,$$

and consequently,

$$\langle x - y, z - y\rangle \leq \frac{\lambda}{2}\|z - y\|^2.$$

Thus, (b) follows by letting $\lambda \to 0$.

Conversely, if $x \in E$ and $y \in S$ satisfy (b), then for every $z \in S$, we have

$$\|x - y\|^2 - \|x - z\|^2 = 2\langle x - y, z - y\rangle - \|z - y\|^2 \leq 0.$$

Thus, x and y satisfy (a).   □

If $E = \mathbb{R}^2$ and S is a closed convex subset of $\mathbb{R}^2$, then condition (b) has a clear geometrical meaning: The angle between the line through x and y and the line through z and y is always obtuse (see Figure 3.3).

Theorem 3.6.6. *If H_1 is a closed subspace of a Hilbert space H, then every element $x \in H$ has a unique decomposition in the form $x = y + z$ where $y \in H_1$ and $z \in H_1^\perp$.*

Proof: If $x \in H_1$, then the obvious decomposition is $x = x + 0$. Suppose now that $x \notin H_1$. Let y be the unique point of H_1 satisfying $\|x - y\| = \inf_{w \in H_1} \|x - w\|$, as in Theorem 3.6.4. We will show that $x = y + (x - y)$ is the desired decomposition.

If $w \in H_1$ and $\lambda \in \mathbb{C}$, then $y + \lambda w \in H_1$ and

$$\|x - y\|^2 \leq \|x - y - \lambda w\|^2 = \|x - y\|^2 - 2\operatorname{Re}\lambda\langle w, x - y\rangle + |\lambda|^2\|w\|^2.$$

Hence,

$$-2\operatorname{Re}\lambda\langle w, x - y\rangle + |\lambda|^2\|w\|^2 \geq 0.$$

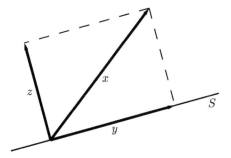

Figure 3.4 Orthogonal decomposition in $\mathbb{R}^2$.

If $\lambda > 0$, then dividing by λ and letting $\lambda \to 0$ gives

$$\text{Re}\langle w, x - y \rangle \leq 0. \tag{3.43}$$

Similarly, replacing λ by $-i\lambda$ ($\lambda > 0$), dividing by λ, and letting $\lambda \to 0$ gives

$$\text{Im}\langle w, x - y \rangle \leq 0. \tag{3.44}$$

Since $y \in H_1$ implies $-y \in H_1$, inequalities (3.43) and (3.44) hold also with $-w$ instead of w. Therefore, $\langle w, x - y \rangle = 0$ for every $w \in H_1$, which means $x - y \in H_1^\perp$.

To prove the uniqueness note that if $x = y_1 + z_1$, $y_1 \in H_1$, and $z_1 \in H_1^\perp$, then $y - y_1 \in H_1$ and $z - z_1 \in H_1^\perp$. Since $y - y_1 = z_1 - z$, we must have $y - y_1 = z_1 - z = 0$. □

The above theorem can be also stated as follows: If H_1 is a closed subspace of a Hilbert space H, then H is the direct sum of H_1 and $H_1^\perp$, that is,

$$H = H_1 \oplus H_1^\perp. \tag{3.45}$$

The representation of H as $H_1 \oplus H_1^\perp$ is called an *orthogonal decomposition* of H. Note that the union of a basis of H_1 and a basis of $H_1^\perp$ is a basis of H.

Theorem 3.6.6 allows us to define a mapping $P_{H_1}(x) = y$, where $y \in H_1$ is the unique element such that $x = y + z$ and $z \in H_1^\perp$. Mapping P_{H_1} is called the *orthogonal projection* onto H_1. Such mappings will be discussed in Section 4.7.

Example 3.6.7. Let $H = \mathbb{R}^2$. Figure 3.4 exhibits the geometric meaning of the orthogonal decomposition in $\mathbb{R}^2$. Here $x \in \mathbb{R}^2$, $x = y + z$, $y \in S$, and $z \in S^\perp$. Note that if s_0 is a unit vector in S, then $y = \langle x, s_0 \rangle s_0$. □

Example 3.6.8. If $H = \mathbb{R}^3$, given a plane P, any vector x can be projected onto the plane P. Figure 3.5 illustrates this situation. □

Theorem 3.6.9. *If S is a closed subspace of a Hilbert space H, then $S^{\perp\perp} = S$.*

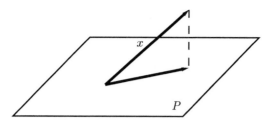

Figure 3.5   Orthogonal projection onto a plane.

Proof: If $x \in S$, then for every $z \in S^{\perp}$, we have $\langle x, z \rangle = 0$, which means $x \in S^{\perp\perp}$. Thus, $S \subset S^{\perp\perp}$. To prove that $S^{\perp\perp} \subset S$ consider an $x \in S^{\perp\perp}$. Since S is closed, $x = y + z$ for some $y \in S$ and $z \in S^{\perp}$. In view of the inclusion $S \subset S^{\perp\perp}$, we have $y \in S^{\perp\perp}$ and thus $z = x - y \in S^{\perp\perp}$, because $S^{\perp\perp}$ is a vector subspace. But $z \in S^{\perp}$, so we must have $z = 0$, which means $x = y \in S$. This shows that $S^{\perp\perp} \subset S$, completing the proof.   □

3.7 Linear Functionals and the Riesz Representation Theorem

In Section 3.5 we remarked that for any fixed vector x_0 in an inner product space E, the formula $f(x) = \langle x, x_0 \rangle$ defines a bounded linear functional on E. It turns out that if E is a Hilbert space, then every bounded linear functional is of this form. Before proving this result, known as the Riesz representation theorem, we discuss some examples and prove a lemma.

Example 3.7.1. Let $H = L^2((a, b))$, $-\infty < a < b < \infty$. Define a linear functional f on H by the formula

$$f(x) = \int_a^b x(t)\,dt.$$

If x_0 denotes the constant function 1 on (a, b), then clearly $f(x) = \langle x, x_0 \rangle$, and thus f is a bounded functional.   □

Example 3.7.2. Let $H = L^2(a, b)$, and let t_0 be a fixed point in (a, b). Let f be a functional on H defined by $f(x) = x(t_0)$. This functional is linear, but it is not bounded.   □

Example 3.7.3. Let $H = \mathbb{C}^n$, and let $n_0 \in \{1, 2, \dots, n\}$. Define f by the formula

$$f\big((x_1, \dots, x_n)\big) = x_{n_0}.$$

We have

$$f\big((x_1, \dots, x_n)\big) = \big\langle (x_1, \dots, x_n), e_{n_0} \big\rangle,$$

where e_{n_0} is the vector which has 1 on the n_0th place and zeros on the remaining places. Thus, f is a bounded linear functional. $\qquad\square$

Example 3.7.4. Let E be the space of all sequences of complex numbers that have only a finite number of nonzero terms with the inner product defined as in l^2, that is,

$$\langle (x_n), (y_n) \rangle = \sum_{n}^{\infty} x_n \overline{y_n}.$$

Then

$$f(x) = \sum_{n}^{\infty} \frac{x_n}{n}$$

defines a bounded linear functional in E. However, there is no $(z_n) \in E$ such that $f(x) = \langle (x_n), (z_n) \rangle$ for every $(x_n) \in E$. Note that E is not a Hilbert space. $\square$

Lemma 3.7.5. *If f is a nontrivial bounded linear functional on a Hilbert space H, then $\dim \mathcal{N}(f)^{\perp} = 1$.*

Proof: Since f is continuous, $\mathcal{N}(f)$ is a closed proper subspace of H and thus $\mathcal{N}(f)^{\perp}$ is not empty. Let $x_1, x_2 \in \mathcal{N}(f)^{\perp}$ be nonzero vectors. Since $f(x_1) \neq 0$ and $f(x_2) \neq 0$, there exists a scalar $a \neq 0$ such that $f(x_1) + af(x_2) = f(x_1 + ax_2) = 0$. Thus, $x_1 + ax_2 \in \mathcal{N}(f)$. On the other hand, since $\mathcal{N}(f)^{\perp}$ is a vector space and $x_1, x_2 \in \mathcal{N}(f)^{\perp}$, we must have $x_1 + ax_2 \in \mathcal{N}(f)^{\perp}$. This is only possible if $x_1 + ax_2 = 0$, which shows that x_1 and x_2 are linearly dependent, because $a \neq 0$. $\qquad\square$

Example 3.7.6. The property in the above lemma need not hold in incomplete inner product spaces. Consider the space E and the functional f in Example 3.7.4. Then f is clearly a nontrivial bounded linear functional on E, but $\mathcal{N}(f)^{\perp} = \{0\}$. $\qquad\square$

Theorem 3.7.7. (Riesz representation theorem) *Let f be a bounded linear functional on a Hilbert space H. There exists exactly one $x_0 \in H$ such that $f(x) = \langle x, x_0 \rangle$ for all $x \in H$. Moreover, we have $\|f\| = \|x_0\|$.*

Proof: If $f(x) = 0$ for all $x \in H$, then $x_0 = 0$ has the desired properties. Assume now that f is a nontrivial functional. Then $\dim \mathcal{N}(f)^{\perp} = 1$, by Lemma 3.7.5. Let z_0 be a unit vector in $\mathcal{N}(f)^{\perp}$. Then, for every $x \in H$, we have

$$x = x - \langle x, z_0 \rangle z_0 + \langle x, z_0 \rangle z_0.$$

Since $\langle x, z_0 \rangle z_0 \in \mathcal{N}(f)^{\perp}$, we must have $x - \langle x, z_0 \rangle z_0 \in \mathcal{N}(f)$, which means that

$$f(x - \langle x, z_0 \rangle z_0) = 0.$$

Consequently,

$$f(x) = f\big(\langle x, z_0 \rangle z_0\big) = \langle x, z_0 \rangle f(z_0) = \big\langle x, \overline{f(z_0)} z_0 \big\rangle.$$

Therefore, if we put

$$x_0 = \overline{f(z_0)} z_0,$$

then $f(x) = \langle x, x_0 \rangle$ for all $x \in H$.

Suppose now that there is another point x_1 such that $f(x) = \langle x, x_1 \rangle$ for all $x \in H$. Then $\langle x, x_0 - x_1 \rangle = 0$ for all $x \in H$, and thus $\langle x_0 - x_1, x_0 - x_1 \rangle = 0$. This is only possible if $x_0 = x_1$.

Finally, we have

$$\|f\| = \sup_{\|x\|=1} |f(x)| = \sup_{\|x\|=1} |\langle x, x_0 \rangle| \le \sup_{\|x\|=1} \big(\|x\| \|x_0\|\big) = \|x_0\|$$

and

$$\|x_0\|^2 = \langle x_0, x_0 \rangle = |f(x_0)| \le \|f\| \|x_0\|.$$

Therefore $\|f\| = \|x_0\|$.   □

The collection H' of all bounded linear functionals on a Hilbert space H is a Banach space (see Theorem 1.5.9). The Riesz representation theorem says that $H' = H$, or more precisely, H' and H are isomorphic. The element x_0 corresponding to a functional f is sometimes called the *representer* of f.

Note that the functional f defined by $f(x) = \langle x_0, x \rangle$, where $x_0 \neq 0$ is a fixed element of a complex Hilbert space H, is not linear. Indeed, we have $f(\alpha x + \beta y) = \overline{\alpha} f(x) + \overline{\beta} f(y)$. Such functionals are often called *antilinear* or *conjugate-linear*.

One of the most important theorems in functional analysis is the Hahn–Banach theorem. It says that a bounded linear functional on a subspace of a normed space E can be extended to a bounded linear functional on E. The proof of this fact is not effective. We can only prove that it is possible, but in general we cannot construct such an extension. The situation in Hilbert spaces is much simpler.

Theorem 3.7.8. *Let f be bounded linear functional defined on a subspace E of a Hilbert space H. Then there exists a bounded linear functional g defined on H such that*

(a) $f(x) = g(x)$ *for all $x \in E$,*

(b) $\|f\| = \|g\|$.

Proof: If E is a closed subspace, then it is a Hilbert space and, by the Riesz representation theorem, there exists an $x_0 \in E$ such that $f(x) = \langle x, x_0 \rangle$ for all $x \in E$. But then g can be defined as $g(x) = \langle x, x_0 \rangle$. Clearly, (a) and (b) are satisfied.

If E is not closed, then we first extend f to a bounded linear functional defined on the closure of E. □

3.8 Exercises

1. Show that
$$\langle x, \alpha y + \beta z \rangle = \overline{\alpha} \langle x, y \rangle + \overline{\beta} \langle x, z \rangle$$
   for all $\alpha, \beta \in \mathbb{C}$, in any inner product space.

2. Prove that the space $C_o(\mathbb{R})$ of all complex-valued continuous functions that vanish outside some finite interval is an inner product space with the inner product
$$\langle f, g \rangle = \int_{-\infty}^{\infty} f(x)\overline{g(x)}dx.$$

3. Verify that the spaces in Examples 3.2.2–3.2.7 are inner product spaces.

4. (a) Let $E = C^1([a, b])$, the space of all continuously differentiable complex-valued functions on $[a, b]$. For $f, g \in E$ define
$$\langle f, g \rangle = \int_a^b f'(x)\overline{g'(x)}dx.$$

   Is $\langle \cdot, \cdot \rangle$ an inner product in E?

   (b) Let $F = \{f \in C^1([a, b]): f(a) = 0\}$. Is $\langle \cdot, \cdot \rangle$ defined in (a) an inner product in F? Is it a Hilbert space?

5. Is the space $C_o^1(\mathbb{R})$ of all continuously differentiable complex-valued continuous functions that vanish outside some finite interval an inner product space if
$$\langle f, g \rangle = \int_{-\infty}^{\infty} f'(x)\overline{g'(x)}dx?$$

   Is it a Hilbert space?

6. Let E be a normed space, $a \in E$, and $r > 0$. Show that the ball $S(a, r) = \{x \in E: \|x - a\| \leq r\}$ is a convex set.

7. Show that, in any inner product space, $\|x - y\| + \|y - z\| = \|x - z\|$ if and only if $y = \alpha x + (1 - \alpha)z$ for some $\alpha \in [0, 1]$. Is the same true in a normed space?

8. Let $E_1, \ldots, E_n$ be inner product spaces. Show that

$$\langle [x_1, \ldots, x_n], [y_1, \ldots, y_n] \rangle = \langle x_1, y_1 \rangle + \cdots + \langle x_n, y_n \rangle$$

defines an inner product in $E = E_1 \times \cdots \times E_n$. If $E_1, \ldots, E_n$ are Hilbert spaces, show that E is a Hilbert space and its norm is defined by

$$\| [x_1, \ldots, x_n] \| = \sqrt{\|x_1\|^2 + \cdots + \|x_n\|^2}.$$

9. Show that, in an inner product space, $x \perp y$ if and only if $\|x + \alpha y\| = \|x - \alpha y\|$ for all scalars α.

10. Show that the *polarization identity*

$$\langle x, y \rangle = \frac{1}{4} \left[\|x + y\|^2 - \|x - y\|^2 + i\|x + iy\|^2 - i\|x - iy\|^2 \right]$$

holds in any inner product space.

11. Show that, for any x in an inner product space, $\|x\| = \sup_{\|y\|=1} |\langle x, y \rangle|$.

12. Prove that any complex normed space with norm $\| \cdot \|$ satisfying the parallelogram law is an inner product space with the inner product defined by

$$\langle x, y \rangle = \frac{1}{4} \left[\|x + y\|^2 - \|x - y\|^2 + i\|x + iy\|^2 - i\|x - iy\|^2 \right],$$

and then $\|x\|^2 = \langle x, x \rangle$.

13. Is $\mathcal{C}([a, b])$ with the norm $\|f\| = \max_{[a,b]} |f(x)|$ an inner product space?

14. Show that $L^2([a, b])$ is the only inner product space among the spaces $L^p([a, b])$.

15. Show that for any elements in an inner product space,

$$\|z - x\|^2 + \|z - y\|^2 = \frac{1}{2} \|x - y\|^2 + 2 \left\| z - \frac{x+y}{2} \right\|^2.$$

The equality is called *Apollonius' identity*.

16. Prove that any finite dimensional inner product space is a Hilbert space.

17. Complete the proof of completeness of $L^{2,p}([a, b])$ (Example 3.3.7) by showing that $\frac{F}{\sqrt{\rho}} \in L^{2,p}([a, b])$ and $f_n \to \frac{F}{\sqrt{\rho}}$ in $L^{2,p}([a, b])$.

18. Let E be an incomplete inner product space. Let H be the completion of E (see Section 1.4). Is it possible to extend the inner product from E onto H such that H would become a Hilbert space?

19. Suppose $x_n \xrightarrow{w} x$ and $y_n \xrightarrow{w} y$ as $n \to \infty$ in a Hilbert space, and $\alpha_n \to \alpha$ in $\mathbb{C}$. Prove or give a counterexample:

    (a) $x_n + y_n \xrightarrow{w} x + y$.

    (b) $\alpha_n x_n \xrightarrow{w} \alpha x$.

    (c) $\langle x_n, y_n \rangle \to \langle x, y \rangle$.

    (d) $\|x_n\| \to \|x\|$.

    (e) If $x_n = y_n$ for all $n \in \mathbb{N}$, then $x = y$.

20. Show that, in a finite dimensional Hilbert space, weak convergence implies strong convergence.

21. If $\sum_{n=1}^{\infty} u_n = u$, show that

$$\sum_{n=1}^{\infty} \langle u_n, x \rangle = \langle u, x \rangle$$

for any x in an inner product space.

22. Let $\{x_1, \ldots, x_n\}$ be a finite orthonormal set in a Hilbert space H. Prove that for any $x \in H$ the vector

$$x - \sum_{k=1}^{n} \langle x, x_k \rangle x_k$$

is orthogonal to x_k for every $k = 1, \ldots, n$.

23. In the inner product space $\mathcal{C}([-\pi, \pi])$, show that the following sequences of functions are orthogonal

    (a) $x_k(t) = \sin kt, k = 1, 2, 3, \ldots,$

    (b) $y_n(t) = \cos nt, n = 0, 1, 2, \ldots.$

24. Show that the application of the Gram–Schmidt process to the sequence of functions

$$f_0(t) = 1, f_1(t) = t, f_2(t) = t^2, \ldots, f_n(t) = t^n, \ldots$$

(as elements of $L^2([-1, 1])$) yields the Legendre polynomials.

25. Show that the application of the Gram–Schmidt process to the sequence of functions

$$f_0(t) = e^{-t^2/2},$$
$$f_1(t) = te^{-t^2/2},$$

$$f_2(t) = t^2 e^{-t^2/2}, \dots,$$
$$f_n(t) = t^n e^{-t^2/2}, \dots$$

(as elements of $L^2(\mathbb{R})$) yields the orthonormal system discussed in Example 3.4.7.

26. Apply the Gram–Schmidt process to the sequence of functions

$$f_0(t) = 1, f_1(t) = t, f_2(t) = t^2, \dots, f_n(t) = t^n, \dots$$

defined on $\mathbb{R}$ with the inner product

$$\langle f, g \rangle = \int_{-\infty}^{\infty} f(t)\overline{g(t)} e^{-t^2} dt.$$

Compare the result with Example 3.4.7.

27. Apply the Gram–Schmidt process to the sequence of functions

$$f_0(t) = 1, f_1(t) = t, f_2(t) = t^2, \dots, f_n(t) = t^n, \dots$$

defined on $[0, \infty)$ with the inner product

$$\langle f, g \rangle = \int_0^{\infty} f(t)\overline{g(t)} e^{-t} dt.$$

The obtained polynomials are called the *Laguerre polynomials* (Edmond Nicolas Laguerre (1834–1886)).

28. Let T_n be the *Chebyshev polynomial* (Pafnuty Lvovich Chebyshev (1821–1894)) of degree n, that is,

$$T_0(x) = 1, \qquad T_n(x) = 2^{1-n} \cos(n \arccos x).$$

Show that the functions

$$\varphi_n(x) = \frac{2^n}{\sqrt{2\pi}} T_n(x), \quad n = 0, 1, 2, \dots,$$

form an orthonormal system in $L^2[(-1, 1)]$ with respect to the inner product

$$\langle f, g \rangle = \int_{-1}^{1} \frac{1}{\sqrt{1 - x^2}} f(x)\overline{g(x)} dx.$$

29. Prove that for any polynomial

$$p_n(x) = x^n + a_{n-1} x^{n-1} + \dots + a_0,$$

we have

$$\max_{[-1,1]}\left|p_n(x)\right| \geq \max_{[-1,1]}\left|T_n(x)\right|,$$

where T_n denotes the Chebyshev polynomial of degree n.

30. Show that the complex functions

$$\varphi_n(z) = \sqrt{\frac{n}{\pi}} z^{n-1}, \quad n = 1, 2, 3, \ldots,$$

form an orthonormal system in the space of continuous complex functions defined in the unit disc $D = \{z \in \mathbb{C}: |z| \leq 1\}$ with respect to the inner product

$$\langle f, g \rangle = \int_D f(z)\overline{g(z)} dz.$$

31. Prove that the complex functions

$$\psi_n(z) = \frac{1}{\sqrt{2\pi}} z^{n-1}, \quad n = 1, 2, 3, \ldots$$

form an orthonormal system in the space of continuous complex functions defined on the unit circle $C = \{z \in \mathbb{C}: |z| = 1\}$ with respect to the inner product

$$\langle f, g \rangle = \int_C f(z)\overline{g(z)} dz.$$

32. With respect to the inner product

$$\langle f, g \rangle = \int_{-1}^{1} f(x)\overline{g(x)}\omega(x) dx,$$

where $\omega(x) = (1 - x)^\alpha (1 + x)^\beta$ and $\alpha, \beta > -1$, show that the *Jacobi polynomials* (Carl Gustav Jacob Jacobi (1804–1851))

$$P_n^{(\alpha\beta)}(x) = \frac{(-1)^n}{n!2^n}(1-x)^{-\alpha}(1+x)^{-\beta}\frac{d^n}{dx^n}\left[(1-x)^\alpha(1+x)^\beta(1-x^2)^n\right]$$

form an orthogonal system.

33. Show that the *Gegenbauer polynomials* (Leopold Bernhard Gegenbauer (1849–1903))

$$C_n^\gamma(x) = \frac{(-1)^n}{n!2^n}(1-x^2)^{1/2-\gamma}\frac{d^n}{dx^n}(1-x^2)^{n+\gamma-1/2},$$

where $\gamma > \frac{1}{2}$, form an orthonormal system with respect to the inner product

$$\langle f, g \rangle = \int_{-1}^{1} f(x)\overline{g(x)}(1 - x^2)^{1/2 - \gamma}\, dx.$$

Note that the Gegenbauer polynomials are a special case of the Jacobi polynomials with $\alpha = \beta = \gamma - \frac{1}{2}$.

34. Find $a, b, c \in \mathbb{C}$, which minimize the value of the integral

$$\int_{-1}^{1} |x^3 - a - bx - cx^2|^2\, dx.$$

35. If x and x_k $(k = 1, \ldots, n)$ belong to a real Hilbert space, show that

$$\left\| x - \sum_{k=1}^{n} \alpha_k x_k \right\|^2 = \|x\|^2 - \sum_{k=1}^{n} \alpha_k \langle x, x_k \rangle + \sum_{k=1}^{n} \sum_{l=1}^{n} \alpha_k \alpha_l \langle x_k, x_l \rangle.$$

Also show that this expression is minimum when $Aa = b$ where $a = (\alpha_1, \ldots, \alpha_n)$, $b = (\langle x, x_1 \rangle, \ldots, \langle x, x_n \rangle)$ and the matrix $A = (\alpha_{kl})$ is defined by $\alpha_{kl} = \langle x_k, x_l \rangle$.

36. If (a_n) is an orthonormal sequence in a Hilbert space H and (α_n) is a sequence in l^2, show that there exists $x \in H$ such that

$$\langle x, a_n \rangle = \alpha_n \quad \text{and} \quad \|(\alpha_n)\| = \|x\|,$$

where $\|(\alpha_n)\|$ denotes the norm in l^2.

37. If α_n and β_n $(n = 1, 2, 3, \ldots)$ are generalized Fourier coefficients of vectors x and y with respect to a complete orthonormal sequence in a Hilbert space, show that

$$\langle x, y \rangle = \sum_{k=1}^{\infty} \alpha_k \overline{\beta_k}.$$

38. Let (x_n) be an orthonormal sequence in a Hilbert space H. Show that (x_n) is complete if and only if $\mathrm{cl}(\mathrm{span}\{x_1, x_2, \ldots\}) = H$. In other words, (x_n) is complete if and only if every element of H can be approximated by a sequence of finite combinations of x_n's.

39. Show that the functions

$$\varphi_n(x) = \frac{e^{-x/2}}{n!} L_n(x), \quad n = 0, 1, 2, \ldots$$

where L_n is the Laguerre polynomial of degree n, that is,

$$L_n(x) = e^x \frac{d^n}{dx^n} \left(x^n e^{-x} \right),$$

form a complete orthonormal system in $L^2(0, \infty)$.

40. Let

$$\varphi_n(x) = \frac{e^{inx}}{\sqrt{2\pi}}, \quad n = 0, \pm 1, \pm 2, \ldots,$$

and let $f \in L^1([-\pi, \pi])$. Define

$$f_n(x) = \sum_{k=-n}^{n} \langle f, \varphi_k \rangle \varphi_k, \quad \text{for } n = 0, 1, 2, \ldots.$$

Show that

$$\frac{f_0(x) + f_1(x) + \cdots + f_n(x)}{n+1} = \sum_{k=-n}^{n} \left(1 - \frac{|k|}{n+1} \right) \langle f, \varphi_k \rangle \varphi_k(x).$$

41. Fill in the details of the proof of Lemma 3.5.1.

42. Let f be a continuous non-negative function defined on $[-\pi, \pi]$ such that $\mathrm{supp} f \subseteq [-\pi + \varepsilon, \pi - \varepsilon]$, for some $0 < \varepsilon < \pi$, and $\int_{-\pi}^{\pi} f(x) dx = 2\pi$. Let g be a 2π-periodic extension of f onto the entire line $\mathbb{R}$. Define

$$k_n(x) = ng(nx) \quad \text{for } n = 1, 2, \ldots.$$

Show that (k_n) is a summability kernel.

43. Show that the sequence of functions

$$\frac{1}{\sqrt{2\pi}}, \frac{\cos x}{\sqrt{\pi}}, \frac{\sin x}{\sqrt{\pi}}, \frac{\cos 2x}{\sqrt{\pi}}, \frac{\sin 2x}{\sqrt{\pi}}, \ldots,$$

is a complete orthonormal sequence in $L^2([-\pi, \pi])$.

44. Show that the following sequence of functions is a complete orthonormal system in $L^2([0, \pi])$:

$$\frac{1}{\sqrt{\pi}}, \sqrt{\frac{2}{\pi}} \cos x, \sqrt{\frac{2}{\pi}} \cos 2x, \sqrt{\frac{2}{\pi}} \cos 3x, \ldots.$$

45. Show that the following sequence of functions is a complete orthonormal system in $L^2([0, \pi])$:

$$\sqrt{\frac{2}{\pi}} \sin x, \sqrt{\frac{2}{\pi}} \sin 2x, \sqrt{\frac{2}{\pi}} \sin 3x, \ldots.$$

46. Give an example of a complete orthonormal sequence in $L^2([a, b])$ for arbitrary $a < b$.

47. What is the orthogonal complement in $L^2(\mathbb{R})$ of the set of all even functions?

48. What is the orthogonal complement in $L^2([-\pi, \pi])$ of the set of all polynomials of odd degree?

49. Let $\mathcal{P}$ be a complete orthonormal system in a Hilbert space H. Show that if $\mathcal{P} = \mathcal{P}_1 \cup \mathcal{P}_2$ and $\mathcal{P}_1 \cap \mathcal{P}_2 = \emptyset$, then $\mathcal{P}_1^\perp = \text{cl}(\text{span}\,\mathcal{P}_2)$.

50. Let S be a subset of an inner product space. Show that $S^\perp = (\text{span}\,S)^\perp$.

51. Prove the following generalization of Theorem 3.6.5 for complex inner product spaces: *Let S be a convex subset of a complex inner product space E, $y \in S$ and let $x \in E$. Then the following conditions are equivalent:*

   (a) $\|x - y\| = \inf_{z \in S} \|x - z\|$,

   (b) $\text{Re}\langle x - y, z - y \rangle \leq 0$ *for all $z \in S$.*

52. Let E be the Banach space $\mathbb{R}^2$ with the norm $\|(x, y)\| = \max\{|x|, |y|\}$. Show that E does not have the closest point property.

53. Let S be a closed subspace of a Hilbert space H and let $(e_1, e_2, \ldots)$ be a complete orthonormal sequence in S. For an arbitrary $x \in H$ there exists $y \in S$ such that $\|x - y\| = \inf_{z \in S} \|x - z\|$. Define y in terms of $(e_1, e_2, \ldots)$.

54. If S is a closed subspace of a Hilbert space H, then $H = S \oplus S^\perp$. Is this true in every inner product space?

55. Show that the functional in Example 3.7.2 is unbounded.

56. The Riesz representation theorem says, that for every bounded linear functional $f \in H'$ on a Hilbert space H, there exists a representer $x_f \in H$ such that $f(x) = \langle x, x_f \rangle$ for all $x \in H$. Let $T : H' \to H$ be the mapping which assigns x_f to f. Prove the following properties of T:

   (a) $T(H') = H$,

   (b) $T(f + g) = T(f) + T(g)$,

   (c) $T(\alpha f) = \overline{\alpha} T(f)$,

   (d) $\|T(f)\| = \|f\|$,

   where $f, g \in H'$ and $\alpha \in \mathbb{C}$.

57. Prove part (b) of Theorem 3.4.27.

58. Show that the space l^2 is separable.

59. Let $\mathcal{P}$ be an uncountable orthonormal system in an inner product space E. Show that, for every $x \in E$, we have $\langle x, e \rangle \neq 0$ for at most countably many $e \in \mathcal{P}$.

Linear Operators on Hilbert Spaces

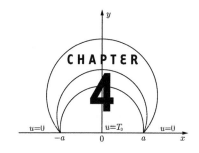

CHAPTER 4

"As long as a branch of science affords an abundance of problems, it is full of life; want of problems means death or cessation of independent development. Just as every human enterprise prosecutes final aims, so mathematical research needs problems. Their solution steels the force of the investigator; thus he discovers new methods and viewpoints and widens his horizon."

David Hilbert

4.1 Introduction

In Section 1.5, we discussed some basic properties of linear mappings between vector spaces. In this chapter, we are interested in the special case when the domain and the range are subspaces of a Hilbert space. In this case, the name *linear operator* or *linear transformation* is usually used. Since nonlinear operators will not be considered here, linear operators will be called simply operators. Operators on Hilbert spaces are widely used to represent physical quantities, and hence their importance is further enhanced in applied mathematics and mathematical physics. The most important operators include differential, integral, and matrix operators. In the next section, we present some examples of operators. Then bilinear functionals and quadratic forms are considered and the Lax–Milgram theorem is proved. This theorem is an important generalization of the Riesz representation theorem. Adjoint and self-adjoint operators are discussed in Section 4.4. Sections 4.5 to 4.8 deal with special linear operators including invertible, normal, isometric, unitary, positive, compact, and projection operators. In Section 4.9, we consider eigenvalues and eigenvectors of linear operators. These concepts play a central role in the theory of operators and their applications, where the spectral decomposition of operators is one of the most important tools. The spectral theorem for compact self-adjoint operators and other related results are discussed in Section 4.10. Basic concepts of the theory of unbounded operators are discussed in Section 4.12.

4.2 Examples of Operators

We begin with some examples of operators. In each case, we are interested whether the operator is bounded. Recall that an operator A is called bounded if there is a number K such that $\|Ax\| \le K\|x\|$ for every x in the domain of A. The norm of A is defined as the infimum of all such numbers K, or equivalently, by

$$\|A\| = \sup_{\|x\|=1} \|Ax\|.$$

We will refer to this norm as the *operator norm*.

In Section 1.6, we proved that an operator A is bounded if and only if it is continuous. It is often much more difficult to find the norm of an operator than just to prove that it is bounded.

Example 4.2.1. (Identity operator and null operator) The simplest examples of operators are the *identity operator* $\mathcal{I}$ and the *null operator*. The identity operator leaves every element unchanged, that is, $\mathcal{I}x = x$ for all $x \in E$. The null operator assigns the zero vector to every element of E. The null operator will be denoted by 0. Obviously, the identity operator and the null operator are bounded and we have $\|\mathcal{I}\| = 1$ and $\|0\| = 0$. A scalar multiple $\alpha\mathcal{I}$ of the identity operator is the operator, which multiplies every element by the scalar α, that is, $(\alpha\mathcal{I})x = \alpha x$. □

Example 4.2.2. Let A be an operator on $\mathbb{C}^N$ and let $\{e_1, \ldots, e_N\}$ be the standard orthonormal base in $\mathbb{C}^N$, that is,

$$e_1 = (1, 0, 0, \ldots, 0),$$
$$e_2 = (0, 1, 0, \ldots, 0),$$
$$\vdots$$
$$e_N = (0, 0, \ldots, 0, 1).$$

Define, for $i, j \in \{1, 2, \ldots, N\}$,

$$\alpha_{ij} = \langle Ae_j, e_i \rangle.$$

Then, for $x = \sum_{j=1}^{N} \lambda_j e_j \in \mathbb{C}^N$, we have $Ax = \sum_{j=1}^{N} \lambda_j Ae_j$, and hence

$$\langle Ax, e_i \rangle = \sum_{j=1}^{N} \lambda_j \langle Ae_j, e_i \rangle = \sum_{j=1}^{N} \alpha_{ij} \lambda_j. \tag{4.1}$$

Thus, every operator on the space $\mathbb{C}^N$ is defined by an $N \times N$ matrix. Conversely, for every $N \times N$ matrix (α_{ij}), formula (4.1) defines an operator on $\mathbb{C}^N$. We thus have a one-to-one correspondence between operators on N-dimensional vector spaces and $N \times N$ matrices. If operator A is defined by the matrix (α_{ij}), then

$$\|A\| \leq \sqrt{\sum_{i=1}^{N} \sum_{j=1}^{N} |\alpha_{ij}|^2}.$$

This implies that every operator on $\mathbb{C}^N$, and thus every operator on any finite dimensional Hilbert space, is bounded. $\qquad\square$

The operator in the next example is defined on a proper subspace of a Hilbert space.

Example 4.2.3. (Differential operator) One of the most important operators in applied mathematics is the *differential operator*

$$(Df)(x) = \frac{df}{dx}(x) = f'(x)$$

defined on a space of differentiable functions. Consider, for example, the differential operator on a subspace of $L^2([-\pi, \pi])$ defined as

$$\mathcal{D}(D) = \{f \in L^2([-\pi, \pi]) : f' \in L^2([-\pi, \pi])\}.$$

If $L^2([-\pi, \pi])$ is equipped with the standard norm $\|f\| = \sqrt{\int_{-\pi}^{\pi} |f(x)|^2 dx}$, then the differential operator is not bounded. Indeed, for $f_n(x) = \sin nx$, $n = 1, 2, 3, \ldots$, we have

$$\|f_n\| = \sqrt{\int_{-\pi}^{\pi} |\sin nx|^2 dx} = \sqrt{\pi}$$

and

$$\|Df_n\| = \sqrt{\int_{-\pi}^{\pi} |n \cos nx|^2 dx} = n\sqrt{\pi}.$$

This example can be easily generalized to an arbitrary interval $[a, b]$ or even $(-\infty, \infty)$. $\qquad\square$

Example 4.2.4. (Integral operator) Another important type of operators is an *integral operator T* defined by

$$(Tx)(s) = \int_a^b K(s, t)x(t)dt,$$

where a and b are finite or infinite, $a < b$, and K is a function defined on the square $(a, b) \times (a, b)$. The function K is called the *kernel* of the operator. The domain of an integral operator depends on K. If

$$\int_a^b \int_a^b |K(s, t)|^2 dt ds < \infty,$$

then T is a bounded operator on $L^2([a, b])$ and

$$\|T\| \leq \sqrt{\int_a^b \int_a^b |K(s, t)|^2 dt ds}.$$

Indeed, for any $x \in L^2([a, b])$, we have

$$\|Tx\|^2 = \int_a^b \left| \int_a^b K(s, t)x(t)dt \right|^2 ds$$

$$\leq \int_a^b \left(\int_a^b |K(s, t)|^2 dt \int_a^b |x(t)|^2 dt \right) ds \quad \text{(by Schwarz's inequality)}$$

$$\leq \int_a^b \int_a^b |K(s, t)|^2 dt ds \int_a^b |x(t)|^2 dt.$$

Thus,

$$\|Tx\| \leq \sqrt{\int_a^b \int_a^b |K(s, t)|^2 dt ds} \|x\|. \qquad \square$$

Example 4.2.5. (Multiplication operator) Let $z \in C([a, b])$. An operator A on $L^2([a, b])$ defined by $(Ax)(t) = z(t)x(t)$ is clearly linear. The function z is called the *multiplier*. Since

$$\|Ax\|^2 = \int_a^b |x(t)|^2 |z(t)|^2 dt \leq \max_{[a,b]} |z(t)|^2 \int_a^b |x(t)|^2 dt,$$

we have

$$\|Ax\| \leq \max_{[a,b]} |z(t)| \|x\|,$$

and thus A is bounded. $\qquad \square$

Two operators A and B are said to be *equal*, $A = B$, if $\mathcal{D}(A) = \mathcal{D}(B)$ and $Ax = Bx$ for every x in that common domain. The sum of two operators is defined by

$$(A + B)x = Ax + Bx$$

and $\mathcal{D}(A + B) = \mathcal{D}(A) \cap \mathcal{D}(B)$. Since we are always assuming that the domain of an operator is a vector subspace, $\mathcal{D}(A) \cap \mathcal{D}(B)$ is never empty because $0 \in \mathcal{D}(A) \cap \mathcal{D}(B)$. However, in the extreme case, we might have $\mathcal{D}(A) \cap \mathcal{D}(B) = \{0\}$. The product of a scalar λ and an operator A is defined by

$$(\lambda A)x = \lambda(Ax).$$

Obviously, $\mathcal{D}(\lambda A) = \mathcal{D}(A)$. The *product AB* of operators A and B is defined by

$$(AB)x = A(Bx).$$

The domain of the product of A and B is

$$\mathcal{D}(AB) = \{x \in \mathcal{D}(B): Bx \in \mathcal{D}(A)\}.$$

The product AB is simply the composition of A and B. Traditionally, in the context of operators, the word *product* is used instead of *composition*.

When an operator A is multiplied by a scalar λ, then the result can also be viewed as the product of the multiplication operator $\lambda \mathcal{I}$ and A, that is, $\lambda A = (\mathcal{I}\lambda)A$. This different interpretation does not change the properties of the operation. In some situations, it may be convenient to identify scalars with multiplication operators. Since $\mathcal{I}A = 1A$, the identity operator is often denoted by 1.

The defined operations have the following obvious properties:

$$
\begin{array}{lll}
A + B = B + A, & (A + B) + C = A + (B + C), & A + 0 = A, \\
\alpha(A + B) = \alpha A + \alpha B, & (\alpha + \beta)A = \alpha A + \beta A, & A0 = 0, \\
A(BC) = (AB)C, & (A + B)C = AC + BC, & A\mathcal{I} = \mathcal{I}A.
\end{array}
$$

It is somewhat unexpected that the equality $A(B + C) = AB + AC$ need not hold in general.

The product of operators is not commutative, that is, AB need not equal BA. Operators A and B for which $AB = BA$ are called *commuting operators*.

Example 4.2.6. (Noncommuting operators)   Let A and B be operators on $\mathbb{C}^2$ defined by the matrices

$$A = \begin{pmatrix} 1 & 0 \\ 0 & 0 \end{pmatrix} \quad \text{and} \quad B = \begin{pmatrix} 0 & 1 \\ 0 & 0 \end{pmatrix}$$

(see Example 4.2.2). Then $AB \neq BA$.   □

Example 4.2.7. (Noncommuting operators) Consider the operators

$$Af(x) = xf(x) \quad \text{and} \quad D = \frac{d}{dx}.$$

It is easy to check that $AD \neq DA$.   □

The square of an operator A is defined as $A^2x = A(Ax)$. By induction, we can define

$$A^nx = A(A^{n-1}x)$$

for any positive integer n. As usual, $A^1 = A$ and $A^0 = \mathcal{I}$.

If A and B are bounded operators, then obviously $A+B$ and λA are bounded (for any scalar λ) and we have

$$\|A + B\| \le \|A\| + \|B\| \quad \text{and} \quad \|\lambda A\| = |\lambda| \|A\|.$$

A similar property holds for the product of operators.

Theorem 4.2.8. *The product AB of bounded operators A and B is bounded and*

$$\|AB\| \le \|A\| \|B\|.$$

Proof: Let A and B be bounded operators on a normed space E, $\|A\| = K_1$ and $\|B\| = K_2$. Then

$$\|ABx\| \le K_1 \|Bx\| \le K_1 K_2 \|x\|,$$

for every $x \in E$.   □

Theorem 4.2.9. *A bounded operator on a separable infinite dimensional Hilbert space can be represented by an infinite matrix.*

Proof: Let A be a bounded operator on a Hilbert space H and let (e_n), $n = 1, 2, 3, \ldots$, be a complete orthonormal sequence in H. For $i, j \in \mathbb{N}$, define

$$\alpha_{ij} = \langle Ae_j, e_i \rangle.$$

For any $x \in H$, we have

$$Ax = A\left(\sum_{j=1}^{\infty} \langle x, e_j \rangle e_j\right) = A\left(\lim_{n \to \infty} \sum_{j=1}^{n} \langle x, e_j \rangle e_j\right)$$

$$= \lim_{n \to \infty} A\left(\sum_{j=1}^{n} \langle x, e_j \rangle e_j\right) \quad \text{(by continuity of } A\text{)}$$

$$= \lim_{n \to \infty} \left(\sum_{j=1}^{n} \langle x, e_j \rangle A e_j \right) \quad \text{(by linearity of } A\text{)}$$

$$= \sum_{j=1}^{\infty} \langle x, e_j \rangle A e_j.$$

Now

$$\langle Ax, e_i \rangle = \left\langle \sum_{j=1}^{\infty} \langle x, e_j \rangle A e_j, e_i \right\rangle = \sum_{j=1}^{\infty} \langle A e_j, e_i \rangle \langle x, e_j \rangle = \sum_{j=1}^{\infty} \alpha_{ij} \langle x, e_j \rangle.$$

Thus, A is represented by the matrix (a_{ij}). □

4.3 Bilinear Functionals and Quadratic Forms

The concepts of a bilinear functional and a quadratic form do not require the structure of an inner product space. They can be defined in any vector space. This discussion is presented in this chapter because of important applications to the theory of operators in Hilbert spaces.

Definition 4.3.1.   (Bilinear functional)
By a *bilinear functional* φ on a complex vector space E, we mean a mapping $\varphi : E \times E \to \mathbb{C}$ satisfying the following two conditions:

(a) $\varphi(\alpha x_1 + \beta x_2, y) = \alpha \varphi(x_1, y) + \beta \varphi(x_2, y),$

(b) $\varphi(x, \alpha y_1 + \beta y_2) = \overline{\alpha} \varphi(x, y_1) + \overline{\beta} \varphi(x, y_2),$

for any scalars α and β and any $x, x_1, x_2, y, y_1, y_2 \in E$.

Bilinear functionals are often called *sesquilinear*. Note that a bilinear functional is linear with respect to the first variable and antilinear with respect to the second variable. Clearly, all bilinear functionals on E constitute a vector space.

Example 4.3.2. Inner product is a bilinear functional. □

Example 4.3.3. Let A and B be operators on an inner product space E. Then $\varphi_1(x, y) = \langle Ax, y \rangle$, $\varphi_2(x, y) = \langle x, By \rangle$, and $\varphi_3(x, y) = \langle Ax, By \rangle$ are bilinear functionals on E. □

Example 4.3.4. Let f and g be linear functionals on a vector space E. Then $\varphi(x, y) = f(x)\overline{g(y)}$ is a bilinear functional on E. □

Definition 4.3.5. (Symmetric, positive, strictly positive, and bounded bilinear functionals)

Let φ be a bilinear functional on E.

(a) φ is called *symmetric* if $\varphi(x, y) = \overline{\varphi(y, x)}$ for all $x, y \in E$.

(b) φ is called *positive* if $\varphi(x, x) \geq 0$ for every $x \in E$.

(c) φ is called *strictly positive* if it is positive and $\varphi(x, x) > 0$ for all $x \neq 0$.

(d) If E is a normed space, then φ is called *bounded* if $|\varphi(x, y)| \leq K\|x\|\|y\|$ for some $K > 0$ and all $x, y \in E$. The *norm* of a bounded bilinear functional is defined by

$$\|\varphi\| = \sup_{\|x\|=\|y\|=1} |\varphi(x, y)|.$$

If $f = g$ in Example 4.3.4, then φ is symmetric and positive. Inner product is strictly positive. If operators A and B in Example 4.3.3 are bounded, then φ_1, φ_2, and φ_3 are bounded. Similarly, if f and g in Example 4.3.4 are bounded, then the defined bilinear functional is also bounded. Note that for a bounded bilinear functional φ on E we have

$$|\varphi(x, y)| \leq \|\varphi\|\|x\|\|y\| \quad \text{for all } x, y \in E.$$

Definition 4.3.6. (Quadratic form)

Let φ be a bilinear functional on a vector space E. The function $\Phi : E \to \mathbb{C}$ defined by $\Phi(x) = \varphi(x, x)$ is called the *quadratic form associated with* φ. A quadratic form Φ on a normed space E is called *bounded* if there exists a constant $K > 0$ such that $|\Phi(x)| \leq K\|x\|^2$ for all $x \in E$. The *norm* of a bounded quadratic form is defined by

$$\|\Phi\| = \sup_{\|x\|=1} |\Phi(x)|.$$

Note that for a bounded quadratic form Φ on a normed space we have $|\Phi(x)| \leq \|\Phi\|\|x\|^2$. A bilinear functional and the associated quadratic form have properties similar to an inner product $\langle x, y \rangle$ and the square of the norm defined by that inner product $\|x\|^2 = \langle x, x \rangle$, respectively.

Theorem 4.3.7. (Polarization identity) *Let φ be a bilinear functional on E, and let Φ be the quadratic form associated with φ. Then*

$$4\varphi(x, y) = \Phi(x + y) - \Phi(x - y) + i\Phi(x + iy) - i\Phi(x - iy) \qquad (4.2)$$

for all $x, y \in E$.

Proof: For any $\alpha, \beta \in \mathbb{C}$, we have

$$\Phi(\alpha x + \beta y) = \varphi(\alpha x + \beta y, \alpha x + \beta y)$$
$$= |\alpha|^2 \Phi(x) + \alpha \bar{\beta} \varphi(x, y) + \bar{\alpha} \beta \varphi(y, x) + |\beta|^2 \Phi(y).$$

Using this equality subsequently for $\alpha - \beta - 1; \alpha = 1$ and $\beta = -1; \alpha = 1$ and $\beta = i; \alpha = 1$ and $\beta = -i$; we get

$$\Phi(x + y) = \Phi(x) + \varphi(x, y) + \varphi(y, x) + \Phi(y),$$
$$-\Phi(x - y) = -\Phi(x) + \varphi(x, y) + \varphi(y, x) - \Phi(y),$$
$$i\Phi(x + iy) = i\Phi(x) + \varphi(x, y) - \varphi(y, x) + i\Phi(y),$$
$$-i\Phi(x - iy) = -i\Phi(x) + \varphi(x, y) - \varphi(y, x) - i\Phi(y).$$

By adding these equalities we obtain (4.2). □

The following simple, but somewhat surprising, result is often useful.

Corollary 4.3.8. *Let φ_1 and φ_2 be bilinear functionals on E. If $\varphi_1(x, x) = \varphi_2(x, x)$ for all $x \in E$, then $\varphi_1 = \varphi_2$, that is, $\varphi_1(x, y) = \varphi_2(x, y)$ for all $x, y \in E$. Similarly, if A and B are operators on E such that $\langle Ax, x \rangle = \langle Bx, x \rangle$ for all $x \in E$, then $A = B$.*

Proof: If $\varphi_1(x, x) = \varphi_2(x, x)$ for all $x \in E$, then the quadratic forms Φ_1 and Φ_2 associated with φ_1 and φ_2, respectively, are equal, and hence, by (4.2), the functionals φ_1 and φ_2 are equal. The proof for operators is obtained by letting $\varphi_1(x, y) = \langle Ax, y \rangle$ and $\varphi_2(x, y) = \langle Bx, y \rangle$. □

Theorem 4.3.9. *A bilinear functional φ on E is symmetric if and only if the associated quadratic form Φ is real.*

Proof: If $\varphi(x, y) = \overline{\varphi(y, x)}$ for all $x, y \in E$, then

$$\Phi(x) = \varphi(x, x) = \overline{\varphi(x, x)} = \overline{\Phi(x)}$$

for every $x \in E$, and thus Φ is real.

Assume now $\Phi(x) = \overline{\Phi(x)}$ for all $x \in E$. Define a bilinear functional ψ on E by

$$\psi(x, y) = \overline{\varphi(y, x)}.$$

Then, for the associated quadratic form Ψ we have

$$\Psi(x) = \overline{\varphi(x, x)} = \overline{\Phi(x)} = \Phi(x).$$

Thus, by Corollary 4.3.8, $\varphi(x, y) = \psi(x, y)$ for all $x, y \in E$. Clearly, this means that $\varphi(x, y) = \overline{\varphi(y, x)}$ for all $x, y \in E$. □

Theorem 4.3.10. *A bilinear functional φ on a normed space E is bounded if and only if the associated quadratic form Φ is bounded. Moreover, we have*

$$\|\Phi\| \le \|\varphi\| \le 2\|\Phi\|. \tag{4.3}$$

Proof: Since

$$\|\Phi\| = \sup_{\|x\|=1} |\Phi(x)| = \sup_{\|x\|=1} |\varphi(x,x)| \le \sup_{\|x\|=\|y\|=1} |\varphi(x,y)| = \|\varphi\|,$$

if φ is bounded, then Φ is bounded and the first inequality follows.

Suppose now that Φ is bounded. In view of (4.2), we have

$$|\varphi(x,y)| = \frac{1}{4}|\Phi(x+y) - \Phi(x-y) + i\Phi(x+iy) - i\Phi(x-iy)|$$

$$\le \frac{1}{4}\|\Phi\|\left(\|x+y\|^2 + \|x-y\|^2 + \|x+iy\|^2 + \|x-iy\|^2\right).$$

Hence, by the parallelogram law,

$$|\varphi(x,y)| \le \|\Phi\|\left(\|x\|^2 + \|y\|^2\right).$$

Consequently,

$$\sup_{\|x\|=\|y\|=1} |\varphi(x,y)| \le \sup_{\|x\|=\|y\|=1} \|\Phi\|\left(\|x\|^2 + \|y\|^2\right) = 2\|\Phi\|.$$

Thus, if Φ is bounded, then φ is bounded and the second inequality in (4.3) follows. $\qquad\square$

Theorem 4.3.11. *Let φ be a bilinear functional on a normed space E and let Φ be the associated quadratic form. If φ is symmetric and bounded, then $\|\varphi\| = \|\Phi\|$.*

Proof: By Theorem 4.3.10, $\|\Phi\| \le \|\varphi\|$. We need to show that the opposite inequality holds as well. Since φ is symmetric, Φ is real, by Theorem 4.3.9. Then, by the polarization identity, we obtain

$$\operatorname{Re}\varphi(x,y) = \frac{1}{4}\left[\Phi(x+y) - \Phi(x-y)\right],$$

and hence

$$|\operatorname{Re}\varphi(x,y)| \le \frac{1}{4}\|\Phi\|\left(\|x+y\|^2 + \|x-y\|^2\right)$$

$$= \frac{1}{2}\|\Phi\|\left(\|x\|^2 + \|y\|^2\right),$$

by the parallelogram law. Let x and y be arbitrary fixed elements of E such that $\|x\| = \|y\| = 1$, and let θ be a complex number such that $|\theta| = 1$ and

$|\varphi(x, y)| = \theta \varphi(x, y)$. Then

$$\left|\varphi(x, y)\right| = \theta \varphi(x, y) = \varphi(\theta x, y) = \left|\text{Re}\, \varphi(\theta x, y)\right|$$

$$\leq \frac{1}{2} \|\Phi\| \left(\|\theta x\|^2 + \|y\|^2 \right) = \|\Phi\|,$$

and thus

$$\|\varphi\| = \sup_{\|x\|=\|y\|=1} \left|\varphi(x, y)\right| \leq \|\Phi\|. \qquad \square$$

Theorem 4.3.12. *Let A be a bounded operator on a Hilbert space H. Then the bilinear functional defined by $\varphi(x, y) = \langle Ax, y \rangle$ is bounded and $\|A\| = \|\varphi\|$.*

Proof: For all $x, y \in H$, by Schwarz's inequality, we have

$$\left|\varphi(x, y)\right| = \left|\langle Ax, y \rangle\right| \leq \|Ax\| \|y\| \leq \|A\| \|x\| \|y\|.$$

Thus, φ is bounded and $\|\varphi\| \leq \|A\|$. On the other hand, we have

$$\|Ax\|^2 = \left|\langle Ax, Ax \rangle\right| = \left|\varphi \langle x, Ax \rangle\right| \leq \|\varphi\| \|Ax\| \|Ax\|.$$

Therefore, for $Ax \neq 0$, we have

$$\|Ax\| \leq \|\varphi\| \|x\|.$$

Since the above inequality is trivially satisfied if $Ax = 0$, we obtain $\|A\| \leq \|\varphi\|$. $\qquad \square$

It turns out that every bounded bilinear functional on a Hilbert space is of the form considered in the above theorem. Note that the theorem remains true if $\varphi(x, y) = \langle Ax, y \rangle$ is replaced by $\varphi(x, y) = \langle x, Ay \rangle$. The same is true about the next theorem. Our reason for choosing $\langle Ax, y \rangle$ in Theorem 4.3.12 and $\langle x, Ay \rangle$ in the theorem that follows is that it is more convenient when adjoint operators are introduced.

Theorem 4.3.13. *Let φ be a bounded bilinear functional on a Hilbert space H. There exists a unique bounded operator A on H such that*

$$\varphi(x, y) = \langle x, Ay \rangle \quad \text{for all } x, y \in H.$$

Proof: For a fixed $y \in H$, $\varphi(x, y)$ is a bounded linear functional on H. Thus, by the Riesz representation theorem, there exists a unique element $Ay \in H$ such that $\varphi(x, y) = \langle x, Ay \rangle$ for all $x \in H$. We have to prove that the mapping $y \to Ay$ is a bounded operator on E. Indeed, for any $x, y_1, y_2 \in H$ and $\alpha, \beta \in \mathbb{C}$ we have

$$\langle x, A(\alpha y_1 + \beta y_2) \rangle = \varphi(x, \alpha y_1 + \beta y_2)$$
$$= \overline{\alpha}\varphi(x, y_1) + \overline{\beta}\varphi(x, y_2)$$
$$= \langle x, \alpha A y_1 + \beta A y_2 \rangle,$$

and thus

$$A(\alpha y_1 + \beta y_2) = \alpha A y_1 + \beta A y_2.$$

Now we show that A is bounded. Since φ is bounded, we have

$$\left| \langle x, Ay \rangle \right| = \left| \varphi(x, y) \right| \le k \|x\| \|y\|$$

for some $k > 0$ and all $x, y \in H$. In particular, for $x = Ay$ we obtain

$$\|Ay\|^2 = \left| \langle Ay, Ay \rangle \right| = \left| \varphi(Ay, y) \right| \le k \|Ay\| \|y\|.$$

Therefore, if $Ay \ne 0$, we get

$$\|Ay\| \le k \|y\|,$$

which is also trivially satisfied if $Ay = 0$. This proves that A is bounded. To prove uniqueness, notice that

$$\langle x, Ay \rangle = \langle x, By \rangle \quad \text{for all } x, y \in H$$

implies $A = B$.   □

Definition 4.3.14.   (Coercive functional)

A bilinear functional φ on a normed space E is called *coercive* (or *elliptic*) if there exists a positive constant K such that

$$\varphi(x, x) \ge K \|x\|^2 \quad \text{for all } x \in E.$$

Example 4.3.15. If z is a continuous real-valued function on $[0, 1]$ such that $\min_{t \in [0,1]} z(t) > 0$, then the bilinear functional φ defined on $L^2([0, 1])$ by

$$\varphi(x, y) = \int_0^1 x(t)\overline{y(t)}z(t)dt$$

is coercive. Indeed, we have

$$\varphi(x, x) = \int_0^1 \left| x(t) \right|^2 z(t)dt \ge K \|x\|^2$$

where $K = \min_{t \in [0,1]} z(t)$.   □

The following theorem, proved by P. Lax and A.N. Milgram in 1954, is an important generalization of the Riesz representation theorem. Applications of this theorem to boundary value problems are discussed in Chapter 6.

Theorem 4.3.16. (Lax–Milgram theorem) *Let φ be a bounded, coercive, bilinear functional on a Hilbert space H. For every bounded linear functional f on H, there exists a unique $x_f \in H$ such that*

$$f(x) = \varphi(x, x_f) \quad \text{for all } x \in H.$$

Proof: By Theorem 4.3.13, there exists a bounded operator A such that

$$\varphi(x, y) = \langle x, Ay \rangle \quad \text{for all } x, y \in H.$$

Since φ is coercive, we have

$$K\|x\|^2 \le \varphi(x, x) = \langle x, Ax \rangle \le \|Ax\| \|x\|,$$

and hence

$$K\|x\| \le \|Ax\| \quad \text{for all } x \in H.$$

Let $x_1, x_2 \in H$. If $Ax_1 = Ax_2$, then $A(x_1 - x_2) = 0$, and thus

$$\|x_1 - x_2\| \le \frac{1}{K} \|A(x_1 - x_2)\| = 0,$$

which implies $x_1 = x_2$. Therefore, A is one-to-one.

Denote the range of A by $\mathcal{R}(A)$. Let (x_n) be a sequence of elements of H. If $\|Ax_n - y\| \to 0$, for some $y \in H$, then

$$\|x_n - x_m\| \le \frac{1}{K} \|Ax_n - Ax_m\| \to 0,$$

as $m, n \to \infty$. Therefore (x_n) is a Cauchy sequence in H. Since H is complete, there exists $x \in H$ such that $\|x_n - x\| \to 0$. Hence $\|Ax_n - Ax\| \to 0$, because A is continuous. Consequently, $Ax = y$, and thus $y \in \mathcal{R}(A)$. This proves that $\mathcal{R}(A)$ is a closed subspace of H. We will prove that $\mathcal{R}(A) = H$.

Suppose $\mathcal{R}(A)$ is a proper subspace of H. Then, by Theorem 3.6.6, there exists a nonzero $x \in H$, which is orthogonal to $\mathcal{R}(A)$, that is,

$$\langle x, Ay \rangle = 0 \quad \text{for all } y \in H.$$

In particular, we have

$$0 = \left| \langle x, Ax \rangle \right| = \left| \varphi(x, x) \right| \ge K\|x\|^2,$$

which contradicts the assumption $x \neq 0$.

If f is a bounded linear functional on H, then there exists a unique $x_0 \in H$ such that

$$f(x) = \langle x, x_0 \rangle \quad \text{for all } x \in H.$$

Since A is a one-to-one mapping and $\mathcal{R}(A) = H$, there exists a unique $x_f \in H$ such that $x_0 = Ax_f$, and hence

$$f(x) = \langle x, Ax_f \rangle = \varphi(x, x_f) \quad \text{for all } x \in H. \qquad \square$$

4.4 Adjoint and Self-Adjoint Operators

Consider a bounded operator A on a Hilbert space H. Since the bilinear functional $\varphi(x, y) = \langle Ax, y \rangle$ is bounded (Theorem 4.3.12), there exists a unique bounded operator A^* such that $\langle Ax, y \rangle = \varphi(x, y) = \langle x, A^*y \rangle$ for all $x, y \in H$ (Theorem 4.3.13).

Definition 4.4.1.   (Adjoint operator)
Let A be a bounded operator on a Hilbert space H. The operator $A^* : H \to H$ defined by

$$\langle Ax, y \rangle = \langle x, A^*y \rangle \quad \text{for all } x, y \in H$$

is called the *adjoint operator* of A.

Adjoint operators, as just defined, are sometimes called *Hilbert-adjoint* to distinguish them from the more general definition considered later in this chapter.

The following properties are direct consequences of Definition 4.4.1:

$$(A + B)^* = A^* + B^*,$$
$$(\alpha A)^* = \overline{\alpha} A^*,$$
$$\left(A^*\right)^* = A,$$
$$\mathcal{I}^* = \mathcal{I},$$
$$(AB)^* = B^* A^*;$$

for arbitrary operators A and B and scalar α.

Theorem 4.4.2. *The adjoint operator A^* of a bounded operator A is bounded. Moreover, we have $\|A\| = \|A^*\|$ and $\|A^*A\| = \|A\|^2$.*

Proof: The argument presented before Definition 4.4.1 shows that the adjoint operator A^* of a bounded operator A is bounded. Since $\langle Ax, y \rangle$

and $\langle x, A^*y \rangle$ define the same bilinear functional, Theorem 4.3.12 gives us $\|A\| = \|A^*\|$.

It follows from Theorem 4.2.8 that

$$\|A^*A\| \le \|A^*\|\|A\| = \|A\|^2.$$

On the other hand, for every $x \in H$, we have

$$\|Ax\|^2 = \langle Ax, Ax \rangle = \langle A^*Ax, x \rangle \le \|A^*Ax\|\|x\| \le \|A^*A\|\|x\|^2.$$

Therefore $\|A^*A\| = \|A\|^2$. $\qquad\qquad\square$

Operators A and A^* need not be equal. For instance, let $H = \mathbb{C}^2$ and let A be defined by

$$A(z_1, z_2) = (0, z_1).$$

Then

$$\langle A(x_1, x_2), (y_1, y_2) \rangle = x_1 \overline{y_2} \quad \text{and} \quad \langle (x_1, x_2), A(y_1, y_2) \rangle = x_2 \overline{y_1}.$$

Operators for which $A = A^*$ are of special interest.

Definition 4.4.3. (Self-adjoint operator)
If $A = A^*$, then A is called *self-adjoint*.

In other words, if A is self-adjoint, then $\langle Ax, y \rangle = \langle x, Ay \rangle$ for all $x, y \in H$.

Example 4.4.4. Let $H = \mathbb{C}^N$ and let $\{e_1, \dots, e_N\}$ be the standard orthonormal base in H. Let A be an operator represented by matrix (a_{ij}), where $a_{ij} = \langle Ae_j, e_i \rangle$, (see Example 4.2.2). Then the adjoint operator A^* is represented by the matrix $b_{kj} = \langle A^*e_j, e_k \rangle$. Consequently

$$b_{kj} = \langle e_j, Ae_k \rangle = \overline{\langle Ae_k, e_j \rangle} = \overline{a_{jk}}.$$

Therefore, the operator A is self-adjoint if and only if $a_{ij} = \overline{a_{ji}}$. A matrix satisfying this condition is often called *Hermitian*. $\qquad\square$

Example 4.4.5. Let H be a separable, infinite dimensional Hilbert space and let $\{e_1, e_2, e_3, \dots\}$ be a complete orthonormal sequence in H. Let A be a bounded operator on H represented by an infinite matrix (a_{ij}) (see Theorem 4.2.9). As in the finite dimensional case, the adjoint operator A^* is represented by the infinite matrix $(\overline{a_{ji}})$. A is self-adjoint if and only if $a_{ij} = \overline{a_{ji}}$ for all $i, j \in \mathbb{N}$. $\qquad\square$

Example 4.4.6. Let T be a Fredholm operator on $L^2([a, b])$ defined by

$$(Tx)(s) = \int_a^b K(s, t)x(t)dt,$$

where K is a function defined on $[a, b] \times [a, b]$ such that

$$\int_a^b \int_a^b |K(s, t)|^2 dsdt < \infty.$$

Note that the condition is satisfied if K is continuous. We have

$$\langle Tx, y \rangle = \int_a^b \int_a^b K(s, t)x(t)\overline{y(s)}dsdt$$

$$= \overline{\int_a^b \int_a^b \overline{K(s, t)x(t)}y(s)dsdt}$$

$$= \overline{\int_a^b x(t)\int_a^b \overline{K(s, t)}y(s)dsdt}.$$

This shows that

$$(T^*x)(s) = \int_a^b \overline{K(t, s)}x(t)dt.$$

Thus, a Fredholm operator is self-adjoint if its kernel satisfies the equality $K(s, t) = \overline{K(t, s)}$. □

Example 4.4.7. Let A be the operator on $L^2([a, b])$ defined by $(Ax)(t) = tx(t)$. Since

$$\langle Ax, y \rangle = \int_a^b tx(t)\overline{y(t)}dt = \int_a^b x(t)\overline{ty(t)}dt = \langle x, Ay \rangle,$$

A is self-adjoint. □

Example 4.4.8. Consider the operator A defined on $L^2(\mathbb{R})$ defined by

$$(Ax)(t) = e^{-|t|}x(t).$$

This is a bounded self-adjoint operator. Boundedness of A can be shown as in Example 4.2.5. Moreover, we have

$$\langle Ax, y \rangle = \int_{-\infty}^{\infty} e^{-|t|}x(t)\overline{y(t)}dt = \int_{-\infty}^{\infty} x(t)\overline{[e^{-|t|}y(t)]}dt = \langle x, Ay \rangle.$$

Thus, A is self-adjoint. □

Now we prove some theorems that can be used to construct more examples of self-adjoint operators.

Theorem 4.4.9. *Let φ be a bounded bilinear functional on H and let A be an operator on H such that $\varphi(x, y) = \langle x, Ay \rangle$ for all $x, y \in H$. Then A is self-adjoint if and only if φ is symmetric.*

Proof: For all $x, y \in H$, we have

$$\langle x, Ay \rangle = \varphi(x, y) = \overline{\varphi(y, x)} = \overline{\langle y, Ax \rangle} = \langle Ax, y \rangle \quad \text{(if φ is symmetric)}$$

and

$$\varphi(x, y) = \langle x, Ay \rangle = \langle Ax, y \rangle = \overline{\langle y, Ax \rangle} = \overline{\varphi(y, x)} \quad \text{(if A is self-adjoint)}. \quad \square$$

Theorem 4.4.10. *Let A be a bounded operator on a Hilbert space H. The operators $T_1 = A^*A$ and $T_2 = A + A^*$ are self-adjoint.*

Proof: For all $x, y \in H$, we have

$$\langle T_1 x, y \rangle = \langle A^*Ax, y \rangle = \langle Ax, Ay \rangle = \langle x, A^*Ay \rangle = \langle x, T_1 y \rangle$$

and

$$\langle T_2 x, y \rangle = \langle (A + A^*)x, y \rangle = \langle x, (A + A^*)^* y \rangle = \langle x, (A + A^*)y \rangle = \langle x, T_2 y \rangle. \quad \square$$

Theorem 4.4.11. *The product of two self-adjoint operators is self-adjoint if and only if the operators commute.*

Proof: Let A and B be self-adjoint operators. Then

$$\langle ABx, y \rangle = \langle Bx, Ay \rangle = \langle x, BAy \rangle.$$

Thus, if $AB = BA$, then AB is self-adjoint. Conversely, if AB is self-adjoint, then the above implies $AB = (AB)^* = BA$. $\quad \square$

Corollary 4.4.12. *If A is self-adjoint, then so is any polynomial of A with real coefficients $\alpha_n, \ldots, \alpha_0$:*

$$\alpha_n A^n + \cdots + \alpha_1 A + \alpha_0 \mathcal{I}.$$

Theorem 4.4.13. *For every bounded operator T on a Hilbert space H, there exist unique self-adjoint operators A and B such that $T = A + iB$ and $T^* = A - iB$.*

Proof: Let T be a bounded operator on H. Define

$$A = \frac{1}{2}(T + T^*) \quad \text{and} \quad B = \frac{1}{2i}(T - T^*).$$

Clearly, A and B are self-adjoint, and $T = A + iB$. Moreover, for any $x, y \in H$, we have

$$\begin{aligned}
\langle Tx, y \rangle &= \langle (A + iB)x, y \rangle \\
&= \langle Ax, y \rangle + i \langle Bx, y \rangle \\
&= \langle x, Ay \rangle + i \langle x, By \rangle \\
&= \langle x, (A - iB)y \rangle.
\end{aligned}$$

Hence, $T^* = A - iB$. Proving the uniqueness is left as an exercise. □

In particular, if T is self-adjoint, then $A = T$ and $B = 0$. Self-adjoint operators are like real numbers in $\mathbb{C}$.

The following property of self-adjoint operators will be useful when we discuss spectral properties of such operators in Sections 4.9 and 4.10.

Theorem 4.4.14. *Let T be a self-adjoint operator on a Hilbert space H. Then*

$$\|T\| = \sup_{\|x\|=1} |\langle Tx, x \rangle|. \tag{4.4}$$

Proof: Let

$$M = \sup_{\|x\|=1} |\langle Tx, x \rangle|.$$

If $\|x\| = 1$, then

$$|\langle Tx, x \rangle| \le \|Tx\| \|x\| = \|Tx\| \le \|T\| \|x\| = \|T\|.$$

Thus,

$$M \le \|T\|. \tag{4.5}$$

On the other hand, for all $x, z \in H$, we have

$$\langle T(x+z), x+z \rangle - \langle T(x-z), x-z \rangle = 2 \big(\langle Tx, z \rangle + \langle Tz, x \rangle \big) = 4 \operatorname{Re} \langle Tx, z \rangle.$$

Therefore,

$$\operatorname{Re} \langle Tx, z \rangle \le \frac{M}{4} \big(\|x+z\|^2 + \|x-z\|^2 \big) = \frac{M}{2} \big(\|x\|^2 + \|z\|^2 \big). \tag{4.6}$$

Now suppose $\|x\| = 1$ and $Tx \ne 0$. If we let $z = \frac{Tx}{\|Tx\|}$, then

$$\operatorname{Re} \langle Tx, z \rangle = \operatorname{Re} \left\langle Tx, \frac{Tx}{\|Tx\|} \right\rangle = \|Tx\|,$$

and, by (4.6),

$$\operatorname{Re}\langle Tx, z\rangle \leq \frac{M}{2}\left(\|x\|^2 + \left\|\frac{Tx}{\|Tx\|}\right\|^2\right) = M.$$

Consequently, $\|T\| \leq M$. $\qquad\qquad\qquad\qquad\qquad\square$

4.5 Invertible, Normal, Isometric, and Unitary Operators

Definition 4.5.1. (Inverse operator)
Let A be an operator defined on a vector subspace of E. An operator B defined on $\mathcal{R}(A)$ is called the *inverse* of A if $ABx = x$ for all $x \in \mathcal{R}(A)$ and $BAx = x$ for all $x \in \mathcal{D}(A)$. An operator which has an inverse is called *invertible*. The inverse of A is denoted by A^{-1}.

If an operator has an inverse, then it is unique. Indeed, suppose B_1 and B_2 are inverses of A. Then

$$B_1 = B_1\mathcal{I} = B_1AB_2 = \mathcal{I}B_2 = B_2.$$

Note also that

$$\mathcal{D}(A^{-1}) = \mathcal{R}(A) \quad \text{and} \quad \mathcal{R}(A^{-1}) = \mathcal{D}(A).$$

In the next theorem, we recall some simple algebraic properties of invertible operators. Easy proofs are left as exercises.

Theorem 4.5.2.

(a) *The inverse of a linear operator is a linear operator.*

(b) *An operator A is invertible if and only if $Ax = 0$ implies $x = 0$.*

(c) *If an operator A is invertible and vectors $x_1, \ldots, x_n$ are linearly independent, then $Ax_1, \ldots, Ax_n$ are linearly independent.*

(d) *If operators A and B are invertible, then the operator AB is invertible and we have $(AB)^{-1} = B^{-1}A^{-1}$.*

It follows from part (c) in the above theorem that, if E is a finite dimensional vector space and A is a linear invertible operator on E, then $\mathcal{R}(A) = E$. As the following example shows, in infinite dimensional vector spaces it is not necessarily true.

Example 4.5.3. Let $E = l^2$. Define an operator A on E by

$$A(x_1, x_2, \ldots) = (0, x_1, x_2, \ldots).$$

Clearly, this is a linear invertible operator on l^2 whose range is a proper subspace of l^2. □

The next example shows that the inverse of a bounded operator is not necessarily bounded.

Example 4.5.4. Let $E = l^2$. Define an operator A on E by

$$A(x_1, x_2, \ldots) = \left(x_1, \frac{x_2}{2}, \frac{x_3}{3}, \ldots, \frac{x_n}{n}, \ldots \right).$$

Since

$$\left\| A(x_1, x_2, \ldots) \right\| = \sqrt{\sum_{n=1}^{\infty} \frac{|x_n|^2}{n^2}} \le \sqrt{\sum_{n=1}^{\infty} |x_n|^2} = \left\| (x_1, x_2, \ldots) \right\|,$$

A is a bounded operator. A is also invertible:

$$A^{-1}(x_1, x_2, \ldots) = (x_1, 2x_2, 3x_3, \ldots, nx_n, \ldots).$$

It is easy to see that A^{-1} is not bounded. □

If E is finite dimensional, then the inverse of any invertible operator on E is bounded, because every operator on a finite dimensional space is bounded.

Theorem 4.5.5. *Let A be a bounded operator on a Hilbert space H such that $\mathcal{R}(A) = H$. If A has a bounded inverse, then the adjoint operator A^* is invertible and $(A^*)^{-1} = (A^{-1})^*$.*

Proof: It suffices to show that

$$(A^{-1})^* A^* x = A^* (A^{-1})^* x = x \tag{4.7}$$

for every $x \in H$. Indeed, for any $y \in H$ we have

$$\langle y, (A^{-1})^* A^* x \rangle = \langle A^{-1} y, A^* x \rangle = \langle A A^{-1} y, x \rangle = \langle y, x \rangle$$

and

$$\langle y, A^* (A^{-1})^* x \rangle = \langle Ay, (A^{-1})^* x \rangle = \langle A^{-1} Ay, x \rangle = \langle y, x \rangle.$$

Thus,

$$\langle y, (A^{-1})^* A^* x \rangle = \langle y, (A^* (A^{-1})^* x \rangle = \langle y, x \rangle \quad \text{for all } y \in H,$$

which implies (4.7). □

Corollary 4.5.6. *If a bounded self-adjoint operator A has a bounded inverse A^{-1}, then A^{-1} is self-adjoint.*

Proof: $(A^{-1})^* = (A^*)^{-1} = A^{-1}$. □

The notion of adjoint operators leads to an important class of operators, called normal operators.

Definition 4.5.7. (Normal operator)
A bounded operator T is called a *normal* operator if it commutes with its adjoint, that is, $TT^* = T^*T$.

Note that T is normal if and only if T^* is normal. Obviously, every self-adjoint operator is normal. The following theorem will help us find examples of normal operators which are not self-adjoint.

Theorem 4.5.8. *A bounded operator T is normal if and only if $\|Tx\| = \|T^*x\|$ for all $x \in H$.*

Proof: For all $x \in H$, we have

$$\langle T^*Tx, x \rangle = \langle Tx, Tx \rangle = \|Tx\|^2.$$

If T is normal, then we also have

$$\langle T^*Tx, x \rangle = \langle TT^*x, x \rangle = \langle T^*x, T^*x \rangle = \|T^*x\|^2,$$

and thus $\|Tx\| = \|T^*x\|$.

Assume now that $\|Tx\| = \|T^*x\|$ for all $x \in H$. By the preceding argument, we have

$$\langle TT^*x, x \rangle = \langle T^*Tx, x \rangle \quad \text{for all } x \in H.$$

Therefore, by Corollary 4.3.8, $TT^* = T^*T$. □

Note that the condition $\|Tx\| = \|T^*x\|$ for all $x \in H$ is much stronger than $\|T\| = \|T^*\|$.

Example 4.5.9. Let H be a Hilbert space and let $Tx = ix$ for all $x \in H$. Since $T^*x = -ix = -Tx$, T is not self-adjoint. On the other hand, $\|Tx\| = \|T^*x\|$ for all $x \in H$, and thus T is normal. □

Theorem 4.5.10. *If A is normal, then $(\alpha I - A)$ is normal for any $\alpha \in \mathbb{C}$.*

Proof: Since $(\alpha I - A)^* = (\bar{\alpha} I - A^*)$, we have

$$(\alpha I - A)(\alpha I - A)^* = |\alpha|^2 I - \bar{\alpha} A - \alpha A^* + AA^* = (\alpha I - A)^*(\alpha I - A).\square$$

Theorem 4.5.11. *Let T be a bounded operator on a Hilbert space H and let A and B be self-adjoint operators on H such that $T = A + iB$. Then T is normal if and only if A and B commute.*

Proof: If $T = A + iB$, then $T^* = A - iB$, and we have

$$TT^* = (A + iB)(A - iB) = A^2 + B^2 - i(AB - BA) \qquad (4.8)$$

and

$$T^*T = (A - iB)(A + iB) = A^2 + B^2 + i(AB - BA). \qquad (4.9)$$

If T is normal, then $AB - BA = 0$, which proves that A and B commute.

On the other hand, if A and B commute, then (4.8) and (4.9) imply $T^*T = A^2 + B^2 = TT^*$. □

Theorem 4.5.12. *If T is a normal operator, then $\|T^n\| = \|T\|^n$ for all $n \in \mathbb{N}$.*

Proof: Note that Theorem 4.2.8 implies that $\|T^n\| \leq \|T\|^n$ for any bounded operator T. To show that $\|T^n\| \geq \|T\|^n$ we fix an x such that $\|x\| = 1$ and use induction to show that

$$\|T^n x\| \geq \|Tx\|^n \qquad (4.10)$$

for all $n \in \mathbb{N}$. Clearly (4.10) holds for $n = 1$. If $Tx = 0$, then the inequality is trivially satisfied for all $n \in \mathbb{N}$. Assume now $Tx \neq 0$ and that (4.10) holds for $n = 1, \dots, m$. First note that

$$\|T^2 x\| = \|T^* Tx\| \geq \langle T^* Tx, x \rangle = \|Tx\|^2 \qquad (4.11)$$

by Theorems 4.5.8 and 4.4.14. Now, from (4.11) and the inductive assumption, we obtain

$$\left\| T^{m+1} x \right\| = \|Tx\| \left\| T^m \frac{Tx}{\|Tx\|} \right\| \geq \|Tx\| \left\| T \frac{Tx}{\|Tx\|} \right\|^m$$
$$= \|Tx\|^{1-m} \left\| T^2 x \right\|^m \geq \|Tx\|^{1-m} \|Tx\|^{2m} = \|Tx\|^{m+1}. \qquad □$$

Definition 4.5.13.   (Isometric operator)
A bounded operator T on a Hilbert space H is called an *isometric operator* if $\|Tx\| = \|x\|$ for all $x \in H$.

Example 4.5.14. Let (e_n), $n \in \mathbb{N}$, be a complete orthonormal sequence in a Hilbert space H. There exists a unique operator A such that $Ae_n = e_{n+1}$ for all $n \in \mathbb{N}$. In fact, if $x = \sum_{n=1}^{\infty} \alpha_n e_n$, then $Ax = \sum_{n=1}^{\infty} \alpha_n e_{n+1}$. Clearly, A is linear and $\|Ax\|^2 = \sum_{n=1}^{\infty} |\alpha_n|^2 = \|x\|^2$. Therefore, A is an isometric operator. Operator A is called a *one-sided shift operator*. □

Theorem 4.5.15. *A bounded operator T on a Hilbert space H is isometric if and only if $T^*T = \mathcal{I}$ on H.*

Proof: If T is isometric, then for every $x \in H$ we have $\|Tx\|^2 = \|x\|^2$, and hence

$$\langle T^*Tx, x \rangle = \langle Tx, Tx \rangle = \|Tx\|^2 = \|x\|^2 = \langle x, x \rangle$$

for all $x \in H$. Thus, $T^*T - \mathcal{I}$, by Corollary 4.3.8. Similarly, if $T^*T = \mathcal{I}$, then

$$\|Tx\| = \sqrt{\langle Tx, Tx \rangle} = \sqrt{\langle T^*Tx, x \rangle} = \sqrt{\langle x, x \rangle} = \|x\|$$

for all $x \in H$. $\qquad \square$

Note that isometric operators "preserve inner product," that is, $\langle Tx, Ty \rangle = \langle x, y \rangle$ for all $x, y \in H$. In particular, $x \perp y$ if and only if $Tx \perp Ty$. An isometric operator is a Hilbert space isomorphism between H and $\mathcal{R}(T)$.

Definition 4.5.16. (Unitary operator)
A bounded operator T on a Hilbert space H is called a *unitary operator* if $T^*T = TT^* = \mathcal{I}$ on H.

In the preceding definition it is essential that the domain and the range of T is the entire space H. The following theorem is an immediate consequence of the definition.

Theorem 4.5.17. *An operator T is unitary if and only if it is invertible and $T^{-1} = T^*$.*

Example 4.5.18. Let H be the Hilbert space of all sequences of complex numbers $x = (\ldots, x_{-1}, x_0, x_1, \ldots)$ such that $\|x\| = \sum_{-\infty}^{\infty} |x_n|^2 < \infty$. The inner product is defined by

$$\langle x, y \rangle = \sum_{-\infty}^{\infty} x_n \overline{y_n}.$$

The operator T defined by $T(x_n) = (x_{n-1})$ is a unitary operator. Indeed, T is invertible and

$$\langle Tx, y \rangle = \sum_{-\infty}^{\infty} x_{n-1} \overline{y_n} = \sum_{-\infty}^{\infty} x_n \overline{y_{n+1}} = \langle x, T^{-1}y \rangle,$$

which implies $T^* = T^{-1}$. $\qquad \square$

Note that unitary operators on H can be defined as isometric operators whose range is the entire space H.

Example 4.5.19. Let $H = L^2([0, 1])$. The operator T on H defined by $(Tx)(t) = x(1 - t)$ is unitary. It is a one-to-one mapping of H onto H and $\|T(x)\| = \|x\|$ for all $x \in H$. □

A unitary operator is obviously normal, but a normal operator need not be unitary. To see that consider any self-adjoint operator A such that $\|A\| \neq 1$.

Theorem 4.5.20. *Let T be a unitary operator. Then T^{-1} and T^* are unitary.*

Proof: Note that

$$\left(T^{-1}\right)^* T^{-1} = T^{**} T^{-1} = TT^{-1} = \mathcal{I}.$$

Similarly, $T^{-1}(T^{-1})^* = \mathcal{I}$, and thus T^{-1} is unitary. Since $T^* = T^{-1}$, by Theorem 4.5.17, T^* is also unitary. □

4.6 Positive Operators

First we consider a partial order defined for self-adjoint operators. If A and B are self-adjoint operators on a Hilbert space H, then we write $A \geq B$ (or $B \leq A$) if $\langle Ax, x \rangle \geq \langle Bx, x \rangle$ for all $x \in H$. This relation has the following natural properties:

If $A \geq B$, then $-A \leq -B$;

If $A \geq B$ and $C \geq D$, then $A + C \geq B + D$;

If $A \geq 0$ and $\alpha \geq 0$ $(\alpha \in \mathbb{R})$, then $\alpha A \geq 0$;

If $A \geq B$ and $B \geq C$, then $A \geq C$.

Proofs of these properties are left as exercises.

Example 4.6.1. Let φ and ψ be non-negative continuous function on $[a, b]$ and let A and B be multiplication operators on $L^2([a, b])$ defined by $Ax = \varphi x$ and $Bx = \psi x$. If $\varphi(t) \geq \psi(t)$ for every $t \in [a, b]$, then $A \geq B$. In fact, for any $x \in L^2([a, b])$, we have

$$\langle Ax, x \rangle = \int_a^b \varphi(t)x(t)\overline{x(t)}dt = \int_a^b \varphi(t)\big|x(t)\big|^2 dt$$

$$\geq \int_a^b \psi(t)\big|x(t)\big|^2 dt = \int_a^b \psi(t)x(t)\overline{x(t)}dt = \langle Bx, x \rangle. \quad \square$$

Theorem 4.6.2. *If A is a self-adjoint operator on H and $\|A\| \leq 1$, then $A \leq \mathcal{I}$.*

Proof: If $\|A\| \leq 1$, then

$$\langle Ax, x \rangle \leq \|A\| \|x\|^2 \leq \langle x, x \rangle = \langle \mathcal{I}x, x \rangle$$

for all $x \in H$.   □

Corollary 4.6.3. *If A is a self-adjoint operator, then there exists $\alpha > 0$ ($\alpha \in \mathbb{R}$) such that $\alpha A \leq \mathcal{I}$.*

Definition 4.6.4. (Positive operator)
An operator A is called *positive* if it is self-adjoint and $\langle Ax, x \rangle \geq 0$ for all $x \in H$.

Clearly, operators A and B in Example 4.6.1 are positive. The next example is a variation of the same idea.

Example 4.6.5. Let K be a positive continuous function defined on $[a, b] \times [a, b]$. The integral operator T on $L^2([a, b])$ defined by

$$(Tx)(s) = \int_a^b K(s, t)x(t)dt$$

is positive. Indeed, we have

$$\langle Tx, x \rangle = \int_a^b \int_a^b K(s, t)x(t)\overline{x(t)}dtds = \int_a^b \int_a^b K(s, t)|x(t)|^2 dtds \geq 0$$

for all $x \in L^2([a, b])$.   □

Theorem 4.6.6. *For any bounded operator A on H, the operators A^*A and AA^* are positive.*

Proof: For any $x \in H$, we have

$$\langle A^*Ax, x \rangle = \langle Ax, Ax \rangle = \|Ax\|^2 \geq 0$$

and

$$\langle AA^*x, x \rangle = \langle A^*x, A^*x \rangle = \|A^*x\|^2 \geq 0.$$ □

Theorem 4.6.7. *If A is an invertible positive operator, then its inverse A^{-1} is positive.*

Proof: If $y \in \mathcal{D}(A^{-1})$, then $y = Ax$ for some $x \in H$, and then

$$\langle A^{-1}y, y \rangle = \langle A^{-1}Ax, Ax \rangle = \langle x, Ax \rangle \geq 0.$$ □

Example 4.6.8. The product of two positive operators is not necessarily positive. Indeed, consider operators on $\mathbb{C}^2$ defined by matrices

$$A = \begin{pmatrix} 1 & 0 \\ 0 & 0 \end{pmatrix} \quad \text{and} \quad B = \begin{pmatrix} 1 & 1 \\ 1 & 1 \end{pmatrix}.$$

It is easy to check that both A and B are positive operators, but the product AB is not. $\square$

Theorem 4.6.9. *Product of two commuting positive operators is a positive operator.*

Proof: Let A and B be commuting positive operators. Without loss of generality, we can assume that $A \neq 0$. Define a sequence of operators

$$A_1 = A/\|A\| \quad \text{and} \quad A_{n+1} = A_n - A_n^2 \quad \text{for } n = 1, 2, \ldots.$$

Note that operators A_n are self-adjoint and commuting. We will show, by induction, that

$$0 \leq A_n \leq \mathcal{I} \tag{4.12}$$

for all $n \in \mathbb{N}$.

For $n = 1$, (4.12) is satisfied by Theorem 4.6.2. Suppose now (4.12) holds for some $k \in \mathbb{N}$. Then

$$\langle A_k^2(\mathcal{I} - A_k)x, x \rangle = \langle A_k(\mathcal{I} - A_k)x, A_k x \rangle = \langle (\mathcal{I} - A_k)A_k x, A_k x \rangle \geq 0$$

and

$$\langle A_k(\mathcal{I} - A_k)^2 x, x \rangle = \langle A_k(\mathcal{I} - A_k)x, (\mathcal{I} - A_k)x \rangle \geq 0,$$

which means

$$A_k^2(\mathcal{I} - A_k) \geq 0 \quad \text{and} \quad A_k(\mathcal{I} - A_k)^2 \geq 0.$$

Consequently,

$$A_{k+1} = A_k^2(\mathcal{I} - A_k) + A_k(\mathcal{I} - A_k)^2 \geq 0$$

and

$$\mathcal{I} - A_{k+1} = (\mathcal{I} - A_k) + A_k^2 \geq 0.$$

This shows that (4.12) holds for $k + 1$, and thus for all $n \in \mathbb{N}$, by induction. We have

$$A_1 = A_1^2 + A_2 = A_1^2 + A_2^2 + A_3 = \cdots = \sum_{k=1}^{n} A_k^2 + A_{n+1}$$

and hence

$$\sum_{k=1}^{n} A_k^2 = A_1 - A_{n+1} \leq A_1.$$

Therefore

$$\sum_{k=1}^{n} \langle A_k x, A_k x \rangle < \langle A_1 x, x \rangle.$$

This shows that the series $\sum_{n=1}^{\infty} \|A_n x\|^2$ converges and $\|A_n x\| \to 0$. Moreover,

$$\left(\sum_{k=1}^{n} A_k^2 \right) x = A_1 x - A_{n+1} x \to A_1 x \quad \text{as } n \to \infty$$

or, equivalently,

$$\sum_{n=1}^{\infty} A_n^2 x = A_1 x.$$

Since B commutes with A_n for all $n \in \mathbb{N}$, we have

$$\langle ABx, x \rangle = \|A\| \langle BA_1 x, x \rangle = \|A\| \sum_{n=1}^{\infty} \langle BA_n^2 x, x \rangle$$

$$= \|A\| \sum_{n=1}^{\infty} \langle BA_n x, A_n x \rangle \geq 0. \qquad \square$$

Corollary 4.6.10. *Let A and B be self-adjoint operators. If $A \leq B$, then $AC \leq BC$ for every positive operator C that commutes with both A and B.*

The next theorem will be useful in studying properties of wavelets (see Chapter 8).

Theorem 4.6.11. *Let A be a positive operator on H such that*

$$\alpha \mathcal{I} \leq A \leq \beta \mathcal{I} \tag{4.13}$$

for some $0 < \alpha < \beta$. Then

(a) *A is invertible,*

(b) *$\mathcal{R}(A) = H$,*

(c) *$\frac{1}{\beta} \mathcal{I} \leq A^{-1} \leq \frac{1}{\alpha} \mathcal{I}$.*

Proof: First note that (4.13) is equivalent to

$$\alpha \|x\|^2 \leq \langle Ax, x \rangle \leq \beta \|x\|^2 \quad \text{for all } x \in H. \tag{4.14}$$

Thus, $Ax = 0$ implies $x = 0$, proving (a).

To prove (b) we will prove that the range $\mathcal{R}(A)$ is closed and that $\mathcal{R}(A)^{\perp} = \{0\}$, and then use Theorem 3.6.9.

Consider a sequence $y_n \in \mathcal{R}(A)$ such that $y_n \to y$ for some $y \in H$. Let $y_n = Ax_n$ for some $x_n \in H$. Since

$$\alpha \|x_n - x_m\|^2 \leq \langle A(x_n - x_m), x_n - x_m \rangle$$
$$= \langle Ax_n - Ax_m, x_n - x_m \rangle$$
$$= \langle y_n - y_m, x_n - x_m \rangle \leq \|y_n - y_m\| \|x_n - x_m\|,$$

we have

$$\alpha \|x_n - x_m\| \leq \|y_n - y_m\|,$$

which implies that the sequence (x_n) is a Cauchy sequence, because (y_n) is. Thus (x_n) has a limit in H, say $x_n \to x$. From continuity of A, we get

$$y_n = Ax_n \to Ax.$$

This implies $y = Ax$, proving that $\mathcal{R}(A)$ is closed.

Now assume that $\langle Ax, y \rangle = 0$ for all $x \in H$. But this means that we must have $\langle Ay, y \rangle = 0$, which implies $y = 0$ by (4.14). Therefore $\mathcal{R}(A)^{\perp} = \{0\}$.

Finally, if

$$\alpha \mathcal{I} \leq A \leq \beta \mathcal{I},$$

then by Theorem 4.6.7 and Corollary 4.6.10,

$$\alpha A^{-1} \leq AA^{-1} \leq \beta A^{-1}$$

and consequently,

$$A^{-1} \leq \frac{1}{\alpha}\mathcal{I} \quad \text{and} \quad \frac{1}{\beta}\mathcal{I} \leq A^{-1}$$

proving (c). $\square$

The following technical theorem will be used in the proof of Theorem 4.6.14.

Theorem 4.6.12. *Let $A_1 \leq A_2 \leq \cdots \leq A_n \leq \cdots$ be self-adjoint operators on H such that $A_nA_m = A_mA_n$ for all $m, n \in \mathbb{N}$. If B is a self-adjoint operator on H such that $A_nB = BA_n$ and $A_n \leq B$ for all $n \in \mathbb{N}$, then there exists a self-adjoint operator A such that*

$$\lim_{n \to \infty} A_nx = Ax \quad \text{for every } x \in H$$

and

$$A_n \leq A \leq B \quad \textit{for every } n \in \mathbb{N}.$$

Proof: Define $C_n = B - A_n$. Operators C_n commute with each other and

$$C_1 \geq C_2 \geq \cdots \geq 0.$$

By Theorem 4.6.9, for $n > m$, the operators

$$(C_m - C_n)C_m \quad \text{and} \quad C_n(C_m - C_n)$$

are positive. Hence,

$$\langle C_m^2 x, x \rangle \geq \langle C_m C_n x, x \rangle \geq \langle C_n^2 x, x \rangle,$$

for every $x \in H$. Since, for an arbitrary fixed $x \in H$, $(\langle C_k^2 x, x \rangle)$ is a nonincreasing sequence of non-negative numbers, it converges and thus

$$\lim_{m,n \to \infty} \langle C_m C_n x, x \rangle = \lim_{n \to \infty} \langle C_n^2 x, x \rangle.$$

Hence,

$$\| C_m x - C_n x \|^2 = \langle (C_m - C_n)^2 x, x \rangle = \langle C_m^2 x, x \rangle - 2 \langle C_m C_n x, x \rangle + \langle C_n^2 x, x \rangle \to 0,$$

as $m, n \to \infty$. Therefore $(C_n x)$ is a Cauchy sequence for every $x \in H$. Consequently, $(C_n x)$, and thus also $(A_n x)$, are convergent for every $x \in H$. It is easy to check that the operator A defined by $Ax = \lim_{n \to \infty} A_n x$ is self-adjoint and that $A_n \leq A \leq B$ for every $n \in \mathbb{N}$. $\qquad \square$

Definition 4.6.13. (Square root)
By a *square root of a positive operator* A we mean a self-adjoint operator B satisfying $B^2 = A$.

Theorem 4.6.14. *Every positive operator A has a unique positive square root B. Moreover, B commutes with every operator commuting with A.*

Proof: Let $A \geq 0$ and let $\alpha > 0$ ($\alpha \in \mathbb{R}$) be such that $\alpha^2 A \leq \mathcal{I}$. Define $T_0 = 0$ and

$$T_{n+1} = T_n + \frac{1}{2}(\alpha^2 A - T_n^2) \tag{4.15}$$

for $n = 0, 1, 2, \ldots$. Note that operators T_n are self-adjoint (as polynomials of A with real coefficients), and they commute with every operator commuting with A. In particular, $T_n T_m = T_m T_n$ for all m and n.

For every $n \in \mathbb{N}$, we have

$$\mathcal{I} - T_{n+1} = \frac{1}{2}(\mathcal{I} - T_n)^2 + \frac{1}{2}(\mathcal{I} - \alpha^2 A) \qquad (4.16)$$

and

$$T_{n+1} - T_n = \frac{1}{2}\big((\mathcal{I} - T_{n-1}) + (\mathcal{I} - T_n)\big)(T_n - T_{n-1}). \qquad (4.17)$$

In view of (4.16), we have $T_n \leq \mathcal{I}$ for all $n \in \mathbb{N}$. Moreover,

$$T_0 \leq T_1 \leq \cdots \leq T_n \leq \cdots.$$

Indeed,

$$T_1 = \frac{1}{2}\alpha^2 A \geq 0 = T_0,$$

and if $T_n - T_{n-1} \geq 0$, then $T_{n+1} - T_n \geq 0$, by (4.17).

By Theorem 4.6.12, the sequence (T_n) converges to a positive self-adjoint operator T. Letting $n \to \infty$ in (4.15) yields

$$T = T + \frac{1}{2}(\alpha^2 A - T^2),$$

that is,

$$\left(\frac{1}{\alpha}T\right)^2 = A.$$

Denote $B = T/\alpha$. The operator B is obviously positive. Since, for each $n \in \mathbb{N}$, T_n commutes with every operator commuting with A, so do T and B.

It remains to prove the uniqueness. Let C be a positive operator such that $C^2 = A$. Since C commutes with A, C commutes with B. Let $x \in H$ and let $y_0 = (B - C)x$. Then

$$\langle By_0, y_0 \rangle + \langle Cy_0, y_0 \rangle = \big\langle (B + C)y_0, y_0 \big\rangle$$
$$= \big\langle (B + C)(B - C)x, y_0 \big\rangle$$
$$= \big\langle (B^2 - C^2)x, y_0 \big\rangle = 0.$$

Since B and C are positive, we have $\langle By_0, y_0 \rangle = \langle Cy_0, y_0 \rangle = 0$. If D is a positive square root of B, then

$$\|Dy_0\|^2 = \langle D^2 y_0, y_0 \rangle = \langle By_0, y_0 \rangle = 0.$$

Hence $Dy_0 = 0$ and also $By_0 = D(Dy_0) = 0$. In a similar fashion, we can prove that $Cy_0 = 0$. Consequently,

$$\|Bx - Cx\|^2 = \langle (B - C)^2 x, x \rangle = \langle (B - C)y_0, x \rangle = 0$$

for arbitrary $x \in H$. This proves $B = C$.   □

Definition 4.6.15.   (Strictly positive operator)
A self-adjoint operator is called *strictly positive* or *positive definite* if $\langle Ax, x \rangle > 0$ for all $x \in H, x \neq 0$.

The operator T in Example 4.6.5 is strictly positive.

4.7 Projection Operators

Theorem 3.6.4 states that if S is a closed subspace of a Hilbert space H, then for every $x \in H$, there exists a unique element $y \in S$ such that $x = y + z$ and $z \in S^\perp$. Thus every closed subspace induces an operator on H, which assigns to x that unique y.

Definition 4.7.1.   (Orthogonal projection operator)
Let S be a closed subspace of a Hilbert space H. The operator P on H defined by

$$Px = y \quad \text{if } x = y + z, y \in S, \text{ and } z \in S^\perp, \tag{4.18}$$

is called the *orthogonal projection operator* onto S, or simply *projection* onto S. The vector y is called the projection of x onto S. Projection onto a subspace S will be usually denoted by P_S.

From the uniqueness of the decomposition $x = y + z$, it follows that projection operators are linear. The Pythagorean formula implies

$$\|Px\|^2 = \|y\|^2 = \|x\|^2 - \|z\|^2 \leq \|x\|^2.$$

Thus, projection operators are bounded and $\|P\| \leq 1$. The zero operator is a projection operator onto the zero subspace. If P_S is a nonzero projection operator, then $\|P_S\| = 1$, because for every $x \in S$, we have $P_S x = x$. The identity operator $\mathcal{I}$ is the projection operator onto the whole space H.

Example 4.7.2. Let S be a closed subspace of a Hilbert space H, and let $\{e_1, e_2, \ldots\}$ be a complete orthonormal system in S. Then the projection operator P_S can be defined by

$$P_S x = \sum_{n=1}^{\infty} \langle x, e_n \rangle e_n.$$

In particular, if S is of dimension 1 and $v \in S$, $\|v\| = 1$, then $P_S x = \langle x, v \rangle v$. $\quad\square$

Example 4.7.3. Let $H = L^2([-\pi, \pi])$. Every $x \in H$ can be represented as $x = y + z$, where y is an even function and z is an odd function. The operator defined by $Px = y$ is the projection operator onto the subspace of all even functions. This operator can be also defined as in Example 4.7.2:

$$Px = \sum_{n=0}^{\infty} \langle x, \varphi_n \rangle \varphi_n,$$

where $\varphi_0(t) = \frac{1}{\sqrt{2\pi}}$ and $\varphi_n(t) = \frac{1}{\sqrt{\pi}} \cos nt$ for $n = 1, 2, \ldots$. $\quad\square$

Example 4.7.4. Let $H = L^2([-\pi, \pi])$ and let P be an operator on H defined by

$$(Px)(t) = \begin{cases} 0 & \text{if } t \leq 0, \\ x(t) & \text{if } t > 0. \end{cases}$$

Then P is the projection operator onto the space of all functions that vanish for $t \leq 0$. $\quad\square$

Note that if P is a projection on H, then

$$\langle Px, y - Py \rangle = 0 \quad \text{for every } x, y \in H. \tag{4.19}$$

Definition 4.7.5.   (Idempotent operator)
An operator T is called *idempotent* if $T^2 = T$.

Every projection operator is idempotent. Indeed, if P is the projection operator onto a subspace S, then P is the identity operator on S. Since $Px \in S$ for every $x \in H$, we have $P^2 x = P(Px) = Px$ for all $x \in H$.

Example 4.7.6. Consider the operator T on $\mathbb{C}^2$ defined by $T(x, y) = (x - y, 0)$. Obviously, T is idempotent. On the other hand, since

$$\langle T(x, y), (x, y) - T(x, y) \rangle = x\bar{y} - |y|^2,$$

$T(x, y)$ need not be orthogonal to $(x, y) - T(x, y)$, and thus T is not a projection. $\quad\square$

Theorem 4.7.7. *A bounded operator is a projection if and only if it is idempotent and self-adjoint.*

Proof: Let P be a projection operator on H. We have already seen that every projection is idempotent. For any $x, y \in H$, we have

$$\langle Px, y \rangle = \langle Px, Py \rangle + \langle Px, y - Py \rangle = \langle Px, Py \rangle$$

$$= \langle Px, Py \rangle + \langle x - Px, Py \rangle = \langle x, Py \rangle,$$

by (4.19). Thus, P is self-adjoint.

Assume now that T is an idempotent and self-adjoint operator on H. Define

$$S = \{x \in H : Tx = x\}.$$

Since T is a bounded operator, S is a closed subspace of H. To prove that T is the projection onto S, we need to show that $Tx \in S$ and $x - Tx \in S^\perp$ for all $x \in H$. The first property follows immediately from the fact that T is idempotent. To prove the second one, note that for any $x \in H$ and $z \in S$, we have

$$\langle x - Tx, z \rangle = \langle x, z \rangle - \langle Tx, z \rangle = \langle x, z \rangle - \langle x, Tz \rangle = \langle x, z \rangle - \langle x, z \rangle = 0. \quad \square$$

Corollary 4.7.8. *If P is a projection operator on H, then $\langle Px, x \rangle = \|Px\|^2$ for all $x \in H$.*

Proof: From Theorem 4.7.7, we have

$$\langle Px, x \rangle = \langle PPx, x \rangle = \big(Px, P^*x\big) = \langle Px, Px \rangle = \|Px\|^2. \qquad \square$$

Example 4.7.9. If P_S is the projection operator onto a closed subspace S, then $\mathcal{I} - P_S$ is the projection operator onto $S^\perp$ and $S^\perp = \{x : P_S x = 0\}$. The operator $\mathcal{I} - P_S$ is sometimes denoted by $P_S^\perp$ and called the *complementary projection*. We thus have $P_S^\perp = P_{S^\perp}$. $\qquad \square$

In general, the sum of two projection operators is not a projection operator. For example, if P is a nonzero projection operator, then $P + P = 2P$ is not a projection operator, because $\|P + P\| = 2$.

Definition 4.7.10.   (Orthogonality of projection operators)
Two projection operators P and Q are called *orthogonal* if $PQ = 0$.

Note that for any two projection operators P and Q we have $PQ = P^*Q^* = (QP)^*$. Thus, $PQ = 0$ if and only if $QP = 0$.

Theorem 4.7.11. *Two projection operators P_R and P_S are orthogonal if and only if $R \perp S$.*

Proof: Assume $P_R P_S = 0$. If $x \in R$ and $y \in S$, then

$$\langle x, y \rangle = \langle P_R x, P_S y \rangle = \langle x, P_R P_S y \rangle = 0.$$

Hence $R \perp S$.

Now, assume $R \perp S$. If $x \in H$, then $P_S x \in S$, and thus $P_S x \perp R$. Hence, $P_R(P_S x) = 0$ for all $x \in H$. This means that P_R and P_S are orthogonal. □

Theorem 4.7.12. *The sum of two projection operators P_R and P_S is a projection operator if and only if $P_R P_S = 0$. In this case, $P_R + P_S = P_{R \oplus S}$.*

Proof: If $P = P_R + P_S$ is a projection operator, then

$$(P_R + P_S)^2 = P_R + P_S$$

or

$$P_R P_S + P_S P_R = 0.$$

Multiplication of this equality by P_R from the left yields

$$P_R P_S + P_R P_S P_R = 0. \tag{4.20}$$

Multiplication of the last equality by P_R from the right gives

$$2 P_R P_S P_R = 0. \tag{4.21}$$

Combining (4.20) and (4.21) yields $P_R P_S = 0$.

Assume now that $P_R P_S = 0$. Then also $P_S P_R = 0$, and thus

$$(P_R + P_S)^2 = P_R + P_S.$$

Therefore, P is idempotent. Since P is also self-adjoint, as a sum of two self-adjoint operators, it is a projection, by Theorem 4.7.7.

Let $P = P_R + P_S$. For any $x \in H$, we have

$$Px = P_R x + P_S x \in R \oplus S.$$

Moreover, if $x = x_1 + x_2$ with $x_1 \in R, x_2 \in S$, then

$$Px = P_R x + P_S x = x_1 + x_2 = x,$$

so P is the identity on $R \oplus S$. □

Theorem 4.7.13. *The product of two projection operators P_R and P_S is a projection operator if and only if P_R and P_S commute. In this case, $P_R P_S = P_{R \cap S}$.*

Proof: Assume that $P = P_R P_S$ is a projection. Then $P^* = P$ and thus

$$P_R P_S = (P_R P_S)^* = P_S^* P_R^* = P_S P_R.$$

Conversely, if $P_R P_S = P_S P_R$, then

$$(P_R P_S)^* = P_S^* P_R^* = P_S P_R = P_R P_S,$$

so $P = P_R P_S$ is self-adjoint. Moreover,

$$P^2 = P_R P_S P_R P_S = P_R^2 P_S^2 = P_R P_S = P,$$

so P is idempotent. Thus, P is a projection.

For $x \in H$, we have $Px = P_R(P_S x) = P_S(P_R x)$, and hence $Px \in R \cap S$. Moreover, for $x \in R \cap S$, we have $Px = P_R(P_S x) = P_R x = x$. Consequently, $P = P_{R \cap S}$. $\qquad \square$

Example 4.7.14. The operators defined by matrices

$$A = \begin{pmatrix} 1 & 0 \\ 0 & 0 \end{pmatrix} \quad \text{and} \quad B = \begin{pmatrix} 1/2 & 1/2 \\ 1/2 & 1/2 \end{pmatrix}$$

are projections in $\mathbb{C}^2$. It is easy to check that AB is not a projection. $\qquad \square$

Theorem 4.7.15. *Let R and S be two closed subspaces of a Hilbert space H, and let P_R and P_S be the respective projections. The following conditions are equivalent:*

(a) $R \subset S$;

(b) $P_S P_R = P_R$;

(c) $P_R P_S = P_R$;

(d) $\|P_R x\| \le \|P_S x\|$ *for all $x \in H$.*

Proof: Assume $R \subset S$. Then $P_R x \in S$ for all $x \in H$. Consequently, $P_S P_R x = P_R x$, and thus (a) implies (b).

If $P_S P_R = P_R$, then

$$P_R = P_R^* = (P_S P_R)^* = P_R^* P_S^* = P_R P_S,$$

so (b) implies (c).

Assume now $P_R P_S = P_R$. Then

$$\|P_R x\| = \|P_R P_S x\| \le \|P_R\| \|P_S x\| \le \|P_S x\|,$$

for all $x \in H$. Thus (c) implies (d).

Finally, suppose (d) holds and (a) does not. Then there exists $x \in R$ such that $x \notin S$. Let $x = y + z$ with $y \in S$ and $z \in S^{\perp}$. Since $x \notin S$, we have $z \neq 0$ and

$$\|P_R x\|^2 = \|y\|^2 + \|z\|^2 > \|y\|^2 = \|P_S x\|^2,$$

which contradicts (d). Thus, $R \subset S$. $\qquad \square$

4.8 Compact Operators

Compact operators constitute an important class of bounded operators. The concept originated from the theory of integral equations of the second kind. They also provide a natural generalization of operators with finite-dimensional range.

Definition 4.8.1. (Compact operator)
An operator A on a Hilbert space H is called a *compact operator* (or *completely continuous operator*) if, for every bounded sequence (x_n) in H, the sequence (Ax_n) contains a convergent subsequence.

Example 4.8.2. Every operator on a finite dimensional Hilbert space is compact. Indeed, if A is an operator on $\mathbb{C}^N$, then it is bounded. Therefore, if (x_n) is a bounded sequence, then (Ax_n) is a bounded sequence in $\mathbb{C}^N$. By the Bolzano–Weierstrass theorem, (Ax_n) contains a convergent subsequence. $\square$

Example 4.8.3. Let y and z be fixed elements of a Hilbert space H. Define

$$Tx = \langle x, y \rangle z.$$

Let (x_n) be a bounded sequence, that is, $\|x_n\| \leq M$ for some $M > 0$ and all $n \in \mathbb{N}$. Since

$$\left|\langle x_n, y \rangle\right| \leq \|x_n\|\|y\| \leq M\|y\|,$$

the sequence $(\langle x_n, y \rangle)$ contains a convergent subsequence $(\langle x_{p_n}, y \rangle)$. Denote the limit of that subsequence by α. Then

$$Tx_{p_n} = \langle x_{p_n}, y \rangle z \to \alpha z \quad \text{as } n \to \infty.$$

Therefore, T is a compact operator. $\square$

Example 4.8.4. Important examples of compact operators are integral operators T on $L^2([a, b])$ defined by

$$(Tx)(s) = \int_a^b K(s, t)x(t)dt,$$

where a and b are finite and K is continuous. We will sketch the proof of compactness of such an operator.

Let $x_n \in L^2([a, b])$ and $\|x_n\| \leq M$ for $n = 1, 2, \ldots$ and some $M > 0$. Then, by Schwarz's inequality, we have

$$\left|(Tx_n)(s)\right| \leq \int_a^b \left|K(s, t)x_n(t)\right| dt \leq M \max\left|K(s, t)\right|\sqrt{b - a}.$$

Thus, the sequence of functions (Tx_n) is uniformly bounded. Moreover, for every $s_1, s_2 \in [a, b]$, we have

$$\left|(Tx_n)(s_1) - (Tx_n)(s_2)\right| \leq \int_a^b \left|K(s_1, t) - K(s_2, t)\right|\left|x_n(t)\right|dt$$

$$\leq \sqrt{\int_a^b \left|K(s_1, t) - K(s_2, t)\right|^2 dt} \sqrt{\int_a^b \left|x_n(t)\right|^2 dt}$$

$$\leq M\sqrt{b - a} \max_{t \in (a,b)} \left|K(s_1, t) - K(s_2, t)\right|.$$

Since K is uniformly continuous, the last inequality implies that the sequence (Tx_n) is equicontinuous. Therefore, by Arzela's theorem (see, for example, Friedman (1982)), (Tx_n) contains a uniformly convergent subsequence. This proves that T is compact, because uniform convergence on $[a, b]$ implies convergence in $L^2([a, b])$.   □

Example 4.8.5. Let S be a finite dimensional subspace of a Hilbert space H. The projection operator P_S is a compact operator.   □

Theorem 4.8.6. *Compact operators are bounded.*

Proof: If an operator A is not bounded, then there exists a sequence (x_n) such that $\|x_n\| = 1$, for all $n \in \mathbb{N}$, and $\|Ax_n\| \to \infty$. Then (Ax_n) does not contain a convergent subsequence, which means that A is not compact.   □

Not every bounded operator is compact.

Example 4.8.7. The identity operator $\mathcal{I}$ on an infinite dimensional Hilbert space H is not compact, although it is bounded. In fact, consider an orthonormal sequence (e_n) in H. Then the sequence $\mathcal{I}e_n = e_n$ does not contain a convergent subsequence.   □

Theorem 4.8.8. *The collection of all compact operators on a Hilbert space H is a vector space.*

The easy proof is left as an exercise.

Theorem 4.8.9. *Let A be a compact operator on a Hilbert space H, and let B be a bounded operator on H. Then AB and BA are compact.*

Proof: Let (x_n) be a bounded sequence in H. Since B is bounded, the sequence (Bx_n) is bounded. Next, since A is compact, the sequence (ABx_n) contains a convergent subsequence, which means that the operator AB is compact. Similarly, since A is compact, the sequence (Ax_n) contains a convergent subsequence (Ax_{p_n}). Now, since B is bounded (and thus continuous), the sequence (BAx_{p_n}) converges. Therefore, the operator BA is compact. $\qquad\square$

Note that the operators in Examples 4.8.2, 4.8.3, and 4.8.5 have finite dimensional ranges.

Definition 4.8.10.   (Finite-dimensional operator)
An operator is called *finite-dimensional* (or a *finite rank operator*) if its range is of finite dimension.

Theorem 4.8.11. *Finite-dimensional bounded operators are compact.*

Proof: Let A be a finite dimensional bounded operator, and let $\{z_1, \ldots, z_k\}$ be an orthonormal basis of the range of A. Define

$$T_n x = \langle Ax, z_n \rangle z_n$$

for $n = 1, \ldots, k$. Since

$$T_n x = \langle Ax, z_n \rangle z_n = \langle x, A^* z_n \rangle z_n,$$

the operators T_n are compact as proved in Example 4.8.3. Since

$$A = \sum_{n=1}^{k} T_n,$$

A is compact, by Theorem 4.8.8. $\qquad\square$

The range of the integral operator considered in Example 4.8.4 is infinite dimensional (unless $K = 0$).

Recall that the space $\mathcal{B}(H, H)$ of all bounded operators on a Hilbert space H is a Banach space with respect to the operator norm (see Theorem 1.5.9). It turns out that the space of all compact operators is a closed subspace of $\mathcal{B}(H, H)$. As we will see later, this has important consequences.

Theorem 4.8.12. *The limit of a uniformly convergent sequence of compact operators is compact. More precisely, if $T_1, T_2, \ldots$ are compact operators on a Hilbert space H and $\|T_n - T\| \to 0$ as $n \to \infty$ for some operator T on H, then T is compact.*

Proof: Let (x_n) be a bounded sequence in H. Since T_1 is compact, there exists a subsequence $(x_{1,n})$ of (x_n) such that $(T_1 x_{1,n})$ is convergent. Similarly, the sequence $(T_2 x_{1,n})$ contains a convergent subsequence $(T_2 x_{2,n})$. In general, for $k \geq 2$, let $(x_{k,n})$ be a subsequence of $(x_{k-1,n})$ such that $(T_k x_{k,n})$ is convergent. Consider the sequence $(x_{n,n})$. Since it is a subsequence of (x_n), we can put $x_{p_n} = x_{n,n}$, where (p_n) is an increasing sequence of positive integers. Obviously, the sequence $(T_k x_{p_n})$ converges for every $k \in \mathbb{N}$. We will show that the sequence $(T x_{p_n})$ converges too.

Let $\varepsilon > 0$. Since $\|T_n - T\| \to 0$, there exists $k \in \mathbb{N}$ such that $\|T_k - T\| < \frac{\varepsilon}{3M}$, where M is a constant such that $\|x_n\| \le M$ for all $n \in \mathbb{N}$. Next, let $k_1 \in \mathbb{N}$ be such that

$$\|T_k x_{p_n} - T_k x_{p_m}\| < \frac{\varepsilon}{3}$$

for all $n, m > k_1$. Then

$$\|T x_{p_n} - T x_{p_m}\| \le \|T x_{p_n} - T_k x_{p_n}\| + \|T_k x_{p_n} - T_k x_{p_m}\| + \|T_k x_{p_m} - T x_{p_m}\|$$
$$< \frac{\varepsilon}{3} + \frac{\varepsilon}{3} + \frac{\varepsilon}{3} = \varepsilon,$$

for sufficiently large n and m. Thus, $(T x_{p_n})$ is a Cauchy sequence in H. Completeness of H implies that $(T x_{p_n})$ converges. $\square$

Corollary 4.8.13. *The limit of a convergent sequence of finite-dimensional operators is a compact operator.*

Theorem 4.8.14. *The adjoint of a compact operator is compact.*

Proof: Let T be a compact operator on a Hilbert space H and let (x_n) be a bounded sequence in H, that is, $\|x_n\| \le M$ for some M for all $n \in \mathbb{N}$. Define $y_n = T^* x_n$, $n = 1, 2, \ldots$. Since T^* is bounded, the sequence (y_n) is bounded. It thus contains a subsequence (y_{k_n}) such that the sequence $(T y_{k_n})$ converges in H. Now, for any $m, n \in \mathbb{N}$, we have

$$\|y_{k_m} - y_{k_n}\|^2 = \left\| T^* x_{k_m} - T^* x_{k_n} \right\|^2$$
$$= \left\langle T^*(x_{k_m} - x_{k_n}), T^*(x_{k_m} - x_{k_n}) \right\rangle$$
$$= \left\langle T T^*(x_{k_m} - x_{k_n}), (x_{k_m} - x_{k_n}) \right\rangle$$
$$\le \left\| T T^*(x_{k_m} - x_{k_n}) \right\| \|x_{k_m} - x_{k_n}\|$$
$$\le 2M \| T y_{k_m} - T y_{k_n} \| \to 0,$$

as $m, n \to \infty$. Therefore (y_{k_n}) is a Cauchy sequence in H, which implies that (y_{k_n}) converges. This proves that T^* is a compact operator. $\square$

In the next theorem, we characterize compactness of operators in terms of weakly convergent sequences. Recall that we write "$x_n \to x$" to denote strong convergence and "$x_n \xrightarrow{w} x$" to denote weak convergence.

Theorem 4.8.15. *An operator T on a Hilbert space H is compact if and only if it maps weakly convergent sequences into strongly convergent sequences. More precisely, T is compact if and only if $x_n \xrightarrow{w} x$ implies $T x_n \to T x$ for any $x_n, x \in H$.*

Proof: Let T be a compact operator. Assume that $x_n \overset{w}{\to} x$ and suppose that $Tx_n \not\to Tx$. Then there exists $\varepsilon > 0$ and a subsequence (x_{p_n}) of (x_n) such that

$$\| Tx_{p_n} - Tx \| > \varepsilon \tag{4.22}$$

for all $n \in \mathbb{N}$. Since the sequence (x_{p_n}) is weakly convergent, it is bounded, by Theorem 3.3.14. Compactness of T implies that the sequence (Tx_{p_n}) has a strongly convergent subsequence (Tx_{q_n}). On the other hand, for every $y \in H$, we have

$$\langle Tx_n, y \rangle = \langle x_n, T^*y \rangle \to \langle x, T^*y \rangle = \langle Tx, y \rangle,$$

so that $Tx_n \overset{w}{\to} Tx$, as well as $Tx_{q_n} \overset{w}{\to} Tx$. Since we already know that the sequence (Tx_{q_n}) is strongly convergent, Theorem 3.3.11 implies $Tx_{q_n} \to Tx$. But this contradicts (4.22).

Assume now that T is such that $Tx_n \to Tx$ whenever $x_n \overset{w}{\to} x$. Let (z_n) be an arbitrary bounded sequence in H. We want to show that (Tz_n) has a convergent subsequence. Let (e_n) be a complete orthonormal sequence in H, and let M be a constant such that $\|z_n\| \le M$ for all $n \in \mathbb{N}$. Since

$$\left| \langle z_n, e_1 \rangle \right| \le M,$$

for all $n \in \mathbb{N}$, the sequence (z_n) has a subsequence $(z_{1,n})$ such that the sequence $(\langle z_{1,n}, e_1 \rangle)$ converges. Similarly, since

$$\left| \langle z_{1,n}, e_2 \rangle \right| \le M,$$

for all $n \in \mathbb{N}$, the sequence $(z_{1,n})$ has a subsequence $(z_{2,n})$ such that the sequence $(\langle z_{2,n}, e_2 \rangle)$ converges. Continuing this procedure, we construct sequences $(z_{m,n})$, $m = 1, 2, 3, \ldots$, such that $(z_{m+1,n})$ is a subsequence of $(z_{m,n})$ for every $m \in \mathbb{N}$, and the limit $\lim_{n\to\infty} \langle z_{m,n}, e_m \rangle$ exists for every $m \in \mathbb{N}$.

Now define

$$x_n = z_{n,n}, \quad n = 1, 2, \ldots.$$

Clearly, (x_n) is a subsequence of (z_n) and the limit $\lim_{n\to\infty} \langle x_n, e_m \rangle$ exists for every $m \in \mathbb{N}$. We will show that (x_n) is weakly convergent. Define

$$\alpha_k = \lim_{n \to \infty} \langle x_n, e_k \rangle, \quad k = 1, 2, \ldots.$$

For any $l, n \in \mathbb{N}$ we have

$$\sum_{k=1}^{l} \left| \langle x_n, e_k \rangle \right|^2 \le \sum_{k=1}^{\infty} \left| \langle x_n, e_k \rangle \right|^2 = \|x_n\|^2 \le M^2.$$

By first letting $n \to \infty$, we obtain

$$\sum_{k=1}^{l} |\alpha_k|^2 \le M^2,$$

and then, by letting $l \to \infty$,

$$\sum_{k=1}^{\infty} |\alpha_k|^2 \le M^2.$$

Define

$$z = \sum_{k=1}^{\infty} \alpha_k e_k.$$

Then, for every $m \in \mathbb{N}$,

$$\langle x_n, e_m \rangle - \langle z, e_m \rangle = \langle x_n, e_m \rangle - \left\langle \sum_{k=1}^{\infty} \alpha_k e_k, e_m \right\rangle$$

$$= \langle x_n, e_m \rangle - \langle \alpha_m e_m, e_m \rangle$$

$$= \langle x_n, e_m \rangle - \alpha_m \to 0, \quad \text{as } n \to \infty.$$

Thus, $\langle x_n, e_m \rangle \to \langle z, e_m \rangle$ as $n \to \infty$ for every $m \in \mathbb{N}$. Since span$\{e_1, e_2, \ldots\}$ is dense in H, it follows that $x_n \overset{w}{\to} z$, by Theorem 3.3.14. Consequently, $Tx_n \to Tz$. □

Corollary 4.8.16. *Compact operators map orthonormal sequences into sequences strongly convergent to 0.*

Proof: Orthonormal sequences are weakly convergent to 0. □

Note that, from the preceding theorem it follows that the inverse of a compact operator on an infinite-dimensional Hilbert space, if it exists, is unbounded.

It has already been noted that compactness of operators is a stronger condition than boundedness. For operators, boundedness is equivalent to continuity: bounded operators are exactly those operators which map strongly convergent sequences into strongly convergent sequences. Theorem 4.8.15 says that compact operators on a Hilbert space can be characterized as those operators which map weakly convergent sequences into strongly convergent sequences. From this point of view, compactness of operators is a stronger type of continuity. For this reason, compact operators are called sometimes *completely continuous operators*. The condition has been used by F. Riesz as the definition of compact operators. Hilbert has used still another (equivalent) definition of compact operators: an operator A defined on a Hilbert space H is compact if $x_n \to x$ weakly and $y_n \to y$ weakly implies $\langle Ax_n, y_n \rangle \to \langle Ax, y \rangle$.

4.9 Eigenvalues and Eigenvectors

Concepts discussed in this section play a central role in the theory of operators and their applications. First we need to introduce a number of new ideas.

Definition 4.9.1.   (Eigenvalue)
Let A be an operator on a complex vector space E. A complex number λ is called an *eigenvalue* of A if there is a nonzero vector $u \in E$ such that

$$Au = \lambda u. \tag{4.23}$$

Every vector u satisfying (4.23) is called an *eigenvector* of A corresponding to the eigenvalue λ. If E is a function space, eigenvectors are often called *eigenfunctions*.

Example 4.9.2. Let S be a linear subspace of an inner product space E and let P_S be the projection on S. The only eigenvalues of P_S are 0 and 1. Indeed, if $P_S u = \lambda u$, for some $\lambda \in \mathbb{C}$ and $0 \neq u \in E$, then

$$\lambda u = \lambda^2 u,$$

because $P_S^2 = P_S$. Therefore $\lambda = 0$ or $\lambda = 1$. The eigenvectors corresponding to 0 are the vectors of E, which are orthogonal to S. The eigenvectors corresponding to 1 are all elements of S. □

It is important to note that every eigenvector corresponds to exactly one eigenvalue, but there is always infinitely many eigenvectors corresponding to one eigenvalue. Indeed, every multiple of an eigenvector is an eigenvector. Moreover, several linearly independent vectors may correspond to the same eigenvalue. We have the following simple theorem.

Theorem 4.9.3. *The collection of all eigenvectors corresponding to an eigenvalue of an operator is a vector space.*

The easy proof is left as an exercise.

Definition 4.9.4.   (Eigenspace)
The set of all eigenvectors corresponding to an eigenvalue λ is called the *eigenspace* of λ. The dimension of that space is called the *multiplicity* of λ. An eigenvalue

of multiplicity one is called *simple* or *nondegenerate*. An eigenvalue of multiplicity greater than one is called *multiple* or *degenerate*. In this case the dimension of the eigenspace is also called the *degree of degeneracy*.

Example 4.9.5. Consider the integral operator A on $L^2([0, 2\pi])$ defined by

$$(Au)(t) = \int_0^{2\pi} \cos(t - y)u(y)dy. \tag{4.24}$$

We will show that A has exactly one nonzero eigenvalue $\lambda = \pi$, and its eigenfunctions are

$$u(t) = a\cos t + b\sin t$$

with arbitrary a and b.

The eigenvalue equation is

$$(Au)(t) = \int_0^{2\pi} \cos(t - y)u(y)dy = \lambda u(t)$$

or

$$\cos t \int_0^{2\pi} \cos y u(y)dy + \sin t \int_0^{2\pi} \sin y u(y)dy = \lambda u(t). \tag{4.25}$$

This means that, for $\lambda \neq 0$, u is a linear combination of cosine and sine functions, that is,

$$u(t) = a\cos t + b\sin t, \tag{4.26}$$

where $a, b \in \mathbb{C}$. Substituting this into (4.25), we obtain

$$\pi a = \lambda a \quad \text{and} \quad \pi b = \lambda b. \tag{4.27}$$

Hence, $\lambda = \pi$, which means that A has exactly one nonzero eigenvalue and its eigenfunctions are given by (4.26). This is a two-dimensional eigenspace, so the multiplicity of the eigenvalue is 2.

Equation (4.25) reveals that $\lambda = 0$ is also an eigenvalue of A. The corresponding eigenfunctions are all functions orthogonal to $\cos t$ and $\sin t$. Therefore, $\lambda = 0$ is an eigenvalue of infinite multiplicity. $\qquad\square$

Note that, if λ is not an eigenvalue of A, then the operator $A - \lambda\mathcal{I}$ is invertible, and conversely. If space E is finite dimensional and λ is not an eigenvalue of A, then the operator $(A - \lambda\mathcal{I})^{-1}$ is bounded, because all operators on a finite dimensional space are bounded. The situation for infinite dimensional spaces is more complicated.

Definition 4.9.6.   (Resolvent and spectrum)
Let A be an operator on a normed space E. The operator

$$A_\lambda = (A - \lambda \mathcal{I})^{-1}$$

is called the *resolvent* of A. The values of λ for which A_λ is defined on the whole space E and is bounded are called *regular values* of A. The set of all regular values of A is called the *resolvent set* and is denoted by $\rho(A)$. The complement of $\rho(A)$ in $\mathbb{C}$ is called the *spectrum* of A and is denoted by $\sigma(A)$. The *spectral radius* of A, denoted by $r(A)$, is defined by

$$r(A) = \sup\{|\lambda|: \lambda \in \sigma(A)\}.$$

Every eigenvalue belongs to the spectrum. The following example shows that the spectrum may contain points that are not eigenvalues. In fact, a non-empty spectrum may contain no eigenvalues at all.

Example 4.9.7. Let E be the space $\mathcal{C}([a, b])$ of continuous functions on the interval $[a, b]$. For a fixed $u \in \mathcal{C}([a, b])$, consider the operator A defined by

$$(Ax)(t) = u(t)x(t).$$

Since

$$(A - \lambda \mathcal{I})^{-1}x(t) = \frac{x(t)}{u(t) - \lambda},$$

the spectrum of A consists of all λ's such that $\lambda - u(t) = 0$ for some $t \in [a, b]$. This means that the spectrum of A is exactly the range of u. If $u(t) = c$ is a constant function, then $\lambda = c$ is an eigenvalue of A. On the other hand, if u is a strictly increasing function, then A has no eigenvalues. The spectrum of A in such a case is the interval $\sigma(A) = [u(a), u(b)]$.   □

The problem of finding eigenvalues and eigenvectors is called the *eigenvalue problem*. One of the main sources of eigenvalue problems in mechanics is the theory of oscillating systems. The state of a given system at a given time t may be represented by an element $u(t) \in H$, where H is an appropriate Hilbert space of functions. The equation of motion in classical mechanics is

$$\frac{d^2u}{dt^2} = Au, \tag{4.28}$$

where A is an operator in H. If the system oscillates, the time dependence of u is sinusoidal, so that $u(t) = v \sin \omega t$, where v is a fixed element of H. If A is linear, then (4.28) becomes

$$Av = (-\omega^2)v. \tag{4.29}$$

This means that $-\omega^2$ is an eigenvalue of A.

Theorem 4.9.8. *If A is a bounded linear operator in a Banach space E and $\|A\| < |\lambda|$, then $\Lambda_\lambda = (\Lambda - \lambda\mathcal{I})^{-1}$ is a bounded operator,*

$$A_\lambda = -\sum_{n=0}^{\infty} \frac{A^n}{\lambda^{n+1}}, \tag{4.30}$$

and

$$\|A_\lambda\| \leq \frac{1}{|\lambda| - \|A\|}. \tag{4.31}$$

Proof: Since $\|A/\lambda\| < 1$, we have

$$\sum_{n=0}^{\infty} \left\| \frac{A^n}{\lambda^n} \right\| \leq \sum_{n=0}^{\infty} \left\| \frac{A}{\lambda} \right\|^n < \infty.$$

Therefore, by the completeness of $\mathcal{B}(E, E)$, there exists a bounded linear operator B on E such that

$$B = \sum_{n=0}^{\infty} \frac{A^n}{\lambda^n}.$$

Moreover,

$$(A - \lambda\mathcal{I})B = (A - \lambda\mathcal{I})\left(\sum_{n=0}^{\infty} \frac{A^n}{\lambda^n} \right) = \sum_{n=0}^{\infty} (A - \lambda\mathcal{I})\frac{A^n}{\lambda^n}$$

$$= \sum_{n=0}^{\infty} \frac{A^{n+1} - \lambda A^n}{\lambda^n} = \lambda \sum_{n=0}^{\infty} \left(\frac{A^{n+1}}{\lambda^{n+1}} - \frac{A^n}{\lambda^n} \right) = -\lambda\mathcal{I}.$$

Similarly, $B(A - \lambda\mathcal{I}) = -\lambda\mathcal{I}$. Thus,

$$A_\lambda = (A - \lambda\mathcal{I})^{-1} = -\frac{B}{\lambda} = -\sum_{n=0}^{\infty} \frac{A^n}{\lambda^{n+1}}.$$

To prove (4.31), we observe that

$$\|A_\lambda\| \leq \frac{1}{|\lambda|} \sum_{n=0}^{\infty} \left\| \frac{A}{\lambda} \right\|^n = \frac{1}{|\lambda|} \frac{1}{1 - \|A/\lambda\|} = \frac{1}{|\lambda| - \|A\|}. \qquad \square$$

The representation (4.30) is usually referred to as the *Neumann series* (Carl Gottfried Neumann (1832–1925)). Note that, if $\|A\| < 1$, then we have $(\mathcal{I} - A)^{-1} = \sum_{n=0}^{\infty} A^n$.

Corollary 4.9.9. *If A is a bounded operator on a Banach space, then $r(A) \leq \|A\|$.*

If the eigenvalues are considered as points in the complex plane, the above result implies that all eigenvalues of a bounded operator A lie in the closed disk of radius $\|A\|$ centered at the origin.

Theorem 4.9.10. *Let T be an invertible operator on a vector space E, and let A be an operator on E. The operators A and TAT^{-1} have the same eigenvalues.*

Proof: Let λ be an eigenvalue of A. This means that there exists a nonzero vector u such that $Au = \lambda u$. Since T is invertible, $Tu \neq 0$ and

$$TAT^{-1}(Tu) = TAu = T(\lambda u) = \lambda Tu.$$

Thus, λ is an eigenvalue of TAT^{-1}.

Assume now that λ is an eigenvalue of TAT^{-1}, that is, $TAT^{-1}u = \lambda u$ for some nonzero vector $u = Tv$. Since $AT^{-1}u = \lambda T^{-1}u$ and $T^{-1}u \neq 0$, λ is an eigenvalue of A. □

The remaining theorems of this section describe properties of operators on Hilbert spaces.

Theorem 4.9.11. *All eigenvalues of a self-adjoint operator on a Hilbert space are real.*

Proof: Let λ be an eigenvalue of a self-adjoint operator A, and let u be an eigenvector of λ, $u \neq 0$. Then

$$\lambda \langle u, u \rangle = \langle \lambda u, u \rangle = \langle Au, u \rangle = \langle u, Au \rangle = \langle u, \lambda u \rangle = \overline{\lambda} \langle u, u \rangle.$$

Since $\langle u, u \rangle \neq 0$, we conclude $\lambda = \overline{\lambda}$. □

Theorem 4.9.12. *If A is a bounded self-adjoint operator on a Hilbert space, then $r(A) = \|A\|$.*

Proof: The theorem is obviously true for $A = 0$. Let A be a nonzero, bounded, self-adjoint operator on a Hilbert space H. In view of Corollary 4.9.9, it suffices to show that there exists a $\lambda \in \sigma(A)$ such that $|\lambda| = \|A\|$. Since, by Theorem 4.4.14, $\|A\| = \sup_{\|x\|=1} |\langle Ax, x \rangle|$, there exists a sequence $x_n \in H$ such that $\|x_n\| = 1$ and $|\langle Ax_n, x_n \rangle| \to \|A\|$, as $n \to \infty$. Without loss of generality, we may assume that $\langle Ax_n, x_n \rangle \to \lambda$, where $|\lambda| = \|A\|$. For every $n \in \mathbb{N}$, we have

$$\|Ax_n - \lambda x_n\|^2 = \|Ax_n\|^2 - 2\lambda \langle Ax_n, x_n \rangle + \lambda^2 \|x_n\|^2$$
$$\leq \|A\|^2 - 2\lambda \langle Ax_n, x_n \rangle + \lambda^2 = 2\lambda \big(\lambda - \langle Ax_n, x_n \rangle\big).$$

Thus,

$$Ax_n - \lambda x_n \to 0, \tag{4.32}$$

as $n \to \infty$. Now, suppose $\lambda \in \rho(A)$. Then, from (4.32) and continuity of $(A - \lambda \mathcal{I})^{-1}$, we obtain

$$1 = \|x_n\| = \left\| (A - \lambda \mathcal{I})^{-1}(A - \lambda \mathcal{I})x_n \right\| \to 0,$$

as $n \to \infty$. This contradiction shows that $\lambda \in \sigma(A)$. $\qquad \square$

Theorem 4.9.13. *All eigenvalues of a positive operator are non-negative. All eigenvalues of a strictly positive operator are positive.*

Proof: Let A be a positive operator, and let $Ax = \lambda x$ for some $x \neq 0$. Since A is self-adjoint, we have

$$0 \leq \langle Ax, x \rangle = \lambda \langle x, x \rangle = \lambda \|x\|^2. \tag{4.33}$$

Thus, $\lambda \geq 0$. The proof of the second part of the theorem is obtained by replacing "$\leq$" by "$<$" in (4.33). $\qquad \square$

Theorem 4.9.14. *All eigenvalues of a unitary operator on a Hilbert space are complex numbers of modulus 1.*

Proof: Let λ be an eigenvalue of a unitary operator A, and let u be an eigenvector of λ, $u \neq 0$. Then

$$\langle Au, Au \rangle = \langle \lambda u, \lambda u \rangle = |\lambda|^2 \|u\|^2.$$

On the other hand,

$$\langle Au, Au \rangle = \langle u, A^*Au \rangle = \langle u, u \rangle = \|u\|^2.$$

Thus, $|\lambda| = 1$. $\qquad \square$

Theorem 4.9.15. *Eigenvectors corresponding to distinct eigenvalues of a self-adjoint or unitary operator on a Hilbert space are orthogonal.*

Proof: Let u_1 and u_2 be eigenvectors corresponding to distinct eigenvalues λ_1 and λ_2 of a self-adjoint operator A, that is, $Au_1 = \lambda_1 u_1$ and $Au_2 = \lambda_2 u_2$, $\lambda_1 \neq \lambda_2$. By Theorem 4.9.11, λ_1 and λ_2 are real. Then

$$\lambda_1 \langle u_1, u_2 \rangle = \langle Au_1, u_2 \rangle = \langle u_1, Au_2 \rangle = \langle u_1, \lambda_2 u_2 \rangle = \overline{\lambda_2} \langle u_1, u_2 \rangle = \lambda_2 \langle u_1, u_2 \rangle,$$

and hence

$$(\lambda_1 - \lambda_2) \langle u_1, u_2 \rangle = 0.$$

Since $\lambda_1 \neq \lambda_2$, we have $\langle u_1, u_2 \rangle = 0$, that is, u_1 and u_2 are orthogonal.

Suppose now A is a unitary operator on a Hilbert space H. Then $AA^* = A^*A = \mathcal{I}$ and $\|Au\| = \|u\|$ for all $u \in H$. First, note that $\lambda_1 \neq \lambda_2$ implies $\lambda_1 \overline{\lambda_2} \neq 1$. Indeed, if $\lambda_1 \overline{\lambda_2} = 1$, then

$$\lambda_2 = \lambda_1 \overline{\lambda_2} \lambda_2 = \lambda_1 |\lambda_2|^2 = \lambda_1,$$

because $|\lambda_2| = 1$, by Theorem 4.9.14. Now

$$\lambda_1 \overline{\lambda_2} \langle u_1, u_2 \rangle = \langle \lambda_1 u_1, \lambda_2 u_2 \rangle = \langle Au_1, Au_2 \rangle = \langle u_1, A^*Au_2 \rangle = \langle u_1, u_2 \rangle.$$

Since $\lambda_1 \overline{\lambda_2} \neq 1$, we get $\langle u_1, u_2 \rangle = 0$, which proves that the eigenvectors u_1 and u_2 are orthogonal. $\square$

In view of Theorem 4.9.12, it is natural to ask whether for any bounded operator A there exists an eigenvalue λ such that $|\lambda| = \|A\|$. In general, the answer is negative, but it is true for compact self-adjoint operators.

Theorem 4.9.16. *If A is a compact, self-adjoint operator on a Hilbert space, then at least one of the numbers $\|A\|$ or $-\|A\|$ is an eigenvalue of A.*

Proof: The theorem is trivially true if $A = 0$. Now assume that A is a non-zero, compact, self-adjoint operator on a Hilbert space H. As shown in the proof of Theorem 4.9.12, there exists a sequence $x_n \in H$ such that $\|x_n\| = 1$ and

$$Ax_n - \lambda x_n \to 0, \tag{4.34}$$

as $n \to \infty$. Compactness of A implies that the sequence (x_n) has a subsequence (x_{p_n}) such that the sequence (Ax_{p_n}) converges. Since $A \neq 0$, it follows from (4.34) that $x_{p_n} \to u$ for some $u \in H$. Note that $\|u\| = 1$, as $\|x_{p_n}\| = 1$ for all $n \in \mathbb{N}$. Finally, from continuity of A and (4.32), we obtain $Au = \lambda u$. $\square$

Corollary 4.9.17. *If A is a compact, self-adjoint operator on a Hilbert space H, then there is a vector $w \in H$ such that $\|w\| = 1$ and*

$$\left| \langle Aw, w \rangle \right| = \sup_{\|x\| \leq 1} \left| \langle Ax, x \rangle \right|.$$

Proof: Let w, $\|w\| = 1$, be an eigenvector corresponding to an eigenvalue λ such that $|\lambda| = \|A\|$. Then

$$\left| \langle Aw, w \rangle \right| = \left| \langle \lambda w, w \rangle \right| = |\lambda| \|w\|^2 = |\lambda| = \|A\| = \sup_{\|x\| \leq 1} \left| \langle Ax, x \rangle \right|$$

by Theorem 4.4.14. $\square$

Theorem 4.9.16 guarantees existence of at least one nonzero eigenvalue. The preceding corollary gives a useful method for finding that eigenvalue by maximizing a quadratic expression.

Theorem 4.9.18. *The eigenspaces corresponding to nonzero eigenvalues of a compact self-adjoint operator are finite dimensional.*

Proof: Let $\lambda \neq 0$ be an eigenvalue of a compact self-adjoint operator A, and let E_λ be the eigenspace corresponding to λ. Suppose E_λ is of infinite dimension. Let $\{x_1, x_2, x_3, \ldots\}$ be an orthonormal basis of E_λ. Then $Ax_n \to 0$ as $n \to \infty$, since the sequence (x_n) is weakly convergent to 0. But this is impossible, because $Ax_n = \lambda x_n$ for all $n \in \mathbb{N}$ and $\lambda \neq 0$. $\qquad\square$

Theorem 4.9.19. *The set of distinct nonzero eigenvalues (λ_n) of a compact self-adjoint operator is either finite or countable with $\lim_{n \to \infty} \lambda_n = 0$.*

Proof: Suppose A is a self-adjoint compact operator that has infinitely many distinct eigenvalues λ_n, $n \in \mathbb{N}$. Let u_n be an eigenvector corresponding to λ_n such that $\|u_n\| = 1$. By Theorem 4.9.15, (u_n) is an orthonormal sequence and thus it converges weakly to 0. Consequently, by Theorem 4.8.15, the sequence (Au_n) converges strongly to 0 and hence

$$|\lambda_n| = \|\lambda_n u_n\| = \|Au_n\| \to 0, \quad \text{as } n \to \infty. \qquad\square$$

Example 4.9.20. We will find the eigenvalues and eigenfunctions of the operator A on $L^2([0, 2\pi])$ defined by

$$(Au)(x) = \int_0^{2\pi} k(x - t)u(t)dt,$$

where k is a periodic function with period 2π, square integrable on $[0, 2\pi]$.

As a trial solution we take

$$u_n(x) = e^{inx}$$

and note that

$$(Au_n)(x) = \int_0^{2\pi} k(x - t)e^{int}dt = e^{inx} \int_{x-2\pi}^{x} k(s)e^{ins}ds.$$

Thus,

$$Au_n = \lambda_n u_n, \quad n \in \mathbb{Z},$$

where

$$\lambda_n = \int_0^{2\pi} k(s)e^{ins}ds.$$

The set of functions $\{u_n\}$, $n \in \mathbb{Z}$, is a complete orthogonal system in $L^2([0, 2\pi])$. Note that A is self-adjoint if $k(x) = k(-x)$ for all x, but the sequence of eigenfunctions is complete even if A is not self-adjoint. $\qquad\square$

Theorem 4.9.21. *Let (P_n) be a sequence of pairwise orthogonal projection operators on a Hilbert space H and let (λ_n) be a sequence of numbers such that $\lambda_n \to 0$ as $n \to \infty$. Then*

(a) *$\sum_{n=1}^{\infty} \lambda_n P_n$ converges in $\mathcal{B}(H, H)$ and thus defines a bounded operator.*

(b) *For each $n \in \mathbb{N}$, λ_n is an eigenvalue of the operator $A = \sum_{n=1}^{\infty} \lambda_n P_n$, and the only other possible eigenvalue of A is 0.*

(c) *If all λ_n's are real, then A is self-adjoint.*

(d) *If all projections P_n are finite-dimensional, then A is compact.*

Proof: (a) In view of the completeness of $\mathcal{B}(H, H)$, it suffices to prove that the sequence of partial sums of $\sum_{n=1}^{\infty} \lambda_n P_n$ is a Cauchy sequence. Let ε be an arbitrary positive number. Since $\lambda_n \to 0$, there exists an $n_0 \in \mathbb{N}$ such that $|\lambda_n| < \varepsilon$ for all $n > n_0$. For every $x \in H$ and every $k, m \in \mathbb{N}$ such that $n_0 < k < m$, we have

$$\left\| \sum_{n=k}^{m} \lambda_n P_n x \right\|^2 = \sum_{n=k}^{m} \|\lambda_n P_n x\|^2 = \sum_{n=k}^{m} |\lambda_n|^2 \|P_n x\|^2 \tag{4.35}$$

$$\leq \varepsilon^2 \sum_{n=k}^{m} \|P_n x\|^2 = \varepsilon^2 \left\| \sum_{n=k}^{m} P_n x \right\|^2 \leq \varepsilon^2 \left\| \sum_{n=k}^{m} P_n \right\|^2 \|x\|^2, \tag{4.36}$$

where the first and the last equalities follow from orthogonality of projections P_n. The sum $\sum_{n=k}^{m} P_n$, being a finite sum of projection operators, is a projection operator and its operator norm is 1. Thus, (4.35) yields

$$\left\| \sum_{n=k}^{m} \lambda_n P_n \right\| = \sup_{\|x\|=1} \left\| \sum_{n=k}^{m} \lambda_n P_n x \right\| \leq \varepsilon,$$

for any $n_0 < k < m$, proving that the sequence of partial sums $\sum_{n=k}^{m} \lambda_n P_n$ is a Cauchy sequence.

(b) Denote the range of P_n by $\mathcal{R}(P_n)$ and let $n_0 \in \mathbb{N}$. If $u \in \mathcal{R}(P_{n_0})$, then $P_{n_0} u = u$ and $P_n u = 0$ for all $n \neq n_0$, because the projections P_n are orthogonal. Thus,

$$Au = \sum_{n=1}^{\infty} \lambda_n P_n u = \lambda_{n_0} u,$$

which shows that λ_{n_0} is an eigenvalue of A.

To prove that there are no other nonzero eigenvalues, suppose u is an eigenvector corresponding to an eigenvalue λ. Set $v_n = P_n u$, $n = 1, 2, \ldots$, and let $w = Qu$, where Q is the projection on the orthogonal complement of $\mathcal{R}(A)$. Then

$$u = \sum_{n=1}^{\infty} v_n + w \qquad (4.37)$$

with $w \perp \mathcal{R}(P_n)$ for all $n \in \mathbb{N}$. Clearly,

$$A\left(\sum_{n=1}^{\infty} v_n + w\right) = A\left(\sum_{n=1}^{\infty} v_n\right) = \sum_{n=1}^{\infty} A v_n$$

since $P_n w = 0$ and A is continuous. Consequently, the eigenvalue equation can be written in the form

$$\sum_{n=1}^{\infty} \lambda_n v_n = \lambda\left(\sum_{n=1}^{\infty} v_n + w\right)$$

or

$$\sum_{n=1}^{\infty} (\lambda - \lambda_n) v_n + \lambda w = 0. \qquad (4.38)$$

Since all vectors in (4.38) are orthogonal, the sum vanishes only if every term vanishes. Hence, $\lambda w = 0$, and for every $n \in \mathbb{N}$ either $\lambda = \lambda_n$ or $v_n = 0$. Finally, if u in (4.37) is a nonzero eigenvector, then either $w \neq 0$ or $v_k \neq 0$ for some $k \in \mathbb{N}$. Therefore, $\lambda = 0$ or $\lambda = \lambda_k$ for some $k \in \mathbb{N}$, by (4.38).

(c) Suppose all λ_n's are real. Since projections are self-adjoint operators, for any $x, y \in H$, we have

$$\langle Ax, y \rangle = \sum_{n=1}^{\infty} \langle \lambda_n P_n x, y \rangle = \sum_{n=1}^{\infty} \lambda_n \langle P_n x, y \rangle$$

$$= \sum_{n=1}^{\infty} \overline{\lambda_n} \langle x, P_n y \rangle = \sum_{n=1}^{\infty} \langle x, \lambda_n P_n y \rangle = \langle x, Ay \rangle.$$

(d) If all projections P_n are finite-dimensional, then A is compact by Corollary 4.8.13. □

In some applications, the following generalization of the notion of eigenvalue is useful.

Definition 4.9.22. (Approximate eigenvalue)
Let T be an operator on a Hilbert space H. A scalar λ is called an *approximate eigenvalue* of T if there exists a sequence of vectors $x_n \in H$ such that $\|x_n\| = 1$ for all $n \in \mathbb{N}$ and $\|Tx_n - \lambda x_n\| \to 0$ as $n \to \infty$.

Obviously, every eigenvalue is an approximate eigenvalue.

Example 4.9.23. Let (e_n) be a orthonormal sequence in a Hilbert space H. Let (λ_n) be a sequence of scalars such that $\lambda_n \to \lambda$ and $\lambda_n \neq \lambda$ for all $n \in \mathbb{N}$. Define an operator T on H by

$$Tx = \sum_{n=1}^{\infty} \lambda_n \langle x, e_n \rangle e_n.$$

It is easy to see that every λ_n is an eigenvalue of T, but λ is not. On the other hand,

$$\| Te_n - \lambda e_n \| = \| \lambda_n e_n - \lambda e_n \| = \left\| (\lambda_n - \lambda) e_n \right\| = |\lambda_n - \lambda| \to 0$$

as $n \to \infty$. Thus, λ is an approximate eigenvalue of T. $\square$

Theorem 4.9.24. *If T is a compact operator, then every nonzero approximate eigenvalue of T is an eigenvalue.*

Proof: Let (x_n) be a sequence of vectors such that $\| x_n \| = 1$ for all $n \in \mathbb{N}$ and $\| Tx_n - \lambda x_n \| \to 0$ as $n \to \infty$, for some $\lambda \neq 0$. Since T is compact, there exists a subsequence (x_{p_n}) of (x_n) such that $Tx_{p_n} \to y$ as $n \to \infty$, for some y. Then

$$\| y - \lambda x_{p_n} \| \leq \| y - Tx_{p_n} \| + \| Tx_{p_n} - \lambda x_{p_n} \| \to 0,$$

as $n \to \infty$. Since $\lambda \neq 0$, we have $x_{p_n} \to \frac{y}{\lambda}$. If we let $u = \frac{y}{\lambda}$, then $\| u \| = 1$ and

$$\| Tu - \lambda u \| \leq \| Tu - Tx_{p_n} \| + \| Tx_{p_n} - y \| \to 0,$$

as $n \to \infty$. Thus, $Tu = \lambda u$. $\square$

For further properties of approximate eigenvalues see the exercises at the end of this chapter.

4.10 Spectral Decomposition

Let H be a finite-dimensional Hilbert space, say $H = \mathbb{C}^N$. It is known from linear algebra that eigenvectors of a self-adjoint operator on H form an orthogonal basis of H. The following theorems generalize this result to infinite-dimensional spaces.

Theorem 4.10.1. (Hilbert–Schmidt theorem) *For every compact, self-adjoint operator A on an infinite-dimensional Hilbert space H, there exists an orthonormal system of eigenvectors (u_n) corresponding to nonzero eigenvalues (λ_n) such* *that every element $x \in H$ has a unique representation in the form*

$$x = \sum_{n=1}^{\infty} \alpha_n u_n + v, \qquad (4.39)$$

where $\alpha_n \in \mathbb{C}$ and $v \in \mathcal{N}(A)$.

Proof: By Theorem 4.9.16 and Corollary 4.9.17, there exists an eigenvalue λ_1 of A such that

$$|\lambda_1| = \sup_{\|x\| \leq 1} |\langle Ax, x \rangle|.$$

Let u_1 be a normalized eigenvector corresponding to λ_1. We set

$$Q_1 = \{x \in H \colon x \perp u_1\},$$

that is, Q_1 is the orthogonal complement of the set $\{u_1\}$. Thus, Q_1 is a closed linear subspace of H. If $x \in Q_1$, then

$$\langle Ax, u_1 \rangle = \langle x, Au_1 \rangle = \lambda_1 \langle x, u_1 \rangle = 0,$$

which means that $x \in Q_1$ implies $Ax \in Q_1$. Therefore, A maps the Hilbert space Q_1 into itself. We can again apply Theorem 4.9.16 and Corollary 4.9.17 with Q_1 in place of H. This gives an eigenvalue λ_2 such that

$$|\lambda_2| = \sup_{\|x\| \leq 1} \left\{ |\langle Ax, x \rangle| \colon x \in Q_1 \right\}.$$

Let u_2 be a normalized eigenvector of λ_2. Clearly $u_1 \perp u_2$. Next we set

$$Q_2 = \{x \in Q_1 \colon x \perp u_2\},$$

and repeat the above argument. Having eigenvalues $\lambda_1, \ldots, \lambda_n$ and the corresponding normalized eigenvectors $u_1, \ldots, u_n$, we define

$$Q_n = \{x \in Q_{n-1} \colon x \perp u_n\}$$

and choose an eigenvalue λ_{n+1} such that

$$|\lambda_{n+1}| = \sup_{\|x\| \leq 1} \left\{ |\langle Ax, x \rangle| \colon x \in Q_n \right\}. \qquad (4.40)$$

For u_{n+1} we choose a normalized vector corresponding to λ_{n+1}.

This procedure can terminate after a finite number of steps. Indeed, it can happen that there is a positive integer k such that $\langle Ax, x \rangle = 0$ for every $x \in Q_k$. Then every element x of H has a unique representation

$$x = \alpha_1 u_1 + \cdots + \alpha_k u_k + v,$$

where $Av = 0$, and

$$Ax = \lambda_1 \alpha_1 u_1 + \cdots + \lambda_k \alpha_k u_k,$$

which proves the theorem in this case.

Now suppose that the described procedure yields an infinite sequence of eigenvalues (λ_n) and eigenvectors (u_n). Let S be the closed space spanned by the vectors $u_1, u_2, \ldots$, that is,

$$S = \left\{ \sum_{n=1}^{\infty} \alpha_n u_n : \sum_{n=1}^{\infty} |\alpha_n|^2 < \infty \right\}.$$

By Theorem 3.6.6, every $x \in H$ has a unique decomposition $x = u + v$ or

$$x = \sum_{n=1}^{\infty} \alpha_n u_n + v,$$

where $v \in S^{\perp}$. It remains to prove that $Av = 0$ for all $v \in S^{\perp}$.

Let $v \in S^{\perp}$, $v \neq 0$. Define $w = v/\|v\|$. Then

$$\langle Av, v \rangle = \|v\|^2 \langle Aw, w \rangle.$$

Since $w \in S^{\perp} \subset Q_n$ for every $n \in \mathbb{N}$, by (4.40), we have

$$\left| \langle Av, v \rangle \right| = \|v\|^2 \left| \langle Aw, w \rangle \right| \leq \|v\|^2 \sup_{\|x\| \leq 1} \left\{ \left| \langle Ax, x \rangle \right| : x \in Q_n \right\}$$

$$= \|v\|^2 |\lambda_{n+1}| \to 0.$$

This implies $\langle Av, v \rangle = 0$ for every $v \in S^{\perp}$. Therefore, by Theorem 4.4.14, the norm of A restricted to $S^{\perp}$ is 0, and thus $Av = 0$ for all $v \in S^{\perp}$.   □

Corollary 4.10.2. (Spectral theorem for compact self-adjoint operators) *Let A be a compact, self-adjoint operator on an infinite-dimensional Hilbert space H. Then H has a complete orthonormal system (an orthonormal basis) $\{v_1, v_2, \ldots\}$ consisting of eigenvectors of A. Moreover, for every $x \in H$,*

$$Ax = \sum_{n=1}^{\infty} \lambda_n \langle x, v_n \rangle v_n, \tag{4.41}$$

where λ_n is the eigenvalue corresponding to v_n.

Proof: To obtain a complete orthonormal system $\{v_1, v_2, \ldots\}$, we need to complement the system $\{u_1, u_2, \ldots\}$, defined in the proof of Theorem 4.10.1, with an arbitrary orthonormal basis of $\mathcal{N}(A)$. The eigenvalues corresponding to the vectors that form $\mathcal{N}(A)$ are all equal zero. Equality (4.41) follows from the continuity of A.   □

Corollary 4.10.3. *Let A be a compact, self-adjoint operator on a Hilbert space H. Then*

$$A = \sum_{n=1}^{\infty} \lambda_n P_n, \tag{4.42}$$

where P_n is a projection on finite-dimensional subspace of H.

Proof: Let $\{v_1, v_2, \ldots\}$ be a complete orthonormal system of eigenvectors of A corresponding to eigenvalues $\{\lambda_1, \lambda_2, \ldots\}$. Let P_n be the projection operator onto the one-dimensional space spanned by v_n. Since $P_n x = \langle x, v_n \rangle v_n$, equation (4.41) can be written as $Ax = \sum_{n=1}^{\infty} \lambda_n P_n x$.

Here is another way of representing A in the form (4.42). Let $\{\lambda_1, \lambda_2, \ldots\}$ be all distinct nonzero eigenvalues of A and let P_n be the projection onto the eigenspace corresponding to λ_n. By Theorem 4.9.18, those eigenspaces are finite-dimensional. $\qquad\square$

Corollary 4.10.3 is just another version of the spectral theorem. This version is important because it has natural extensions to more general classes of operators. It is also useful because it leads to an elegant expression for powers and other functions of an operator.

Let A, λ_n, and P_n be as in Corollary 4.10.3. Then

$$A^2 = A \left(\sum_{n=1}^{\infty} \lambda_n P_n \right) = \sum_{n=1}^{\infty} \lambda_n A P_n = \sum_{n=1}^{\infty} \lambda_n^2 P_n,$$

because $A P_n x = \lambda_n P_n x$ for every $x \in H$. Similarly, for any $k \in \mathbb{N}$, we get

$$A^k = \sum_{n=1}^{\infty} \lambda_n^k P_n, \tag{4.43}$$

and, hence, for any polynomial $p(t) = \alpha_n t^n + \cdots + \alpha_1 t$, we have

$$p(A) = \sum_{n=1}^{\infty} p(\lambda_n) P_n.$$

The constant term in p must be zero, because otherwise the sequence $(p(\lambda_n))$ would not converge to zero. To deal with polynomials with a nonzero constant term α_0, we add $\alpha_0 \mathcal{I}$ to the series. Note that, in such a case, $p(A)$ is not a compact operator.

This method can be generalized in the following way.

Definition 4.10.4.　(Function of an operator)
Let f be a real-valued function on $\mathbb{R}$ such that

$$f(\lambda) \to 0 \quad \text{as } \lambda \to 0 \quad \text{and} \quad f(0) = 0.$$

For a compact, self-adjoint operator $A = \sum_{n=1}^{\infty} \lambda_n P_n$, we define

$$f(A) = \sum_{n=1}^{\infty} f(\lambda_n) P_n. \tag{4.44}$$

Theorem 4.9.21 ensures that the series in (4.44) converges and that $f(A)$ is self-adjoint and compact.

Example 4.10.5.　Let $A = \sum_{n=1}^{\infty} \lambda_n P_n$ be a compact, self-adjoint operator such that $\lambda_n \geq 0$ for all $n \in \mathbb{N}$. For any $\alpha > 0$, we can define A^α by

$$A^\alpha x = \sum_{n=1}^{\infty} \lambda_n^\alpha P_n x.$$

Note that in the case $\alpha = \frac{1}{2}$, the defined operator $A^{1/2}$ is the same as the square root of A introduced in Definition 4.6.13. Indeed, by (4.43), we have

$$\left(\sqrt{A}\right)^2 = \sum_{n=1}^{\infty} \left(\sqrt{\lambda_n}\right)^2 P_n = \sum_{n=1}^{\infty} \lambda_n P_n = A,$$

because all λ_n's are non-negative.　□

Example 4.10.6.　Let $A = \sum_{n=1}^{\infty} \lambda_n P_n$ be a compact, self-adjoint operator. We can define sine of A by

$$\sin A = \sum_{n=1}^{\infty} (\sin \lambda_n) P_n.$$

The condition that $f(\lambda) \to 0$ as $\lambda \to 0$ in Definition 4.10.1 can be replaced by boundedness of f in a neighborhood of the origin. Indeed, if $A = \sum_{n=1}^{\infty} \lambda_n P_n$ and $P_n x = \langle x, v_n \rangle v_n$, then, for any $x \in H$, we have

$$\left(f(A)\right)x = \sum_{n=1}^{\infty} f(\lambda_n) \langle x, v_n \rangle v_n,$$

where convergence of the series is justified by Theorem 3.4.10, because

$$\left| f(\lambda_n) \langle x, v_n \rangle \right|^2 \leq M \left| \langle x, v_n \rangle \right|^2$$

for some constant M, and hence $(f(\lambda_n)\langle x, \lambda_n\rangle) \in l^2$. Clearly, in this case, we cannot expect $f(A)$ to be a compact operator. $\qquad\square$

Theorem 4.10.7. *If eigenvectors $u_1, u_2, \ldots$ of a self-adjoint operator T on a Hilbert space H form a complete orthonormal system in H and all eigenvalues are positive (or non-negative), then T is strictly positive (or positive).*

Proof: Suppose $u_1, u_2, \ldots$ is a complete orthonormal system of eigenvalues of T corresponding to eigenvalues $\lambda_1, \lambda_2, \ldots$. Then any nonzero vector $u \in H$ can be represented as $u = \sum_{n=1}^{\infty} \alpha_n u_n$, and we have

$$\langle Tu, u \rangle = \left\langle Tu, \sum_{n=1}^{\infty} \alpha_n u_n \right\rangle = \sum_{n=1}^{\infty} \overline{\alpha_n}\langle Tu, u_n \rangle = \sum_{n=1}^{\infty} \overline{\alpha_n}\langle u, Tu_n \rangle$$

$$= \sum_{n=1}^{\infty} \overline{\alpha_n}\langle u, \lambda_n u_n \rangle = \sum_{n=1}^{\infty} \lambda_n \overline{\alpha_n}\langle u, u_n \rangle = \sum_{n=1}^{\infty} \lambda_n \overline{\alpha_n} \alpha_n$$

$$= \sum_{n=1}^{\infty} \lambda_n |\alpha_n|^2 \geq 0$$

if all eigenvalues are non-negative. If all λ_n's are positive, then the last inequality becomes strict. $\qquad\square$

Theorem 4.10.8. *For any two commuting, compact, self-adjoint operators A and B on a Hilbert space H, there exists a complete orthonormal system in H of common eigenvectors of A and B.*

Proof: Let λ be an eigenvalue of A, and let S be the corresponding eigenspace. For any $x \in S$, we have

$$ABx = BAx = B(\lambda x) = \lambda Bx.$$

This means that Bx is an eigenvector of A corresponding to λ provided $Bx \neq 0$. In any case, $Bx \in S$ and hence B maps S into itself. Since B is a compact, self-adjoint operator, by Corollary 4.10.2, S has an orthonormal basis consisting of eigenvalues of B, but these vectors are also eigenvectors of A, because they belong to S. If we repeat the same with every eigenspace of A, then the union of all these eigenvectors will be an orthonormal basis of H. $\qquad\square$

4.11 Unbounded Operators

Boundedness of an operator was an essential assumption in almost every theorem proved in this chapter. Methods used were developed with boundedness or continuity in mind. However, in the most important applications of the theory of Hilbert spaces we often have to deal with operators which are not bounded. In this section, we will briefly discuss some basic problems, concepts, and methods in the theory of unbounded operators. Some concepts, like the adjoint operator, will require a new approach in this setting.

An operator A defined in a Hilbert space H, that is, $\mathcal{R}(A) \subset H$, is called *unbounded* if it is not bounded. Therefore, to show that an operator A is unbounded it suffices to find a sequence of elements $x_n \in H$ such that $\|x_n\| \leq M$ (for some M and all $n \in \mathbb{N}$) and $\|Ax_n\| \to \infty$. Since for linear operators boundedness is equivalent to continuity, unboundedness is equivalent to lack of continuity (at any point). Consequently, we can show that an operator A is unbounded by finding a sequence (x_n) convergent to 0 such that the sequence (Ax_n) does not converge to 0.

One of the most important unbounded operators is the differential operator on $L^2(\mathbb{R})$ (see Example 4.2.3). Other important unbounded operators arise from quantum mechanics and will be discussed in Chapter 7.

Example 4.11.1. Let A be a compact operator on an infinite dimensional Hilbert space H. If A is invertible, then A^{-1} is unbounded. Indeed, let $v_n \in H$ be an orthonormal sequence and let $z_n = Av_n$. Then $z_n \to 0$, but $A^{-1}z_n \not\to 0$. □

It will be convenient to adopt the following convention: When we say "A is an operator on a Hilbert space H," we mean that the domain of A is the whole space H, and when we say "A is an operator in a Hilbert space H," we mean that the domain of A is a subset of H.

If the domain of a bounded operator A is a proper subspace of a Hilbert space H, then A has a unique extension to a bounded operator defined on the closure of $\mathcal{D}(A)$. More precisely, there exists a bounded operator B defined on $\mathrm{cl}\,\mathcal{D}(A)$ such that $Ax = Bx$ for every $x \in \mathcal{D}(A)$. Such an operator can be defined by

$$Bx = \lim_{n\to\infty} Ax_n, \quad \text{where } x_n \in \mathcal{D}(A) \text{ and } x_n \to x.$$

In this case $\|B\| = \|A\|$. Then B can be extended to a bounded operator C defined on all of H by

$$C = BP_{\mathcal{D}(B)},$$

where $P_{\mathcal{D}(B)}$ is the orthogonal projection operator onto the subspace $\mathcal{D}(B)$. Again we have $\|C\| = \|B\|$. We may thus always assume that the domain of a bounded operator is the entire space H, or at least a closed subspace of H. An unbounded operator defined on a proper subspace of a Hilbert space usually does not have a natural extension onto the closure of its domain. For instance, the differential operator is defined on a dense subspace of $L^2(\mathbb{R})$, but it does not have a natural extension onto $L^2(\mathbb{R})$. On the other hand, it may be still possible to extend the domain of an unbounded operator in such a way that, although the domain of the extension is not the whole space, it has better properties.

Definition 4.11.2.   (Extension of operators)
Let A and B be operators in a vector space E. If

$$\mathcal{D}(A) \subset \mathcal{D}(B)$$

and

$$Ax = Bx \quad \text{for every } x \in \mathcal{D}(A),$$

then B is called an *extension* of A and we write $A \subset B$.

When performing typical operations on operators defined on subspaces of a common vector space E, we have to keep track of the domains. For instance, the operator $A + B$ is defined for all $x \in \mathcal{D}(A) \cap \mathcal{D}(B)$, that is, $\mathcal{D}(A + B) = \mathcal{D}(A) \cap \mathcal{D}(B)$. It may happen that $\mathcal{D}(A) \cap \mathcal{D}(B) = \{0\}$, and then the sum $A + B$ does not make much sense. Similarly, $\mathcal{D}(AB) = \{x \in \mathcal{D}(B) \colon Bx \in \mathcal{D}(A)\}$. The usual properties need not hold. For example, we have the equality $(A + B)C = AC + BC$, but, in general, the inclusion $AB + AC \subset A(B + C)$ cannot be replaced by equality.

Definition 4.11.3.   (Densely defined operator)
An operator A defined in a normed space E is called *densely defined* if its domain is a dense subset of E, that is, cl $\mathcal{D}(A) = E$.

The differential operator $D = \frac{d}{dx}$ is densely defined in $L^2(\mathbb{R})$, because the subspace of differentiable functions is dense in $L^2(\mathbb{R})$.

Theorem 4.11.4. *Let A be a densely defined operator in a Hilbert space H and let E be the set of all $y \in H$ for which $\langle Ax, y \rangle$ is a continuous functional on $\mathcal{D}(A)$. There exists a unique operator B defined on E such that*

$$\langle Ax, y \rangle = \langle x, By \rangle \quad \text{for all } x \in \mathcal{D}(A) \text{ and } y \in E.$$

Proof: For any $y \in E$, the functional $f_y(x) = \langle Ax, y \rangle$, being continuous on a dense subspace of H, has a unique extension to a continuous functional $\tilde{f}_y$ on H. By the Riesz representation theorem, there exists a unique $z_y \in H$ such that $\tilde{f}_y(x) = \langle x, z_y \rangle$ for all $x \in H$. If we define $B(y) = z_y$, then we will have

$$\langle Ax, y \rangle = f_y(x) = \tilde{f}_y(x) = \langle x, z_y \rangle = \langle x, By \rangle$$

for all $x \in \mathcal{D}(A)$ and $y \in E$. Proof of linearity of B is left as an exercise.   □

The preceding result justifies the following definition.

Definition 4.11.5.   (Adjoint of a densely defined operator)
Let A be a densely defined operator in a Hilbert space H. The *adjoint* A^* of A is the operator defined on the set of all $y \in H$ for which $\langle Ax, y \rangle$ is a continuous functional on $\mathcal{D}(A)$ and such that

$$\langle Ax, y \rangle = \langle x, A^*y \rangle \quad \text{for all } x \in \mathcal{D}(A) \text{ and } y \in \mathcal{D}(A^*).$$

Note that for a bounded operator on H this definition is equivalent to Definition 4.4.1.

Example 4.11.6. Let $\mathcal{C}_0^1(\mathbb{R})$ denote the space of all continuously differentiable functions on $\mathbb{R}$ with compact support. This is a dense subspace of $L^2(\mathbb{R})$. Consider the differential operator D defined on $\mathcal{C}_0^1(\mathbb{R})$. Since

$$\langle Dx, y \rangle = \int_{-\infty}^{\infty} \left(\frac{d}{dt}x(t) \right) \overline{y(t)} dt = -\int_{-\infty}^{\infty} x(t) \left(\frac{d}{dt}\overline{y(t)} \right) dt,$$

$\langle Dx, y \rangle$ is a continuous functional on $\mathcal{C}_0^1(\mathbb{R})$ (with respect to the norm of $L^2(\mathbb{R})$) for every $y \in L^2(\mathbb{R})$ such that $y' \in L^2(\mathbb{R})$. Moreover,

$$\langle Dx, y \rangle = -\int_{-\infty}^{\infty} x(t) \left(\frac{d}{dt}\overline{y(t)} \right) dt = \int_{-\infty}^{\infty} x(t) \overline{\left(-\frac{d}{dt}y(t) \right)} dt.$$

Note that it is not correct to write $D^* = -D$, since the domain of D^* is not $\mathcal{C}_0^1(\mathbb{R})$.   □

Theorem 4.11.7. *Let A and B be densely defined operators in a Hilbert space H.*

(a) *If $A \subset B$, then $B^* \subset A^*$.*

(b) *If $\mathcal{D}(B^*)$ is dense in H, then $B \subset B^{**}$.*

Proof: To prove (a), consider a $y \in \mathcal{D}(B^*)$. Then $\langle Bx, y \rangle$, as a function of x, is a continuous functional on $\mathcal{D}(B)$. Since $\mathcal{D}(A) \subset \mathcal{D}(B)$, $\langle Bx, y \rangle$ is a continuous functional on $\mathcal{D}(A)$. But $Bx = Ax$ for $x \in \mathcal{D}(A)$, so $\langle Ax, y \rangle$ is a continuous functional on $\mathcal{D}(A)$. This proves that $y \in \mathcal{D}(A^*)$. The equality $A^*y = B^*y$ for $y \in \mathcal{D}(B^*)$ follows from uniqueness of the adjoint operator.
To prove (b), observe that the condition

$$\langle Bx, y \rangle = \langle x, B^*y \rangle \quad \text{for all } x \in \mathcal{D}(B) \text{ and all } y \in \mathcal{D}(B^*)$$

can be rewritten as

$$\langle B^*y, x\rangle = \langle y, Bx\rangle \quad \text{for all } y \in \mathcal{D}(B^*) \text{ and all } x \in \mathcal{D}(B). \tag{4.45}$$

Since $\mathcal{D}(B^*)$ is dense in H, B^{**} exists and we have

$$\langle B^*y, x\rangle = \langle y, B^{**}x\rangle \quad \text{for all } y \in \mathcal{D}(B^*) \text{ and all } x \in \mathcal{D}(B^{**}). \tag{4.46}$$

Now, by an argument similar to that in part (a), we can show that $\mathcal{D}(B) \subset \mathcal{D}(B^{**})$ and $B(x) = B^{**}(x)$ for any $x \in \mathcal{D}(B)$. $\quad\square$

Theorem 4.11.8. *If A is a one-to-one operator in a Hilbert space and both A and its inverse A^{-1} are densely defined, then A^* is also one-to-one and*

$$\left(A^*\right)^{-1} = \left(A^{-1}\right)^*. \tag{4.47}$$

Proof: Let $y \in \mathcal{D}(A^*)$. Then for every $x \in \mathcal{D}(A^{-1})$, we have $A^{-1}x \in \mathcal{D}(A)$ and hence

$$\langle A^{-1}x, A^*y\rangle = \langle AA^{-1}x, y\rangle = \langle x, y\rangle.$$

This means that $A^*y \in \mathcal{D}((A^{-1})^*)$ and

$$\left(A^{-1}\right)^* A^*y = \left(AA^{-1}\right)^*y = y. \tag{4.48}$$

Next, take an arbitrary $y \in \mathcal{D}(A^{-1})^*$. Then, for each $x \in \mathcal{D}(A)$, we have $Ax \in \mathcal{D}(A^{-1})$. Hence

$$\langle Ax, \left(A^{-1}\right)^*y\rangle = \langle A^{-1}Ax, y\rangle = \langle x, y\rangle.$$

This shows that $(A^{-1})^*y \in \mathcal{D}(A^*)$ and

$$A^*\left(A^{-1}\right)^*y = \left(A^{-1}A\right)^*y = y. \tag{4.49}$$

Now equality (4.47) follows from (4.48) and (4.49). $\quad\square$

Theorem 4.11.9. *If A, B, and AB are densely defined operators in H, then $B^*A^* \subset (AB)^*$.*

Proof: Suppose $x \in \mathcal{D}(AB)$ and $y \in \mathcal{D}(B^*A^*)$. Since $x \in \mathcal{D}(B)$ and $A^*y \in \mathcal{D}(B^*)$, it follows that

$$\langle Bx, A^*y\rangle = \langle x, B^*A^*y\rangle.$$

On the other hand, since $Bx \in \mathcal{D}(A)$ and $y \in \mathcal{D}(A^*)$, we have

$$\langle ABx, y\rangle = \langle Bx, A^*y\rangle.$$

Hence

$$\langle ABx, y\rangle = \langle x, B^*A^*y\rangle.$$

Since this holds for all $x \in \mathcal{D}(AB)$, we have $y \in \mathcal{D}((AB)^*)$ and $(B^*A^*)y = (AB)^*y$. $\qquad\qquad\qquad\qquad\qquad\qquad\qquad\qquad\qquad\qquad\qquad\qquad\square$

Self-adjoint operators have been already discussed in Section 4.4. In that section we limited our discussion to bounded operators. Without the boundedness condition the matter is more delicate.

Definition 4.11.10.   (Self-adjoint operator)

Let A be a densely defined operator in a Hilbert space H. A is called *self-adjoint* if $A = A^*$.

Remember that $A = A^*$ means that $\mathcal{D}(A^*) = \mathcal{D}(A)$ and $A(x) = A^*(x)$ for all $x \in \mathcal{D}(A)$. If A is a bounded, densely defined operator in H, then A has a unique extension to a bounded operator on H, and then its domain as well as the domain of its adjoint, is the whole space H. In the case of unbounded operators, it is possible that a densely defined operator A has an adjoint A^* such that $A(x) = A^*(x)$ whenever $x \in \mathcal{D}(A) \cap \mathcal{D}(A^*)$, but $\mathcal{D}(A^*) \neq \mathcal{D}(A)$, and thus A is not self-adjoint. The following relaxation of the conditions in Definition 4.11.10 seems reasonable in the case of unbounded operators.

Definition 4.11.11.   (Symmetric operator)

A densely defined operator A in a Hilbert space H is said to be *symmetric* if

$$\langle Ax, y \rangle = \langle x, Ay \rangle \quad \text{for every } x, y \in \mathcal{D}(A).$$

Clearly, every self-adjoint operator is symmetric. In the first of the following two examples, we present an unbounded self-adjoint operator. The second one shows that a symmetric operator need not be self-adjoint.

Example 4.11.12. Consider a bounded operator on $H = l^2$ defined by

$$A(x_n) = \left(\frac{x_n}{n} \right).$$

Note that A is self-adjoint and one-to-one. The subspace $\mathcal{R}(A) = \mathcal{D}(A^{-1})$ consists of all sequences $(y_n) \in l^2$ such that

$$\sum_{n=1}^{\infty} n^2 |y_n|^2 < \infty$$

and it is dense in H. The inverse A^{-1} is defined by

$$A^{-1}(y_n) = (ny_n).$$

Clearly, A^{-1} is an unbounded operator. By Theorem 4.11.8,

$$\left(A^{-1}\right)^* = \left(A^*\right)^{-1} = A^{-1}.$$

Thus, A^{-1} is self-adjoint.   □

Example 4.11.13. Consider the operator $A = id/dt$ with the domain

$$\mathcal{D}(A) = \{f \in L^2([a, b]): f' \text{ is continuous and } f(a) = f(b) = 0\}.$$

Since

$$\langle Af, g \rangle = \int_a^b if'(t)\overline{g(t)}dt = \int_a^b f(t)\overline{ig'(t)}dt = \langle f, Ag \rangle$$

for all $f, g \in \mathcal{D}(A)$, A is symmetric. However, $\langle Af, g \rangle$ is a continuous functional on $\mathcal{D}(A)$ for any function g continuously differentiable on $[a, b]$, not necessarily satisfying $g(a) = g(b)$. Consequently, $\mathcal{D}(A^*) \neq \mathcal{D}(A)$ and A is not self-adjoint.   □

The following theorem gives a nice characterization of symmetric operators.

Theorem 4.11.14. *A densely defined operator A in a Hilbert space H is symmetric if and only if $A \subset A^*$.*

Proof: Suppose $A \subset A^*$. Since

$$\langle Ax, y \rangle = \langle x, A^*y \rangle \quad \text{for all } x \in \mathcal{D}(A) \text{ and all } y \in \mathcal{D}(A^*), \qquad (4.50)$$

we have

$$\langle Ax, y \rangle = \langle x, Ay \rangle \quad \text{for all } x, y \in \mathcal{D}(A). \qquad (4.51)$$

Thus, A is symmetric.

If A is symmetric, then (4.50) and (4.51) hold, which implies $A \subset A^*$.   □

Corollary 4.11.15. *If A is symmetric, then A^* is the maximal symmetric extension of A.*

Proof: Let B be a symmetric operator such that $A \subset B$. Then, by Theorem 4.11.9, $B^* \subset A^*$. Hence $A \subset B \subset B^* \subset A^*$.   □

Corollary 4.11.16. *If A is symmetric and $\mathcal{D}(A) = H$, then A is self-adjoint.*

Proof: $A \subset A^*$ and $\mathcal{D}(A) = H$ implies $A^* = A$.   □

Now we are going to consider the concept of a closed operator. It can be thought of as a weaker version of boundedness.

Recall that by the graph of an operator A, defined on $\mathcal{D}(A) \subset E_1$ and with the range $\mathcal{R}(A) \subset E_2$, we mean the set $\mathcal{G}(A) = \{(x, Ax): x \in \mathcal{D}(A)\}$. The graph of $A : \mathcal{D}(A) \to \mathcal{R}(A)$ is thus a subset of $E_1 \times E_2$. Under our usual assumptions (A is linear and $\mathcal{D}(A)$ is a vector space) $\mathcal{G}(A)$ is a vector space. Note that, if $A \subset B$, then $\mathcal{G}(A) \subset \mathcal{G}(B)$.

Definition 4.11.17.   (Closed operator)

An operator A from a normed space E_1 into a normed space E_2 is called *closed* if its graph $\mathcal{G}(A)$ is a closed subspace of $E_1 \times E_2$, that is,

$$x_n \in \mathcal{D}(A), x_n \to x, \text{ and } Ax_n \to y \quad \text{implies} \quad x \in \mathcal{D}(A) \text{ and } Ax = y.$$

The domain $\mathcal{D}(A)$ of a closed operator A need not be closed. It can be proved, that every closed operator from a Banach space into a Banach space is bounded (this is known as the closed graph theorem). Thus, the domain of an unbounded closed operator in a Hilbert space H cannot be closed, and thus it cannot be the entire space H.

Theorem 4.11.18.   *If A is closed and invertible, then A^{-1} is closed.*

Proof: If $\mathcal{G}(A) = \{(x, Ax) : x \in \mathcal{D}(A)\}$ is closed, then, obviously,

$$\mathcal{G}(A^{-1}) = \{(Ax, x) : x \in \mathcal{D}(A)\}$$

is closed.   □

Theorem 4.11.19.   *If A is densely defined, then A^* is closed.*

Proof: If $y_n \in \mathcal{D}(A^*)$, $y_n \to y$, and $A^* y_n \to z$, then for any $x \in \mathcal{D}(A)$, we have

$$\langle Ax, y \rangle = \lim_{n \to \infty} \langle Ax, y_n \rangle = \lim_{n \to \infty} \langle x, A^* y_n \rangle = \langle x, z \rangle.$$

Hence $y \in \mathcal{D}(A^*)$ and $A^* y = z$.   □

The property of being a closed operator is a desirable one. If a given operator A is not closed, is it possible to extend A to a closed operator? A possible approach to this problem is to use the closure of $\mathcal{G}(A)$ in $H \times H$ to define an operator. If $\operatorname{cl} \mathcal{G}(A)$ defines an operator, then it is a closed extension of A.

Theorem 4.11.20.   *Let A be an operator in a Hilbert space H. Then there exists an operator B such that $\mathcal{G}(B) = \operatorname{cl} \mathcal{G}(A)$ if and only if the following condition is satisfied:*

$$x_n \in \mathcal{D}(A), x_n \to 0, Ax_n \to y \quad \text{implies} \quad y = 0. \tag{4.52}$$

Proof: Suppose $\mathcal{G}(B) = \mathrm{cl}\,\mathcal{G}(A)$ for some operator B. If $x_n \in \mathcal{D}(A)$, $x_n \to 0$, and $Ax_n \to y$, then $(0, y) \in \mathcal{G}(B)$, because $\mathcal{G}(B)$ is a closed set. Since $(0, 0) \in \mathcal{G}(B)$, we must have $y = 0$.

Now assume that (4.52) is satisfied and let $(x, y_1), (x, y_2) \in \mathrm{cl}\,\mathcal{G}(A)$. Then there exist $(x_n, Ax_n), (z_n, Az_n) \in \mathcal{G}(A)$ such that $x_n \to x$, $z_n \to x$, $Ax_n \to y_1$, and $Az_n \to y_2$. Since $x_n - z_n \in \mathcal{G}(A)$, $x_n - z_n \to 0$, and $A(x_n - z_n) \to y_1 - y_2$, (4.52) implies $y_1 - y_2 = 0$. This shows that B can be defined by

$$B(x) = y \quad \text{if } (x, y) \in \mathrm{cl}\,\mathcal{G}(A).$$

Proof of linearity of B is left as an exercise. $\qquad\square$

Note that the operator B described in Theorem 4.11.20 is the minimal closed extension of A.

Theorem 4.11.21. *Every symmetric, densely defined operator has a closed symmetric extension.*

Proof: Let A be a symmetric, densely defined operator in a Hilbert space H. First we will show that condition (4.52) is satisfied. Let $x_n \in \mathcal{D}(A)$, $x_n \to 0$, and $Ax_n \to y$. Since A is symmetric, we have

$$\langle y, z \rangle = \lim_{n \to 0} \langle Ax_n, z \rangle = \lim_{n \to 0} \langle x_n, Az \rangle = 0,$$

for any $z \in \mathcal{D}(A)$. This implies $y = 0$, because $\mathcal{D}(A)$ is dense in H. Now, by Theorem 4.11.20, there exists a closed operator B such that $\mathcal{G}(B) = \mathrm{cl}\,\mathcal{G}(A)$, and hence $A \subset B$. It remains to prove that B is symmetric.

If $x, y \in \mathcal{D}(B)$, then there exist $x_n, y_n \in \mathcal{D}(A)$ such that

$$x_n \to x, \quad Ax_n \to Bx,$$
$$y_n \to y, \quad Ay_n \to By.$$

Since A is symmetric, we have

$$\langle Ax_n, y_n \rangle = \langle x_n, Ay_n \rangle.$$

By letting $n \to \infty$, we get

$$\langle Bx, y \rangle = \langle x, By \rangle.$$

Therefore, B is symmetric. $\qquad\square$

The following is an example of a theorem on solvability of linear equations involving closed densely defined operators. Note the role of these two conditions in the proof of the theorem.

Theorem 4.11.22. *Let A be a densely defined closed operator in a Hilbert space H.*

(a) *For any $u, v \in H$, there exist unique $x \in D(A)$ and $y \in D(A^*)$ such that $Ax + y = u$ and $x - A^*y = v$.*

(b) *For any $v \in H$, there exists a unique $x \in D(A^*A)$ such that $A^*Ax + x = v$.*

Proof: (a) Consider the Hilbert space $H_1 = H \times H$. Since A is closed, $G(A)$ is a closed subspace of H_1. Thus, by Theorem 3.6.6,

$$H_1 = G(A) \oplus G(A)^\perp,$$

with $G(A) \cap G(A)^\perp = \{0\}$.

Now, $(z, y) \in G(A)^\perp$ if and only if $\langle (x, Ax), (z, y) \rangle = 0$ for all $x \in D(A)$ or, equivalently,

$$\langle x, z \rangle + \langle Ax, y \rangle = 0 \quad \text{for all } x \in D(A).$$

Thus,

$$(z, y) \in G(A)^\perp \quad \text{if and only if} \quad \langle Ax, y \rangle = \langle x, -z \rangle \quad \text{for all } x \in D(A).$$

In other words,

$$(z, y) \in G(A)^\perp \quad \text{if and only if} \quad y \in D(A^*) \text{ and } z = -A^*y.$$

Consequently, if $(v, u) \in H \times H$, then there exist unique $x \in D(A)$ and $y \in D(A^*)$ such that

$$(v, u) = (x, Ax) + \left(-A^*y, y\right),$$

which completes the proof of (a).

(b) If $u = 0$ in (a), then it follows that there are unique $x \in D(A)$ and $y \in D(A^*)$ such that $Ax + y = 0$ and $x - A^*y = v$. Thus, $x - A^*(-Ax) = v$ or $A^*(Ax) + x = v$. $\qquad\square$

We close this section with an interesting observation. Let A be a closed operator in a Hilbert space H. We know that this does not imply boundedness of A. On the other hand, it is always possible to redefine the inner product on $D(A)$ in such a way that $D(A)$ becomes a Hilbert space and A becomes a bounded operator on $D(A)$. In fact, for $x, y \in D(A)$ define

$$\langle x, y \rangle_1 = \langle x, y \rangle + \langle Ax, Ay \rangle,$$

where $\langle \cdot, \cdot \rangle$ denotes the inner product in H. The proof of completeness of $D(A)$ with respect to the norm

$$\|x\|_1 = \|x\|^2 + \|Ax\|^2,$$

and the boundedness of A in this new Hilbert space is left as an exercise.

4.12 Exercises

1. Let $E = \{f \in L^2(\mathbb{R}): f' \in L^2(\mathbb{R})\}$. Show that the differential operator is not bounded in E.

2. If A is an operator on a complex Hilbert space H such that $Ax \perp x$ for every $x \in H$, show that $A = 0$.

3. Give an example of a bounded operator A such that $\|A^2\| \neq \|A\|^2$.

4. Let A be a bounded operator defined on a proper subspace of a Hilbert space H.

   (a) Define an operator A_1 on the closure $\operatorname{cl}\mathcal{D}(A)$ of the domain of A by $A_1 x = \lim_{n \to \infty} Ax_n$, where $x_n \in \mathcal{D}(A)$ and $x_n \to x$. Show that A_1 is well defined, that is, $A_1 x$ does not depend on a particular choice of the sequence (x_n). Show that A_1 is a linear and bounded operator defined on $\operatorname{cl}\mathcal{D}(A)$.

   (b) Define an operator B on H by $Bx = A_1 x_1$, where x_1 is the projection of x onto $\operatorname{cl}\mathcal{D}(A)$. Show that B is a bounded operator on H.

   (c) Show that $\|A\| = \|B\|$.

   Since $A = B$ on $\mathcal{D}(A)$, B is an extension of A.

5. Let φ be a symmetric, positive, bilinear functional on a vector space E. Show that

   $$|\varphi(x, y)|^2 \leq \varphi(x, x)\varphi(y, y).$$

6. Let (e_n) be a complete orthonormal sequence in a Hilbert space H and let (λ_n) be a sequence of scalars.

   (a) Show that there exists a unique operator T on H such that $Te_n = \lambda_n e_n$.

   (b) Show that T is bounded if and only if the sequence (λ_n) is bounded.

   (c) For a bounded sequence (λ_n), find the norm of T.

7. Let $A: \mathbb{R}^2 \to \mathbb{R}^2$ be defined by $A(x, y) = (x + 2y, 3x + 2y)$. Find the eigenvalues and eigenvectors of A.

8. Let $T: \mathbb{R}^2 \to \mathbb{R}^2$ be defined by $T(x, y) = (x + 3y, 2x + y)$. Show that $T^* \neq T$.

9. Let $A: \mathbb{R}^3 \to \mathbb{R}^3$ be given by $A(x, y, z) = (3x - z, 2y, -x + 3z)$. Show that A is self-adjoint.

10. If A is a self-adjoint operator and B is a bounded operator, show that B^*AB is self-adjoint.

11. Prove that the representation $T = A + iB$ in Theorem 4.4.13 is unique.

12. If $A^*A + B^*B = 0$, show that $A = B = 0$.

13. Let A be an operator on H. Show the following:

    (a) A is anti-Hermitian if and only if iA is self-adjoint.

    (b) $A - A^*$ is anti-Hermitian.

14. If T is self-adjoint and $T \neq 0$, show that $T^n \neq 0$ for all $n \in \mathbb{N}$.

15. Let A be a self-adjoint operator. Show the following:

    (a) $\|Ax + ix\|^2 = \|Ax\|^2 + \|x\|^2$.

    (b) The operator $U = (A - i\mathcal{I})(A + i\mathcal{I})^{-1}$ is unitary. (U is called the *Cayley transform* of A after the name of Arthur Cayley (1821–1895).)

16. Show that the limit of a convergent sequence of self-adjoint operators is a self-adjoint operator.

17. If T is a bounded operator on H with one-dimensional range, show that there exist vectors $y, z \in H$ such that $Tx = \langle x, z \rangle y$ for all $x \in H$. Hence, show the following:

    (a) $T^*x = \langle x, y \rangle z$ for all $x \in H$.

    (b) $T^2 = \lambda T$, λ is a scalar.

    (c) $\|T\| = \|y\| \|z\|$.

    (d) $T^* = T$ if and only if $y = \alpha z$ for some real scalar α.

18. Let A be a bounded self-adjoint operator on a Hilbert space H such that $\|A\| < 1$. Prove that $\langle x, (\mathcal{I} - A)x \rangle \geq (1 - \|A\|)\|x\|^2$ for all $x \in H$.

19. Show that the product of isometric operators is an isometric operator.

20. Let (e_n) be a complete orthonormal sequence in a Hilbert space H. Show that a bounded operator A on H is unitary if and only if (Ae_n) is a complete orthonormal sequence in H.

21. Let (e_n), $n \in \mathbb{Z} = \{\ldots, -2, -1, 0, 1, 2, \ldots\}$, be a complete orthonormal system in a Hilbert space H. Show that there exists a unique bounded operator A on H such that $Ax_n = x_{n+1}$ for all $n \in \mathbb{Z}$. Operator A is called a *two-sided shift operator*. Show that A is isometric and unitary.

22. Show that the product of two unitary operators is a unitary operator.

23. Let A be a bounded operator on a Hilbert space. Define the exponential operator by

$$e^A = \sum_{n=0}^{\infty} \frac{A^n}{n!} \quad (A^0 = \mathcal{I}).$$

Show that e^A is a well-defined operator. Prove the following:

(a) $(e^A)^n = e^{nA}$ for any $n \in \mathbb{N}$.

(b) $e^0 = \mathcal{I}$.

(c) e^A is invertible (even if A is not) and its inverse is e^{-A}.

(d) $e^A e^B = e^{A+B}$ for any commuting operators A and B.

(e) If A is self-adjoint, then e^{iA} is unitary.

24. If T is a normal operator on H and λ is a scalar, show that

$$\left\| T^* x - \bar{\lambda} x \right\| = \left\| Tx - \lambda x \right\| \quad \text{for all } x \in H.$$

25. Show that, if the kernel $K(x, y)$ satisfies $K(x, y) = \overline{K(y, x)}$, then for any real α the operator

$$(Tu)(x) = \alpha u(x) + i \int_a^b K(x, y) u(y) dy$$

on $L^2([a, b])$ is normal.

26. Show that, for any invertible operator T, the operator $T^* T$ is also invertible.

27. Let T and S be commuting operators. Show that if both T and S are normal, then $S + T$ and ST are normal.

28. If $T^* T = \mathcal{I}$, is it true that $TT^* = \mathcal{I}$?

29. Let A, B, C, and D be positive operators on a Hilbert space. Prove the following:

(a) If $A \geq B$ and $C \geq D$, then $A + C \geq B + D$.

(b) If $A \geq 0$ and $\alpha \geq 0$ $(\alpha \in \mathbb{R})$, then $\alpha A \geq 0$.

(c) If $A \geq B$ and $B \geq C$, then $A \geq C$.

30. If A is a positive operator and B is a bounded operator, show that $B^* AB$ is positive.

31. If A and B are positive operators and $A + B = 0$, show that $A = B = 0$.

32. Show that for any self-adjoint operator A, there exists positive operators S and T such that $A = S - T$ and $ST = 0$.

33. Show that if A is a positive definite operator, then it is invertible and its inverse is positive definite.

34. Find all operators $T: \mathbb{R}^2 \to \mathbb{R}^2$ such that $T^2 = \mathcal{I}$. Which one is the positive square root of $\mathcal{I}$?

35. Find the positive square root of the operator T on $L^2([a, b])$ defined by $(Tf)(t) = g(t)f(t)$, where g is a positive continuous function on $[a, b]$.

36. Show that $\|\sqrt{A}\| = \sqrt{\|A\|}$.

37. Let A and B be positive operators on a Hilbert space. Show that $A^2 = B^2$ implies $A = B$.

38. Let A and B be commuting positive operators. Show that $\sqrt{AB} = \sqrt{A}\sqrt{B}$.

39. If P is self-adjoint and P^2 is a projection operator, is P a projection operator?

40. Let T be a multiplication operator on $L^2([a, b])$. Find necessary and sufficient conditions for T to be a projection.

41. Give an example of two noncommuting projection operators.

42. Show that P is a projection if and only if $P = P^*P$.

43. If P, Q, and $P + Q$ are projections, show that $PQ = 0$.

44. Show that every projection P is a positive operator and $0 \leq P \leq \mathcal{I}$.

45. If T is an isometric operator, show that TT^* is a projection.

46. Show that for projections P and Q, the operator $P + Q - PQ$ is a projection if and only if $PQ = QP$.

47. Prove that the collection of all compact operators on a Hilbert space H is a vector space.

48. Show that the projection onto a closed subspace F of a Hilbert space H is a compact operator if and only if F is finite-dimensional.

49. Show that the operator $T: l^2 \to l^2$ defined by $T(x_n) = (2^{-n}x_n)$ is compact.

50. Show that a self-adjoint operator T is compact if and only if there exists a sequence of finite-dimensional operators strongly convergent to T.

51. Show that the space of all eigenvectors corresponding to a nonzero eigenvalue of a compact operator is finite-dimensional.

52. Show that eigenvalues of a symmetric operator are real and eigenvectors corresponding to different eigenvalues are orthogonal.

53. Show that shift operators have no eigenvalues.

54. Give an example of a self-adjoint operator which has no eigenvalues.

55. Give an example of a normal operator which has no eigenvalues.

56. Show that a nonzero vector x is an eigenvector of an operator A if and only if $|\langle Ax, x \rangle| = \|Ax\| \|x\|$.

57. Show that if the eigenvectors of a self-adjoint operator A form a complete orthogonal system and all eigenvalues are non-negative (or positive), then A is positive (or strictly positive).

58. Prove the spectral theorem for the finite-dimensional case: If $T : \mathbb{R}^N \to \mathbb{R}^N$ is a self-adjoint operator, then there exists an orthonormal system of vectors $\varphi_1, \dots, \varphi_N \in \mathbb{R}^N$ and scalars $\lambda_1, \dots, \lambda_N \in \mathbb{C}$ such that $T\varphi_k = \lambda_k \varphi_k$, $k = 1, \dots, N$. Hence, the matrix corresponding to T relative to the basis $(\varphi_1, \dots, \varphi_N)$ is

$$
\begin{pmatrix}
\lambda_1 & 0 & \cdots & 0 \\
0 & \lambda_2 & \cdots & 0 \\
\vdots & \vdots & \ddots & \vdots \\
0 & 0 & \cdots & \lambda_N
\end{pmatrix}.
$$

59. If λ is an approximate eigenvalue of an operator T, show that $|\lambda| \leq \|T\|$.

60. Show that if T has an approximate eigenvalue λ such that $|\lambda| = \|T\|$, then $\sup_{\|x\| \leq 1} |\langle Tx, x \rangle| = \|T\|$.

61. Show that for every bounded self-adjoint operator T on a Hilbert space, at least one of the numbers $\|T\|$ or $-\|T\|$ is an approximate eigenvalue.

62. If λ is an approximate eigenvalue of T, show that $\lambda + \mu$ is an approximate eigenvalue of $T + \mu I$ and $\lambda\mu$ is an approximate eigenvalue of μT.

63. Show that for every approximate eigenvalue λ of an isometric operator, we have $|\lambda| = 1$.

64. Show that every approximate eigenvalue of a self-adjoint operator is real.

65. Show that if λ is an approximate eigenvalue of a normal operator T, then $\bar{\lambda}$ is an approximate eigenvalue of T^*.

66. For arbitrary unbounded operators A, B, and C, prove the following:

    (a) $(A + B)C = AC + BC$.

    (b) $AB + AC \subset A(B + C)$.

    Give an example of operators A, B, and C for which $AB + AC \neq A(B + C)$.

67. Show that $(A + B)^* \supset A^* + B^*$.

68. Give an example of a closed operator whose domain is not a closed set.

69. Show that A^{**} is symmetric whenever A is symmetric.

70. If A is an operator on a Hilbert space H and there exists an operator B on H such that $\langle Ax, y \rangle = \langle x, By \rangle$ for all $x, y \in H$, show that A is bounded and $B = A^*$.

71. Let A be a closed operator in a Hilbert space H. Prove that $\mathcal{D}(A)$ is a Hilbert space with respect to the inner product defined by

$$\langle x, y \rangle_1 = \langle x, y \rangle + \langle Ax, Ay \rangle,$$

where $\langle \cdot, \cdot \rangle$ denotes the inner product in H. Prove that A is a bounded operator on $\mathcal{D}(A)$ with the defined inner product.

Applications to Integral and Differential Equations

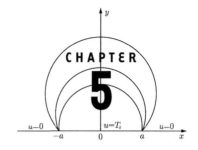

CHAPTER

5

"Since a general solution must be judged impossible from want of analysis, we must be content with the knowledge of some special cases, and that all the more, since the development of various cases seems to be the only way to bringing us at last to a more perfect knowledge."

Leonhard Euler

"The profound study of nature is the most fertile source of mathematical discoveries."

Joseph Fourier

5.1 Introduction

In this chapter we discuss some applications of the theory of Hilbert spaces to integral and differential equations. The goal is to illustrate possibilities of applications of techniques developed in Chapters 3 and 4. In Section 5.2, we give a number of important existence and uniqueness theorems for operator equations. In subsequent sections, particular applications are discussed. More specifically, Sections 5.3 to 5.7 deal with different kinds of integral equations and their applications. Ordinary differential equations and differential operators are discussed in Section 5.8. Section 5.9 is devoted to Sturm–Liouville systems. Section 5.10 deals with inverse differential operators and Green's functions. Section 5.11 is devoted to an important operator on $L^2(\mathbb{R})$: the Fourier transform. Applications of the Fourier transform to ordinary and integral equations are discussed in Section 5.12.

Equations with the unknown function under the integral sign are called *integral equations*. The equation of the form

$$\varphi(x) = \int_a^x K(x, t) f(t) dt,$$

where φ is given and f is the unknown function, is called a *Volterra equation of the first kind* (Vito Volterra (1860–1940)). The function K is called the *kernel*. If the unknown function appears also outside the integral, that is,

$$f(x) = \int_a^x K(x, t) f(t) dt + \varphi(x),$$

then the equation is called a *Volterra equation of the second kind*.

Similarly, the equations with fixed limits of integration

$$\varphi(x) = \int_a^b K(x, t) f(t) dt$$

and

$$f(x) = \int_a^b K(x, t) f(t) dt + \varphi(x),$$

are called *Fredholm equations of the first* and *second kind*, respectively (Erik Ivar Fredholm (1866–1927)). If $\varphi = 0$, the equation is called *homogeneous*, otherwise it is called *nonhomogeneous*.

5.2 Basic Existence Theorems

Equations in applied mathematics can often be written as operator equations of the form

$$Tx = x, \tag{5.1}$$

where T is an operator in a Hilbert space and x is the unknown. It is easy to recognize that (5.1) is an eigenvalue equation with the eigenvalue $\lambda = 1$.

As in Section 5.6, solutions of Equation (5.1) will be called *fixed points* of the mapping T. Fixed points are thus elements of the space that are unchanged by the action of T. There is a large body of mathematics dealing with the existence and determination of fixed points.

Recall from Chapter 1 (see Definition 1.6.2) that by a contraction mapping we mean a mapping $T: E \to E$, where E is a subset of a normed space, for which there exists a positive number $\alpha < 1$ such that

$$\|Tx - Ty\| \leq \alpha \|x - y\| \quad \text{for all } x, y \in E. \tag{5.2}$$

Now we are going to use the Banach fixed point theorem (Theorem 1.6.4) to obtain the existence and uniqueness of solutions of certain operator equations. For convenience, we restate the theorem in a slightly different form geared for applications:

Theorem 5.2.1. (Contraction mapping theorem) *Let S be a closed subset of a Banach space, and let $T : S \to S$ be a contraction mapping. Then*

(a) *the equation $Tx = x$ has one and only one solution in S, and*

(b) *the unique solution x can be obtained as the limit of the sequence (x_n) of elements of S defined by $x_n = Tx_{n-1}$, $n = 1, 2, \ldots$, where x_0 is an arbitrary element of S:*

$$x = \lim_{n \to \infty} T^n x_0. \tag{5.3}$$

This theorem not only represents an existence and uniqueness result but also gives an algorithm for finding the solution by an iterative procedure, which is known as the *method of successive approximations*. Theorem 5.2.1 can be used to prove existence and to find solutions of algebraic, differential, and integral equations.

The following is a useful generalization of Theorem 5.2.1. It will be essential in the proof of the existence and uniqueness of solutions of a certain type of integral equations; see Theorem 5.5.1.

Theorem 5.2.2. *Let E be a Banach space, and let $T : E \to E$. If T^m is a contraction for some $m \in \mathbb{N}$, then T has a unique fixed point $x_0 \in E$ and $x_0 = \lim_{n \to \infty} T^n x$ for any $x \in E$.*

Proof: In view of Theorem 5.2.1, T^m has a unique fixed point $x_0 \in E$. Then $T^m x_0 = x_0$ and hence

$$T^m(Tx_0) = T(T^m x_0) = Tx_0.$$

By uniqueness of the fixed point of T^m, we must have $Tx_0 = x_0$. The proof of the fact that $x_0 = \lim_{n \to \infty} T^n x$ for any $x \in E$ is left as an exercise. $\square$

Theorem 5.2.3. *If A is a bounded linear operator on a Banach space E, and φ is an arbitrary element of E, then the operator defined by*

$$Tf = \alpha Af + \varphi \tag{5.4}$$

has a unique fixed point for any sufficiently small $|\alpha|$. More precisely, if k is a positive constant such that

$$\|Af\| \le k\|f\| \quad \text{for all } f \in E,$$

then $Tf = f$ has a unique solution whenever $|\alpha|k < 1$.

Proof: Since A is bounded, there exists a constant k such that

$$\|Af_1 - Af_2\| \le k\|f_1 - f_2\| \quad \text{for all } f_1, f_2 \in E.$$

Thus,

$$\|Tf_1 - Tf_2\| = |\alpha|\|Af_1 - Af_2\| \le |\alpha|k\|f_1 - f_2\|,$$

and hence T is a contraction whenever $|\alpha| < 1/k$. In such a case T has a unique fixed point by Theorem 5.2.1. □

When the iterative process is applied in the case described in the previous theorem, we obtain the following sequence approximating the solution:

$$f_0 = \text{an arbitrary element of } E,$$

$$f_1 = Tf_0 = \alpha Af_0 + \varphi,$$

$$f_2 = T(\alpha Af_0 + \varphi) = \alpha^2 A^2 f_0 + \alpha A\varphi + \varphi,$$

$$\vdots$$

$$f_n = \alpha^n A^n f_0 + \alpha^{n-1} A^{n-1}\varphi + \cdots + \alpha^2 A^2\varphi + \alpha A\varphi + \varphi,$$

$$\vdots \ .$$

Therefore, the solution f can be written as

$$f = \varphi + \alpha A\varphi + \alpha^2 A^2\varphi + \cdots + \alpha^n A^n\varphi + \cdots. \tag{5.5}$$

Note that the choice of f_0 is irrelevant for the solution. However, since some choices of f_0 give faster convergence of the series, in applications it may be important to make a good "first guess."

Formally, the expansion (5.5) can be obtained directly from the equation

$$f - \alpha Af = \varphi$$

by expanding $(\mathcal{I} - \alpha A)^{-1}$ into a geometric series

$$(\mathcal{I} - \alpha A)^{-1} = \mathcal{I} + \alpha A + \alpha^2 A^2 + \cdots. \tag{5.6}$$

Theorem 4.9.8 provides a rigorous justification for this equality.

Corollary 5.2.4. *Let A be a bounded linear operator in a Banach space. Then the equation*

$$x = x_0 + \alpha Ax$$

has a unique solution given by

$$x = \sum_{n=0}^{\infty} \alpha^n A^n x_0$$

whenever $|\alpha|\,\|A\| < 1$.

The following theorem, attributed to Charles Emile Picard (1856–1941), is an important example of the application of the fixed point theorem to ordinary differential equations.

Theorem 5.2.5. (Picard's existence and uniqueness theorem) *Consider the initial value problem for the ordinary differential equation*

$$\frac{dy}{dx} = f(x, y) \tag{5.7}$$

with the initial condition

$$y(x_0) = y_0, \tag{5.8}$$

where f is a continuous function in some closed domain

$$R = \{(x, y): a \le x \le b, c \le y \le d\}$$

containing the point (x_0, y_0) in its interior. If f satisfies the Lipschitz condition

$$\left|f(x, y_1) - f(x, y_2)\right| \le K|y_1 - y_2| \tag{5.9}$$

for some $K \in R$ and all $(x, y_1), (x, y_2) \in R$, then there exists a unique solution $y = \varphi(x)$ of the problem (5.7) to (5.8) defined in some neighborhood of x_0.

Proof: Observe that every solution of the integral equation

$$y(x) = y_0 + \int_{x_0}^{x} f\big(t, y(t)\big)dt \tag{5.10}$$

satisfies (5.7) and (5.8), and conversely. Consider the operator T defined on $\mathcal{C}([a, b])$ by

$$(T\varphi)(x) = y_0 + \int_{x_0}^{x} f\big(t, \varphi(t)\big)dt. \tag{5.11}$$

Let

$$M = \sup\{\left|f(x, y)\right|: (x, y) \in R\},$$

and select $\varepsilon > 0$ such that $K\varepsilon < 1$ and $[x_0 - \varepsilon, x_0 + \varepsilon] \subset [a, b]$. If

$$S = \{\varphi(x) \in \mathcal{C}([x_0 - \varepsilon, x_0 + \varepsilon]): \left|\varphi(x) - y_0\right| \le M\varepsilon$$
$$\text{for all } x \in [x_0 - \varepsilon, x_0 + \varepsilon]\},$$

then S is a closed subset of the Banach space $\mathcal{C}([x_0 - \varepsilon, x_0 + \varepsilon])$ with the sup-norm

$$\|\varphi\| = \sup_{[x_0 - \varepsilon, x_0 + \varepsilon]} \left|\varphi(x)\right|.$$

Furthermore, if $\varphi \in S$ and $x \in [x_0 - \varepsilon, x_0 + \varepsilon]$, then

$$\left|(T\varphi)(x) - y_0\right| = \left|\int_{x_0}^{x} f(t, \varphi(t))\,dt\right| \leq M\varepsilon,$$

and thus T maps S onto itself. Finally, for any $\varphi_1, \varphi_2 \in S$, we have

$$\|T\varphi_1 - T\varphi_2\| = \sup_{[x_0-\varepsilon,x_0+\varepsilon]} \left|\int_{x_0}^{x} \big(f(t, \varphi_1(t)) - f(t, \varphi_2(t))\big)\,dt\right| \leq K\varepsilon\|\varphi_1 - \varphi_2\|.$$

Thus, since $K\varepsilon < 1$, T is a contraction. Therefore, in view of Theorem 5.2.1, there is a unique solution φ of the equation $T\varphi = \varphi$, that is, $y = \varphi$ is a unique solution of (5.10).   □

We close this section with a version of the so-called *Fredholm alternative*. It is one of the most important theorems in applied mathematics. It gives a definite criterion for existence of solutions of linear operator equations. Roughly speaking, it replaces the question of uniqueness of solution of a nonhomogeneous equation by the question of existence of a nontrivial solution of the corresponding homogeneous equation.

Theorem 5.2.6. (Fredholm alternative for self-adjoint compact operators) *Let A be a self-adjoint compact operator on a Hilbert space H. Then the nonhomogeneous operator equation*

$$f = Af + \varphi \tag{5.12}$$

has a unique solution for every $\varphi \in H$ if and only if the homogeneous equation

$$g = Ag \tag{5.13}$$

has only trivial solution $g = 0$. Moreover, if Equation (5.12) has a solution, then $\langle \varphi, g \rangle = 0$ for every solution g of (5.13).

Proof: In view of Corollary 4.10.2, H has an orthonormal basis (v_n) consisting of eigenvectors of A with corresponding eigenvalues (λ_n). Let

$$\varphi = \sum_{n=1}^{\infty} c_n v_n. \tag{5.14}$$

We seek a solution of (5.12) in the form

$$f = \sum_{n=1}^{\infty} a_n v_n.$$

Thus,

$$\sum_{n=1}^{\infty} a_n v_n = \sum_{n=1}^{\infty} a_n \lambda_n v_n + \sum_{n=1}^{\infty} c_n v_n,$$

and hence,

$$a_n = \frac{c_n}{1 - \lambda_n} \quad \text{for all } n \in \mathbb{N}, \tag{5.15}$$

provided $\lambda_n \neq 1$. If (5.13) has no nonzero solutions, 1 is not an eigenvalue of A, and hence (5.15) is valid. Therefore, if (5.12) has a solution, it has to be of the form

$$f = \sum_{n=1}^{\infty} \frac{c_n}{1 - \lambda_n} v_n. \tag{5.16}$$

This shows that if (5.12) has a solution, then it is unique.

To prove that (5.12) has a solution, it suffices to show that the series (5.16) is always convergent. By Theorem 4.9.19, we have $\lambda_n \to 0$, and therefore,

$$\frac{1}{1 - \lambda_n} \leq M$$

for some constant M and all $n \in \mathbb{N}$. Consequently,

$$\sum_{n=1}^{\infty} \left| \frac{c_n}{1 - \lambda_n} \right|^2 \leq M^2 \sum_{n=1}^{\infty} |c_n|^2 < \infty.$$

Thus, by Theorem 3.4.10, series (5.16) converges and its sum is a solution of (5.12).

If (5.13) has a nontrivial solution g and f is a solution of (5.12), then also $f + cg$ is a solution of (5.12) for any $c \in \mathbb{C}$. Therefore, (5.12) has infinitely many solutions in that case.

Finally, suppose f and g are solutions of (5.12) and (5.13), respectively. Then

$$\langle f, g \rangle = \langle Af, g \rangle + \langle \varphi, g \rangle = \langle f, Ag \rangle + \langle \varphi, g \rangle = \langle f, g \rangle + \langle \varphi, g \rangle,$$

since $Ag = g$. This means $\langle \varphi, g \rangle = 0$. Hence, if (5.12) has a solution, then φ is orthogonal to every solution of (5.13). $\qquad \square$

Theorem 5.2.6 is called an *alternative* because of the following version: Either the equation

$$f = Af + \varphi$$

has a solution for every φ or the corresponding homogeneous equation

$$f = Af$$

has a nontrivial solution.

5.3 Fredholm Integral Equations

In this and the following sections, we study solvability of integral equations.

Theorem 5.3.1. (Existence and uniqueness of solution of the Fredholm non-homogeneous linear integral equation of the second kind)   *The equation*

$$f(x) = \alpha \int_a^b K(x, y) f(y) dy + \varphi(x) \tag{5.17}$$

has a unique solution $f \in L^2([a, b])$ *provided the kernel K is continuous in* $[a, b] \times [a, b]$, $\varphi \in L^2([a, b])$, *and* $|\alpha| k < 1$, *where*

$$k = \sqrt{\int_a^b \int_a^b |K(x, y)|^2 dx dy}.$$

Proof: Consider the operator

$$(Tf)(x) = \alpha \int_a^b K(x, y) f(y) dy + \varphi(x).$$

Since $\varphi \in L^2([a, b])$, $Tf \in L^2([a, b])$ if

$$\int_a^b K(x, y) f(y) dy \in L^2([a, b]). \tag{5.18}$$

By Schwarz's inequality, we find

$$\left| \int_a^b K(x, y) f(y) dy \right| \leq \int_a^b |K(x, y) f(y)| dy$$

$$\leq \left(\int_a^b |K(x, y)|^2 dy \right)^{1/2} \left(\int_a^b |f(y)|^2 dy \right)^{1/2}.$$

Therefore,

$$\left| \int_a^b K(x, y) f(y) dy \right|^2 \leq \left(\int_a^b |K(x, y)|^2 dy \right) \left(\int_a^b |f(y)|^2 dy \right)$$

and

$$\int_a^b \left| \int_a^b K(x, y) f(y) dy \right|^2 dx \leq \int_a^b \left(\int_a^b |K(x, y)|^2 dy \int_a^b |f(y)|^2 dy \right) dx$$

$$\leq \int_a^b \int_a^b |K(x, y)|^2 dy dx \int_a^b |f(y)|^2 dy.$$

Since

$$\int_a^b \int_a^b |K(x,y)|^2 dy dx < \infty \quad \text{and} \quad \int_a^b |f(y)|^2 dy < \infty,$$

(5.18) is satisfied and thus, T maps $L^2([a,b])$ into itself.

Note that the above shows also that the operator defined by

$$(Af)(x) = \int_a^b K(x,y)f(y)dy$$

is bounded. Therefore, by Theorem 5.2.3, the equation $Tf = f$ has a unique solution whenever $|\alpha|k < 1$. $\qquad\square$

Example 5.3.2. Consider the integral equation

$$f(x) = \alpha \int_a^b e^{(x-y)/2} f(y) dy + \varphi(x), \tag{5.19}$$

where φ is a given function. Since

$$\int_a^b \int_a^b \left(e^{(x-y)/2}\right)^2 dx dy = \frac{(e^b - e^a)^2}{e^{a+b}},$$

Equation (5.19) has a unique solution whenever

$$|\alpha| < \frac{e^{(a+b)/2}}{e^b - e^a}. \qquad\square$$

Theorem 5.3.3. (Existence and uniqueness of the solution of the nonlinear Fredholm integral equation) *Suppose*

(a) $\left\| \int_a^b K(x,y,f(y))dy \right\| \leq M\|f\| \quad \text{for all } f \in L^2([a,b]),$

(b) $|K(x,y,z_1) - K(x,y,z_2)| \leq N(x,y)|z_1 - z_2| \quad \text{for all } x, y, z_1, z_2 \in [a,b],$

(c) $\int_a^b \int_a^b |N(x,y)|^2 dx dy = k^2 < \infty.$

Then the nonlinear Fredholm equation

$$f(x) = \alpha \int_a^b K(x,y,f(y))dy + \varphi(x) \tag{5.20}$$

has a unique solution $f \in L^2([a,b])$ for every $\varphi \in L^2([a,b])$ and every α such that $|\alpha|k < 1$.

Proof: Consider the operator

$$Tf = \alpha Af + \varphi,$$

where

$$(Af)(x) = \int_a^b K(x, y, f(y))dy.$$

Then

$$\|Tf_1 - Tf_2\| = |\alpha| \left\| \int_a^b \left(K(x, y, f_1(y)) - K(x, y, f_2(y)) \right) dy \right\|$$

$$\leq |\alpha| \left(\int_a^b \left(\int_a^b |K(x, y, f_1(y)) - K(x, y, f_2(y))| dy \right)^2 dx \right)^{1/2}$$

$$\leq |\alpha| \left(\int_a^b \left(\int_a^b N(x, y)|f_1(y) - f_2(y)| dy \right)^2 dx \right)^{1/2}$$

$$\leq |\alpha| k \|f_1 - f_2\|.$$

Clearly, if $|\alpha| k < 1$, then T is a contraction operator and thus it has a unique fixed point. That fixed point is a solution of Equation (5.20). $\square$

Note that when $K(x, y, f(y)) = K_1(x, y)f(y)$, Equation (5.20) reduces to (5.17). So Theorem 5.3.3 includes the linear case as well.

5.4 Method of Successive Approximations

Consider an operator equation

$$f = \varphi + \alpha Tf. \tag{5.21}$$

If T is an integral operator with a kernel K, that is,

$$Tf(x) = \int_a^b K(x, t)f(t)dt,$$

then (5.21) leads to a Fredholm integral equation of the second kind:

$$f(x) = \varphi(x) + \alpha \int_a^b K(x, t)f(t)dt. \tag{5.22}$$

In such a case, we have

$$(T^2f)(x) = T\left(\int_a^b K(x, t)f(t)dt \right)$$

$$= \int_a^b K(x, z) \left(\int_a^b K(z, t) f(t) dt \right) dz$$

$$= \int_a^b \left(\int_a^b K(x, z) K(z, t) dz \right) f(t) dt.$$

Therefore, T^2 is an integral operator whose kernel is

$$\int_a^b K(x, z) K(z, t) dz.$$

Similarly,

$$T^n f(x) = \int_a^b K_n(x, t) f(t) dt \quad \text{for } n \geq 2,$$

where the kernel K_n of T^n is given by

$$K_n(x, t) = \int_a^b K(x, \xi) K_{n-1}(\xi, t) d\xi \quad \text{for } n > 2.$$

The kernel can be also written as

$$K_n(x, t) = \int_a^b \cdots \int_a^b K(x, \xi_{n-1}) K(\xi_{n-1}, \xi_{n-2}) \cdots K(\xi_1, t) d\xi_{n-1} d\xi_{n-2} \ldots d\xi_1.$$

We next apply Corollary 5.2.4 to obtain the following result regarding the solvability of (5.21) and hence also the integral equation (5.22).

If $|\alpha| \|T\| < 1$, then Equation (5.21) has a unique solution given by the Neumann series

$$f = \varphi + \sum_{n=1}^{\infty} \alpha^n T^n \varphi. \tag{5.23}$$

Hence, the integral equation (5.22) has a unique solution f given by

$$f(x) = \varphi(x) + \alpha \int_a^b \left[\sum_{n=1}^{\infty} \alpha^{n-1} K_n(x, t) \right] \varphi(t) dt. \tag{5.24}$$

If we adopt the notation

$$\Gamma(x, t; \alpha) = \sum_{n=1}^{\infty} \alpha^{n-1} K_n(x, t),$$

then the solution can be written in the form

$$f(x) = \varphi(x) + \alpha \int_a^b \Gamma(x, t; \alpha) \varphi(t) dt. \tag{5.25}$$

The function Γ is often called the *resolvent kernel*.

Example 5.4.1. We obtain the Neumann series solution of the integral equation

$$f(x) = x + \frac{1}{2} \int_{-1}^{1} (t - x)f(t)dt.$$

First we set $f_0(x) = x$. Then

$$f_1(x) = x + \frac{1}{2} \int_{-1}^{1} (t - x)t\,dt = x + \frac{1}{3}.$$

Substituting f_1 back into the original equation, we find

$$f_1(x) = x + \frac{1}{2} \int_{-1}^{1} (t - x)\left(t + \frac{1}{3}\right)dt = x + \frac{1}{3} - \frac{1}{3}x.$$

Continuing this process, we obtain

$$f_3(x) = x + \frac{1}{3} - \frac{x}{3} - \frac{1}{3^2},$$

$$f_4(x) = x + \frac{1}{3} - \frac{x}{3} - \frac{1}{3^2} + \frac{x}{3^2},$$

$$\vdots$$

$$f_{2n}(x) = x + \sum_{m=1}^{n} (-1)^{m-1} 3^{-m} - x \sum_{m=1}^{n} (-1)^{m-1} 3^{-m}.$$

By letting $n \to \infty$, we get

$$f(x) = \frac{3}{4}x + \frac{1}{4}.$$

It is easy to verify this solution by direct substitution into the original equation. $\qquad\square$

5.5 Volterra Integral Equations

The Volterra equations

$$\int_{a}^{x} K(x, y)f(y)dy = \varphi(x), \tag{5.26}$$

and

$$f(x) - \alpha \int_{a}^{x} K(x, y)f(y)dy = \varphi(x) \tag{5.27}$$

are special cases of the Fredholm equations. The latter reduces to the corresponding Volterra equation if $K(x, y) = 0$ for $y > x$. However, Volterra equations have many important and interesting properties that cannot be obtained from the theory of Fredholm equations, so a separate study is definitely needed.

Theorem 5.5.1. (Volterra equation of the second kind) *Suppose $\varphi \in L^2([a, b])$ and the kernel K satisfies the condition*

$$\int_a^b \int_a^b |K(x, y)|^2 dxdy < \infty. \tag{5.28}$$

Then the equation

$$f(x) = \alpha \int_a^x K(x, y)f(y)dy + \varphi(x) \tag{5.29}$$

has a unique solution in $L^2([a, b])$ for arbitrary $\alpha \in \mathbb{C}$. The solution can be written in the form

$$f(x) = \varphi(x) + \sum_{n=1}^{\infty} \alpha^n \int_a^x K_n(x, t)\varphi(t)dt, \tag{5.30}$$

where the kernels $K_n(x, t)$ satisfy the recurrence relation

$$K_1(x, t) = K(x, t),$$

$$K_n(x, t) = \int_a^x K(x, \xi)K_{n-1}(\xi, t)d\xi, \quad \text{for } n \geq 2. \tag{5.31}$$

Proof: We set

$$A(x) = \int_a^x |K(x, y)|^2 dy \quad \text{and} \quad B(y) = \int_y^b |K(x, y)|^2 dx.$$

By (5.28), A and B are integrable functions, so that there exists a constant M such that

$$\int_a^b A(x)dx \leq M \quad \text{and} \quad \int_a^b B(y)dy \leq M.$$

We also introduce the function λ on $[a, b]$ defined by

$$\lambda(x) = \int_a^x A(t)dt.$$

Clearly, $0 \leq \lambda(x) \leq M$ for all $x \in [a, b]$.

Consider the operator

$$(Tf)(x) = \alpha \int_a^x K(x, y)f(y)dy + \varphi(x).$$

We shall show that T^n is a contraction for some $n \in \mathbb{N}$, and then use Theorem 5.2.2 to conclude that T has a fixed point. That fixed point must be a unique solution of (5.29).

If we write

$$Tf = \alpha Wf + \varphi,$$

where

$$(Wf)(x) = \int_a^x K(x, y)f(y)dy,$$

then

$$T^n f = \varphi + \alpha W\varphi + \alpha^2 W^2\varphi + \cdots + \alpha^n W^n f.$$

The operators W^m can be written in the form

$$(W^m g)(x) = \int_a^x K_m(x, y)g(y)dy,$$

where the kernels K_n are defined by (5.31). Indeed, for $m = 2$ we have

$$(W^2 g)(x) = \int_a^x K(x, z) \int_a^z K(z, y)g(y)dydz.$$

This integral can be considered as a double integral over the triangular region $\{(y, z): a \le y \le z \text{ and } a \le z \le x\}$ (see Figure 5.1). After interchanging the order of integration, we obtain

$$(W^2 g)(x) = \int_a^x \int_y^x K(x, z)K(z, y)dzg(y)dy.$$

If we denote

$$K_2(x, y) = \int_y^x K(x, z)K(z, y)dz,$$

then by a similar argument we get

$$(W^3 g)(x) = \int_a^x \int_y^x K(x, z)K_2(z, y)dzg(y)dy,$$

and so on, as stated earlier.

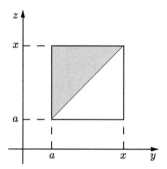

Figure 5.1   The triangular region $\{(y, z): a \le y \le z \text{ and } a \le z \le x\}$.

To estimate $\|W^m\|$, we examine K_m. For $m = 2$, application of Schwarz's inequality gives

$$\left|K_2(x, y)\right|^2 = \left|\int_y^x K(x, z)K(z, y)dz\right|^2$$

$$\le \int_y^x \left|K(x, z)\right|^2 dz \int_y^x \left|K_1(z, y)\right|^2 dz \le A(x)B(y).$$

Similarly,

$$\left|K_3(x, y)\right|^2 \le \int_y^x \left|K(x, z)\right|^2 dz \int_y^x \left|K_2(z, y)\right|^2 dz$$

$$\le A(x)B(y) \int_y^x A(z)dz = A(x)B(y)\big(\lambda(x) - \lambda(y)\big).$$

By induction, we can show that

$$\left|K_m(x, y)\right|^2 \le A(x)B(y) \frac{(\lambda(x) - \lambda(y))^{m-2}}{(m-2)!} \quad \text{for } m \ge 2.$$

Therefore,

$$\left|T^m f_1(x) - T^m f_2(x)\right|^2 = |\alpha|^{2m} \left|\int_a^x K_m(x, y)\big(f_1(y) - f_2(y)\big)dy\right|^2$$

$$\le |\alpha|^{2m} \int_a^x \frac{A(x)B(y)[\lambda(x) - \lambda(y)]^{m-2}}{(m-2)!}dy \int_a^x \left|f_1(y) - f_2(y)\right|^2 dy$$

$$\le \frac{|\alpha|^{2m}A(x)(\lambda(x))^{m-2}}{(m-2)!} \int_a^x B(y)dy\|f_1 - f_2\|^2$$

$$\le \frac{|\alpha|^{2m}A(x)(\lambda(x))^{m-2}M}{(m-2)!}\|f_1 - f_2\|^2.$$

Integrating with respect to x in $[a, b]$, we obtain

$$\left\| T^m f_1 - T^m f_2 \right\|^2 \leq \frac{|\alpha|^{2m} M^m}{(m-1)!} \|f_1 - f_2\|^2 \quad \text{for } m \geq 2.$$

Therefore, since there exists $n \in \mathbb{N}$ such that

$$\frac{|\alpha|^{2n} M^n}{(n-1)!} < 1,$$

T^n is a contraction. In view of Theorem 5.2.2, Equation (5.29) has a unique solution that can be written in the form

$$\lim_{n \to \infty} T^n f = \varphi + \alpha W \varphi + \alpha^2 W^2 \varphi + \alpha^3 W^3 \varphi + \cdots,$$

or, equivalently,

$$f(x) = \varphi(x) + \sum_{n=1}^{\infty} \alpha^n \int_a^x K_n(x, t) \varphi(t) dt. \qquad \square$$

Note that the restriction on α imposed in the result of Section 5.4 is not needed in Theorem 5.5.1.

Theorem 5.5.2. (Homogeneous Volterra equation)   *The homogeneous Volterra equation*

$$f(x) = \alpha \int_0^x K(x, t) f(t) dt, \quad x \in [0, 1] \tag{5.32}$$

has only the trivial solution $f = 0$.

Proof: From (5.32), we have

$$\left| f(x) \right| \leq |\alpha| \int_0^x \left| K(x, t) \right| \left| f(t) \right| dt \leq |\alpha| M p, \tag{5.33}$$

where

$$p = \int_0^1 \left| f(t) \right| dt$$

and M is a constant such that $|K(x, t)| \leq M$ for all $x, t \in [0, 1]$. Hence, by using (5.33) in (5.32), we obtain

$$\left| f(x) \right| \leq |\alpha| \int_0^x \left| K(x, t) \right| |\alpha| M p \, dt \leq |\alpha|^2 M^2 p x.$$

By continuing the process, we get

$$\left| f(x) \right| \leq |\alpha|^n M^n p \frac{x^{n-1}}{(n-1)!} \leq \frac{|\alpha|^n M^n p}{(n-1)!} \to 0 \quad \text{as } n \to \infty.$$

This shows $f(x) = 0$ for all $x \in [0, 1]$. □

5.6 Method of Solution for a Separable Kernel

It is of interest to examine the solvability of the Fredholm integral equation of the second kind with separable kernel. The kernel K is called *separable* if it is of the form

$$K(s, t) = \sum_{k=1}^{n} M_k(s)N_k(t), \tag{5.34}$$

where the functions M_k and N_k are assumed to belong to $L^2([a, b])$, so that the kernel is square integrable. Such kernels are also called *degenerate*. They include all polynomials and many transcendental functions, for instance

$$\cos(s + t) = \cos s \cos t - \sin s \sin t.$$

For separable kernels, the Fredholm equation of the second kind can be put into the form

$$f(x) = \varphi(x) + \alpha \sum_{k=1}^{n} M_k(x) \int_a^b N_k(t)f(t)dt \tag{5.35}$$

or

$$f(x) = \varphi(x) + \alpha \sum_{k=1}^{n} c_k M_k(x), \tag{5.36}$$

where

$$c_k = \int_a^b N_k(t)f(t)dt, \quad k = 1, \ldots, n. \tag{5.37}$$

The constants c_k depend on the unknown solution f and therefore, they are not known. On the other hand, we know that the solution of Equation (5.35) is of the form (5.36), and thus the problem is reduced to determination of the constants $c_1, \ldots, c_n$. This can be done in the following way.

First, we multiply Equation (5.36) by $N_m(x)$ and integrate with respect to x to eliminate the x-dependence. Consequently, it follows by (5.37) that

$$c_m = \alpha \sum_{k=1}^{n} a_{mk}c_k + b_m, \tag{5.38}$$

where

$$a_{mk} = \int_a^b N_m(x)M_k(x)dx \quad \text{and} \quad b_m = \int_a^b N_m(x)\varphi(x)dx.$$

Equation (5.38) can be written in the matrix form

$$(\mathcal{I} - \alpha A)c = b$$

or

$$c = (\mathcal{I} - \alpha A)^{-1}b,$$

where $A = (a_{mk})$, $b = (b_1, \ldots, b_n)$, and $c = (c_1, \ldots, c_n)$. This equation is equivalent to a system of linear simultaneous algebraic equations of the form

$$
\begin{aligned}
(1 - \alpha a_{11})c_1 - \alpha a_{12}c_2 &\quad - \alpha a_{13}c_3 - \cdots &\quad - \alpha a_{1n}c_n = b_1, \\
-\alpha a_{21}c_1 + (1 - \alpha a_{22})c_2 &\quad - \alpha a_{23}c_3 - \cdots &\quad - \alpha a_{2n}c_n = b_2, \\
\vdots &\quad \ddots &\quad \vdots \\
-\alpha a_{n1}c_1 - \alpha a_{n2}c_2 &\quad - \alpha a_{n3}c_3 - \cdots &\quad + (1 - \alpha a_{nn})c_n = b_n.
\end{aligned}
$$

This system has a unique solution if $\det(\mathcal{I} - \alpha A) \neq 0$.

If the original equation is homogeneous, that is, if $\varphi(x) = 0$, then $b = 0$. Then the system of homogeneous equations

$$(\mathcal{I} - \alpha A)c = 0 \tag{5.39}$$

has a nontrivial solution if and only if

$$\det(\mathcal{I} - \alpha A) = 0.$$

We can use the roots of this equation to find solutions of (5.35). Note that in this case we have infinitely many solutions.

Example 5.6.1. We use the above method to solve the integral equation

$$f(x) = \alpha \int_0^1 (1 - 3xt)f(t)dt. \tag{5.40}$$

We have

$$K(x, t) = 1 - 3xt = \sum_{k=1}^{2} M_k(x)N_k(t),$$

where

$$M_1(x) = 1, \quad N_1(t) = 1, \quad M_2(x) = -3x, \quad N_2(t) = t.$$

Since

$$a_{mk} = \int_0^1 N_m(x)M_k(x)dx,$$

we have

$$a_{11} = 1, \quad a_{12} = -\frac{3}{2}, \quad a_{21} = \frac{1}{2}, \quad a_{22} = -1.$$

The solution of (5.40) has the form

$$f(x) = \alpha \sum_{k=1}^{2} c_k M_k(x),$$

where

$$c_k = \int_0^1 N_k(t)f(t)dt, \quad k = 1, 2.$$

The system of algebraic equations associated with this problem is

$$(1 - \alpha)c_1 + \alpha\frac{3}{2}c_2 = 0,$$

$$-\alpha\frac{1}{2}c_1 + (1 + \alpha)c_2 = 0.$$

The roots of the equation

$$\begin{vmatrix} 1 - \alpha & \frac{3}{2}\alpha \\ -\frac{1}{2}\alpha & 1 + \alpha \end{vmatrix} = 0$$

are $\alpha = 2$ and $\alpha = -2$. Substituting them into the algebraic system, we obtain

$$\begin{matrix} c_1 = 3c_2 & \text{for } \alpha = 2, \\ c_1 = c_2 & \text{for } \alpha = -2. \end{matrix}$$

Hence the eigenfunctions are

$$f_1(x) = 2\big(c_1 M_1(x) + c_2 M_2(x)\big) = 2(c_1 - 3c_2 x) = a(1 - x) \qquad \text{for } \alpha = 2;$$
$$f_2(x) = -2\big(c_1 M_1(x) + c_2 M_2(x)\big) = -2(c_1 - 3c_2 x) = b(1 - 3x) \quad \text{for } \alpha = -2.$$

Therefore, Equation (5.40) has nontrivial solution only if $\alpha = 2$ or $\alpha = -2$ and the solutions are $f(x) = a(1 - x)$ and $f(x) = b(1 - 3x)$, respectively; a and b are arbitrary scalars. $\qquad \square$

Example 5.6.2. Consider now a nonhomogeneous equation

$$f(x) = \varphi(x) + \alpha \int_0^1 (1 - 3xt)f(t)dt. \tag{5.41}$$

If

$$\begin{vmatrix} 1 - \alpha & \frac{3}{2}\alpha \\ -\frac{1}{2}\alpha & 1 + \alpha \end{vmatrix} \neq 0,$$

then the equation has a unique solution of the form

$$f(x) = \varphi(x) + \alpha \sum_{k=1}^{2} c_k M_k(x),$$

where c_1 and c_2 are solutions of the linear system

$$(1-\alpha)c_1 + \alpha\frac{3}{2}c_2 = b_1 = \int_0^1 N_1(x)\varphi(x)dx = \int_0^1 \varphi(x)dx,$$

$$-\alpha\frac{1}{2}c_1 + (1+\alpha)c_2 = b_2 = \int_0^1 N_2(x)\varphi(x)dx = \int_0^1 x\varphi(x)dx.$$

On the other hand, if $\alpha = 2$, then the equation has a solution if and only if

$$\int_0^1 \varphi(x)(1-x)dx = 0,$$

and in this case the solution is given by

$$f(x) = \varphi(x) - 2\int_0^1 \varphi(\xi)d\xi + a(1-x).$$

Finally, if $\alpha = -2$, the equation has a solution if and only if

$$\int_0^1 \varphi(x)(1-3x)dx = 0,$$

and in this case the solution is

$$f(x) = \varphi(x) - 2\int_0^1 \xi\varphi(\xi)d\xi + b(1-3x). \qquad \square$$

5.7 Volterra Integral Equations of the First Kind and Abel's Integral Equation

In general, it is difficult to deal with Volterra integral equations of the first kind. In this section, we consider some special cases which can be solved without much difficulty. We consider the equation

$$\int_0^x K(x,t)f(t)dt = \varphi(x), \quad x \in [0,1], \tag{5.42}$$

where K and φ are assumed to be differentiable. We differentiate (5.42) with respect to x to obtain

$$K(x,x)f(x) + \int_0^x \frac{\partial}{\partial x}K(x,t)f(t)dt = \varphi'(x).$$

If $K(x, x) \neq 0$, this equation can be transformed into a Volterra integral equation of the second kind:

$$f(x) + \int_0^x \frac{\frac{\partial}{\partial x} K(x, t)}{K(x, x)} f(t) dt = \frac{\varphi'(x)}{K(x, x)}. \tag{5.43}$$

If the kernel and the function on the right-hand side of (5.43) are square integrable, then (5.43) has a unique solution in $L^2([0, 1])$ by the general theory of integral equations.

Example 5.7.1. (Abel's equation) An equation of the form

$$\int_0^x \frac{f(t)}{(x - t)^\alpha} dt = \varphi(x), \tag{5.44}$$

where $0 \leq \alpha < 1$, φ is continuous and $\varphi(0) = 0$, is called *Abel's equation*. We denote the kernel of (5.44) by K_α. If we apply the Volterra integral operator with kernel K_β $(0 \leq \beta < 1)$ to both sides of (5.44), we obtain

$$\int_0^x \left[\int_t^x \frac{dz}{(x - z)^\beta (z - t)^\alpha} \right] f(t) dt = \int_0^x \frac{\varphi(t)}{(x - t)^\beta} dt. \tag{5.45}$$

Next we put

$$z = t + (x - t)v$$

inside the square bracket on the left-hand side of (5.45) to obtain

$$\int_t^x \frac{dz}{(x - z)^\beta (z - t)^\alpha} = (x - t)^{1-\alpha-\beta} \int_0^1 \frac{dv}{(1 - v)^\beta v^\alpha}$$

$$= (x - t)^{1-\alpha-\beta} \frac{\Gamma(1 - \alpha)\Gamma(1 - \beta)}{\Gamma(2 - \alpha - \beta)},$$

where Γ denotes the *Euler gamma function* (Leonhard Euler (1707–1783)):

$$\Gamma(p) = \int_0^\infty x^{p-1} e^{-x} dx, \quad p > 0.$$

Consequently, the integral Equation (5.45) becomes

$$\int_0^x \frac{f(t)}{(x - t)^{\alpha+\beta-1}} dt = \frac{\Gamma(2 - \alpha - \beta)}{\Gamma(1 - \alpha)\Gamma(1 - \beta)} \int_0^x \frac{\varphi(t)}{(x - t)^\beta} dt.$$

In particular, if we set $a + \beta - 1 = 0$, this takes a simple form

$$\int_0^x f(t) dt = \frac{\Gamma(1)}{\Gamma(\alpha)\Gamma(1 - \alpha)} \int_0^x \frac{\varphi(t)}{(x - t)^{1-\alpha}} dt = \frac{\sin \pi \alpha}{\pi} \int_0^x \frac{\varphi(t)}{(x - t)^{1-\alpha}} dt, \tag{5.46}$$

because $\Gamma(1) = 1$ and $\Gamma(\alpha)\Gamma(1 - \alpha) = \pi / \sin \pi \alpha$.

If we assume that the right-hand side of (5.46) is differentiable, we can differentiate both sides to obtain

$$f(x) = \frac{\sin \pi \alpha}{\pi} \frac{d}{dx} \int_0^x \frac{\varphi(t)}{(x - t)^{1-\alpha}} dt. \tag{5.47}$$

This is the desired solution of (5.44).

In particular, if $\alpha = \frac{1}{2}$, then Abel's equation (5.44) has the solution

$$f(x) = \frac{1}{\pi} \frac{d}{dx} \int_0^x \frac{\varphi(t)}{\sqrt{x - t}} dt.$$

In 1825, Niels Henrik Abel (1802–1829) solved this equation in connection with the problem of tautochronous motion. This problem deals with the motion of a heavy particle sliding under the action of gravitational acceleration g, without a friction, down a curve which passes through the origin. We want to find a curve for which the particle descends to its lowest point in a constant time independent of the initial position.

The velocity of the particle at any point (ξ, η) satisfies the following equation:

$$v = \frac{ds}{dt} = \sqrt{2g}\sqrt{y - \eta}, \tag{5.48}$$

where the initial position of the particle is (x, y), s is the arc length of the curve from the initial point (x, y) to (ξ, η), and t is the time. Integrating (5.48), we obtain

$$\sqrt{2g} \int_0^T dt = - \int_y^0 \frac{ds}{\sqrt{y - \eta}} = \int_0^y \frac{ds}{\sqrt{y - \eta}}.$$

Assuming $s = f(\eta)$ with $f(0) = 0$, we get

$$\sqrt{2g}T = \int_0^y \frac{f'(\eta)}{(y - \eta)^{1/2}} d\eta. \tag{5.49}$$

This is Abel's equation with $\alpha = \frac{1}{2}$, and so its solution is

$$f'(y) = \frac{1}{\pi} \frac{d}{dy} \int_0^y \frac{\sqrt{2g}T}{(y - \eta)^{1/2}} d\eta = \frac{T\sqrt{2g}}{\pi} \frac{d}{dy} \int_0^y \frac{d\eta}{(y - \eta)^{1/2}} = \frac{T\sqrt{2g}}{\pi \sqrt{y}}.$$

Thus,

$$f(y) = 2\sqrt{2ay}, \quad a = \frac{gT^2}{\pi}. \tag{5.50}$$

This plane curve represents a cycloid with vertex at the origin and the x-axis as the tangent at the vertex. $\square$

5.8 Ordinary Differential Equations and Differential Operators

In this section, we consider operators arising from boundary value problems associated with second-order ordinary differential equations. This type of an ordinary differential equation is most common in applied mathematics, science, and engineering.

Consider the second-order linear differential equation of the form

$$Lu = f \tag{5.51}$$

over $[a, b]$ subject to two nonhomogeneous boundary conditions

$$B_1(u) = b_1, \tag{5.52}$$

$$B_2(u) = b_2, \tag{5.53}$$

where B_1 and B_2 are functions of the unknown solution u and its derivatives, and b_1, b_2 are constants. Finally, assume that the operator L has the form

$$(Lu)(x) = a_2(x)\frac{d^2}{dx^2}u(x) + a_1(x)\frac{d}{dx}u(x) + a_0(x)u(x), \tag{5.54}$$

where

$$a_2(x) \neq 0 \quad \text{on } [a, b].$$

In this section, we consider only those functions B_1 and B_2 for which boundary conditions (5.52) and (5.53) can be written as

$$B_1(u) = \alpha_{11}u(a) + \alpha_{12}u'(a) + \beta_{11}u(b) + \beta_{12}u'(b) = b_1, \tag{5.55}$$

$$B_2(u) = \alpha_{21}u(a) + \alpha_{22}u'(a) + \beta_{21}u(b) + \beta_{22}u'(b) = b_2, \tag{5.56}$$

where α_{ij} and β_{ij} are real constants. We assume that the row vectors

$$(\alpha_{11}, \alpha_{12}, \beta_{11}, \beta_{12}) \quad \text{and} \quad (\alpha_{21}, \alpha_{22}, \beta_{21}, \beta_{22})$$

are linearly independent so that (5.55) and (5.56) represent two essentially different boundary conditions.

There are two common kinds of boundary data. The first type is called the *separated boundary conditions*: $\beta_{11} = \beta_{12} = \alpha_{21} = \alpha_{22} = 0$, that is,

$$B_1(u) = \alpha_{11}u(a) + \alpha_{12}u'(a) = b_1, \tag{5.57}$$

$$B_2(u) = \beta_{21}u(a) + \beta_{22}u'(a) = b_2. \tag{5.58}$$

Another kind of boundary condition, which often arises in practice, is called the *periodic boundary conditions*:

$$u(a) = u(b), \tag{5.59}$$

$$u'(a) = u'(b). \tag{5.60}$$

Example 5.8.1. Consider the operator $L = D^2 + \omega^2$ with the separated boundary conditions $u(0) = u(1) = 0, 0 \le x \le 1$. $\qquad\qquad$ □

Example 5.8.2. Consider the deflection of a string stretched under a constant tension T_0 with fixed ends at $x = a$ and $x = b$, and subject to a force distribution $f(x)$. The differential equation and the boundary conditions for the vertical displacement $u(x)$ of the string of density $\rho(x)$ are

$$\frac{d^2 u}{dx^2} = f(x), \quad a \le x \le b;$$

$$u(a) = u(b) = 0,$$

where $f(x) = g\rho(x)/T_0$ and g is the acceleration due to gravity.

In the context of boundary value problems when we deal with a differential operator we actually have in mind the whole system: the differential operator itself and the given boundary conditions. This convention is evident in the following definition. $\qquad\qquad$ □

Definition 5.8.3.   (Domain of a differential operator)

Let L be a differential operator in $L^2([a, b])$. The *domain* of L, denoted by $\mathcal{D}(L)$, is the subset of $L^2([a, b])$ consisting of all functions for which the highest derivative in L is square integrable and satisfying the homogeneous boundary conditions.

For example, if L is a second-order differential operator, then $\mathcal{D}(L)$ consists of all functions in $L^2([a, b])$ which have square integrable second derivatives and satisfy the boundary conditions $B_1(u) = B_2(u) = 0$.

Since differential operators are unbounded, adjoint operators are defined as in Section 4.11.

Definition 5.8.4.   (Adjoint of a differential operator)

Let L be a differential operator in $L^2([a, b])$. An operator L^* is called the *adjoint* of L if

$$\langle Lu, v \rangle = \langle u, L^* v \rangle \tag{5.61}$$

for all $u \in \mathcal{D}(L)$ and $v \in \mathcal{D}(L^*)$.

The adjoint of the differential operator in (5.54) can be found as follows:

$$
\langle Lu, v \rangle = \int_a^b (Lu)(x)v(x)dx
$$

$$
= \int_a^b \big(a_2(x)u''(x) + a_1(x)u'(x) + a_0(x)u(x)\big)v(x)dx
$$

$$
= \big[a_2(x)\big(u'(x)v(x) - u(x)v'(x)\big) + \big(a_1(x) - a_2'(x)\big)u(x)v(x)\big]_a^b
$$

$$
+ \int_a^b u(x)\big(\big(a_2(x)v(x)\big)'' - \big(a_1(x)v(x)\big)' + a_0(x)v(x)\big)dx.
$$

If the right-hand side of this result is equal to $\langle u, L^*v \rangle$, we have

$$
\langle u, L^*v \rangle = \int_a^b u(x)\big(\big(a_2(x)v(x)\big)'' - \big(a_1(x)v(x)\big)' + a_0(x)v(x)\big)dx
$$

$$
+ \big[a_2(x)\big(u'(x)v(x) - u(x)v'(x)\big) + \big(a_1(x) - a_2'(x)\big)u(x)v(x)\big]_a^b.
$$

Thus, L^* consists of the second-order differential operator

$$
\frac{d^2}{dx^2}\big(a_2(x)v(x)\big) - \frac{d}{dx}\big(a_1(x)v(x)\big) + a_0(x)v(x)
$$

and some boundary terms. Therefore, the adjoint of

$$
L = a_2\frac{d^2}{dx^2} + a_1\frac{d}{dx} + a_0
$$

is

$$
L^* = a_2\frac{d^2}{dx^2} + \big(2a_2' - a_1\big)\frac{d}{dx} + \big(a_2'' - a_1' + a_0\big). \tag{5.62}
$$

The expression

$$
J(u, v) = a_2\big(vu' - uv'\big) + \big(a_1 - a_2'\big)uv \tag{5.63}
$$

is called the *bilinear concomitant* of u and v. We can say that $v \in \mathcal{D}(L^*)$ if it is twice differentiable and $[J(u, v)]_a^b = 0$ for every $u \in \mathcal{D}(L)$. From the definition of J, it follows that $[J(u, v)]_a^b = 0$ for all $u \in \mathcal{D}(L)$ only if v satisfies two homogeneous boundary conditions $B_1^*(v) = B_2^*(v) = 0$. These are referred to as *adjoint boundary conditions*. In other words, corresponding to the homogeneous boundary value problem

$$
Lu = f, \qquad B_1(u) = B_2(u) = 0,
$$

there is an adjoint boundary value problem

$$
L^*v = f, \qquad B_1^*(v) = B_2^*(v) = 0.
$$

Definition 5.8.5.   (Self-adjoint and formally self-adjoint differential operator) A differential operator L is called *self-adjoint* if $L = L^*$ and $\mathcal{D}(L) = \mathcal{D}(L^*)$. A differential operator L is called *formally self-adjoint* if $L = L^*$.

It follows from (5.62) that if L is self-adjoint, then $a_2' = a_1$. Consequently, a self-adjoint differential operator L can be written as

$$L = \frac{d}{dx}\left(a_2\frac{d}{dx}\right) + a_0, \tag{5.64}$$

and $J(u, v)$ is then given by

$$J(u, v) = a_2\left(u'v - uv'\right).$$

Example 5.8.6. We will consider the differential operator defined by

$$Lu = \frac{d^2u}{dx^2}$$

and the following boundary value problem:

$$Lu = 0, \quad a \leq x \leq b,$$
$$B_1(u) = u(a) = 0 \quad \text{and} \quad B_2(u) = u'(a) = 0.$$

Clearly, $L = L^*$. To determine the adjoint boundary conditions, we compute

$$\begin{aligned}
\left[J(u, v)\right]_a^b &= \left[u'(x)v(x) - u(x)v'(x)\right]_a^b \\
&= u'(b)v(b) - u(b)v'(b) - v(a)u'(a) + u(a)v'(b) \\
&= u'(b)v(b) - u(b)v'(b).
\end{aligned}$$

Hence $[J(u, v)]_a^b = 0$ for all $u \in \mathcal{D}(L)$ if

$$B_1^*(v) = v(b) = 0 \quad \text{and} \quad B_2^*(v) = -v'(b) = 0.$$

Although this operator is formally self-adjoint, it is not self-adjoint because $\mathcal{D}(L) \neq \mathcal{D}(L^*)$. □

Example 5.8.7. Consider the boundary value problem

$$(Lu)(x) = a_2(x)\frac{d^2}{dx^2}u(x) + a_1(x)\frac{d}{dx}u(x) + a_0(x)u(x),$$

$$u(0) = 0, \quad u'(1) = 0,$$

where $a_2(x) \neq 0$ on $[0, 1]$. Then

$$L^* = a_2 \frac{d^2}{dx^2} + (2a_2' - a_1)\frac{d}{dx} + (a_2'' - a_1' + a_0).$$

To find the adjoint boundary conditions, we compute

$$[J(u, v)]_0^1 = [a_2(x)(u'(x)v(x) - u(x)v'(x)) + u(x)v(x)(a_1(x) \quad a_2'(x))]_0^1$$
$$= (a_1(1) - a_2'(1))u(1)v(1) - a_2(1)v'(1)u(1) - a_2(0)v(0)u'(0).$$

Hence, $[J(u, v)]_0^1 = 0$ for all $u \in \mathcal{D}(L)$, if

$$B_1^* = v(0) = 0 \quad \text{and} \quad B_2^*(v) = v'(1) - \frac{(a_1(1) - a_2'(1))v(1)}{a_2(1)} = 0.$$

Thus, in general, $\mathcal{D}(L) \neq \mathcal{D}(L^*)$. However, if $a_2' = a_1$, then $L = L^*$ and

$$B_1^*(v) = v(0) = 0 \quad \text{and} \quad B_2^*(v) = v'(1) = 0,$$

so that $\mathcal{D}(L) = \mathcal{D}(L^*)$. That means that the boundary value problem is self-adjoint. □

If an operator L of the form

$$(Lu)(x) = a_2(x)\frac{d^2}{dx^2}u(x) + a_1(x)\frac{d}{dx}u(x) + a_0(x)u(x),$$

where $a_2(x) \neq 0$ on $[a, b]$, is not self-adjoint, then it can be transformed into a self-adjoint form. We multiply L by

$$\frac{1}{a_2(x)} \exp\left[\int \frac{a_1(x)}{a_2(x)}dx\right]$$

to obtain

$$\frac{1}{a_2(x)} \exp\left[\int \frac{a_1(x)}{a_2(x)}dx\right](Lu)(x)$$
$$= \frac{d}{dx}\left\{\exp\left[\int \frac{a_1(x)}{a_2(x)}dx\right]\frac{d}{dx}u(x)\right\} + \frac{a_0(x)}{a_2(x)}\exp\left[\int \frac{a_1(x)}{a_2(x)}dx\right]u(x),$$

which is a formally self-adjoint operator.

To study the second-order differential equations, it is convenient to express them in the form

$$(Lu)(x) + \lambda \omega(x)u(x) = 0, \quad a \leq x \leq b, \tag{5.65}$$

where L is self-adjoint, that is,

$$L = \frac{d}{dx}\left(p\frac{d}{dx}\right) + q,$$

λ is a constant parameter, and ω is a given function, called the *weight function*. We assume $\omega(x) > 0$ except possibly at isolated points where $\omega(x) = 0$.

For a given value of the parameter λ, a function u_λ satisfying (5.65) and prescribed boundary conditions is called an *eigenfunction* corresponding to λ, which is then called an *eigenvalue*. Clearly, there is no guarantee that for every value of λ there exists an eigenfunction u_λ.

Example 5.8.8. (Legendre operator and Legendre polynomials) The Legendre differential equation is

$$\frac{d}{dx}\left[(1-x^2)\frac{du}{dx}\right] + \lambda u = 0, \quad -1 \le x \le 1$$

($p(x) = 1 - x^2$, $q(x) = 0$, $\omega(x) = 1$). The eigenfunctions of this equation are called the *Legendre polynomials*:

$$P_n(x) = \frac{1}{2^n n!}\frac{d^n}{dx^n}(x^2 - 1)^n,$$

and the corresponding eigenvalues are $\lambda_n = n(n+1)$, $n = 1, 2, 3, \ldots$. By the *Legendre operator*, we mean

$$Lu = -\frac{d}{dx}\left[(1-x^2)\frac{du}{dx}\right]. \qquad \square$$

Example 5.8.9. (Associated Legendre operator and associated Legendre functions) The eigenfunctions of the associated Legendre differential equation

$$\frac{d}{dx}\left[(1-x^2)\frac{du}{dx}\right] - \frac{m^2 u}{1-x^2} + \lambda u = 0, \quad -1 \le x \le 1$$

($p(x) = 1 - x^2$, $q(x) = -m^2/(1-x^2)$, $\omega(x) = 1$) are the *associated Legendre functions*:

$$P_n^m(x) = (1-x^2)^{m/2}\frac{d^m}{dx^m}P_n(x), \quad n > m,$$

and the corresponding eigenvalues are $\lambda_n = n(n+1)$, $n = 1, 2, 3, \ldots$. Note that the operator

$$Lu = -\frac{d}{dx}\left[(1-x^2)\frac{du}{dx}\right] + \frac{m^2 u}{1-x^2}$$

is positive definite. $\qquad \square$

Example 5.8.10. (Chebyshev operator and Chebyshev polynomials) The Chebyshev differential equation is

$$\sqrt{1-x^2}\frac{d}{dx}\left[\sqrt{1-x^2}\frac{du}{dx}\right] + \lambda u = 0, \quad -1 \le x \le 1$$

$(p(x) = \sqrt{1-x^2}, q(x) = 0, \omega(x) = (1-x^2)^{-1/2})$. The eigenfunctions

$$T_n(x) = \cos\left(n\cos^{-1}x\right)$$

are called the *Chebyshev polynomials*. The corresponding eigenvalues are $\lambda_n = n^2$. The *Chebyshev operator* is defined by

$$Lu = -\frac{d}{dx}\left[\sqrt{1-x^2}\frac{du}{dx}\right]. \qquad \square$$

Example 5.8.11. (Jacobi operator and Jacobi polynomials) The differential equation

$$(1-x)^{-\alpha}(1+x)^{-\beta}\frac{d}{dx}\left[(1-x)^{\alpha+1}(1+x)^{\beta+1}\frac{du}{dx}\right] + \lambda u = 0, \quad -1 \le x \le 1,$$

with $\alpha > -1$ and $\beta > -1$, is called the *Jacobi differential equation* $(p(x) = (1-x)^{\alpha+1}(1+x)^{\beta+1}, q(x) = 0, \omega(x) = (1-x)^\alpha (1+x)^\beta)$. The functions

$$P_n^{(\alpha,\beta)}(x) = \frac{(-1)^n}{2^n n!}\left[(1-x)^{\alpha+n}(1+x)^{\beta+n}\right],$$

called the *Jacobi polynomials*, are the eigenfunctions of the Jacobi equation. The eigenvalues are $\lambda_n = n(n+\alpha+\beta+1)$. The *Jacobi operator* is defined by

$$Lu = -\frac{d}{dx}\left[(1-x)^{\alpha+1}(1+x)^{\beta+1}\frac{du}{dx}\right].$$

If $\alpha = \beta = 0$, the Jacobi equation becomes the Legendre equation. For $\alpha = \beta = -\frac{1}{2}$, it reduces to the Chebyshev equation. If $\alpha = \beta = \gamma - \frac{1}{2}$, then we get the Gegenbauer polynomials $C_n^\gamma(x)$. $\qquad \square$

Example 5.8.12. (Laguerre operator and Laguerre polynomials) The differential equation

$$e^x\frac{d}{dx}\left[xe^{-x}\frac{du}{dx}\right] + \lambda u = 0, \quad 0 < x < \infty,$$

is called the *Laguerre differential equation*. The *Laguerre operator* is

$$Lu = -\frac{d}{dx}\left[xe^{-x}\frac{du}{dx}\right].$$

Its eigenvalues are $\lambda_n = n$ and the corresponding eigenfunctions, called the *Laguerre polynomials*, are defined by

$$L_n(x) = \frac{1}{n!} e^x \frac{d^n}{dx^n} \left(x^n e^{-x} \right) = \sum_{k=1}^{n} \binom{n}{k} \frac{x^k}{k!}. \qquad \square$$

Example 5.8.13. (Associated Laguerre operator and associated Laguerre functions) The equation

$$x^{-\alpha} e^x \frac{d}{dx} \left[x^{\alpha+1} e^{-x} \frac{du}{dx} \right] + \lambda u = 0, \qquad 0 \leq x \leq \infty,$$

is called the *associated Laguerre equation*. The *associated Laguerre functions*

$$L_n^\alpha(x) = (-1)^n \frac{d^\alpha}{dx^\alpha} \left[L_{n+\alpha}(x) \right] = \frac{e^x x^{-\alpha}}{n!} \frac{d^n}{dx^n} \left(e^{-x} x^{n+\alpha} \right)$$

are the eigenfunctions of the differential operator

$$Lu = -\frac{d}{dx} \left[x^{\alpha+1} e^{-x} \frac{du}{dx} \right]$$

with the eigenvalues $\lambda_n = n - \alpha$. $\qquad \square$

Example 5.8.14. (Bessel operator and Bessel functions) By the *Bessel equation* we mean

$$\frac{1}{x} \frac{d}{dx} \left[x \frac{du}{dx} \right] - \frac{v^2}{x^2} u + \lambda u = 0, \qquad 0 < x < a.$$

The corresponding operator is

$$Lu = -\frac{d}{dx} \left[x \frac{du}{dx} \right] + \frac{v^2}{x} u,$$

and its eigenfunctions are the Bessel functions $J_v(k_n x)$ with the eigenvalues $\lambda_n = k_n^2$, $n = 1, 2, 3, \ldots$, where $a k_n$ are positive zeros of $J_v(t)$. The Bessel operator is positive definite, and its eigenfunctions form a complete orthogonal system in $L^2[(0, a])$ with respect to the inner product with the weight function $\omega(x) = x$. $\qquad \square$

Example 5.8.15. (Hermite operator and Hermite polynomials) The differential equation

$$e^{x^2} \frac{d}{dx} \left[e^{-x^2} \frac{du}{dx} \right] + \lambda u = 0, \qquad -\infty < x < \infty,$$

and the operator

$$Lu = -\frac{d}{dx}\left[e^{-x^2}\frac{du}{dx}\right]$$

are called the *Hermite equation* and *Hermite operator*, respectively. The eigen-functions

$$H_n(x) = (-1)^n e^{x^2}\frac{d^n}{dx^n}\left(e^{-x^2}\right)$$

are called the *Hermite polynomials*; the eigenvalues are $\lambda_n = 2n$. □

5.9 Sturm–Liouville Systems

In this section we are going to concentrate on a special class of boundary value problems referred to as the *Sturm–Liouville system* (Jacques Charles François Sturm (1803–1855), Joseph Liouville (1809–1882)).

Definition 5.9.1. (Regular Sturm–Liouville system)
A *regular Sturm–Liouville system* consists of an ordinary differential equation

$$\frac{d}{dx}\left[p(x)\frac{du}{dx}\right] + \big(q(x) + \lambda\omega(x)\big)u = 0, \quad a \leq x \leq b, \tag{5.66}$$

with the boundary conditions

$$a_1 u(a) + a_2 u'(a) = 0, \tag{5.67a}$$

$$b_1 u(b) + b_2 u'(b) = 0, \tag{5.67b}$$

where p, q, and ω are continuous real-valued functions on $[a, b]$, p and ω are positive in $[a, b]$, $p'(x)$ exists and is continuous in $[a, b]$, a_1, a_2, b_1, b_2 are given real numbers such that $a_1^2 + a_2^2 > 0$ and $b_1^2 + b_2^2 > 0$.

Definition 5.9.2. (Singular Sturm–Liouville system)
Suppose p, q, and ω are functions defined in $[a, b]$ satisfying all the conditions in Definition 5.9.1 except that p is only assumed to be positive in (a, b), and vanishes at one or both end points of the interval $[a, b]$. By the *singular Sturm–Liouville system*, we mean the system consisting of the differential equation (5.66) in the open or semiopen interval with boundary conditions

(a) u is bounded on (a, b),

(b) if p does not vanish at an end point, then p satisfies boundary conditions of the form (5.67a) and (5.67b) at that end point.

Example 5.9.3. Consider the regular Sturm–Liouville system

$$u'' + \lambda u = 0, \quad 0 \le x \le \pi,$$

$$u(0) = u(\pi) = 0.$$

Suppose $\lambda < 0$ and let $\nu = \sqrt{|\lambda|}$. Then, the general solution of the equation is

$$u(x) = Ae^{\nu x} + Be^{-\nu x}.$$

This solution satisfies the given boundary conditions if and only if

$$A + B = 0,$$

$$Ae^{\nu \pi} + Be^{-\nu \pi} = 0.$$

Since $\nu > 0$, $e^{\nu \pi} \ne e^{-\nu \pi}$ and therefore the only solution is $A = B = 0$. This means that the system has no nonzero solutions if $\lambda < 0$. In other words, there are no negative eigenvalues. A similar argument shows that $\lambda = 0$ is not an eigenvalue.

However, when $\lambda > 0$, then the solutions of the equation are

$$u(x) = A \cos \sqrt{\lambda} x + B \sin \sqrt{\lambda} x.$$

The boundary conditions give

$$A = 0 \quad \text{and} \quad B \sin \sqrt{\lambda} \pi = 0.$$

Since $\lambda \ne 0$ and $B = 0$ yields trivial solution, we must have $B \ne 0$ and

$$\sin \sqrt{\lambda} \pi = 0.$$

Hence, the eigenvalues are $\lambda_n = n^2$, $n = 1, 2, \ldots$, and the eigenfunctions are

$$u_n(x) = \sin nx.$$

Note that $\lambda_n \to \infty$ as $n \to \infty$, unlike the case of self-adjoint compact operators when the eigenvalues converge to 0; see Theorem 4.9.19. Section 5.10, particularly Theorem 5.10.4, will explain this.   □

Another type of problem that often occurs in practice is the *periodic Sturm–Liouville system*:

$$\frac{d}{dx}\left[p(x)\frac{du}{dx}\right] + \big(q(x) + \lambda \omega(x)\big)u = 0, \quad a \le x \le b,$$

$$p(a) = p(b), \qquad u(a) = u(b), \qquad u'(a) = u'(b).$$

Example 5.9.4. Find the eigenvalues and eigenfunctions of the following periodic Sturm–Liouville system

$$u'' + \lambda u = 0, \quad -\pi \leq x \leq \pi,$$

$$u(-\pi) = u(\pi), \qquad u'(-\pi) = u'(\pi).$$

Note that here $p(x) = 1$, and hence $p(-\pi) = p(\pi)$. When $\lambda > 0$ the general solution of the equation is

$$u(x) = A \cos \sqrt{\lambda} x + B \sin \sqrt{\lambda} x.$$

Using the boundary conditions, we get

$$2B \sin \sqrt{\lambda} \pi = 0,$$

$$2A\sqrt{\lambda} \sin \sqrt{\lambda} \pi = 0.$$

Thus, for nontrivial solutions, we must have

$$\sin \sqrt{\lambda} \pi = 0.$$

The equation is satisfied if

$$\lambda = \lambda_n = n^2, \quad n = 1, 2, 3, \ldots.$$

For every eigenvalue $\lambda_n = n^2$, we have two linearly independent solutions $\cos nx$ and $\sin nx$.

It can be readily shown that the system has no negative eigenvalues. However, $\lambda = 0$ is an eigenvalue and the corresponding eigenfunction is the constant function $u(x) = 1$. Thus, the eigenvalues are

$$0, 1, 4, \ldots, n^2, \ldots,$$

and the corresponding eigenfunctions are

$$1, \cos x, \sin x, \cos 2x, \sin 2x, \ldots, \cos nx, \sin nx, \ldots.$$

Throughout the remainder of this section, L denotes the differential operator in the Sturm–Liouville differential equation, that is,

$$Lu = \frac{d}{dx}\left[p(x)\frac{du}{dx} \right] + q(x)u.$$

For the regular Sturm–Liouville system, we denote by $\mathcal{D}(L)$ the domain of L, that is, $\mathcal{D}(L)$ is the space of all complex-valued functions u defined on $[a, b]$

for which u'' belongs to $L^2([a, b])$ and that satisfy boundary conditions (5.67a) and (5.67b). Then, we have

$$L : \mathcal{D}(L) \rightarrow L^2([a, b]).$$

For the singular Sturm–Liouville system, we need only to replace (5.67a) and (5.67b) by

(a) u is bounded on (a, b),

(b) $b_1 u(b) + b_2 u'(b) = 0,$

where b_1 and b_2 are real constants such that $b_1^2 + b_2^2 > 0$.   □

The following identity is attributed to Joseph-Louis Lagrange (1736–1813).

Theorem 5.9.5. (Lagrange's identity)   *For any $u, v \in \mathcal{D}(L)$, we have*

$$uLv - vLu = \frac{d}{dx}\left[p\left(u\frac{dv}{dx} - v\frac{du}{dx}\right)\right]. \tag{5.68}$$

Proof: We have

$$uLv - vLu = u\frac{d}{dx}\left[p\frac{dv}{dx}\right] + quv - v\frac{d}{dx}\left[p\frac{du}{dx}\right] - quv$$

$$= \frac{d}{dx}\left[p\left(u\frac{dv}{dx} - v\frac{du}{dx}\right)\right]. \qquad \square$$

Theorem 5.9.6. (Abel's formula)   *If u and v are two solutions of the equation*

$$Lu + \lambda\omega u = 0 \tag{5.69}$$

in $[a, b]$, then

$$p(x)W(x; u, v) = constant,$$

where W is the Wronskian defined by

$$W(x; u, v) = \begin{vmatrix} u(x) & u'(x) \\ v(x) & v'(x) \end{vmatrix}.$$

The Wronskian is named after Josef Hoëné de Wronski (1778–1853).

Proof: Since u and v are solutions of (5.69), we have

$$\frac{d}{dx}\left[p(x)\frac{du}{dx}\right] + \left(q(x) + \lambda\omega(x)\right)u = 0,$$

$$\frac{d}{dx}\left[p(x)\frac{dv}{dx}\right] + \left(q(x) + \lambda\omega(x)\right)v = 0.$$

Multiplying the first equation by v and the second by u, and then subtracting, we obtain

$$u\frac{d}{dx}\left[p\frac{dv}{dx}\right] - v\frac{d}{dx}\left[p\frac{du}{dx}\right] = 0.$$

By integrating this equation from a to x, we find

$$p(x)\left[u(x)v'(x) - u'(x)v(x)\right] - p(a)\left[u(a)v'(a) - u'(a)v(a)\right] = \text{constant}.$$

This is Abel's formula. $\qquad\qquad\square$

Theorem 5.9.7. *Eigenfunctions of a regular Sturm–Liouville system are unique except for a constant factor.*

Proof: Suppose u and v are eigenfunctions corresponding to the same eigenvalue λ. According to Abel's formula, we have

$$p(x)W(x; u, v) = \text{constant}.$$

Since $p > 0$, if $W(x; u, v)$ vanishes at a point in $[a, b]$, then it vanishes everywhere in $[a, b]$. From the boundary conditions, we have

$$a_1 u(a) + a_2 u'(a) = 0,$$
$$a_1 v(a) + a_2 v'(a) = 0.$$

Since a_1 and a_2 are not both zero, we get

$$W(a; u, v) = \begin{vmatrix} u(a) & u'(a) \\ v(a) & v'(a) \end{vmatrix} = 0.$$

Therefore, $W(x; u, v) = 0$ for all $x \in [a, b]$, which proves linear dependence of u and v. $\qquad\qquad\square$

Theorem 5.9.8. *For any $u, v \in \mathcal{D}(L)$, we have*

$$\langle Lu, v \rangle = \langle u, Lv \rangle,$$

where $\langle , \rangle$ denotes the inner product of $L^2([a, b])$. In other words, L is a self-adjoint operator.

Proof: Since all constants involved in the boundary conditions of a Sturm–Liouville system are real, if $v \in \mathcal{D}(L)$, then $\bar{v} \in \mathcal{D}(L)$. Also, since p, q, and ω are real-valued, $\overline{Lv} = L\bar{v}$. Consequently,

$$\langle Lu, v \rangle - \langle u, Lv \rangle = \int_a^b (\bar{v}Lu - uL\bar{v})dx = \left[p(u'\bar{v} - u\bar{v}')\right]_a^b, \qquad (5.70)$$

by Lagrange's identity (5.68). We will show that the last term in the preceding equality vanishes for both the regular and singular systems. If $p(a) = 0$, the result follows immediately. If $p(a) > 0$, then u and v satisfy boundary conditions of the form (5.67a) and (5.67b) at $x = a$. That is,

$$\begin{bmatrix} u(a) & u'(a) \\ \overline{v}(a) & \overline{v}'(a) \end{bmatrix} \begin{bmatrix} a_1 \\ a_2 \end{bmatrix} = 0.$$

Since a_1 and a_2 are not both zero, we have

$$u(a)\overline{v}'(a) - \overline{v}(a)u'(a) = 0.$$

A similar argument can be applied to the other end point $x = b$, so that we conclude

$$\left[p(u\overline{v}' - \overline{v}u') \right]_a^b = 0. \qquad \square$$

Theorem 5.9.9. *Eigenvalues of a Sturm–Liouville system are real.*

Proof: Let λ be an eigenvalue of a Sturm–Liouville system, and let u be the corresponding eigenfunction. This means that $u \neq 0$ and $Lu = -\lambda \omega u$. Then

$$0 = \langle Lu, u \rangle - \langle u, Lu \rangle = \langle -\lambda \omega u, u \rangle - \langle u, -\lambda \omega u \rangle$$

$$= \left(\overline{\lambda} - \lambda \right) \int_a^b \omega(x) |u(x)|^2 dx.$$

Since $\omega(x) > 0$ in $[a, b]$ and $u \neq 0$, the integral is a positive number. Therefore $\overline{\lambda} = \lambda$, completing the proof. $\qquad \square$

This theorem states that all eigenvalues of a regular Sturm–Liouville system are real, but it does not guarantee that an eigenvalue exists. It is proved in Section 5.10 that a regular Sturm–Liouville system has an infinite sequence of eigenvalues.

Theorem 5.9.10. *Eigenfunctions corresponding to distinct eigenvalues of a Sturm–Liouville system are orthogonal with respect to the inner product with the weight function $\omega(x)$.*

Proof: Suppose u_1 and u_2 are eigenfunctions corresponding to eigenvalues λ_1 and λ_2, $\lambda_1 \neq \lambda_2$. Thus,

$$Lu_1 = -\lambda \omega u_1 \quad \text{and} \quad Lu_2 = -\lambda_2 \omega u_2.$$

Hence,

$$u_1 Lu_2 - u_2 Lu_1 = -\lambda_2 \omega u_1 u_2 + \lambda_1 \omega u_1 u_2 = (\lambda_1 - \lambda_2)\omega u_1 u_2. \qquad (5.71)$$

By Theorem 5.9.5, we have

$$u_1 L u_2 - u_2 L u_1 = \frac{d}{dx} \left[p \left(u_1 \frac{du_2}{dx} - u_2 \frac{du_1}{dx} \right) \right]. \tag{5.72}$$

Combining (5.71) and (5.72) and integrating from a to b, we get

$$(\lambda_1 - \lambda_2) \int_a^b \omega(x) u_1(x) u_2(x) dx$$

$$= \left[p(x) \left(u_1(x) \frac{du_2(x)}{dx} - u_2(x) \frac{du_1(x)}{dx} \right) \right]_a^b = 0,$$

by boundary conditions (5.67a) and (5.67b). Since $\lambda_1 \neq \lambda_2$, we conclude

$$\int_a^b \omega(x) u_1(x) u_2(x) dx = 0. \qquad \square$$

5.10 Inverse Differential Operators and Green's Functions

A typical boundary value problem for an ordinary differential equation can be written in the operator form as

$$Lu = f. \tag{5.73}$$

We seek a solution u, which satisfies this equation and the given boundary conditions. If $\mathcal{D}(L)$ is defined as the space of functions satisfying those boundary conditions, then the problem reduces to finding a solution of (5.73) in $\mathcal{D}(L)$. One way to approach the problem is by looking for the inverse operator L^{-1}. If it is possible to find L^{-1}, then the solution of (5.73) can be obtained as $u = L^{-1}(f)$. It turns out that it is possible in many important cases, and the inverse operator is an integral operator of the form

$$u = (L^{-1}f)(x) = \int_a^b G(x, t) f(t) dt.$$

The function G is called the *Green's function* (George Green (1793–1841)) of the operator L. Existence of the Green's function and its determination is not a simple problem. However, we will examine the question more closely in the case of Sturm–Liouville systems.

Theorem 5.10.1. *Suppose $\lambda = 0$ is not an eigenvalue of the following regular Sturm–Liouville system:*

$$Lu = \frac{d}{dx} \left[p(x) \frac{du}{dx} \right] + q(x) u = f(x), \quad a \leq x \leq b, \tag{5.74}$$

with the homogeneous boundary conditions

$$a_1 u(a) + a_2 u'(a) = 0, \tag{5.75}$$

$$b_1 u(b) + b_2 u'(b) = 0, \tag{5.76}$$

where p and q are continuous real-valued functions on $[a, b]$, p is positive in $[a, b]$, $p'(x)$ exists and is continuous in $[a, b]$, a_1, a_2, b_1, b_2 are given real numbers such that $a_1^2 + a_2^2 > 0$ and $b_1^2 + b_2^2 > 0$. Then, for any $f \in C([a, b])$, the system has a unique solution

$$u(x) = \int_a^b G(x, t) f(t) dt,$$

where $G(x, t)$ is the Green's function given by

$$G(x, t) = \begin{cases} \dfrac{u_2(x)u_1(t)}{p(t)W(t)} & \text{for } a \leq t < x, \\[2mm] \dfrac{u_1(x)u_2(t)}{p(t)W(t)} & \text{for } x < t \leq b, \end{cases}$$

where u_1 and u_2 are nonzero solutions of the homogeneous system

$$\frac{d}{dx}\left[p(x)\frac{du}{dx} \right] + q(x)u = 0$$

with boundary conditions (5.75) and (5.76), respectively, and

$$W(t) = u_1(t)u_2'(t) - u_2(t)u_1'(t)$$

is the Wronskian.

Proof: According to the theory of ordinary differential equations, the general solution of (5.74) is of the form

$$u(x) = c_1 u_1(x) + c_2 u_2(x) + u_p(x), \tag{5.77}$$

where c_1 and c_2 are constants, u_1 and u_2 are two linearly independent solutions of the homogeneous equation $Lu = 0$, and u_p is any particular solution of (5.74).

The particular solution u_p can be found by the method of variation of parameters. Thus, we are looking for a solution in the form

$$u_p(x) = v_1(x)u_1(x) + v_2(x)u_2(x), \tag{5.78}$$

where v_1 and v_2 are functions to be determined. Since there are infinitely many pairs of functions v_1 and v_2 for which u_p satisfies (5.74), we impose a second condition:

$$v_1' u_1 + v_2' u_2 = 0. \tag{5.79}$$

We now have

$$u_p' = v_1 u_1' + v_2 u_2'$$

and

$$u_p'' = v_1 u_1'' + v_2 u_2'' \mid v_1' u_1' + v_2' u_2'.$$

Substituting into (5.74), we get

$$v_1 \left(p u_1'' + p' u_1' + q u_1 \right) + v_2 \left(p u_2'' + p' u_2' + q u_2 \right) + p \left(v_1' u_1' + v_2' u_2' \right) = f.$$

Since u_1 and u_2 are solutions of the homogeneous equation, the first two terms vanish so that the above result becomes

$$v_1' u_1' + v_2' u_2' = \frac{f}{p}. \tag{5.80}$$

Solving (5.79) and (5.80) for v_1' and v_2', we obtain

$$v_1'(x) = -\frac{f(x) u_2(x)}{p(x) W(x; u_1, u_2)} \quad \text{and} \quad v_2'(x) = \frac{f(x) u_1(x)}{p(x) W(x; u_1, u_2)}. \tag{5.81}$$

We will show that the Wronskian does not vanish at any point of $[a, b]$. Indeed, suppose that

$$W = u_1 u_2' - u_2 u_1' = \begin{vmatrix} u_1 & u_2 \\ u_1' & u_2' \end{vmatrix}$$

vanishes at some $\xi \in [a, b]$. Then the system of equations

$$\alpha u_1(\xi) + \beta u_2(\xi) = 0,$$
$$\alpha u_1'(\xi) + \beta u_2'(\xi) = 0$$

has a nontrivial solution such that α and β are not both zero. Then the function $g = \alpha u_1 + \beta u_2$ is a solution of the initial value problem

$$\frac{d}{dx} \left[p(x) \frac{du}{dx} \right] + q(x) u = 0, \qquad g(\xi) = g'(\xi) = 0.$$

But we know that the above problem has only the trivial solution, and thus $g = 0$. This means that u_1 and u_2 are linearly dependent, contrary to the assumption.

By Abel's formula (Theorem 5.9.6), $p(x) W(x; u_1, u_2)$ is a constant. Since W does not vanish in $[a, b]$ and p is assumed to be positive, the constant is not zero. Denote

$$c = \frac{1}{p(x)W(x; u_1, u_2)}.$$

Now, by integrating differential equations (5.81), we get

$$v_1(x) = -\int cf(x)u_2(x)dx \quad \text{and} \quad v_2(x) = \int cf(x)u_1(x)dx,$$

and finally,

$$u_p(x) = -cu_1(x)\int_b^x f(t)u_2(t)dt + cu_2(x)\int_a^x f(t)u_1(t)dt \quad (5.82)$$

$$= \int_a^x cu_2(x)u_1(t)f(t)dt + \int_x^b cu_1(x)u_2(t)f(t)dt. \quad (5.83)$$

Consequently, if we define the Green's function as

$$G(x, t) = \begin{cases} cu_2(x)u_1(t) & \text{for } a \le t < x, \\ cu_1(x)u_2(t) & \text{for } x < t \le b, \end{cases} \quad (5.84)$$

we can write

$$u_p(x) = \int_a^b G(x, t)f(t)dt$$

provided the integral exists. This follows immediately from the continuity of G. The proof of continuity of G is left as an exercise. □

Denote by T the integral operator defined in Theorem 5.10.1, that is,

$$(Tf)(x) = \int_a^b G(x, t)f(t)dt. \quad (5.85)$$

We are going to examine properties of T.

Theorem 5.10.2. *The operator T defined by (5.85) is a self-adjoint compact operator from $L^2([a, b])$ into $C([a, b])$.*

Proof: Compactness of integral operators was discussed in Example 4.8.4. In Example 4.4.6 we prove that an integral operator of the form (5.85) is self-adjoint if $G(x, t) = \overline{G(t, x)}$. It is easy to see that, in this case, the condition is satisfied. Finally, continuity of Tf follows from the continuity of G. □

Operator T, as a compact, self-adjoint operator, admits spectral representation (see Corollary 4.10.2). The following two theorems describe the connection between eigenvalues and eigenfunctions of the regular Sturm–Liouville operator L and the corresponding integral operator T.

Theorem 5.10.3. *If $\lambda = 0$ is not an eigenvalue of the Sturm–Liouville system defined in Theorem 5.10.1, then $\lambda = 0$ is not an eigenvalue of the integral operator T defined by (5.85).*

Proof: Suppose $Tf = 0$. Then

$$0 = (Tf)'(x) = \frac{d}{dx}\left[-cu_1(x) \int_b^x f(t)u_2(t)dt + cu_2(x) \int_a^x f(t)u_1(t)dt \right]$$

$$= c\left(-u_1'(x) \int_b^x f(t)u_2(t)dt + u_1(x)u_2(x)f(x) \right.$$

$$\left. + u_2'(x) \int_a^x f(t)u_1(t)dt - u_1(x)u_2(x)f(x) \right)$$

$$= c\left(-u_1'(x) \int_b^x f(t)u_2(t)dt + u_2'(x) \int_a^x f(t)u_1(t)dt \right).$$

Therefore, we have the following system of equations:

$$u_1(x) \int_b^x f(t)u_2(t)dt - u_2(x) \int_a^x f(t)u_1(t)dt = 0,$$

$$u_1'(x) \int_b^x f(t)u_2(t)dt - u_2'(x) \int_a^x f(t)u_1(t)dt = 0.$$

Since the determinant

$$\begin{vmatrix} u_1 & u_2 \\ u_1' & u_2' \end{vmatrix}$$

does not vanish at any point of $[a, b]$ (see the proof of Theorem 5.10.1), we conclude

$$\int_b^x f(t)u_2(t)dt = 0 \quad \text{and} \quad \int_a^x f(t)u_1(t)dt = 0$$

for all $x \in [a, b]$. This implies $f = 0$, and thus, the equation $Tf = 0$ has only the trivial solution. □

Theorem 5.10.4. *Under the assumptions of Theorem 5.10.1, λ is an eigenvalue of L if and only if $1/\lambda$ is an eigenvalue of T. Moreover, if f is an eigenfunction of L corresponding to the eigenvalue λ, then f is an eigenfunction of T corresponding to the eigenvalue $1/\lambda$.*

Proof: Suppose $Lf = \lambda f$ for some nonzero f in the domain of L. By Theorem 5.10.1, we have

$$f = T(\lambda f)$$

or, equivalently, since $\lambda \neq 0$,

$$Tf = \frac{1}{\lambda}f.$$

This shows that $1/\lambda$ is an eigenvalue of T and f is the corresponding eigenfunction.

Conversely, if f is an eigenfunction of T corresponding to λ, $f \neq 0$ and $\lambda \neq 0$, then

$$Tf = \lambda f,$$

and hence,

$$f = L(Tf) = L(\lambda f) = \lambda L f.$$

Therefore, $1/\lambda$ is an eigenvalue of L and the corresponding eigenfunction is f. $\qquad\qquad\square$

5.11 The Fourier Transform

In this section, we introduce the Fourier transform in $L^2(\mathbb{R})$ and discuss its basic properties. The definition of the transform in $L^2(\mathbb{R})$ is not trivial. The integral

$$\frac{1}{\sqrt{2\pi}} \int_{-\infty}^{\infty} e^{-i\omega x} f(x)dx$$

cannot be used as a definition of the Fourier transform in $L^2(\mathbb{R})$ because not all functions in $L^2(\mathbb{R})$ are integrable. It is, however, possible to extend the Fourier transform from $L^1(\mathbb{R}) \cap L^2(\mathbb{R})$ onto $L^2(\mathbb{R})$. In the first part of this section, we discuss properties of the Fourier transform in $L^1(\mathbb{R})$. Then, we show that the extension onto $L^2(\mathbb{R})$ is possible and study properties of that extension.

Let f be an integrable function on $\mathbb{R}$. Consider the integral

$$\int_{-\infty}^{\infty} e^{-i\omega x} f(x)dx, \quad \omega \in \mathbb{R}. \tag{5.86}$$

Since the function $g(x) = e^{-i\omega x}$ is continuous, the product $e^{-i\omega x} f(x)$ is a locally integrable function for every $\omega \in \mathbb{R}$ (see Theorem 2.9.4). Moreover, since $|e^{-i\omega x}| = 1$ for all $\omega, x \in \mathbb{R}$, we have $|e^{-i\omega x} f(x)| = |f(x)|$ and thus, by Theorem 2.9.5, the integral (5.86) exists for all $\omega \in \mathbb{R}$.

Definition 5.11.1.   (Fourier transform in $L^1(\mathbb{R})$)
Let $f \in L^1(\mathbb{R})$. The function $\hat{f}$ defined by

$$\hat{f}(\omega) = \frac{1}{\sqrt{2\pi}} \int_{-\infty}^{\infty} e^{-i\omega x} f(x)dx \tag{5.87}$$

is called the *Fourier transform* of f.

In some books the Fourier transform is defined without the factor $\frac{1}{\sqrt{2\pi}}$. Another variation is the definition without the "$-$" sign in the exponent, that is, $\int_{-\infty}^{\infty} e^{i\omega x} f(x)\,dx$. These details do not change the theory of the Fourier transforms at all.

Instead of "$\hat{f}$" the notation "$\mathcal{F}\{f(x)\}$" is also used. The latter is especially convenient if instead of a letter "f" or "g" we want to use a formula describing a function, for example, $\mathcal{F}(e^{-x^2})$. We will use both symbols freely.

Example 5.11.2.

(a) Let $\alpha > 0$. Then

$$\mathcal{F}\{e^{-\alpha|x|}\} = \frac{1}{\sqrt{2\pi}} \int_{-\infty}^{\infty} e^{-i\omega x} e^{-\alpha|x|}\,dx$$

$$= \frac{1}{\sqrt{2\pi}} \int_{-\infty}^{0} e^{x(\alpha - i\omega)}\,dx + \frac{1}{\sqrt{2\pi}} \int_{0}^{\infty} e^{-x(\alpha + i\omega)}\,dx$$

$$= \frac{1}{\sqrt{2\pi}} \left[\frac{1}{\alpha - i\omega} + \frac{1}{\alpha + i\omega} \right] = \sqrt{\frac{2}{\pi}} \frac{\alpha}{\alpha^2 + \omega^2}.$$

(b)

$$\mathcal{F}\{e^{-\alpha x^2}\} = \frac{1}{\sqrt{2\pi}} \int_{-\infty}^{\infty} \exp\left[-\alpha \left(x + \frac{i\omega}{2\alpha} \right)^2 - \frac{\omega^2}{4\alpha} \right] dx$$

$$= \frac{1}{\sqrt{2\pi}} e^{-\omega^2/(4\alpha)} \int_{-\infty}^{\infty} e^{-\alpha u^2}\,du = \frac{1}{\sqrt{2\alpha}} e^{-\omega^2/(4\alpha)}. \qquad \square$$

The following theorem is an immediate consequence of Definition 5.11.1.

Theorem 5.11.3. (Linearity) *Let $f, g \in L^1(\mathbb{R})$ and $\alpha, \beta \in \mathbb{C}$. Then*

$$\mathcal{F}(\alpha f + \beta g) = \alpha \mathcal{F}(f) + \beta \mathcal{F}(g).$$

Theorem 5.11.4. *The Fourier transform of an integrable function is a continuous function.*

Proof: Let $f \in L^1(\mathbb{R})$. For any $\omega, \lambda \in \mathbb{R}$, we have

$$\left| \hat{f}(\omega + \lambda) - \hat{f}(\omega) \right| = \left| \frac{1}{\sqrt{2\pi}} \int_{-\infty}^{\infty} e^{-i\omega x} \left(e^{-i\lambda x} - 1 \right) f(x)\,dx \right|$$

$$\leq \frac{1}{\sqrt{2\pi}} \int_{-\infty}^{\infty} \left| e^{-i\lambda x} - 1 \right| |f(x)|\,dx. \qquad (5.88)$$

Since

$$\left|e^{-i\lambda x} - 1\right|\left|f(x)\right| \le 2\left|f(x)\right|$$

and

$$\lim_{\lambda \to 0}\left|e^{-i\lambda x} - 1\right| = 0 \quad \text{for all } x \in \mathbb{R},$$

we conclude

$$\lim_{\lambda \to 0}\frac{1}{\sqrt{2\pi}}\int_{-\infty}^{\infty}\left|e^{-i\lambda x} - 1\right|\left|f(x)\right|dx = 0$$

by the Dominated Convergence Theorem. This proves continuity of $\hat{f}$. (Actually, since the inequality in (5.88) is independent of ω, we have proved that $\hat{f}$ is uniformly continuous.) $\qquad\square$

Recall from Section 2.6 that the integral $\int_{-\infty}^{\infty}|f(x)|dx$ defines a norm in $L^1(\mathbb{R})$. This norm will be denoted by $\|\cdot\|_1$, that is,

$$\|f\|_1 = \int_{-\infty}^{\infty}\left|f(x)\right|dx \quad \text{for } f \in L^1(\mathbb{R}).$$

Theorem 5.11.5. *If $f_1, f_2, \ldots \in L^1(\mathbb{R})$ and $\|f_n - f\|_1 \to 0$ as $n \to \infty$, then $\hat{f}_n \to \hat{f}$ uniformly on $\mathbb{R}$.*

Proof: First notice that

$$\left|\hat{f}(\omega)\right| \le \frac{1}{\sqrt{2\pi}}\int_{-\infty}^{\infty}\left|e^{-i\omega x}f(x)\right|dx \le \frac{1}{\sqrt{2\pi}}\int_{-\infty}^{\infty}\left|f(x)\right|dx,$$

for every $\omega \in \mathbb{R}$. Thus,

$$\sup_{\omega \in \mathbb{R}}\left|\hat{f}_n(\omega) - \hat{f}(\omega)\right| \le \frac{1}{\sqrt{2\pi}}\int_{-\infty}^{\infty}\left|f_n(x) - f(x)\right|dx = \|f_n - f\|_1 \to 0,$$

proving the theorem. $\qquad\square$

Theorem 5.11.6. (Riemann–Lebesgue lemma) *If $f \in L^1(\mathbb{R})$, then*

$$\lim_{|\omega| \to \infty}\left|\hat{f}(\omega)\right| = 0.$$

Proof: Since $e^{-i\omega x} = -e^{-i\omega x - i\pi}$, we have

$$\hat{f}(\omega) = -\frac{1}{\sqrt{2\pi}}\int_{-\infty}^{\infty}e^{-i\omega(x + \pi/\omega)}f(x)dx = -\frac{1}{\sqrt{2\pi}}\int_{-\infty}^{\infty}e^{-i\omega x}f\left(x - \frac{\pi}{\omega}\right)dx.$$

Hence,

$$\hat{f}(\omega) = \frac{1}{2}\left[\frac{1}{\sqrt{2\pi}}\int_{-\infty}^{\infty}e^{-i\omega x}f(x)dx - \frac{1}{\sqrt{2\pi}}\int_{-\infty}^{\infty}e^{-i\omega x}f\left(x - \frac{\pi}{\omega}\right)dx\right]$$

$$= \frac{1}{2}\frac{1}{\sqrt{2\pi}}\int_{-\infty}^{\infty}e^{-i\omega x}\left(f(x) - f\left(x - \frac{\pi}{\omega}\right)\right)dx.$$

We thus have

$$\left|\hat{f}(\omega)\right| \le \frac{1}{2\sqrt{2\pi}}\int_{-\infty}^{\infty}\left|f(x) - f\left(x - \frac{\pi}{\omega}\right)\right|dx.$$

Since

$$\lim_{|\omega|\to\infty}\int_{-\infty}^{\infty}\left|f(x) - f\left(x - \frac{\pi}{\omega}\right)\right|dx = 0,$$

by Theorem 2.4.3, we have $\lim_{|\omega|\to\infty}|\hat{f}(\omega)| = 0$. $\qquad\square$

Note that the space $\mathcal{C}_0(\mathbb{R})$ of all continuous functions on $\mathbb{R}$, which vanish at infinity (that is, such that $\lim_{|x|\to\infty}f(x) = 0$), is a normed space with respect to the norm defined by

$$\|f\| = \sup_{x\in\mathbb{R}}\left|f(x)\right|.$$

Theorems 5.11.3 to 5.11.6 show that the Fourier transform is a continuous linear operator from $L^1(\mathbb{R})$ into $\mathcal{C}_0(\mathbb{R})$.

Theorem 5.11.7. *Let $f \in L^1(\mathbb{R})$. Then*

(a) $\mathcal{F}\{e^{i\alpha x}f(x)\} = \hat{f}(\omega - \alpha)$ *(translation)*,

(b) $\mathcal{F}\{f(x - x_0)\} = \hat{f}(\omega)e^{-i\omega x_0}$ *(shifting)*,

(c) $\mathcal{F}\{f(\alpha x)\} = \dfrac{1}{\alpha}\hat{f}\left(\dfrac{\omega}{\alpha}\right),\quad \alpha > 0$ *(scaling)*,

(d) $\mathcal{F}\{\bar{f}(x)\} = \overline{\mathcal{F}\{f(-x)\}}$ *(conjugate)*.

Proof follows easily from Definition 5.11.1.

Example 5.11.8. (Modulated Gaussian function) If

$$f(x) = e^{i\omega_0 x - \frac{1}{2}x^2},$$

then

$$\hat{f}(\omega) = e^{-\frac{1}{2}(\omega - \omega_0)^2}.$$

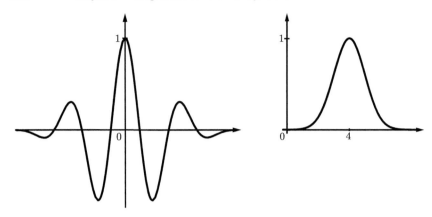

Figure 5.2   Modulated Gaussian and its Fourier transform ($\omega_0 = 4$).

This readily follows from Example 5.11.2(b) combined with the translation property of the Fourier transform. The graphs of $\operatorname{Re} f(x)$ and $\hat{f}(\omega)$ are drawn in Figure 5.2. The Gaussian function is named after Johann Carl Friedrich Gauss (1777–1855).   □

Theorem 5.11.9. *If f is a continuous piecewise differentiable function, $f, f' \in L^1(\mathbb{R})$, and $\lim_{|x|\to\infty} f(x) = 0$, then*

$$\mathcal{F}\{f'\} = i\omega \mathcal{F}\{f\}.$$

Proof: Simple integration by parts gives

$$\frac{1}{\sqrt{2\pi}} \int_{-\infty}^{\infty} f'(x)e^{-i\omega x}\,dx = \frac{1}{\sqrt{2\pi}}\left[f(x)e^{-i\omega x}\right]_{-\infty}^{\infty} + \frac{i\omega}{\sqrt{2\pi}} \int_{-\infty}^{\infty} f(x)e^{-i\omega x}\,dx$$

$$= i\omega \hat{f}(\omega).$$ □

Corollary 5.11.10. *If f is a continuous function, n-times piecewise differentiable, and $f, f', \dots, f^{(n)} \in L^1(\mathbb{R})$, and*

$$\lim_{|x|\to\infty} f^{(k)}(x) = 0 \quad \text{for } k = 0, \dots, n-1,$$

then

$$\mathcal{F}\{f^{(n)}\} = (i\omega)^n \mathcal{F}\{f\}.$$

Because of our definition of the Fourier transform, it is better to redefine the convolution of two functions $f, g \in L^1(\mathbb{R})$ as follows:

$$(f * g)(x) = \frac{1}{\sqrt{2\pi}} \int_{-\infty}^{\infty} f(x - u)g(u)\,du.$$

The main reason is the simplicity of the formula in the next theorem.

Theorem 5.11.11. (Convolution theorem) *Let $f, g \in L^1(\mathbb{R})$. Then*

$$\mathcal{F}(f * g) = \mathcal{F}(f)\mathcal{F}(g).$$

Proof: Let $f, g \in L^1(\mathbb{R})$ and $h = f * g$. Then $h \in L^1(\mathbb{R})$, by Theorem 2.15.1, and we have

$$\hat{h}(\omega) = \frac{1}{\sqrt{2\pi}} \int_{-\infty}^{\infty} h(x) e^{-i\omega x} dx$$

$$= \frac{1}{\sqrt{2\pi}} \int_{-\infty}^{\infty} e^{-i\omega x} \frac{1}{\sqrt{2\pi}} \int_{-\infty}^{\infty} f(x-u)g(u)\,du\,dx$$

$$= \frac{1}{2\pi} \int_{-\infty}^{\infty} g(u) \int_{-\infty}^{\infty} e^{-i\omega x} f(x-u)\,dx\,du$$

$$= \frac{1}{2\pi} \int_{-\infty}^{\infty} g(u) \int_{-\infty}^{\infty} e^{-i\omega(x+u)} f(x)\,dx\,du$$

$$= \frac{1}{\sqrt{2\pi}} \int_{-\infty}^{\infty} g(u) e^{-i\omega u}\,du \frac{1}{\sqrt{2\pi}} \int_{-\infty}^{\infty} e^{-i\omega x} f(x)\,dx$$

$$= \hat{g}(\omega)\hat{f}(\omega). \qquad \square$$

We will now discuss the extension of the Fourier transform onto $L^2(\mathbb{R})$. In the following theorem, and in the remaining part of this section, $\| \cdot \|_2$ denotes the norm in $L^2(\mathbb{R})$, that is,

$$\|f\|_2 = \sqrt{\int_{-\infty}^{\infty} |f(x)|^2 dx} \quad \text{for } f \in L^2(\mathbb{R}).$$

Theorem 5.11.12. *Let f be a continuous function on $\mathbb{R}$ vanishing outside a bounded interval. Then $\hat{f} \in L^2(\mathbb{R})$ and*

$$\left\|\hat{f}\right\|_2 = \|f\|_2.$$

Proof: Suppose first that f vanishes outside the interval $[-\pi, \pi]$. Using Parseval's formula for the orthonormal sequence of functions on $[-\pi, \pi]$,

$$\varphi_n(x) = \frac{1}{\sqrt{2\pi}} e^{-inx}, \quad n = 0, \pm 1, \pm 2, \ldots,$$

we get

$$\|f\|_2^2 = \sum_{n=-\infty}^{\infty} \left| \frac{1}{\sqrt{2\pi}} \int_{-\infty}^{\infty} e^{-inx} f(x)\,dx \right|^2 = \sum_{n=-\infty}^{\infty} |\hat{f}(n)|^2.$$

Since this inequality holds also for $g(x) = e^{-i\xi x} f(x)$ instead of $f(x)$, we obtain

$$\|f\|_2^2 = \sum_{n=-\infty}^{\infty} |\hat{f}(n+\xi)|^2,$$

in view of $\|f\|_2^2 = \|g\|_2^2$. Integration of both sides with respect to ξ from 0 to 1 yields

$$\|f\|_2^2 = \sum_{n=-\infty}^{\infty} \int_0^1 |\hat{f}(n+\xi)|^2 d\xi = \int_{-\infty}^{\infty} |\hat{f}(\xi)|^2 d\xi = \|\hat{f}\|_2^2.$$

If f does not vanish outside $[-\pi, \pi]$, then we take a positive number λ for which the function $g(x) = f(\lambda x)$ vanishes outside $[-\pi, \pi]$. Then

$$\hat{g}(x) = \frac{1}{\lambda} \hat{f}\left(\frac{x}{\lambda}\right),$$

and thus

$$\|f\|_2^2 = \lambda \|g\|_2^2 = \lambda \|\hat{g}\|_2^2 = \lambda \int_{-\infty}^{\infty} \left|\frac{1}{\lambda} \hat{f}\left(\frac{\xi}{\lambda}\right)\right|^2 d\xi = \int_{-\infty}^{\infty} |\hat{f}(\xi)|^2 dx = \|\hat{f}\|_2^2.$$

$\square$

The space of all continuous functions on $\mathbb{R}$ with compact support is dense in $L^2(\mathbb{R})$. Theorem 5.11.12 shows that the Fourier transform is a continuous mapping from that space into $L^2(\mathbb{R})$. Since the mapping is linear, it has a unique extension to a linear mapping from $L^2(\mathbb{R})$ into itself. This extension will be called the *Fourier transform* on $L^2(\mathbb{R})$.

Definition 5.11.13.   (Fourier transform in $L^2(\mathbb{R})$)
Let $f \in L^2(\mathbb{R})$, and let (φ_n) be a sequence of continuous functions with compact support convergent to f in $L^2(\mathbb{R})$, that is, $\|f - \varphi_n\|_2 \to 0$. The Fourier transform of f is defined by

$$\hat{f} = \lim_{n \to \infty} \hat{\varphi}_n, \qquad (5.89)$$

where the limit is with respect to the norm in $L^2(\mathbb{R})$.

Theorem 5.11.12 guarantees that the limit exists and is independent of a particular sequence approximating f. It is important to remember that the convergence in $L^2(\mathbb{R})$ does not imply pointwise convergence, and therefore, the Fourier transform of a square integrable function is not defined at every point, unlike the Fourier transform of an integrable function. The Fourier transform of a square integrable function is defined almost everywhere. For this reason we cannot say that if $f \in L^1(\mathbb{R}) \cap L^2(\mathbb{R})$, then the Fourier transform defined by (5.87) and the one defined by (5.89) are equal. To be precise, we should say that the function defined by (5.87) belongs to the equivalence class of square integrable functions defined by (5.89). In spite of this difference, we will use the same symbol to denote both transforms. It will not cause any misunderstanding.

The following theorem is an immediate consequence of Definition 5.11.13 and Theorem 5.11.12.

Theorem 5.11.14. (Parseval's relation) *If $f \in L^2(\mathbb{R})$, then*

$$\|\hat{f}\|_2 = \|f\|_2.$$

In physical problems, the quantity $\|f\|_2$ is a measure of energy, and $\|\hat{f}\|_2$ represents the power spectrum of f.

Theorem 5.11.15. *Let $f \in L^2(\mathbb{R})$. Then*

$$\hat{f}(\omega) = \lim_{n \to \infty} \frac{1}{\sqrt{2\pi}} \int_{-n}^{n} e^{-i\omega x} f(x) dx, \tag{5.90}$$

where the convergence is with respect to the norm in $L^2(\mathbb{R})$.

Proof: For $n = 1, 2, 3, \ldots$ define

$$f_n(x) = \begin{cases} f(x) & \text{if } |x| < n, \\ 0 & \text{if } |x| \geq n. \end{cases}$$

Then $\|f - f_n\|_2 \to 0$, and thus, $\|\hat{f} - \hat{f}_n\|_2 \to 0$ as $n \to \infty$. $\qquad\square$

Theorem 5.11.16. (Weak Parseval's relation) *If $f, g \in L^2(\mathbb{R})$, then*

$$\int_{-\infty}^{\infty} f(x)\hat{g}(x) dx = \int_{-\infty}^{\infty} \hat{f}(x)g(x) dx. \tag{5.91}$$

Proof: For $n = 1, 2, 3, \ldots$ define

$$f_n(x) = \begin{cases} f(x) & \text{if } |x| < n, \\ 0 & \text{if } |x| \geq n, \end{cases}$$

and

$$g_n(x) = \begin{cases} g(x) & \text{if } |x| < n, \\ 0 & \text{if } |x| \geq n. \end{cases}$$

Since

$$\hat{f}_m(x) = \frac{1}{\sqrt{2\pi}} \int_{-\infty}^{\infty} e^{-ix\xi} f_m(\xi) d\xi,$$

we have

$$\int_{-\infty}^{\infty} \hat{f}_m(x) g_n(x) dx = \frac{1}{\sqrt{2\pi}} \int_{-\infty}^{\infty} g_n(x) \int_{-\infty}^{\infty} e^{-ix\xi} f_m(\xi) d\xi \, dx.$$

The function

$$e^{-ix\xi} g_n(x) f_m(\xi)$$

is integrable over $\mathbb{R}^2$, and thus the Fubini theorem can be applied. Consequently,

$$\int_{-\infty}^{\infty} \hat{f}_m(x) g_n(x) dx = \frac{1}{\sqrt{2\pi}} \int_{-\infty}^{\infty} f_m(\xi) \int_{-\infty}^{\infty} e^{-ix\xi} g_n(x) dx d\xi$$

$$= \int_{-\infty}^{\infty} f_m(\xi) \hat{g}_n(\xi) d\xi.$$

Since $\|g - g_n\|_2 \to 0$ and $\|\hat{g} - \hat{g}_n\|_2 \to 0$ as $n \to \infty$, we obtain

$$\int_{-\infty}^{\infty} \hat{f}_m(x) g(x) dx = \int_{-\infty}^{\infty} f_m(x) \hat{g}(x) dx,$$

by continuity of inner product. For the same reason, by letting $m \to \infty$, we get

$$\int_{-\infty}^{\infty} \hat{f}(x) g(x) dx = \int_{-\infty}^{\infty} f(x) \hat{g}(x) dx. \qquad \square$$

The following technical lemma will be useful in the proof of the important inversion theorem for the Fourier transform in $L^2(\mathbb{R})$.

Lemma 5.11.17. *Let $f \in L^2(\mathbb{R})$ and let $g = \overline{\hat{f}}$. Then $f = \overline{\hat{g}}$.*

Proof: From Theorems 5.11.14 and 5.11.16 and equality $g = \overline{\hat{f}}$, we obtain

$$\langle f, \overline{\hat{g}} \rangle = \langle \hat{f}, \overline{g} \rangle = \langle \hat{f}, \hat{f} \rangle = \|\hat{f}\|_2^2 = \|f\|_2^2. \tag{5.92}$$

Hence also

$$\overline{\langle f, \overline{\hat{g}} \rangle} = \|f\|_2^2. \tag{5.93}$$

Finally, by Parseval's equality,

$$\|\hat{g}\|_2^2 = \|g\|_2^2 = \|\hat{f}\|_2^2 = \|f\|_2^2. \tag{5.94}$$

Using (5.92) to (5.94), we get

$$\left\| f - \overline{\hat{g}} \right\|_2^2 = \langle f - \overline{\hat{g}}, f - \overline{\hat{g}} \rangle = \|f\|_2^2 - \langle f, \overline{\hat{g}} \rangle - \overline{\langle f, \overline{\hat{g}} \rangle} + \|\hat{g}\|_2^2 = 0.$$

This shows that $f = \bar{\hat{g}}$. □

Theorem 5.11.18. (Inversion of the Fourier transform in $L^2(\mathbb{R})$) *Let* $f \in L^2(\mathbb{R})$. *Then*

$$f(x) = \lim_{n \to \infty} \frac{1}{\sqrt{2\pi}} \int_{-n}^{n} e^{i\omega x} \hat{f}(\omega) d\omega,$$

where the convergence is with respect to the norm in $L^2(\mathbb{R})$.

Proof: Let $f \in L^2(\mathbb{R})$. If $g = \bar{\hat{f}}$, then, by Lemma 5.11.17,

$$f(x) = \bar{\hat{g}}(x) = \lim_{n \to \infty} \frac{1}{\sqrt{2\pi}} \int_{-n}^{n} \overline{e^{-i\omega x} g(\omega)} d\omega$$

$$= \lim_{n \to \infty} \frac{1}{\sqrt{2\pi}} \int_{-n}^{n} e^{i\omega x} \overline{g(\omega)} d\omega$$

$$= \lim_{n \to \infty} \frac{1}{\sqrt{2\pi}} \int_{-n}^{n} e^{i\omega x} \hat{f}(\omega) d\omega.$$ □

Corollary 5.11.19. *If* $f \in L^1(\mathbb{R}) \cap L^2(\mathbb{R})$, *then the equality*

$$f(x) = \frac{1}{\sqrt{2\pi}} \int_{-\infty}^{\infty} e^{i\omega x} \hat{f}(\omega) d\omega \tag{5.95}$$

holds almost everywhere in $\mathbb{R}$.

The transform defined by (5.95) is called the *inverse Fourier transform*. One of the main reasons for introducing the factor $\frac{1}{\sqrt{2\pi}}$ in the definition of the Fourier transform is the symmetry of the transform and its inverse:

$$\mathcal{F}\{f(x)\} = \frac{1}{\sqrt{2\pi}} \int_{-\infty}^{\infty} e^{-i\omega x} f(x) dx,$$

$$\mathcal{F}^{-1}\{f(\omega)\} = \frac{1}{\sqrt{2\pi}} \int_{-\infty}^{\infty} e^{i\omega x} f(\omega) d\omega.$$

Corollary 5.11.20. (Duality) *If* $f \in L^1(\mathbb{R}) \cap L^2(\mathbb{R})$, *then* $\mathcal{F}\{\mathcal{F}\{f(x)\}\} = f(-x)$ *almost everywhere in* $\mathbb{R}$.

Proof: It suffices to replace x by $-x$ in (5.95). □

Theorem 5.11.21. (General Parseval's relation) *If* $f, g \in L^2(\mathbb{R})$, *then*

$$\int_{-\infty}^{\infty} f(x) \overline{g(x)} dx = \int_{-\infty}^{\infty} \hat{f}(\omega) \overline{\hat{g}(\omega)} d\omega.$$

Proof: The polarization identity

$$\langle f, g \rangle = \frac{1}{4}\left(|f + g|^2 - |f - g|^2 + i|f + ig|^2 - i|f - ig|^2\right)$$

implies that every isometry preserves inner product. Since the Fourier transform is an isometry on $L^2(\mathbb{R})$, we have $\langle f, g \rangle = \langle \hat{f}, \hat{g} \rangle$. ☐

The following theorem summarizes the results of this section. It is known as the Plancherel theorem (Michel Plancherel (1885–1967)).

Theorem 5.11.22. (Plancherel theorem) *For every $f \in L^2(\mathbb{R})$, there exists $\hat{f} \in L^2(\mathbb{R})$ such that the following holds:*

(a) *If $f \in L^1(\mathbb{R}) \cap L^2(\mathbb{R})$,   then* $\hat{f}(\omega) = \dfrac{1}{\sqrt{2\pi}} \displaystyle\int_{-\infty}^{\infty} e^{-i\omega x} f(x)dx,$

(b) $\left\| \hat{f}(\omega) - \dfrac{1}{\sqrt{2\pi}} \displaystyle\int_{-n}^{n} e^{-i\omega x} f(x)dx \right\|_2 \to 0$ *as $n \to \infty$,*

(c) $\left\| f(x) - \dfrac{1}{\sqrt{2\pi}} \displaystyle\int_{-n}^{n} e^{i\omega x} \hat{f}(\omega)d\omega \right\|_2 \to 0$ *as $n \to \infty$,*

(d) $\|f\|_2^2 = \|\hat{f}\|_2^2,$

(e) *The map $f \mapsto \hat{f}$ is a Hilbert space isomorphism of $L^2(\mathbb{R})$ onto $L^2(\mathbb{R})$.*

Proof: The only part of this theorem that remains to be proved is the fact that the Fourier transform is "onto." Let $f \in L^2(\mathbb{R})$, and define

$$h = \overline{f} \quad \text{and} \quad g = \overline{\hat{h}}.$$

Then, by Lemma 5.11.17, $\overline{f} = h = \overline{\hat{g}}$, and hence $f = \hat{g}$. This shows that every square integrable function is the Fourier transform of a square integrable function. ☐

Theorem 5.11.23. *The Fourier transform is a unitary operator on $L^2(\mathbb{R})$.*

Proof: First note that, for any $g \in L^2(\mathbb{R})$, we have

$$\mathcal{F}\{\overline{g}\}(\omega) = \frac{1}{\sqrt{2\pi}} \int_{-\infty}^{\infty} e^{-i\omega x} \overline{g(x)}dx = \overline{\frac{1}{\sqrt{2\pi}} \int_{-\infty}^{\infty} e^{i\omega x} g(x)dx} = \overline{\mathcal{F}^{-1}\{g\}(\omega)}.$$

Now, using Theorem 5.11.16, we obtain

$$\langle \mathcal{F}\{f\}, g \rangle = \int_{-\infty}^{\infty} \mathcal{F}\{f\}(x)\overline{g(x)}dx$$

$$= \int_{-\infty}^{\infty} f(x)\mathcal{F}\{\bar{g}\}(x)dx$$

$$= \int_{-\infty}^{\infty} f(x)\overline{\mathcal{F}^{-1}\{g(x)\}}dx$$

$$= \langle f, \mathcal{F}^{-1}\{g\}\rangle.$$

This shows that $\mathcal{F}^{-1} - \mathcal{F}^*$, and thus $\mathcal{F}$ is unitary.   □

The Fourier transform can be defined for functions in $L^1(\mathbb{R}^N)$ by

$$\hat{f}(\omega) = \frac{1}{(2\pi)^{N/2}} \int_{\mathbb{R}^N} e^{-i\omega\cdot x} f(x)dx$$

where $\omega = (\omega_1, \ldots, \omega_N)$, $x = (x_1, \ldots, x_N)$, and $\omega\cdot x = \omega_1 x_1 + \cdots + \omega_N x_N$. The theory of the Fourier transform in $L^1(\mathbb{R}^N)$ is similar to the one-dimensional case. Moreover, the extension to $L^2(\mathbb{R}^N)$ is possible, and it has similar properties, including the inversion theorem and the Plancherel theorem.

We close this section with some examples of Fourier transforms. We chose these particular functions because of their connection with the wavelet transform (see Chapter 8).

Example 5.11.24. (Second derivative of a Gaussian function) If

$$f(x) = (1 - x^2)e^{-\frac{1}{2}x^2},$$

then

$$\hat{f}(\omega) = \omega^2 e^{-\frac{1}{2}\omega^2}.$$

Indeed, we have

$$\mathcal{F}\{(1 - x^2)e^{-\frac{1}{2}x^2}\} = -\mathcal{F}\left\{\frac{d^2}{dx^2}e^{-\frac{1}{2}x^2}\right\}$$

$$= -(i\omega)^2\mathcal{F}\{e^{-\frac{1}{2}x^2}\}$$

$$= \omega^2 e^{-\frac{\omega^2}{2}}.$$

The functions $f(x)$ and $\hat{f}(\omega)$ are plotted in Figure 5.3.   □

Example 5.11.25. (The Haar function) The Haar function is defined by

$$f(x) = \begin{cases} 1 & \text{if } 0 \leq x < \frac{1}{2}, \\ -1 & \text{if } \frac{1}{2} \leq x < 1, \\ 0 & \text{otherwise.} \end{cases}$$

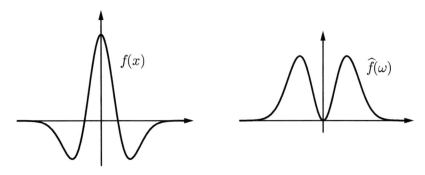

Figure 5.3   Second derivative of a Gaussian function and its Fourier transform.

Clearly,

$$\hat{f}(\omega) = \frac{1}{\sqrt{2\pi}}\left(\int_0^{1/2} e^{-i\omega x}dx - \int_{1/2}^1 e^{-i\omega x}dx\right)$$

$$= \frac{1}{\sqrt{2\pi}}\frac{1}{i\omega}\left(1 - 2e^{-\frac{i\omega}{2}} + e^{-i\omega}\right)$$

$$= \frac{1}{\sqrt{2\pi}}\frac{e^{-\frac{i\omega}{2}}}{i\omega}\left(e^{\frac{i\omega}{2}} - 2 + e^{-\frac{i\omega}{2}}\right)$$

$$= \frac{i}{\sqrt{2\pi}}e^{-\frac{i\omega}{2}}\frac{\sin^2(\frac{\omega}{4})}{\frac{\omega}{4}}.$$

The Haar function is named after Alfréd Haar (1885–1933). The Haar function and its Fourier transform are drawn in Figure 8.2 in Chapter 8.   □

Example 5.11.26. (The Shannon function)   The Shannon function is defined by

$$f(x) = \frac{\sin 2\pi x - \sin \pi x}{\pi x}.$$

The Fourier transform of $f(x)$ is

$$\hat{f}(\omega) = \begin{cases} \frac{1}{\sqrt{2\pi}} & \text{if } \pi < |\omega| < 2\pi, \\ 0 & \text{otherwise.} \end{cases}$$

The graph of the Shannon function and its Fourier transform are shown on Figure 5.4. The Shannon function is named after Claude Elwood Shannon (1916–2001).   □

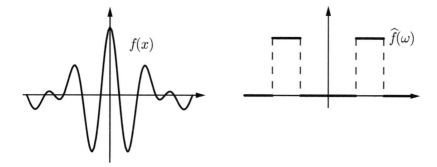

Figure 5.4   The Shannon function and its Fourier transform.

5.12 Applications of the Fourier Transform to Ordinary Differential Equations and Integral Equations

In this section, we discuss some examples of applications of the Fourier transform to ordinary differential equations and integral equations.

Consider the nth order linear ordinary differential equation with constant coefficients

$$Ly(t) = f(t), \tag{5.96}$$

where L is the nth order differential operator given by

$$L = a_n D^n + a_{n-1} D^{n-1} + \cdots + a_1 D + a_0, \tag{5.97}$$

$a_0, a_1, \ldots, a_n$ are constants, $D = \frac{d}{dt}$, and $f \in L^1(\mathbb{R})$ or $f \in L^2(\mathbb{R})$.

Application of the Fourier transform to both sides of (5.96) gives

$$\left[a_n (i\omega)^n + a_{n-1}(i\omega)^{n-1} + \cdots + a_1(i\omega) + a_0 \right] \hat{y}(\omega) = \hat{f}(\omega)$$

or

$$\hat{p}(i\omega)\hat{y}(\omega) = \hat{f}(\omega),$$

where $\hat{p}(z) = a_n z^n + a_{n-1}z^{n-1} + \cdots + a_1 z + a_0$.

Thus,

$$\hat{y}(\omega) = \frac{\hat{f}(\omega)}{\hat{p}(i\omega)} = \hat{f}(\omega)\hat{g}(\omega), \tag{5.98}$$

where

$$\hat{g}(\omega) = \frac{1}{\hat{p}(i\omega)}.$$

Now, the convolution Theorem 5.11.11 gives the solution

$$y(t) = \frac{1}{\sqrt{2\pi}} \int_{-\infty}^{\infty} f(\xi)g(t - \xi)d\xi, \tag{5.99}$$

provided $g(t) = \mathcal{F}^{-1}\{\hat{g}(\omega)\}$ is known explicitly.

To give a physical interpretation of the result, we consider the differential equation associated with a sudden impulse function $f(t) = \delta(t)$:

$$LG(t) = \delta(t). \tag{5.100}$$

(For a rigorous discussion of the Dirac delta distribution δ, see Section 6.2.) Application of the Fourier transform to (5.100) yields the solution

$$G(t) = \mathcal{F}^{-1}\left\{ \frac{1}{\sqrt{2\pi}}\hat{g}(\omega) \right\} = \frac{1}{\sqrt{2\pi}}g(t). \tag{5.101}$$

Now, the solution (5.99) can be written as

$$y(t) = \int_{-\infty}^{\infty} f(\xi)G(t - \xi)d\xi. \tag{5.102}$$

Clearly, $G(t)$ behaves like a Green's function, that is, it is the response to a *unit impulse*. In any physical system, $f(t)$ is usually called the *input function*, while $y(t)$ is called the *output* obtained by the superposition principle. The Fourier transform of $\sqrt{2\pi}\,G(t)$ is called the *admittance* $\hat{g}(\omega) = (\hat{p}(i\omega))^{-1}$. To determine the response to a given input, we first find the Fourier transform of the input, multiply the result by the admittance, and then apply the inverse Fourier transform to the product. We illustrate these ideas by a simple electrical circuit problem.

Example 5.12.1. The electric current $I(t)$ in the circuit is governed by the equation

$$L\frac{dI}{dt} + RI = E, \tag{5.103}$$

where L is the inductance, R is the resistance, and E is the applied electromagnetic force. With $E(t) = E_0 e^{-|t|}$, application of the Fourier transform (with respect to t) to Equation (5.103) gives

$$(i\omega L + R)\hat{I}(\omega) = \sqrt{\frac{2}{\pi}}\frac{E_0}{1 + \omega^2}$$

or

$$\hat{I}(\omega) = \sqrt{\frac{2}{\pi}}\frac{E_0}{(i\omega L + R)(1 + \omega^2)}.$$

The inverse Fourier transform yields

$$I(t) = \frac{E_0}{\pi} \int_{-\infty}^{\infty} \frac{e^{i\omega t} d\omega}{(i\omega L + R)(1 + \omega^2)}.$$

This integral can readily be evaluated by the theory of residues. For $t > 0$,

$$I(t) = \frac{E_0}{\pi} 2\pi i \left([\text{residue at } \omega = i] + \left[\text{residue at } \omega = \frac{iR}{L} \right] \right)$$

$$= 2iE_0 \left(\frac{e^{-t}}{2i(R - L)} + \frac{e^{-Rt/L}}{iL(1 - R^2/L^2)} \right)$$

$$= E_0 \left(\frac{e^{-t}}{R - L} - \frac{2Le^{-Rt/L}}{R^2 - L^2} \right).$$

Similarly, for $t < 0$, we obtain

$$I(t) = -\frac{E_0}{\pi} \cdot 2\pi i [\text{residue at } \omega = -i] = \frac{E_0 e^t}{L + R}. \tag{5.104}$$

At $t = 0$, the current is continuous, hence

$$I(0) = \lim_{t \to 0} I(t) = \frac{E_0}{R + L}. \qquad \square$$

Example 5.12.2. (Synthesis and resolution of a pulse — physical interpretation of convolution) A time-dependent electric, optical, or electromagnetic pulse can be regarded as a superposition of plane waves of all real frequencies so that the total pulse can be represented by the inverse Fourier transform

$$f(t) = \frac{1}{2\pi} \int_{-\infty}^{\infty} F(\omega) e^{i\omega t} d\omega, \tag{5.105}$$

where the factor $1/2\pi$ is introduced because the angular frequency ω is related to the linear frequency ν by $\omega = 2\pi \nu$, and negative frequencies are introduced for mathematical convenience so that we can avoid dealing with the cosine and sine functions separately. Clearly, $F(\omega)$ can be represented by the Fourier transform of $f(t)$ as

$$F(\omega) = \int_{-\infty}^{\infty} f(t) e^{-i\omega t} dt. \tag{5.106}$$

This represents the *resolution* of the pulse $f(t)$ into its angular frequency components, and (5.105) gives a *synthesis* of the pulse from its individual components.

Consider a simple electrical device such as amplifier with an input function $f(t)$ and an output function $g(t)$. For an input of a single frequency ω, we have

$f(t) = e^{i\omega t}$. The amplifier will change the amplitude and may also change the phase so that the output can be expressed in terms of the input, the amplitude, and phase modifying function $\Phi(\omega)$ as

$$g(t) = \Phi(\omega)f(t), \qquad (5.107)$$

where $\Phi(\omega)$ is usually called the *transfer function*, and it is, in general, a complex function of the real variable ω. This function is generally independent of the absence or presence of any other frequency components. Thus, the total output may be obtained by integrating over the entire input as modified by the amplifier

$$g(t) = \frac{1}{2\pi} \int_{-\infty}^{\infty} \Phi(\omega)F(\omega)e^{i\omega t} d\omega. \qquad (5.108)$$

Therefore, the total output $g(t)$ can readily be calculated from any given input $f(t)$ and known transfer function $\Phi(\omega)$. On the other hand, the transfer function is obviously characteristic of the amplifier and can, in general, be obtained as the Fourier transform of some function $\varphi(t)$:

$$\Phi(\omega) = \int_{-\infty}^{\infty} \varphi(t)e^{-i\omega t} dt. \qquad (5.109)$$

The convolution Theorem 5.11.11 allows us to rewrite (5.108) as

$$g(t) = \mathcal{F}^{-1}\{\Phi(\omega)F(\omega)\} = \int_{-\infty}^{\infty} f(\tau)\varphi(t - \tau)d\tau. \qquad (5.110)$$

Physically, the result represents an output signal (effect) as the integral superposition of an input signal (cause) $f(t)$ modified by $\varphi(t - \tau)$. Indeed, (5.110) is the most general mathematical representation of an output in terms of an input modified by the amplifier where t is the time variable. Assuming the principle of causality, that is, every effect has a cause, we must require $\tau < t$. The principle of causality is imposed by requiring

$$\varphi(t - \tau) = 0 \quad \text{for } \tau > t. \qquad (5.111)$$

Consequently, (5.110) reduces to the form

$$g(t) = \int_{-\infty}^{t} f(\tau)\varphi(t - \tau)d\tau. \qquad (5.112)$$

To determine the significance of $\varphi(t)$, we use a sudden impulse function $f(\tau) = \delta(\tau)$ so that (5.112) becomes

$$g(t) = \int_{-\infty}^{t} \delta(\tau)\varphi(t - \tau)d\tau = \varphi(t)H(t). \qquad (5.113)$$

This recognizes $\varphi(t)$ as the output corresponding to a unit impulse at $t = 0$, and the Fourier transform $\Phi(\omega)$ of $\varphi(t)$ is given by

$$\Phi(\omega) = \int_0^\infty \varphi(t)e^{-i\omega t}\,dt \qquad (5.114)$$

with $\varphi(t) = 0$ for $t < 0$. ☐

Example 5.12.3. We will use the Fourier transform to solve the following ordinary differential equation:

$$-\frac{d^2u}{dx^2} + \alpha^2 u = f(x), \qquad (5.115)$$

where $f \in L^2(\mathbb{R})$.

Applying the Fourier transform to both sides of (5.115), we obtain

$$\left(\omega^2 + \alpha^2\right)\hat{u}(\omega) = \hat{f}(\omega).$$

Hence,

$$\hat{u}(\omega) = \frac{1}{\omega^2 + \alpha^2}\hat{f}(\omega).$$

Since

$$\frac{1}{\sqrt{2\pi}}\int_{-\infty}^\infty \frac{e^{i\omega x}}{\omega^2 + \alpha^2}\,d\omega = \frac{1}{2\alpha}e^{-\alpha|x|},$$

we have

$$\frac{1}{\omega^2 + \alpha^2} = \mathcal{F}\left\{\frac{1}{2\alpha}e^{-\alpha|x|}\right\}.$$

From the convolution Theorem 5.11.11, we get

$$\mathcal{F}\{u\} = \mathcal{F}\left\{\frac{1}{2\alpha}e^{-\alpha|x|}\right\}\mathcal{F}\{f\} = \mathcal{F}\left\{\left\{\frac{1}{2\alpha}e^{-\alpha|x|}\right\} * f\right\}$$

and hence

$$u(x) = \left\{\frac{1}{2\alpha}e^{-\alpha|x|}\right\} * f.$$

This means that the solution of the equation can be written as

$$u(x) = \frac{1}{\sqrt{2\pi}}\int_{-\infty}^\infty \frac{e^{-\alpha|x-t|}}{2\alpha}f(t)\,dt. \qquad ☐$$

Example 5.12.4. Consider an infinite beam on an elastic foundation under a prescribed vertical load $W(x)$. The vertical deflection $u(x)$ is governed by the following differential equation

$$EIu^{(iv)} + ku = W(x), \tag{5.116}$$

where EI is the flexural rigidity, and k is the foundation modulus of the beam. We solve this equation assuming that $W(x)$ has compact support so that $u, u', u'',$ and u''' all tend to zero as $|x| \to \infty$.

We first rewrite the equation:

$$u^{(iv)} + \alpha^4 u = w(x), \tag{5.117}$$

where $\alpha^4 = k/EI$, and $w(x) = W(x)/EI$. Application of the Fourier transform to (5.117) gives

$$\hat{u}(\omega) = \frac{\hat{w}(\omega)}{\omega^4 + \alpha^4}.$$

The inverse Fourier transform yields the solution

$$u(x) = \frac{1}{\sqrt{2\pi}} \int_{-\infty}^{\infty} \frac{\hat{w}(\omega)}{\omega^4 + \alpha^4} e^{i\omega x} d\omega \tag{5.118}$$

$$= \frac{1}{2\pi} \int_{-\infty}^{\infty} \frac{e^{i\omega x} d\omega}{\omega^4 + \alpha^4} \int_{-\infty}^{\infty} w(\xi) e^{-i\omega\xi} d\xi \tag{5.119}$$

$$= \int_{-\infty}^{\infty} w(\xi) G(\xi, x) d\xi, \tag{5.120}$$

where

$$G(\xi, x) = \frac{1}{2\pi} \int_{-\infty}^{\infty} \frac{e^{i\omega(x-\xi)}}{\omega^4 + \alpha^4} d\omega = \frac{1}{\pi} \int_{0}^{\infty} \frac{\cos(x-\xi)\omega}{\omega^4 + \alpha^4} d\omega. \tag{5.121}$$

This integral can be evaluated by complex contour integration (or with the aid of tables of Fourier integrals). We simply state the result:

$$G(\xi, x) = \frac{1}{2\alpha^3} e^{-\alpha|x-\xi|/\sqrt{2}} \sin\left(\frac{\alpha|x-\xi|}{\sqrt{2}} + \frac{\pi}{4}\right). \tag{5.122}$$

In particular, consider a point load of unit strength acting at some point x_0, that is, $w(x) = \delta(x - x_0)$. Then the solution (5.120) becomes

$$u(x) = \int_{-\infty}^{\infty} \delta(\xi - x_0) G(\xi, x) d\xi = G(x_0, x). \tag{5.123}$$

Thus, the kernel $G(\xi, x)$ in (5.120) has the physical significance of being the deflection, as a function of x, due to a point unit load at ξ. So the deflection due to a point load of strength $w(\xi)d\xi$ at ξ is $w(\xi)d\xi G(\xi, x)$, and hence (5.120) represents the superposition of all such incremental deflections.   □

Example 5.12.5. We will solve the integral equation of the convolution type:

$$\int_{-\infty}^{\infty} K(x-t)u(t)dt + \lambda u(x) = f(x),$$

where $K, f \in L^1(\mathbb{R})$. This is a *Fredholm integral equation with convolution kernel*.

Application of the Fourier transform gives

$$\sqrt{2\pi}\hat{K}(\omega)\hat{u}(\omega) + \lambda\hat{u}(\omega) = \hat{f}(\omega).$$

The inverse Fourier transform leads to a formal solution in the form

$$u(x) = \frac{1}{\sqrt{2\pi}} \int_{-\infty}^{\infty} \frac{\hat{f}(\omega)e^{i\omega x}}{\sqrt{2\pi}\hat{K}(\omega)+\lambda} d\omega. \tag{5.124}$$

As an example, consider $K(x) = 1/x$. Then

$$\hat{K}(\omega) = \frac{1}{\sqrt{2\pi}} \int_{-\infty}^{\infty} \frac{e^{-i\omega x}}{x} dx = i\sqrt{\frac{\pi}{2}} \operatorname{sgn}(-\omega) = -i\sqrt{\frac{\pi}{2}} \operatorname{sgn}\omega.$$

Thus, the solution (5.124) becomes

$$u(x) = \frac{1}{\sqrt{2\pi}} \int_{-\infty}^{\infty} \frac{\hat{f}(\omega)e^{i\omega x}}{\lambda - i\pi \operatorname{sgn}\omega} d\omega. \qquad \Box$$

Example 5.12.6. (The Hilbert transform) We solve the integral equation

$$\frac{1}{\pi} \int_{-\infty}^{\infty} \frac{f(t)}{x-t} dt = f_H(x), \tag{5.125}$$

where $f_H \in L^2(\mathbb{R})$ and the above integral is the Cauchy principal value:

$$\int_{-\infty}^{\infty} \frac{f(t)}{x-t} dt = \lim_{\varepsilon \to 0} \left(\int_{-\infty}^{-\varepsilon} + \int_{\varepsilon}^{\infty} \right) \frac{f(t)}{x-t} dt.$$

The function $f_H(x)$ is called the *Hilbert transform* of $f(t)$ and is often denoted by $\mathcal{H}\{f(t)\} = f_H(x)$. Our problem is thus to find the inverse Hilbert transform.

First we rewrite Equation (5.125) as

$$\frac{1}{\sqrt{2\pi}} \int_{-\infty}^{\infty} f(t)g(x-t)dt = f_H(x),$$

where $g(x) = \sqrt{2/\pi}(1/x)$. Application of the Fourier transform with respect to x gives

$$\hat{f}(\omega) = \frac{\hat{f}_H(\omega)}{\hat{g}(\omega)}, \qquad \hat{g}(\omega) = -i \operatorname{sgn}\omega.$$

The inverse Fourier transform yields the solution:

$$f(x) = -\frac{1}{\sqrt{2\pi}} \int_{-\infty}^{\infty} (i\,\mathrm{sgn}\,\omega)\hat{f}_H(\omega)e^{i\omega x}\,d\omega.$$

By convolution theorem, this reduces to the form

$$f(x) = -\frac{1}{\pi} \int_{-\infty}^{\infty} \frac{f_H(\xi)}{x - \xi}\,d\xi = -\mathcal{H}\{f_H(\xi)\}. \tag{5.126}$$

Consequently, $\mathcal{H}^2 f = -f$ and thus $\mathcal{H}^{-1} = -\mathcal{H}$.   □

Example 5.12.7. Consider the integral equation

$$\int_{-\infty}^{\infty} e^{-|x-t|}u(t)\,dt = -\frac{1}{4}u(x) + e^{-|x|}, \quad -\infty < x < \infty. \tag{5.127}$$

Application of the Fourier transform with respect to x yields

$$\frac{2}{1+\omega^2}\hat{u}(\omega) = -\frac{1}{4}\hat{u}(\omega) + \sqrt{\frac{2}{\pi}}\frac{1}{1+\omega^2},$$

so that

$$\hat{u}(\omega) = \frac{1}{\sqrt{2\pi}}\frac{8}{\omega^2 + 9}.$$

By the inverse Fourier transform, we obtain

$$u(x) = \frac{4}{\pi} \int_{-\infty}^{\infty} \frac{e^{i\omega x}}{\omega^2 + 9}\,d\omega.$$

To evaluate this integral for $x > 0$, we use a semicircular closed contour in the lower half of the complex plane. It turns out that

$$u(x) = \frac{4}{3}e^{-3x}, \quad x > 0.$$

Similarly, for $x < 0$, we use a closed semicircular contour in the upper half of the complex plane to obtain

$$u(x) = \frac{4}{3}e^{3x}, \quad x < 0.$$

Hence, the solution of (5.127) is

$$u(x) = \frac{4}{3}e^{-3|x|}. \tag{5.128}$$

□

5.13 Exercises

1. Determine the fixed points, if any, of the following operators:

   (a) $T(x) = x + a$ on any vector space.

   (b) $T : \mathbb{R}^2 \to \mathbb{R}^2$ defined by $T(x, y) = (x, 0)$.

   (c) $T : \mathbb{R}^2 \to \mathbb{R}^2$ defined by $T(x, y) = (y, y)$.

   (d) $T : \mathbb{R}^2 \to \mathbb{R}^2$ defined by

$$T(x, y) = (x \cos \varphi + y \sin \varphi, -x \sin \varphi + y \cos \varphi),$$

   where φ is a fixed real number.

2. Suppose T is an operator on $\mathcal{C}([0, 1])$ defined by

$$(Tu)(t) = \int_0^t (u(x))^2 dx.$$

   Show that T is not a contraction on the closed unit ball in $\mathcal{C}([0, 1])$, but that it is one on the closed ball of radius $\frac{1}{4}$ in $\mathcal{C}([0, 1])$.

3. Show that the operator $T : \mathcal{C}([0, 1]) \to \mathcal{C}([0, 1])$ defined by

$$(Tx)(t) = x(0) + \lambda \int_0^t x(\tau) d\tau, \quad \lambda \in \mathbb{R},$$

   is a contraction provided $|\lambda| < 1$.

4. Complete the proof of Theorem 5.2.2 by showing that $x_0 = \lim_{n \to \infty} T^n x$ for any $x \in E$.

5. Show that the nonlinear integral equation

$$f(x) = \int_0^1 e^{-sx} \cos(\alpha f(s)) ds, \quad 0 \le x \le 1, \ 0 < \alpha < 1,$$

   has a unique solution.

6. Consider a system of ordinary differential equations

$$\frac{d}{dx} \varphi_k(x) = f_k\big(x_0, \varphi_1(x), \varphi_2(x), \dots, \varphi_N(x)\big)$$

   with the initial data

$$\varphi_k(x_0) = y_{0k},$$

   where $k = 1, 2, \dots, N$; and the functions $f_k(x_0, y_1, y_2, \dots, y_N)$ are continuous in some domain $\Omega \subset \mathbb{R}^{N+1}$, and $(x_0, y_{01}, y_{02}, \dots, y_{0N}) \in \Omega$.

Moreover, we assume that the functions f_k satisfy the Lipschitz condition

$$\left| f_k(x, y_1, y_2, \ldots, y_N) - f_k(x, z_1, z_2, \ldots, z_N) \right| \leq \sup_x \max_{1 \leq m \leq N} |y_m - z_m|$$

in Ω. Prove that this system has a unique system of solutions $y_k = \varphi_k(x)$ in some interval $|x - x_0| < d$.

7. Use the method presented in Section 5.6 to solve the following homogeneous Fredholm equation:

$$f(x) = \lambda \int_{-1}^{1} (x + t) f(t) dt.$$

8. Use the method presented in Section 5.6 to solve the following nonhomogeneous equation:

$$f(x) = \varphi(x) + \lambda \int_{0}^{1} \left(\pi x \sin \pi t + 2\pi x^2 \sin 2\pi t \right) f(t) dt.$$

9. Express the solution of the integral equation

$$f(x) = \varphi(x) + \lambda \int_{0}^{2\pi} \cos(x + t) f(t) dt$$

in the resolvent form

$$f(x) = \varphi(x) + \lambda \int_{0}^{2\pi} \Gamma(x, t; \lambda) \varphi(t) dt,$$

where λ is not an eigenvalue. Obtain the general solution, if it exists, for $\varphi(x) = \sin x$.

10. Show that the solution of the differential equation

$$\frac{d^2 f}{dx^2} + xf = 1, \qquad f(0) = f'(0) = 0,$$

satisfies the nonhomogeneous Volterra equation

$$f(x) = \frac{x^2}{2} + \int_{0}^{x} t(t - x) f(t) dt.$$

11. Transform the problems

(a) $\dfrac{d^2 f}{dx^2} + f = x, \qquad f(0) = 0, \qquad f'(1) = 0,$

(b) $\dfrac{d^2f}{dx^2} + f = x,$ $f(0) = 1,$ $f'(1) = 0,$

into Fredholm integral equations.

12. Discuss the solutions of the integral equation

$$f(x) = \varphi(x) + \lambda \int_0^1 (x + t) f(t) dt.$$

13. When do the following integral equations have solutions?

(a) $f(x) = \varphi(x) + \lambda \int_0^1 (1 - 3xt) f(t) dt.$

(b) $f(x) = \varphi(x) + \lambda \int_0^{2\pi} \sin(x + t) f(t) dt.$

(c) $f(x) = \varphi(x) + \lambda \int_0^1 xt f(t) dt.$

(d) $f(x) = \varphi(x) + \lambda \int_{-1}^1 \sum_{n=1}^m P_n(x) P_n(t) f(t) dt$, where P_n is the nth degree Legendre polynomial.

(e) $f(x) = x + \frac{1}{2} \int_{-1}^1 (x + t) f(t) dt.$

14. Find the eigenvalues and eigenfunctions of the following integral equations:

(a) $f(x) = \lambda \int_0^{2\pi} \cos(x - t) f(t) dt.$

(b) $f(x) = \lambda \int_{-1}^1 (t - x) f(t) dt.$

(c) $f(x) = \varphi(x) + \lambda \int_0^{2\pi} \cos(x + t) f(t) dt.$

15. Solve the integral equations

(a) $f(x) = \varphi(x) + \lambda \int_0^1 tf(t) dt.$

(b) $f(x) = x + \lambda \int_0^{1/2} f(t) dt.$

(c) $f(x) = \dfrac{5x}{6} + \frac{1}{2} \int_0^1 xt f(t) dt.$

(d) $f(x) = x + \int_0^1 (1 + xt) f(t) dt.$

(e) $f(x) = e^x + \lambda \int_0^1 2e^{x+t} f(t) dt.$

16. Use the separable kernel method to show that

$$f(x) = \lambda \int_0^1 \cos x \sin t f(t) dt$$

has no solution except the trivial solution $f = 0$.

17. Obtain the Neumann series solutions of the following equations:

(a) $f(x) = x + \frac{1}{2}\int_{-1}^{1}(t+x)f(t)dt.$

(b) $f(x) = x + \int_{0}^{x}(t-x)f(t)dt.$

(c) $f(x) = x - \int_{0}^{x}(t-x)f(t)dt.$

(d) $f(x) = 1 - 2\int_{0}^{x}tf(t)dt.$

18. If $Lu = u'' + \omega^2 u$, show that L is formally self-adjoint and the concomitant is $J(u, v) = vu' - uv'$. Moreover, if u is a solution of $Lu = 0$ and v is a solution of $L^*v = 0$, then the concomitant of u and v is a constant.

19. Let L be a self-adjoint differential operator given by (5.64). If u_1 and u_2 are two solutions of $Lu = 0$, and $J(u_1, u_2) = 0$ for some x for which $a_2(x) \neq 0$, then show that u_1 and u_2 are linearly independent.

20. Consider the differential operator

$$L = e^x D^2 + e^x D, \quad D = \frac{d}{dx}, \quad 0 \leq x \leq 1,$$

$$u'(0) = 0, \qquad u(1) = 0.$$

Show that L is formally self-adjoint.

21. Prove continuity of the Green's function defined in Theorem 5.10.1.

22. Find eigenvalues and eigenfunctions of the following Sturm–Liouville system:

$$u'' + \lambda u = 0, \quad 0 \leq x \leq \pi,$$

$$u(0) = u'(\pi) = 0.$$

23. Transform the Euler equation

$$x^2 u'' + xu' + \lambda u = 0, \quad 1 \leq x \leq e,$$

with the boundary conditions

$$u(1) = u(e) = 0$$

into the Sturm–Liouville system

$$\frac{d}{dx}\left[x\frac{du}{dx}\right] + \frac{1}{x}\lambda u = 0,$$

$$u(1) = u(e) = 0.$$

Find the eigenvalues and eigenfunctions.

24. Prove that $\lambda = 0$ is not an eigenvalue of the system defined in Example 5.9.3.

25. Show that the Sturm–Liouville operator $L = DpD + q$, $D = d/dx$, is positive if $p(x) > 0$ and $q(x) \geq 0$ for all $x \in [a, b]$.

26. Show that the Sturm–Liouville operator L in $L^2([a, b])$ given by

$$L = \frac{1}{r(x)}(DpD + q)$$

is not symmetric.

27. Provide a detailed proof for Corollary 5.11.19.

28. Prove Theorem 5.11.7.

29. Find the Fourier transform of

(a) $f(x) = \begin{cases} 1 & \text{if } x \in [-a, a], \\ 0 & \text{otherwise}, \end{cases}$

(b) $f(x) = \begin{cases} 1 - \frac{|x|}{2} & \text{if } x \in [-2, 2], \\ 0 & \text{otherwise}, \end{cases}$

(c) $f(x) = (1 - x^2)e^{-\frac{1}{2}x^2}$,

(d) $f(x) = \int_{-\infty}^{\infty} g(x + t)\overline{g(t)}\,dt$.

30. Show that under appropriate conditions

(a) $\hat{f}'(\omega) = -i\mathcal{F}\{xf(x)\}$,

(b) $\hat{f}^{(r)}(\omega) = (-i)^r \mathcal{F}\{x^r f(x)\}$.

31. Use the Parseval relation to evaluate

(a) $\int_{-\infty}^{\infty} \left(\frac{\sin x}{x}\right)^2 dx$,

(b) $\int_{-\infty}^{\infty} \left(\frac{\sin x}{x}\right)^3 dx$,

(c) $\int_{-\infty}^{\infty} \left(\frac{\sin x}{x}\right)^4 dx$.

32. Show that $\widehat{f}(\omega) * \widehat{g}(\omega) = \mathcal{F}\{f(x)g(x)\}$.

33. Show that the eigenvalues of the Fourier transform are $\lambda = 1, -1, i, -i$.

34. The Hermite functions are given by

$$h_n(x) = \frac{1}{n!} e^{-x^2/2} H_n(x),$$

where $H_n(x)$ is the Hermite polynomial of degree n (see Example 5.8.15). Show that

$$\widehat{h}_n(\omega) = (-i)^n h_n(\omega).$$

35. Use the Fourier transform to solve the forced linear harmonic oscillator

$$\ddot{x} + \omega^2 x = a \sin \Omega t, \quad t > 0, \ \omega \neq \Omega, \qquad x(0+) = 0 = \dot{x}(0+).$$

Examine the case when $\omega = \Omega$.

36. Solve the problem discussed in Example 5.12.1 with $E(t) = E_0 e^{-\alpha t} \sin \omega t \cdot H(t)$ and $I(0+) = I_0$.

37. If there is a capacitor in the circuit discussed in Example 5.12.1, then the current $I(t)$ satisfies the following integro-differential equation:

$$L\frac{dI}{dt} + RI + \frac{1}{C}\left[q_0 + \int_0^t I(t)dt\right] = E(t),$$

where q_0 is the initial charge on the capacitor so that

$$q = q_0 + \int_0^t I(t)dt$$

is the charge and $dq/dt = I$.

   Solve this problem using the Fourier transform and the following conditions:

$$I = q = E = 0 \quad \text{for } t < 0,$$
$$I(0+) = I_0 \quad \text{and} \quad q(0+) = q_0.$$

Examine the special case when $E(t) = H(t)$.

38. Use the Fourier transform to solve the following problem:

$$y'' + 3y' + 2y = e^{-x}, \quad x > 0, \qquad y(0+) = y_0 \quad \text{and} \quad y'(0+) = y_{00}.$$

39. Use the Fourier transform to solve the following pair of coupled differential systems for $t > 0$:

$$x' + y' - x + 3y = e^{-t},$$

$$x' + y' + 2x + y = e^{-2t},$$

$$x(0+) = x_0 \quad \text{and} \quad y(0+) = y_0.$$

40. Show that the formal solution of

$$\ddot{x} + 2k\dot{x} + \sigma^2 x = f(t)$$

is

$$x(t) = \frac{1}{\sqrt{2\pi}} \int_{-\infty}^{\infty} \frac{\hat{f}(\omega)e^{i\omega t}d\omega}{\sigma^2 - \omega^2 + 2i\omega k}.$$

41. (a) Show that the solution of the integral equation

$$u(x) - \lambda \int_{-\infty}^{\infty} e^{-|x-t|}u(t)dt = e^{-|x|}$$

is

$$u(x) = \frac{e^{-\sqrt{1-2\lambda}|x|}}{\sqrt{1-2\lambda}} \quad \text{for } \lambda > \frac{1}{2}.$$

(b) If

$$Tu = \int_{-\infty}^{\infty} e^{-|x-t|}u(t)dt,$$

show that $\|T\| \leq 2$; $\|\cdot\|$ denotes the norm in $L^2(\mathbb{R})$.

42. Prove the following properties of the Hilbert transform $\tilde{\varphi}(x) = \mathcal{H}\{\varphi(t); x\}$:

(a) $\mathcal{H}\{\varphi(t+a); x\} = \mathcal{H}\{\varphi(t); x+a\}$.

(b) $\mathcal{H}\{\varphi(at); x\} = \mathcal{H}\{\varphi(t); ax\}, a > 0$.

(c) $\mathcal{H}\{\varphi(-t); x\} = -\mathcal{H}\{\varphi(t); -x\}$.

(d) $\mathcal{H}\{\varphi'(t); x\} = (d/dx)\tilde{\varphi}(x)$.

(e) $\mathcal{H}\{t\varphi(t); x\} = x\tilde{\varphi}(x) + (1/\pi)\int_{-\infty}^{\infty} \varphi(t)dt$.

43. Show that if $f \in L^2(\mathbb{R})$, then $\mathcal{H}\{f\} \in L^2(\mathbb{R})$ and $\|\mathcal{H}\{f\}\| = \|f\|$.

44. Show that $\mathcal{F}\{\mathcal{H}\{f\}\} = (-i\,\mathrm{sgn}\,\omega)\mathcal{F}\{f\}$.

Generalized Functions and Partial Differential Equations

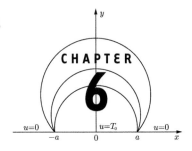

CHAPTER 6

"However varied may be the imagination of man, nature is still a thousand times richer. . . . Each of the theories of physics . . . presents (partial differential) equations under a new aspect . . . without these theories, we should not know partial differential equations."

Henri Poincaré

"Between 1930 and 1940 several mathematicians began to investigate systematically the concept of a "weak" solution of a linear partial differential equation, which appeared episodically (and without a name) in Poincaré's work . . . .

It was one of the main contributions of Laurent Schwartz when he saw, in 1945, that the concept of distribution introduced by Sobolev (which he had rediscovered independently) could give a satisfactory generalization of the Fourier transform including all the preceding ones."

Jean Dieudonné

6.1 Introduction

In this chapter, we shall first discuss briefly the basic concepts and properties of distributions. The theory of distributions was initiated by S.L. Sobolev in 1936. The concept of distributions was independently introduced by Laurent Schwartz (1915–2002) in the 1950s. Since Schwartz was the one who developed the theory almost to its present form, distributions are often called *Schwartz distributions*. Distributions have found applications in many areas of mathematics and physics, including differential and integral equations.

In Section 6.3, we discuss Sobolev spaces. The definitions and some basic properties are presented. The remainder of this chapter deals with Green's functions for partial differential equations of most common interest. This is followed by the form of the Green's identity associated with partial differential operators. Section 6.5 discusses weak solutions of elliptic boundary value problems. The final section is devoted to examples of applications of the Fourier transform to partial differential equations of physical interest.

6.2 Distributions

Consider a partial differential operator L of order m in N variables

$$L = \sum_{|\alpha| \le m} a_\alpha D^\alpha, \tag{6.1}$$

where $\alpha = (\alpha_1, \ldots, \alpha_N)$ is a multi-index, α_n's are non-negative integers, $|\alpha| = \alpha_1 + \cdots + \alpha_N$, $a_\alpha = a_{\alpha_1, \alpha_2, \ldots, \alpha_N}$ are constant coefficients, and

$$D^\alpha = \left(\frac{\partial}{\partial x_1} \right)^{\alpha_1} \cdots \left(\frac{\partial}{\partial x_N} \right)^{\alpha_N} = \frac{\partial^{|\alpha|}}{\partial x_1^{\alpha_1} \ldots \partial x_N^{\alpha_N}}.$$

Let $f \in C(\mathbb{R}^N)$ and let u be a function satisfying the equation

$$Lu = \sum_{|\alpha| \le m} a_\alpha D^\alpha u = f. \tag{6.2}$$

This equation makes sense only if all derivatives of u appearing in L exist. This puts a rather severe restriction on the family of solutions and, as we will see later, quite often there is no such solution.

Let φ be an infinitely differentiable function on $\mathbb{R}^N$ vanishing outside of some bounded set. First we multiply both sides of Equation (6.2) by $\varphi(x)$ to obtain

$$\sum_{|\alpha| \le m} a_\alpha \left(D^\alpha u(x) \right) \varphi(x) = f(x) \varphi(x). \tag{6.3}$$

Next we integrate (6.3) over $\mathbb{R}^N$:

$$\int_{\mathbb{R}^N} \left(\sum_{|\alpha| \le m} a_\alpha \left(D^\alpha u(x) \right) \varphi(x) \right) dx = \int_{\mathbb{R}^N} f(x) \varphi(x) dx. \tag{6.4}$$

Since $\varphi(x) = 0$ outside a bounded subset of $\mathbb{R}^N$, integration by parts yields

$$\int_{\mathbb{R}^N} \left(\sum_{|\alpha| \le m} a_\alpha u(x)(-1)^{|\alpha|} D^\alpha \varphi(x) \right) dx = \int_{\mathbb{R}^N} f(x) \varphi(x) dx. \tag{6.5}$$

If we put

$$L^* = \sum_{|\alpha| \le m} (-1)^{|\alpha|} a_\alpha D^\alpha, \tag{6.6}$$

then (6.5) can be written as

$$\int_{\mathbb{R}^N} u(x)\bigl(L^*\varphi\bigr)(x)dx = \int_{\mathbb{R}^N} f(x)\varphi(x)dx. \tag{6.7}$$

This analysis shows that every function u satisfying (6.2) also satisfies (6.7). On the other hand, a function u may satisfy (6.7) without being differentiable. This observation becomes the fundamental idea behind the theory of generalized functions.

Let f and g be continuous functions on $\mathbb{R}^N$. The condition

$$f(x) = g(x) \quad \text{for every } x \in \mathbb{R}^N$$

follows if

$$\int_{\mathbb{R}^N} f(x)\varphi(x)dx = \int_{\mathbb{R}^N} g(x)\varphi(x)dx$$

for every $\varphi \in C^\infty(\mathbb{R}^N)$ vanishing outside of some bounded set. The essential difference is that instead of comparing the values at every point $x \in \mathbb{R}^N$, we test functions f and g using smooth functions with bounded support. For this reason, those functions are called *test functions*. They play the crucial role in the theory of distributions.

Definition 6.2.1.   (Test function)
By a *test function*, we mean an infinitely differentiable function on $\mathbb{R}^N$ vanishing outside of some bounded set. The space of all test functions is denoted by $\mathcal{D}(\mathbb{R}^N)$ or simply by $\mathcal{D}$.

Instead of an infinitely differentiable function, we will often say a *smooth function*.

Example 6.2.2. The existence of nontrivial test functions is not obvious. In the case when the number of variables $N = 1$, an example of a test function is

$$\varphi(x) = \begin{cases} e^{(x^2-1)^{-1}} & \text{if } |x| < 1, \\ 0 & \text{otherwise} \end{cases}$$

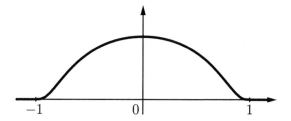

Figure 6.1　The graph of $\varphi(x) = e^{(x^2-1)^{-1}}$ for $|x| < 1$.

(see Figure 6.1). Using this test function, we can easily generate a number of examples. The following are test functions:

$$\varphi(ax + b), \quad \text{where } a, b \text{ are constants, } a \neq 0,$$
$$f(x)\varphi(x), \quad \text{where } f \text{ is an arbitrary smooth function,}$$
$$\varphi^{(k)}(x), \quad \text{where } k \text{ is a positive integer.}$$

The function

$$\varphi(x) = \begin{cases} e^{(x_1^2 + \cdots + x_N^2 - 1)^{-1}} & \text{if } x_1^2 + \cdots + x_N^2 < 1, \\ 0 & \text{otherwise} \end{cases}$$

is a test function on $\mathbb{R}^N$. Another way to obtain examples of test functions on $\mathbb{R}^N$ is to take arbitrary test functions $\varphi_1, \ldots, \varphi_N$ defined on $\mathbb{R}$ and then define

$$\varphi(x) = \varphi_1(x_1)\varphi_2(x_2)\cdots\varphi_N(x_N). \qquad \square$$

Theorem 6.2.3. *The space of test functions $\mathcal{D}$ is a vector space. Moreover, if $\varphi, \psi \in \mathcal{D}$, then*

(a) *$f\varphi \in \mathcal{D}$ for every smooth function f,*

(b) *$\varphi \circ A \in \mathcal{D}$ for every affine transformation A of $\mathbb{R}^N$ onto $\mathbb{R}^N$,*

(c) *$\varphi * \psi \in \mathcal{D}$.* $\qquad \square$

The proof is left as an exercise.

To define distributions, we need to introduce a convergence in the space of test functions.

Definition 6.2.4.　(Convergence of test functions)
Let $\varphi_1, \varphi_2, \ldots$ and φ be test functions. We say that the sequence (φ_n) converges to φ in $\mathcal{D}$, denoted by $\varphi_n \xrightarrow{\mathcal{D}} \varphi$, if the following two conditions are satisfied:

(a)　$\varphi_1, \varphi_2, \ldots$ and φ vanish outside some bounded set $S \subset \mathbb{R}^N$,

(b) $D^\alpha \varphi_n \to D^\alpha \varphi$ uniformly on $\mathbb{R}^N$ for every multi-index α.

Example 6.2.5. Let $\varphi \in \mathcal{D}$ and let $\{v_n\}$ be a sequence of vectors in $\mathbb{R}^N$ convergent to 0. Define $\varphi_n(x) = \varphi(x - v_n)$. Then $\varphi_n \xrightarrow{\mathcal{D}} \varphi$. In other words, translation is a continuous operation in $\mathcal{D}$.

Let $\varphi \in \mathcal{D}$, and let (a_n) be a sequence of scalars convergent to a scalar a. Then $a_n \varphi \xrightarrow{\mathcal{D}} a\varphi$. In other words, multiplication by scalars is a continuous operation on $\mathcal{D}$.

Let φ be a nonzero test function, and let (a_n) and (b_n) be sequences of positive scalars convergent to 0. Define $\varphi_n(x) = b_n \varphi(a_n x)$. Then

$$D^\alpha \varphi_n \to 0 \quad \text{uniformly on } \mathbb{R}^N \text{ for every multi-index } \alpha.$$

In this case, however, the sequence $\{\varphi_n\}$ is not convergent in $\mathcal{D}$ because condition (a) is not satisfied. The supports of φ_n's expand without a bound as $n \to \infty$. □

The properties of the convergence in $\mathcal{D}$ listed in the following theorem are direct consequences of the definition. The easy proofs are left as an exercise.

Theorem 6.2.6. *Let* $\varphi_n \xrightarrow{\mathcal{D}} \varphi$ *and* $\psi_n \xrightarrow{\mathcal{D}} \psi$. *Then*

(a) $a\varphi_n + b\psi_n \xrightarrow{\mathcal{D}} a\varphi + b\psi$, *for any scalars* a, b,

(b) $f\varphi_n \xrightarrow{\mathcal{D}} f\varphi$, *for any smooth function* f *defined on* $\mathbb{R}^N$,

(c) $\varphi_n \circ A \xrightarrow{\mathcal{D}} \varphi \circ A$, *for any affine transformation* A *of* $\mathbb{R}^N$ *onto* $\mathbb{R}^N$,

(d) $D^\alpha \varphi_n \xrightarrow{\mathcal{D}} D^\alpha \varphi$, *for any multi-index* α. □

Definition 6.2.7. (Distribution)
By a *distribution* F on $\mathbb{R}^N$, we mean a continuous linear functional on $\mathcal{D}(\mathbb{R}^N)$. In other words, a mapping $F : \mathcal{D}(\mathbb{R}^N) \to \mathbb{C}$ is called a distribution if

(a) $F(a\varphi + b\psi) = aF(\varphi) + bF(\psi)$ for every $a, b \in \mathbb{C}$ and $\varphi, \psi \in \mathcal{D}(\mathbb{R}^N)$,

(b) $F(\varphi_n) \to F(\varphi)$ (in $\mathbb{C}$) whenever $\varphi_n \xrightarrow{\mathcal{D}} \varphi$.

The space of all distributions is denoted by $\mathcal{D}'(\mathbb{R}^N)$ or simply by $\mathcal{D}'$. It will be convenient to write $\langle F, \varphi \rangle$ instead of $F(\varphi)$. Note that $\langle F, \varphi \rangle$ is not an inner product. Since it is always clear from the context what is meant, this inconsistency does not lead to any problems.

Distributions generalize the concept of a function. Formally, a function on $\mathbb{R}^N$ is not a distribution because its domain is not $\mathcal{D}$. However, every locally integrable function f on $\mathbb{R}^N$ can be identified with a distribution F in the following way:

$$\langle F, \varphi \rangle = \int_{\mathbb{R}^N} f\varphi.$$

Definition 6.2.8.   (Regular and singular distributions)
A distribution $F \in \mathcal{D}'$ is called a *regular distribution* if there exists a locally integrable function f such that

$$\langle F, \varphi \rangle = \int_{\mathbb{R}^N} f\varphi, \tag{6.8}$$

for every $\varphi \in \mathcal{D}$. A distribution which is not regular is called a *singular distribution*.

The proof of the fact that (6.8) defines a distribution is left as an exercise. Note that there is no problem with integrability of the product $f\varphi$ because it vanishes outside a bounded set.

The integral $\int_{\mathbb{R}^N} f\varphi$ can be interpreted, at least for some test functions φ, as the average value of f with respect to probability whose density function is φ. Thus, one can think of $\langle F, \varphi \rangle$ as an average value of F, and of distributions as objects that have average values in the neighborhood of every point. In general, distributions do not have values at points. This interpretation of distributions is very natural from the point of view of physics: when a quantity is measured, the result is not the exact value at a single point.

Example 6.2.9. Let Ω be an open (or just measurable) set in $\mathbb{R}^N$. The functional F defined by

$$\langle F, \varphi \rangle = \int_{\Omega} \varphi$$

is a distribution. Observe that it is a regular distribution since

$$\langle F, \varphi \rangle = \int_{\mathbb{R}^N} \chi_{\Omega} \varphi,$$

where χ_{Ω} is the characteristic function of Ω.

In particular, if $\Omega = (0, \infty) \times \cdots \times (0, \infty)$, we obtain a distribution

$$\langle H, \varphi \rangle = \int_0^{\infty} \cdots \int_0^{\infty} \varphi(x) dx_1 \ldots dx_N,$$

which is called the *Heaviside function* (Oliver Heaviside (1850–1925)). The letter H will be used to denote this distribution, as well as the characteristic function of $\Omega = (0, \infty) \times \cdots \times (0, \infty)$.   □

Example 6.2.10. (Dirac delta distribution) One of the most important examples of distributions is the so called *Dirac delta* (Paul Adrien Maurice Dirac (1902–1984)). It is denoted by δ and defined by

$$\langle \delta, \varphi \rangle = \varphi(0). \tag{6.9}$$

The linearity of δ is obvious. To prove continuity note that $\varphi_n \xrightarrow{\mathcal{D}} \varphi$ implies $\varphi_n \to \varphi$ uniformly on $\mathbb{R}^N$, and thus $\varphi_n(x) \to \varphi(x)$ for every $x \in \mathbb{R}^N$. The Dirac delta is a singular distribution. $\square$

Example 6.2.11. Let α be a multi-index. The functional F on $\mathcal{D}$ defined by

$$\langle F, \varphi \rangle = D^\alpha \varphi(0) \tag{6.10}$$

is a distribution. The proof is similar to the proof of the fact that the Dirac delta is a distribution. Actually, distributions defined by (6.10) are closely related to δ (see Example 6.2.14). $\square$

The success of the theory of distributions is largely due to the fact that most concepts of calculus can be defined for distributions. When adopting definitions for distributions, we expect that new definitions agree with classical ones when applied to regular distributions. The following approach will ensure that. When looking for an extension of some operation A, that is defined for functions, we first consider regular distributions:

$$\langle F, \varphi \rangle = \int f\varphi.$$

Since we expect AF to be the same as Af, it is natural to define

$$\langle AF, \varphi \rangle = \int Af\varphi.$$

If we can find a continuous operation A^* which maps $\mathcal{D}$ into $\mathcal{D}$ such that

$$\int Af\varphi = \int fA^*\varphi,$$

then it makes sense to define, for an arbitrary distribution F,

$$\langle AF, \varphi \rangle = \langle F, A^*\varphi \rangle.$$

For example, if the described method is used to find a natural definition of the derivative of a distribution, it suffices to note that

$$\int_{\mathbb{R}^N} \frac{\partial}{\partial x_k} f(x)\varphi(x)dx = -\int_{\mathbb{R}^N} f(x)\frac{\partial}{\partial x_k}\varphi(x)dx.$$

Definition 6.2.12.   (Derivatives of distributions)
Let F be a distribution. Then the derivative $\partial F / \partial x_k$ is defined by

$$\left\langle \frac{\partial F}{\partial x_k}, \varphi \right\rangle = -\left\langle F, \frac{\partial \varphi}{\partial x_k} \right\rangle. \tag{6.11}$$

More generally, if α is a multi-index, then by $D^\alpha F$ we denote the functional defined by

$$\langle D^\alpha F, \varphi \rangle = (-1)^{|\alpha|} \langle F, D^\alpha \varphi \rangle. \tag{6.12}$$

The fact that $\langle D^\alpha F, \varphi \rangle$ is well defined for every $\varphi \in \mathcal{D}$ does not mean that $D^\alpha F$ is a distribution. That has to be proved.

Theorem 6.2.13. *If F is a distribution, then $D^\alpha F$ is a distribution for any multi-index α.*

Proof: We need to show that $D^\alpha F$ is linear and continuous. If F is a distribution, then

$$
\begin{aligned}
\langle D^\alpha F, a\varphi + b\psi \rangle &= \langle F, a(-1)^{|\alpha|} D^\alpha \varphi + b(-1)^{|\alpha|} D^\alpha \psi \rangle \\
&= a \langle F, (-1)^{|\alpha|} D^\alpha \varphi \rangle + b \langle F, (-1)^{|\alpha|} D^\alpha \psi \rangle \\
&= a \langle D^\alpha F, \varphi \rangle + b \langle D^\alpha F, \psi \rangle,
\end{aligned}
$$

so that $D^\alpha F$ is linear. Moreover, since $\varphi_n \xrightarrow{\mathcal{D}} \varphi$ implies $D^\alpha \varphi_n \xrightarrow{\mathcal{D}} D^\alpha \varphi$, the functional $D^\alpha F$ is continuous. Thus, $D^\alpha F$ is a distribution.   □

Example 6.2.14. Let H denote the Heaviside function of a single variable x. Then

$$\left\langle \frac{d}{dx} H, \varphi \right\rangle = -\left\langle H, \frac{d}{dx}\varphi \right\rangle = -\int_0^\infty \varphi'(x)dx = \varphi(0) = \langle \delta, \varphi \rangle.$$

Therefore,

$$\frac{d}{dx} H = \delta. \tag{6.13}$$

Moreover,

$$\left\langle \frac{d}{dx}\delta, \varphi \right\rangle = \langle \delta', \varphi \rangle = -\varphi'(0) \tag{6.14}$$

and similarly,

$$\left\langle \frac{d^n}{dx^n}\delta, \varphi \right\rangle = \langle \delta^{(n)}, \varphi \rangle = (-1)^n \varphi^{(n)}(0). \tag{6.15}$$

In general,

$$\langle D^\alpha \delta, \varphi \rangle = (-1)^{|\alpha|} D^\alpha \varphi(0). \tag{6.16}$$

□

Definition 6.2.15.   (Weak distributional convergence)
A sequence of distributions (F_n) is convergent to a distribution F if

$$\langle F_n, \varphi \rangle \to \langle F, \varphi \rangle \quad \text{for every } \varphi \in \mathcal{D}.$$

This type of convergence is called the *weak distributional convergence*. Since this is the only convergence of distributions considered in this book, the notation $F_n \to F$ or $\lim_{n\to\infty} F_n = F$ will be used.

Example 6.2.16. Let $f_1, f_2, \ldots$ be a sequence of continuous functions on $\mathbb{R}^N$. Suppose, for some continuous function f, $f_n \to f$ uniformly on every compact subset of $\mathbb{R}^N$. Define regular distributions

$$\langle F_n, \varphi \rangle = \int_{\mathbb{R}^N} f_n \varphi, \quad n = 1, 2, \ldots,$$

and

$$\langle F, \varphi \rangle = \int_{\mathbb{R}^N} f \varphi.$$

Then $F_n \to F$. Indeed, if Ω denotes the support of a test function φ, then

$$\langle F_n, \varphi \rangle = \int_{\mathbb{R}^N} f_n \varphi = \int_\Omega f_n \varphi \to \int_\Omega f \varphi = \langle F, \varphi \rangle.$$

□

Example 6.2.17. Let $f_1, f_2, \ldots \in L^1(\mathbb{R}^N)$. Suppose, for some $f \in L^1(\mathbb{R}^N)$, $f_n \to f$ in $L^1(\mathbb{R}^N)$, that is, $\int_{\mathbb{R}^N} |f_n - f| \to 0$. Define regular distributions F_n and F as in the previous example. Then $F_n \to F$. In fact, we have

$$\left| \langle F_n, \varphi \rangle - \langle F, \varphi \rangle \right| = \left| \int_{\mathbb{R}^N} (f_n - f)\varphi \right| \le \int_{\mathbb{R}^N} |f_n - f||\varphi| \le \sup_{\mathbb{R}^N} |\varphi| \int_{\mathbb{R}^N} |f_n - f| \to 0,$$

for every $\varphi \in \mathcal{D}$.

□

Instead of saying "the sequence of regular distributions generated by a sequence of functions (f_n) converges to a distribution F," we will say simply "the sequence of functions (f_n) converges to a distribution F in the distributional sense" or just "the sequence (f_n) is distributionally convergent to F."

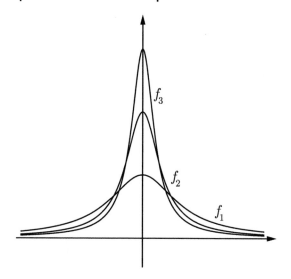

Figure 6.2   Functions $f_n(x) = \frac{n}{\pi(1+n^2x^2)}$ for $n = 1, 2, 3$.

Example 6.2.18. Consider the sequence of functions on $\mathbb{R}$ defined by

$$f_n(x) = \frac{n}{\pi(1 + n^2x^2)}, \quad n = 1, 2, \ldots$$

(see Figure 6.2). We will show that the sequence (f_n) is distributionally convergent to the Dirac delta distribution δ.

Let φ be a test function with the support contained in the interval $[-a, a]$. We need to show that

$$\int_{-\infty}^{\infty} f_n(x)\varphi(x)dx \to \varphi(0) \quad \text{as } n \to \infty,$$

or, equivalently,

$$\int_{-\infty}^{\infty} f_n(x)\varphi(x)dx - \varphi(0) \to 0 \quad \text{as } n \to \infty.$$

Since

$$\int_{-\infty}^{\infty} f_n(x)dx = \int_{-\infty}^{\infty} \frac{ndx}{\pi(1 + n^2x^2)} = 1 \quad \text{for all } n \in \mathbb{N},$$

we have

$$\left| \int_{-\infty}^{\infty} f_n(x)\varphi(x)dx - \varphi(0) \right| = \left| \int_{-\infty}^{\infty} f_n(x)\big(\varphi(x) - \varphi(0)\big)dx \right|$$

$$\leq \left| \varphi(0) \int_{-\infty}^{-a} f_n(x)dx \right| + \left| \int_{-a}^{a} f_n(x)\big(\varphi(x) - \varphi(0)\big)dx \right| + \left| \varphi(0) \int_{a}^{\infty} f_n(x)dx \right|.$$

Direct integration shows that

$$\lim_{n\to\infty}\left|\varphi(0)\int_{-\infty}^{-a}f_n(x)dx\right| = \lim_{n\to\infty}\left|\varphi(0)\int_{a}^{\infty}f_n(x)dx\right| = 0.$$

It thus remains to prove that

$$\lim_{n\to\infty}\left|\int_{-a}^{a}f_n(x)\big(\varphi(x)-\varphi(0)\big)dx\right| = 0. \tag{6.17}$$

Note that

$$\left|\int_{-a}^{a}f_n(x)\big(\varphi(x)-\varphi(0)\big)dx\right| \le \int_{-a}^{a}\left|f_n(x)\big(\varphi(x)-\varphi(0)\big)\right|dx$$

$$\le \max\left|\varphi'(x)\right|\int_{-a}^{a}\left|xf_n(x)\right|dx,$$

since $|\varphi(x)-\varphi(0)| \le \max|\varphi'(x)||x|$, by the mean value theorem, and

$$\lim_{n\to\infty}\int_{-a}^{a}\left|xf_n(x)\right|dx = \lim_{n\to\infty}\frac{\ln(1+n^2a^2)}{\pi n} = 0. \qquad \square$$

Theorem 6.2.19. *If $F_n \to F$ in $\mathcal{D}'(\mathbb{R})$, then*

$$D^\alpha F_n \to D^\alpha F$$

for every multi-index α.

Proof: We have

$$\langle D^\alpha F_n, \varphi\rangle = (-1)^{|\alpha|}\langle F_n, D^\alpha\varphi\rangle \to (-1)^{|\alpha|}\langle F, D^\alpha\varphi\rangle = \langle D^\alpha F, \varphi\rangle$$

for every test function φ. $\qquad\square$

Note that Theorem 6.2.19 allows us to differentiate convergent sequences or series of distributions term by term.

Definition 6.2.20. (Antiderivative of a distribution)
Let $F \in \mathcal{D}'(\mathbb{R})$. A distribution G on $\mathbb{R}$ is called an *antiderivative* of F if $G' = F$.

Theorem 6.2.21. *Every distribution has an antiderivative.*

Proof: Let $\varphi_0 \in \mathcal{D}(\mathbb{R})$ be a fixed test function such that

$$\int_{-\infty}^{\infty}\varphi_0(x)dx = 1. \tag{6.18}$$

Then, for every test function $\varphi \in \mathcal{D}(\mathbb{R})$, there exists a test function $\varphi_1 \in \mathcal{D}(\mathbb{R})$ such that

$$\varphi = K\varphi_0 + \varphi_1,$$

where

$$K = \int_{-\infty}^{\infty} \varphi(x)dx \quad \text{and} \quad \int_{-\infty}^{\infty} \varphi_1(x)dx = 0.$$

Let $F \in \mathcal{D}'(\mathbb{R})$. Define a functional G on $\mathcal{D}$ by

$$\langle G, \varphi \rangle = \langle G, K\varphi_0 + \varphi_1 \rangle = KC_0 - \langle F, \psi \rangle, \tag{6.19}$$

where C_0 is a constant and ψ is the test function defined by

$$\psi(x) = \int_{-\infty}^{x} \varphi_1(t)dt.$$

Then G is a distribution and $G' = F$. $\qquad\qquad\qquad\qquad\qquad\qquad \square$

It can be proved that if G_1 and G_2 are antiderivatives of a distribution F, then $G_1 - G_2$ is a constant function, or, more precisely, there exists a constant C such that

$$\langle G_1 - G_2, \varphi \rangle = C \int_{-\infty}^{\infty} \varphi(x)dx$$

for every test function φ. This follows easily from the following theorem.

Theorem 6.2.22. *If $F \in \mathcal{D}'(\mathbb{R})$ and $F' = 0$, then F is a constant function.*

Proof: Let $\varphi \in \mathcal{D}(\mathbb{R})$. Using the notation of the proof of Theorem 6.2.21, we have

$$\langle F, \varphi \rangle = \langle F, K\varphi_0 + \varphi_1 \rangle = \langle F, K\varphi_0 \rangle + \langle F, \varphi_1 \rangle = \langle F, \varphi_0 \rangle \int_{-\infty}^{\infty} \varphi(x)dx,$$

because

$$\langle F, \varphi_1 \rangle = -\langle F', \psi \rangle = 0.$$

Thus, F is the regular distribution generated by the constant function $C = \langle F, \varphi_0 \rangle$. $\qquad\qquad\qquad\qquad\qquad\qquad\qquad \square$

In applications we often need to consider distributions on an open subset of $\mathbb{R}^N$.

Definition 6.2.23.   (The space of test functions on an open subset of $\mathbb{R}^N$)
Let Ω be an open subset of $\mathbb{R}^N$. By the space of test functions on Ω, we mean the

space of all smooth functions defined on Ω with support contained in some compact subset of Ω. Elements of $\mathcal{D}(\Omega)$ are called *test functions on Ω*.

Note that the function

$$\varphi(x) = \begin{cases} e^{(x^2-1)^{-1}} & \text{if } |x| < 1, \\ 0 & \text{otherwise,} \end{cases}$$

is not a member of $\mathcal{D}((-1, 1))$. On the other hand, the function $\psi(x) = \varphi(\alpha x)$ is a member of $\mathcal{D}((-1, 1))$ for every $\alpha > 1$.

Definition 6.2.24.   (Convergence in $\mathcal{D}(\Omega)$)
Let $\varphi_1, \varphi_2, \ldots$ and φ be test functions on Ω. We say that the sequence $\{\varphi_n\}$ converges to φ in $\mathcal{D}(\Omega)$, denoted by $\varphi_n \to \varphi$ in $\mathcal{D}(\Omega)$, if the following two conditions are satisfied:

(a) $\varphi_1, \varphi_2, \ldots$ and φ vanish outside some compact set $S \subset \Omega$,

(b) $D^\alpha \varphi_n \to D^\alpha \varphi$ uniformly on Ω for every multi-index α.

Definition 6.2.25.   (The space of distributions on an open subset of $\mathbb{R}^N$)
By $\mathcal{D}'(\Omega)$ we denote the space of all continuous linear functions on $\mathcal{D}(\Omega)$. Members of $\mathcal{D}'(\Omega)$ are called *distributions on Ω*.

Definition 6.2.26.   (Multiplication of distributions by smooth functions)
Let $F \in \mathcal{D}'(\Omega)$, and let g be a smooth function on Ω. By the *product of g and F*, we mean the distribution defined by $\langle gF, \varphi \rangle = \langle F, g\varphi \rangle$.

It is easy to show that the product of a distribution and a smooth function is well defined.

Theorem 6.2.27. (Leibniz formula)   *Let $F \in \mathcal{D}'(\Omega)$, and let g be a smooth function on Ω. Then*

$$D^\alpha(gF) = \sum_{\beta \leq \alpha} \binom{\alpha_1}{\beta_1} \cdots \binom{\alpha_N}{\beta_N} D^\beta g D^{\alpha-\beta} F,$$

where $\beta \leq \alpha$ means $\beta_j \leq \alpha_j$ for $j = 1, \ldots, N$.

We leave the proof as an exercise.

In the following sections, distributions are often used in a less precise fashion and ideas not mentioned in this section are used. The arguments can be made precise and formal definitions can be given, but it goes beyond the scope of this book.

6.3 Sobolev Spaces

The study of partial differential equations naturally involves function spaces which are defined not only in terms of the functions themselves, but also of their derivatives. The Banach space of all bounded continuous functions on cl Ω, where Ω is an open set in $\mathbb{R}^N$, with the norm of uniform convergence, is rather unsuitable for the study of partial differential equations. For example, if

$$\frac{\partial^2 u}{\partial x_1^2} + \cdots + \frac{\partial^2 u}{\partial x_N^2} = f$$

with f continuous, it is generally not true that $u \in C^2(\Omega)$. Sobolev spaces turn out to be much more useful for the study of partial differential equations. In this section, we give the basic definitions and discuss some properties of Sobolev spaces. Several results are stated without proofs. An extensive treatment of Sobolev spaces can be found in Adams and Fournier (2003) or Ziemer (1989).

Throughout this section, Ω stands for an open set in $\mathbb{R}^N$.

Definition 6.3.1.   (Sobolev space $W^{m,p}(\Omega)$)
Let $m \geq 0$ be an integer and $1 \leq p \leq \infty$. The *Sobolev space* $W^{m,p}(\Omega)$ is defined to be the set of all functions $u \in L^p(\Omega)$ such that $D^\alpha u \in L^p(\Omega)$ for all $|\alpha| \leq m$.

Since a function $u \in L^p(\Omega)$ need not be differentiable, the meaning of the condition $D^\alpha u \in L^p(\Omega)$ requires an explanation. If $u \in L^p(\Omega)$, then u can be identified with a regular distribution $G_u \in \mathcal{D}'(\Omega)$. The distribution G_u has derivatives of all orders and $D^\alpha G_u \in D'(\Omega)$. If $D^\alpha G_u$ is a regular distribution corresponding to a function in $L^p(\Omega)$, then we simply write $D^\alpha u \in L^p(\Omega)$.

It is easy to see that $W^{m,p}(\Omega)$ is a vector space. The functional

$$\|u\|_{m,p,\Omega} = \left(\sum_{|\alpha| \leq m} \int_\Omega |D^\alpha u|^p \right)^{1/p} \tag{6.20}$$

is a norm in $W^{m,p}(\Omega)$ if $1 \leq p < \infty$. For $p = \infty$ the norm is defined by

$$\|u\|_{m,\infty,\Omega} = \max_{0 \leq |\alpha| \leq m} \|D^\alpha u\|_\infty. \tag{6.21}$$

Example 6.3.2. For $1 \leq p < \infty$, we have

$$\|u\|_{1,p,\Omega} = \left(\int_\Omega |u|^p + \sum_{k=1}^N \int_\Omega \left| \frac{\partial u}{\partial x_k} \right|^p \right)^{1/p},$$

$$\|u\|_{2,p,\Omega} = \left(\int_\Omega |u|^p + \sum_{k=1}^N \int_\Omega \left| \frac{\partial u}{\partial x_k} \right|^p + \sum_{k,l=1}^N \int_\Omega \left| \frac{\partial^2 u}{\partial x_k \partial x_l} \right|^p \right)^{1/p}. \qquad \square$$

Note that, if $m_1 > m_2$, then

$$W^{m_1,p}(\Omega) \subset W^{m_2,p}(\Omega) \subset L^p(\Omega) \tag{6.22}$$

and

$$\|u\|_{m_1,p,\Omega} \geq \|u\|_{m_2,p,\Omega} \geq \|u\|_{L^p(\Omega)} \tag{6.23}$$

for any $u \in W^{m_1,p}(\Omega)$.

Theorem 6.3.3. *The Sobolev space $W^{m,p}(\Omega)$ is a Banach space. If $p < \infty$, it is separable.*

The following property of Sobolev spaces is of fundamental importance for applications.

Theorem 6.3.4. *The differential operator D^α is a continuous mapping from $W^{m,p}(\Omega)$ to $W^{m-|\alpha|,p}(\Omega)$, for any $|\alpha| \leq m$.*

Proof: Let D_j denote the derivative with respect to the jth variable. Then, for any $u \in W^{m,p}(\Omega)$, we have

$$\|D_j u\|_{m-1,p,\Omega}^p = \sum_{|\alpha| \leq m-1} \|D^\alpha D_j u\|_{L^p(\Omega)}^p \leq \sum_{|\beta| \leq m} \|D^\beta u\|_{L^p(\Omega)}^p = \|u\|_{m,p,\Omega}^p,$$

$$\tag{6.24}$$

which proves the theorem in the case of differential operators of order one. The general case follows easily. $\qquad \square$

Theorem 6.3.5. *Let f be a function on Ω such that $D^\alpha f$ is bounded on Ω for every $|\alpha| \leq m$. Then multiplication by f is a continuous mapping from $W^{m,p}(\Omega)$ into $W^{m,p}(\Omega)$.*

A mapping $\varphi : \Omega \to \mathbb{R}^N$ is called a C^r-*diffeomorphism* $(r \in \mathbb{N})$, if the following conditions are satisfied:

(a) $D^\alpha \varphi$ is continuous in Ω for all $|\alpha| \leq r$;

(b) φ is one-to-one in Ω;

(c) The Jacobian of φ does not vanish in Ω.

If φ is a C^r-diffeomorphism in Ω, then it is invertible and φ^{-1} is a C^r-diffeomorphism in $\varphi(\Omega)$ (see, for example, Mikusiński and Taylor (2001)).

Theorem 6.3.6. *Let Ω be bounded, $m \in \mathbb{N}$, and let φ be a C^m-diffeomorphism in Ω defined in a neighborhood of $\mathrm{cl}\,\Omega$. Then the mapping $u \mapsto u \circ \varphi^{-1}$ is a continuous linear operator from $W^{m,p}(\Omega)$ into $W^{m,p}(\varphi(\Omega))$.*

The case when $p = 2$ is of special interest to us, because $W^{m,2}(\Omega)$ is an inner product space. The space $W^{m,2}(\Omega)$ is defined as a subspace of $L^2(\Omega)$. However, it is not a closed subspace of $L^2(\Omega)$ and therefore, it is not a Hilbert space with respect to the inner product of $L^2(\Omega)$. It turns out that $W^{m,2}(\Omega)$ is a Hilbert space if the inner product is defined by

$$\langle u, v \rangle_{m,\Omega} = \sum_{|\alpha| \le m} \int_\Omega D^\alpha u(x)\overline{D^\alpha v(x)}dx. \qquad (6.25)$$

The space $W^{m,2}(\Omega)$ is isomorphic to the space $H^m(\Omega)$ (defined in Example 3.3.8) and the inner product (6.25) yields the norm (6.20) for $p = 2$.

It is natural to consider the space $L^p(\Omega)$ as a special case of the Sobolev space with $m = 0$. We denote the $L^p(\Omega)$-norm of a function by $\|\cdot\|_{0,p,\Omega}$.

When $\Omega = \mathbb{R}^N$, the space $H^m(\mathbb{R}^N)$ can also be defined via the Fourier transform. If $u \in H^m(\mathbb{R}^N)$, then $D^\alpha u \in L^2(\mathbb{R}^N)$ for all $|\alpha| \le m$. Thus, the Fourier transform of $D^\alpha u$ is well defined and given by

$$\mathcal{F}\{D^\alpha u\}(\omega) = i^{|\alpha|}\omega^\alpha \widehat{u}(\omega)$$

and so $\{\omega^\alpha \widehat{u}(\omega)\} \in L^2(\mathbb{R}^N)$ for all $|\alpha| \le m$. Conversely, if $u \in L^2(\mathbb{R}^N)$ and $\{\omega^\alpha \widehat{u}(\omega)\} \in L^2(\mathbb{R}^N)$ for all $|\alpha| \le m$, then $u \in H^m(\mathbb{R}^N)$. We can express this in a more convenient form using the following lemma.

Lemma 6.3.7. *For every $m \in \mathbb{N}$, there exist positive constants K_1 and K_2 such that*

$$K_1\left(1 + \|\omega\|^2\right)^m \le \sum_{|\alpha| \le m}\left|\omega^\alpha\right|^2 \le K_2\left(1 + \|\omega\|^2\right)^m, \qquad (6.26)$$

for all $\omega \in \mathbb{R}^N$.

Proof: We recall $\|\omega\|^2 = \omega_1^2 + \omega_2^2 + \cdots + \omega_N^2$ and $|\omega|^\alpha = |\omega_1|^{\alpha_1}\cdots|\omega_N|^{\alpha_N}$. Using a simple induction argument on m, it follows that the same powers of ω occur in $(1 + \|\omega\|^2)^m$ and $\sum_{|\alpha| \le m}|\omega^\alpha|^2$ with different coefficients. Since the number of terms is finite and depends only on m, the inequalities (6.26) follow. $\qquad\square$

In view of this lemma, we can define the space $H^m(\mathbb{R}^N)$ as follows:

$$H^m(\mathbb{R}^N) = \left\{u \in L^2(\mathbb{R}^N) \colon \left(1 + \|\omega\|^2\right)^m\widehat{u}(\omega) \in L^2(\mathbb{R}^N)\right\}. \qquad (6.27)$$

It follows from the Plancherel theorem that the norm $\| \cdot \|_{m,\mathbb{R}^N}$ in $H^m(\mathbb{R}^N)$ is equivalent to the norm

$$\|u\|_{H^m(\mathbb{R}^N)} = \left\{ \int_\Omega \left(1 + \|\omega\|^2\right)^m |\widehat{u}(\omega)|^2 d\omega \right\}^{1/2}. \tag{6.28}$$

The advantage of this definition is that (6.27) can be used to define $H^s(\mathbb{R}^N)$ and the norm for all real numbers $s \geq 0$.

We return to the Sobolev spaces $W^{m,p}(\Omega)$. The map

$$u \to \left(u, \frac{\partial u}{\partial x_1}, \ldots, \frac{\partial u}{\partial x_N} \right) \tag{6.29}$$

is an isometry of $W^{1,p}(\Omega)$ into $(L^p(\Omega))^{N+1}$, where the latter space is equipped with the norm

$$\|u\| = \left(\sum_{r=1}^{N+1} \|u_r\|_{0,p,\Omega}^p \right)^{1/p} \tag{6.30}$$

for $u = (u_r) \in (L^p(\Omega))^{N+1}$.

If $1 \leq p < \infty$, the space of test functions $\mathcal{D}(\Omega)$ is dense in $L^p(\Omega)$. Moreover, if $\phi \in \mathcal{D}(\Omega)$, then $D^\alpha \phi \in \mathcal{D}(\Omega)$ for every α, and thus $\mathcal{D}(\Omega) \subset W^{m,p}(\Omega)$ for any m and p.

Theorem 6.3.8. $\mathcal{D}(\mathbb{R}^N)$ *is a dense subspace of* $W^{m,p}(\mathbb{R}^N)$ *for any integer* $m \geq 0$ *and* $1 \leq p < \infty$.

If Ω is a proper subspace of $\mathbb{R}^N$, then this theorem need not hold. However, we have the following result.

Theorem 6.3.9. $C^\infty(\Omega)$ *is a dense subspace of* $W^{m,p}(\Omega)$ *for any integer* $m \geq 0$ *and* $1 \leq p < \infty$.

6.4 Fundamental Solutions and Green's Functions for Partial Differential Equations

Consider a partial differential operator L of order m in N variables

$$L = \sum_{|\alpha| \leq m} A_\alpha D^\alpha, \tag{6.31}$$

where $\alpha = (\alpha_1, \ldots, \alpha_N)$ is a multi-index, the α_n's are non-negative integers, $|\alpha| = \alpha_1 + \cdots + \alpha_N, A_\alpha = A_{\alpha_1, \alpha_2, \ldots, \alpha_N}(x_1, x_2, \ldots, x_N)$ are functions in $\mathbb{R}^N$ (possibly constant), and

$$D^\alpha = \left(\frac{\partial}{\partial x_1}\right)^{\alpha_1} \cdots \left(\frac{\partial}{\partial x_N}\right)^{\alpha_N} = \frac{\partial^{|\alpha|}}{\partial x_1^{\alpha_1} \cdots \partial x_N^{\alpha_N}}.$$

The formal adjoint of L is

$$L^*v = \sum_{|\alpha| \le m} (-1)^{|\alpha|} D^\alpha (A_\alpha v). \tag{6.32}$$

For example, the most general linear partial differential operator of order two in two independent variables is

$$L = \sum_{|\alpha| \le 2} A_\alpha D^\alpha \tag{6.33}$$

$$= A_{2,0}(x_1, x_2)\frac{\partial^2}{\partial x_1^2} + A_{1,1}(x_1, x_2)\frac{\partial^2}{\partial x_1 \partial x_2} + A_{0,2}(x_1, x_2)\frac{\partial^2}{\partial x_2^2}$$

$$+ A_{1,0}(x_1, x_2)\frac{\partial}{\partial x_1} + A_{0,1}(x_1, x_2)\frac{\partial}{\partial x_2} + A_{0,0}(x_1, x_2). \tag{6.34}$$

When seeking a solution of the equation $L(x) = Y$, we may be interested in a solution which is a differentiable function, or just a function with D^α understood as generalized derivatives, or, finally, we may seek a solution which is a distribution. For this reason, the solutions of such an equation are classified as follows:

Definition 6.4.1.

(a) (Classical solution). Let f be a function on $\mathbb{R}^N$. Every function u on $\mathbb{R}^N$ which is sufficiently differentiable so that

$$\sum_{|\alpha| \le m} A_\alpha D^\alpha u$$

is well defined as a function and such that the equation

$$Lu = \sum_{|\alpha| \le m} A_\alpha D^\alpha u = f \tag{6.35}$$

is satisfied is called a *classical solution* of (6.35).

(b) (Weak solution). By a *weak solution* of (6.35), we mean a function on $\mathbb{R}^N$ which need not be sufficiently differentiable to make Lu meaningful in the classical sense. In this case, f may be a function or a distribution.

(c) (Distributional solution). Let $f \in \mathcal{D}'(\mathbb{R}^N)$. Every $u \in \mathcal{D}'(\mathbb{R}^N)$ satisfying (6.35) is called a *distributional solution* of (6.35).

Note that if f in (6.35) is a singular distribution, then the equation cannot have a classical solution. The remarkable fact is that including distributions one can generate new solutions of classical equations (equations where f is a function). Some classical equations may not even have a classical solution but can have distributional solutions.

Example 6.4.2. The differential equation

$$xu' = 0$$

has a weak solution of the form $u(x) = c_1 H(x) + c_2$, where H is the Heaviside function and c_1 and c_2 are arbitrary constants. If $c_1 \neq 0$, this is not a classical solution because H is not differentiable. However, $u(x) = c_2$ is a classical solution. □

Example 6.4.3. The differential equation

$$x^2 u' = 0 \tag{6.36}$$

has the same classical and weak solutions as $xu' = 0$. Equation (6.36) has also distributional solutions

$$u(x) = c_1 \delta(x) + c_2 H(x) + c_3,$$

where c_1, c_2, and c_3 are arbitrary constants. Clearly, the distributional solutions have no analogues in the classical theory of differential equations. □

Example 6.4.4. The nonhomogeneous equation

$$u''(x) = \delta''(x)$$

has distributional solutions of the form

$$u(x) = \delta(x) + c_1 x + c_2,$$

where c_1 and c_2 are arbitrary constants. □

It may be noted that, for a large class of problems, all generalized solutions of (6.35) turn out to be classical solutions.

Equations of the form

$$LG = \delta \tag{6.37}$$

are of particular interest. Suppose G is a distribution satisfying (6.37). Then, for any distribution f with compact support, the convolution $f * G$ is well defined and

$$L(f * G) = \sum_{|\alpha| \leq m} A_\alpha D^\alpha (f * G)$$

$$= \sum_{|\alpha| \leq m} A_\alpha (f * D^\alpha G)$$

$$= f * \left(\sum_{|\alpha| \leq m} A_\alpha D^\alpha G \right)$$

$$= f * \delta = f.$$

Thus, if G is a solution of $LG = \delta$, then $f * G$ is a solution of $Lu = f$. This explains the importance of the equations $Lu = \delta$, at least in the context of existence of solutions of partial differential equations.

Definition 6.4.5.   (Fundamental solution)
By a *fundamental solution* of a differential operator L, we mean a distributional solution of the equation $Lu = \delta$.
  Clearly, a fundamental solution is not unique.

Example 6.4.6. It is easy to check that the Heaviside H function is a fundamental solution of the operator $L = d/dx$. If c is any constant, then $H(x) + c$ is also a fundamental solution of L. $\qquad\square$

  Consider the following boundary value problem

$$Lu = f \quad \text{in } \Omega \tag{6.38}$$

with one of the following boundary conditions:

$$u = 0 \quad \text{on } \partial\Omega, \tag{6.39}$$

$$\frac{\partial u}{\partial n} = 0 \quad \text{on } \partial\Omega, \tag{6.40}$$

or

$$\frac{\partial u}{\partial n} + au = 0 \quad \text{on } \partial\Omega, \tag{6.41}$$

where Ω is a bounded open set in $\mathbb{R}^N$ and $\partial\Omega$ is the boundary of Ω. The problem of finding the solution of (6.38) subject to the boundary condition (6.39) is called the *homogeneous Dirichlet problem* (Johann Peter Gustav Lejeune Dirichlet (1805–1859)). The problem of solving (6.38) satisfying the boundary condition (6.40) is known as the *homogeneous Neumann problem*. Finally, the problem of finding the solution of (6.38) subject to the boundary condition (6.41) is called the *Robin* (or *mixed*) *problem*.

A fundamental solution G of the differential operator L satisfies (6.37), but G need not satisfy the given boundary conditions. A fundamental solution which satisfies the homogeneous conditions is known as the *Green's function*.

Definition 6.4.7. (Green's function)
By the *Green's function* G of the operator L with homogeneous boundary conditions, we mean a fundamental solution of the equation $LG = \delta$ satisfying those boundary conditions.

The remarkable fact is that the Green's function can be used to solve the boundary value problems given by (6.38) to (6.41).

The most important linear partial differential equations arise in problems of applied mathematics, mathematical physics, and engineering sciences. Included here are only examples of three-dimensional equations of most common interest:

Example 6.4.8. (The Laplace equation)

$$\nabla^2 u = 0, \tag{6.42}$$

where

$$\nabla^2 u = \frac{\partial^2 u}{\partial x^2} + \frac{\partial^2 u}{\partial y^2} + \frac{\partial^2 u}{\partial z^2}.$$

The operator ∇^2 is known as the *Laplace operator* (Pierre-Simon Laplace (1749–1827)). It is also denoted by Δ.

This equation is satisfied by the electrostatic potential in the absence of charges, by the gravitational potential in the absence of mass, by the equilibrium displacement of a membrane with a given displacement of its boundary, by the velocity potential for an inviscid, incompressible, irrotational homogeneous fluid in the absence of sources and sinks, by the temperature in steady-state heat flow in the absence of sources and sinks, and in many other situations. □

Example 6.4.9. (The Poisson equation)

$$\nabla^2 u = -f(x, y, z), \tag{6.43}$$

where f is a given function.

The Poisson equation (Siméon Denis Poisson (1781–1840)) is satisfied by the electrostatic potential in the presence of charge, by the gravitational potential in the presence of distributed matter, by the equilibrium displacement of a membrane under distributed forces, by the velocity potential for an inviscid, incompressible, irrotational, homogeneous fluid in the presence of distributed sources or sinks, by the steady-state temperature in the presence of thermal sources or sinks, and in many other physical situations.   □

Example 6.4.10. (The nonhomogeneous wave equation)

$$\frac{\partial^2 u}{\partial t^2} - \nabla^2 u = -f(x, y, z). \tag{6.44}$$

This equation and the corresponding homogeneous form arise in a large number of physical situations. Some of these problems include the vibrating string, vibrating membrane, acoustic problems for the velocity potential for the fluid flow through which sound can be transmitted, longitudinal vibrations of an elastic rod or beam, and both electric and magnetic fields in the absence of charge and dielectric.   □

Example 6.4.11. (The nonhomogeneous heat (or diffusion) equation)

$$\frac{\partial u}{\partial t} - \nabla^2 u = f(x, y, z). \tag{6.45}$$

□

Example 6.4.12. (Telegrapher's equation)

$$\frac{\partial^2 \varphi}{\partial t^2} + a\frac{\partial \varphi}{\partial t} + b\varphi = \frac{\partial^2 \varphi}{\partial x^2}, \tag{6.46}$$

where a and b are constants. This equation arises in the study of propagation of electrical signals in a cable transmission line. Both the current I and the voltage V satisfy an equation of the form (6.46). This equation also arises in the propagation of pressure waves in the study of pulsatile blood flow in arteries and in the one-dimensional random motion of bugs along a hedge.   □

Example 6.4.13. (The inhomogeneous Helmholtz's equation) (Hermann Ludwig Ferdinand von Helmholtz (1821–1894))

$$\nabla^2 \psi + \lambda \psi = -f(x, y, z), \tag{6.47}$$

where λ is a constant. This is essentially a time-independent inhomogeneous wave equation (6.44) with λ as the separation constant.   □

Example 6.4.14. (The biharmonic wave equation)

$$\nabla^4 \psi - \frac{1}{c^2}\frac{\partial^2 \psi}{\partial t^2} = 0. \tag{6.48}$$

In elasticity theory, the displacement of a thin elastic plate in small vibrations satisfies this equation. When ψ is independent of time t, (6.48) reduces to the so called *biharmonic equation*

$$\nabla^4 \psi = 0. \tag{6.49}$$

This is the equilibrium equation for the distribution of stress in an elastic medium satisfied by Airy's stress function ψ (George Biddell Airy (1801–1892)). In fluid dynamics, the equation is satisfied by the stream function ψ in an incompressible viscous fluid flow. □

Example 6.4.15. (The time-independent Schrödinger equation in quantum mechanics) (Erwin Rudolf Josef Alexander Schrödinger (1887–1961))

$$\frac{\hbar^2}{2m} \nabla^2 \psi + (E - V)\psi = 0, \tag{6.50}$$

where m is the mass of the particle whose wave function is ψ, $h = 2\pi\hbar$ is the universal Planck's constant, V is the potential energy, and E is a constant. If $V = 0$, (6.50) reduces to the Helmholtz's equation. □

Example 6.4.16. (The Klein–Gordon equation)

$$\Box u + d^2 u = 0, \tag{6.51}$$

where

$$\Box = \nabla^2 - \frac{1}{c^2} \frac{\partial^2}{\partial t^2} \tag{6.52}$$

is the d'Alembertian operator and d is a constant. □

The Laplacian has many remarkable features similar to a Sturm–Liouville operator in one spatial dimension. The following differential identity can be verified by direct computation:

$$u\nabla^2 v = \nabla \cdot (u\nabla v) - (\nabla u) \cdot (\nabla v). \tag{6.53}$$

The integration over a domain Ω yields

$$\int_\Omega u\nabla^2 v\, d\tau = \int_\Omega \nabla \cdot (u\nabla u)\, d\tau - \int_\Omega (\nabla u) \cdot (\nabla v)\, d\tau. \tag{6.54}$$

Application of the divergence theorem to the first integral on the right with the vector field $(u\nabla v)$ and using the fact that $\nabla v \cdot n$ is the directional derivative $\partial v/\partial n$, we obtain the *Green's first identity*

$$\int_\Omega u\nabla^2 v\, d\tau = \int_{\partial\Omega} u\frac{\partial v}{\partial n}\, ds - \int_\Omega (\nabla u) \cdot (\nabla v)\, d\tau. \tag{6.55}$$

In particular, if $u = v$, (6.55) becomes

$$\int_\Omega u\nabla^2 u d\tau = \int_{\partial\Omega} u\frac{\partial u}{\partial n}ds - \int_\Omega |\nabla u|^2 d\tau. \tag{6.56}$$

Note that this is the higher dimensional analogue of the integration by parts formula

$$\int_a^b uu''dx = [uu']_a^b - \int_a^b (u')^2 dx. \tag{6.57}$$

Interchanging u and v in (6.53) and subtracting from (6.53) gives

$$u\nabla^2 v - v\nabla^2 u = \nabla \cdot (u\nabla v - v\nabla u). \tag{6.58}$$

Integrating this equality over Ω and using the divergence theorem yields the *Green's second identity*

$$\int \left(u\nabla^2 v - v\nabla^2 u\right) d\tau = \int_{\partial\Omega} \left(u\frac{\partial v}{\partial n} - v\frac{\partial u}{\partial n}\right) ds. \tag{6.59}$$

This formula can be also interpreted as the higher dimensional analogue of the integration by parts formula

$$\int_a^b (uv'' - vu'')dx = [uv' - vu']_a^b \tag{6.60}$$

or the integral form of the Lagrange identity as stated in Theorem 5.9.5.

From (6.56) we note that if the boundary conditions on u are such that the integral over $\partial\Omega$ vanishes, then the operator $-\nabla^2$ is positive definite. It also follows from (6.59) that ∇^2 is a formally self-adjoint operator. Integrating once with respect to t, it is easy to check that the operator $\partial/\partial t$ has an adjoint $-\partial/\partial t$. Consequently, the adjoint of the operator $(\nabla^2 - \partial/\partial t)$ is $(\nabla^2 + \partial/\partial t)$, and $(\nabla^2 - \partial^2/\partial t^2)$ is formally self-adjoint.

Finally, the Green's identities (6.56) and (6.59) can be generalized for operators more general than the Laplace operator by using the following identity:

$$v(\nabla^2 u + \mathbf{a}\cdot\nabla u + bu) - u(\nabla^2 v - \nabla\cdot(\mathbf{a}v) + bv) = \nabla\cdot(v\nabla u - u\nabla v + \mathbf{a}uv), \tag{6.61}$$

where u and v are arbitrary differentiable functions, $\mathbf{a}$ is a vector field, and b is a constant scalar. This identity can be written in terms of differential operators L and L^* as

$$vLu - uL^*v = \nabla\cdot(v\nabla u - u\nabla v + \mathbf{a}uv). \tag{6.62}$$

Application of the divergence theorem gives

$$\int_\Omega \left(vLu - uL^*v\right) d\tau = \int_\Omega \nabla\cdot(v\nabla u - u\nabla v + \mathbf{a}uv) d\tau$$

$$= \int_{\partial\Omega} (v\nabla u - u\nabla v + \mathbf{a}uv) \cdot \mathbf{n}\,ds$$

$$= \int_{\partial\Omega} \left(v\frac{\partial u}{\partial n} - u\frac{\partial v}{\partial n} + a_n uv \right) ds, \tag{6.63}$$

where $\partial u/\partial n = \mathbf{n} \cdot \nabla u$ and $a_n = \mathbf{a} \cdot \mathbf{n}$.

This is an obvious extension of (6.59) and includes a special case when L is self-adjoint. It follows from the definition of L and L^* that $Lv = L^*v$ only if

$$\mathbf{a} \cdot \nabla v = -\nabla \cdot (\mathbf{a}v) = -\mathbf{a} \cdot \nabla v - v\nabla \cdot \mathbf{a}.$$

This gives $\nabla \cdot \mathbf{a} = 0$ when $v = 1$. If we assume v to take values x, y, and z successively, then $a_k = 0$, $k = 1, 2, 3$. Thus, L is self-adjoint if and only if $\mathbf{a} = \mathbf{0}$. We can also conclude that every self-adjoint operator in the sense of the definition (6.62) has the form

$$Lu = \nabla^2 u + bu.$$

Finally, it also follows from the vector identity

$$\nabla u \cdot \nabla v + u\nabla^2 v = \nabla \cdot (u\nabla v) \tag{6.64}$$

that

$$v(\nabla^2 u + bu) + \nabla u \cdot \nabla v - buv = \nabla \cdot (v\nabla u). \tag{6.65}$$

Therefore, if L is self-adjoint, application of the divergence theorem gives

$$\int_{\Omega} vLu\,d\tau + \int_{\Omega} (\nabla u \cdot \nabla v - buv)d\tau = \int_{\partial\Omega} v\frac{\partial u}{\partial n}\,ds. \tag{6.66}$$

This is clearly a generalization of (6.55).

The second integral on the left-hand side of (6.66) expressed as a bilinear form

$$E(u, v) = \int_{\Omega} (\nabla u \cdot \nabla v - buv)d\tau \tag{6.67}$$

is called the *Dirichlet integral* of the operator L. If L is elliptic and $b < 0$, then $E(u, v)$ is strictly positive since

$$E(u, u) = \int_{\Omega} \left((\nabla u)^2 - bu^2 \right)d\tau > 0 \tag{6.68}$$

for any differentiable function u which does not vanish identically in Ω. This property of the Dirichlet integral plays a fundamental role in the theory of elliptic partial differential equations. However, this feature is not universally applicable to all elliptic equations. For example, the inhomogeneous Helmholtz's equation (6.47) with $\lambda \geq 0$ has an associated Dirichlet integral, which is not positive.

In general, if u and v are functions with continuous derivatives of order m, that is, $D^\alpha u$ and $D^\alpha v$ are continuous functions for every multi-index α with $|\alpha| \leq m$, then it can be shown that

$$vLu - uL^*v = \operatorname{div} J(u, v), \tag{6.69}$$

where L and L^* are given by (6.31) and (6.32), and J is a vector bilinear form in u and v involving only derivatives of u and v of order $m - 1$ or less. Equation (6.69) is an extension of the one-dimensional version of Green's formula for partial differential operators. The integral form of (6.69) is the most general Green's identity

$$\int_\Omega \left(vLu - uL^*v\right)d\tau = \int_{\partial\Omega} \mathbf{n} \cdot J ds, \tag{6.70}$$

where Ω is a bounded region in $\mathbb{R}^N$ with sufficiently smooth boundary $\partial\Omega$, and $\mathbf{n}$ is the unit outward normal drawn to the surface element ds.

We now illustrate the form of the Green's second identity associated with several important differential operators.

For the Helmholtz operator, we have

$$L = L^* = \nabla^2 + k^2,$$

and the Green's identity (6.63) takes the form

$$\int_\Omega (vLu - uL^*v)d\tau = \int_{\partial\Omega} \left(v\frac{\partial u}{\partial n} - u\frac{\partial v}{\partial n}\right)ds. \tag{6.71}$$

For the wave operator

$$L = L^* = c^2\frac{\partial^2}{\partial x^2} - \frac{\partial^2}{\partial t^2}$$

and the Green's identity, (6.63) assumes the form

$$\int_\Omega (vLu - uL^*v)d\tau = \int_{\partial\Omega}\left[c^2\left(v\frac{\partial u}{\partial x} - u\frac{\partial v}{\partial n}\right)\hat{\mathbf{i}} - \left(v\frac{\partial u}{\partial t} - u\frac{\partial v}{\partial t}\right)\hat{\mathbf{j}}\right]\cdot\hat{\mathbf{n}}ds. \tag{6.72}$$

Finally, for the diffusion operator

$$L = K\frac{\partial}{\partial t} - \frac{\partial^2}{\partial x^2}, \qquad L^* = -K\frac{\partial}{\partial t} - \frac{\partial^2}{\partial x^2},$$

and the Green's identity, (6.63) takes the form

$$\int_\Omega (vLu - uL^*v)d\tau = \int_{\partial\Omega}\left[\left(u\frac{\partial u}{\partial x} - v\frac{\partial v}{\partial n}\right)\hat{\mathbf{i}} + Kuv\hat{\mathbf{j}}\right]\cdot\hat{\mathbf{n}}ds. \tag{6.73}$$

Without a rigorous discussion, we shall illustrate the use of the Green's function in finding the solution of the boundary value problems (6.38) to (6.41) for a linear inhomogeneous partial differential equation.

Without loss of generality, we consider the problem in $\mathbb{R}^3$. As before, the Green's function $G(\mathbf{x}, \boldsymbol{\xi})$ of this problem satisfies the equation

$$LG(\mathbf{x}, \boldsymbol{\xi}) = \delta(\mathbf{x} - \boldsymbol{\xi}). \tag{6.74}$$

Physically, the Green's function $G(\mathbf{x}, \boldsymbol{\xi})$ represents the effect at the point $\mathbf{x}$ of a Dirac delta function source at the point $\mathbf{x} = \boldsymbol{\xi}$.

Multiplying (6.74) by $f(\boldsymbol{\xi})$ and integrating over the volume V of the $\boldsymbol{\xi}$ space so that $d\mathbf{v} = d\xi\, d\eta\, d\zeta$, we find

$$L\left\{ \int_V G(\mathbf{x}, \boldsymbol{\xi}) f(\boldsymbol{\xi}) d\mathbf{v} \right\} = \int_V \delta(\mathbf{x} - \boldsymbol{\xi}) f(\boldsymbol{\xi}) d\mathbf{v} = f(\mathbf{x}). \tag{6.75}$$

Comparing (6.38) and (6.75), we see that the solution of (6.38) can be written as

$$u(x) = \int_V G(\mathbf{x}, \boldsymbol{\xi}) f(\boldsymbol{\xi}) d\mathbf{v}. \tag{6.76}$$

Clearly, this is valid no matter how many components $\mathbf{x}$ may have. Accordingly, the Green's function method can be applied, in principle, to any linear constant coefficient inhomogeneous partial differential equation in any number of independent variables. However, although a neat formulation has been developed, in practice, the construction of a Green's function is not an easy problem.

We next illustrate by the following examples the application of the Green's function to linear constant coefficient inhomogeneous partial differential equations.

Example 6.4.17. The solution of the Poisson equation

$$-\nabla^2 u = f(x, y, z) \tag{6.77}$$

is given by

$$u(x, y, z) = \int_{\mathbb{R}^3} G(\mathbf{x}, \boldsymbol{\xi}) f(\boldsymbol{\xi}) d\boldsymbol{\xi}, \tag{6.78}$$

where the Green's function $G(\mathbf{x}, \boldsymbol{\xi})$ of $-\nabla^2$ is

$$G(\mathbf{x}, \boldsymbol{\xi}) = \frac{1}{4\pi} \frac{1}{|\mathbf{x} - \boldsymbol{\xi}|}. \tag{6.79}$$

To find the fundamental solution for $-\nabla^2$, we need to solve the equation

$$-\nabla^2 G(\mathbf{x}, \boldsymbol{\xi}) = \delta(x - \xi)\delta(y - \eta)\delta(z - \zeta), \tag{6.80}$$

where $\mathbf{x} \neq \boldsymbol{\xi}$.

We apply the three-dimensional Fourier transform with respect to x, y, and z

$$\tilde{G}(\boldsymbol{\kappa}, \boldsymbol{\xi}) = \frac{1}{(2\pi)^{3/2}} \int_{\mathbb{R}^3} G(\mathbf{x}, \boldsymbol{\xi}) e^{-i\boldsymbol{\kappa}\cdot\mathbf{x}} d\mathbf{x} \tag{6.81}$$

to Equation (6.80) to obtain

$$\kappa^2 \tilde{G} = \frac{1}{(2\pi)^{3/2}} e^{-i\boldsymbol{\kappa}\cdot\boldsymbol{\xi}} \tag{6.82}$$

where $\boldsymbol{\kappa} = (k, l, m)$ is the Fourier transform variable.

The inverse Fourier transform gives the solution

$$G(\mathbf{x}, \boldsymbol{\xi}) = \frac{1}{(2\pi)^3} \int_{\mathbb{R}^3} e^{i\boldsymbol{\kappa}\cdot(\mathbf{x}-\boldsymbol{\xi})} \frac{d\boldsymbol{\kappa}}{\kappa^2}$$

$$= \frac{1}{(2\pi)^3} \int_{\mathbb{R}^3} e^{i\boldsymbol{\kappa}\cdot\mathbf{r}} \frac{d\boldsymbol{\kappa}}{\kappa^2}, \tag{6.83}$$

where $\mathbf{r} = |\mathbf{x} - \boldsymbol{\xi}|$.

Integral (6.83) can be evaluated using spherical coordinates in the $\boldsymbol{\kappa}$ variables. We choose the polar axis along the $\mathbf{r}$ direction and denote the spherical coordinates by κ, θ, φ. Then $\boldsymbol{\kappa}\cdot\mathbf{r} = \kappa r \cos\theta$ where $r = |\mathbf{r}|$. Then (6.83) becomes

$$G(\mathbf{x}, \boldsymbol{\xi}) = \frac{1}{(2\pi)^3} \int_0^\infty \kappa^2 d\kappa \int_0^\pi \sin\theta \, d\theta \int_0^{2\pi} e^{i\kappa r \cos\theta} \frac{d\varphi}{\kappa^2}$$

$$= \frac{1}{(2\pi)^2} \int_0^\infty \frac{2\sin\kappa r}{\kappa r} d\kappa$$

$$= \frac{1}{4\pi r} = \frac{1}{4\pi |\mathbf{x} - \boldsymbol{\xi}|}, \tag{6.84}$$

provided $r > 0$. This completes the proof.

In electrodynamics, the fundamental solution (6.84) has a well-known interpretation. It is essentially the potential at the point $\mathbf{x}$ produced by a unit point charge at the point $\boldsymbol{\xi}$. This is what can be expected from a physical point of view because $\delta(\mathbf{x} - \boldsymbol{\xi})$ is the charge-density corresponding to a unit point charge at $\boldsymbol{\xi}$.

The solution of (6.77) is

$$u(x, y, z) = \int_{\mathbb{R}^3} G(\mathbf{x}, \boldsymbol{\xi}) f(\boldsymbol{\xi}) d\boldsymbol{\xi} = \frac{1}{4\pi} \int_{\mathbb{R}^3} \frac{f(\xi, \eta, \zeta)}{|\mathbf{x} - \boldsymbol{\xi}|} d\xi \, d\eta \, d\zeta. \tag{6.85}$$

The integrand in (6.85) consists of the given charge distribution $f(\mathbf{x})$ at $\mathbf{x} = \boldsymbol{\xi}$ and the Green's function $G(\mathbf{x}, \boldsymbol{\xi})$. Physically, $G(\mathbf{x}, \boldsymbol{\xi}) f(\boldsymbol{\xi})$ represents the resulting potentials due to elementary point charges, and the total potential due to a given charge distribution $f(\mathbf{x})$ is then obtained by the integral super-position of the resulting potentials. This is the so-called *principle of superposition*.

□

Example 6.4.18. The fundamental solution of the two-dimensional Helm-holtz equation

$$-\nabla^2 G + \lambda^2 G = \delta(x - \xi)\delta(y - \eta), \quad -\infty < x, y < \infty \tag{6.86}$$

is

$$G(\mathbf{x}, \boldsymbol{\xi}) = \frac{1}{2\pi} \int_0^\infty \frac{r J_0[r\{(x - \xi)^2 + (y - \eta)^2\}^{1/2}]dr}{r^2 + \lambda^2}. \tag{6.87}$$

It is convenient to introduce the change of variables $x^* = x - \xi$ and $y^* = y - \eta$. Consequently, (6.86) assumes the form, dropping the asterisks,

$$G_{xx} + G_{yy} - \lambda^2 G = -\delta(x)\delta(y). \tag{6.88}$$

Now we apply the double Fourier transform

$$\tilde{G}(\boldsymbol{\kappa}) = \frac{1}{2\pi} \int_{\mathbb{R}^2} e^{-i\boldsymbol{\kappa}\cdot\mathbf{x}} G(x, y)d\mathbf{x} \tag{6.89}$$

to (6.88) to obtain the solution

$$\tilde{G}(\boldsymbol{\kappa}) = \frac{1}{2\pi} \frac{1}{(\kappa^2 + \lambda^2)}, \tag{6.90}$$

where $\boldsymbol{\kappa} = (k, l)$.

The inverse Fourier transform gives the solution

$$G(x, y) = \frac{1}{4\pi^2} \int_{\mathbb{R}^2} e^{i\boldsymbol{\kappa}\cdot\mathbf{x}} \left(\kappa^2 + \lambda^2\right)^{-1} d\boldsymbol{\kappa}. \tag{6.91}$$

In terms of the polar coordinates $x = \rho\cos\theta, y = \rho\sin\theta, k = r\cos\varphi, l = r\sin\varphi$, this integral has the representation

$$G(x, y) = \frac{1}{4\pi^2} \int_0^\infty \frac{r\,dr}{r^2 + \lambda^2} \int_0^{2\pi} e^{ir\rho\cos(\varphi-\theta)} d\varphi.$$

The second integral can be expressed in terms of the Bessel function as

$$\int_0^{2\pi} e^{ir\rho\cos\varphi} d\varphi = 2\pi J_0(r\rho).$$

Hence, the solution becomes

$$G(x, y, \xi, \eta) = \frac{1}{2\pi} \int_0^\infty \frac{r J_0(r\rho) dr}{r^2 + \lambda^2}. \tag{6.92}$$

Thus, in terms of the original coordinates, the fundamental solution is

$$G(\mathbf{x}, \boldsymbol{\xi}) = \frac{1}{2\pi} \int_0^\infty \frac{r J_0[r\{(x-\xi)^2 + (y-\eta)^2\}^{1/2}] dr}{r^2 + \lambda^2}.$$

Accordingly, the solution of the Helmholtz equation $(\nabla^2 - \lambda^2)u = -f(x, y)$ is

$$u(x, y) = \int_{\mathbb{R}^2} G(\mathbf{x}, \boldsymbol{\xi}) f(\boldsymbol{\xi}) d\boldsymbol{\xi}, \tag{6.93}$$

where $G(\mathbf{x}, \boldsymbol{\xi})$ is given by (6.87).

Since the integral in (6.92) does not exist for $\lambda = 0$, the Green's function for the two-dimensional Poisson equation (6.86) cannot be obtained from (6.92). Instead, we differentiate (6.92) with respect to ρ to obtain

$$\frac{\partial G}{\partial \rho} = \frac{1}{2\pi} \int_0^\infty \frac{r^2 J_0'(r\rho) dr}{r^2 + \lambda^2}$$

which is, for $\lambda = 0$,

$$\frac{\partial G}{\partial \rho} = \frac{1}{2\pi} \int_0^\infty J_0'(r\rho) dr = -\frac{1}{2\pi\rho}.$$

It then follows that

$$G(\rho, \theta) = -\frac{1}{2\pi} \ln \rho.$$

In terms of the original coordinates, we have

$$G(x, y, \xi, \eta) = -\frac{1}{4\pi} \ln\left[(x-\xi)^2 + (y-\eta)^2\right]. \tag{6.94}$$

This is the Green's function for the two-dimensional Poisson equation $\nabla^2 = -f(x, y)$. Consequently, the solution of this equation is

$$u(x, y) = -\frac{1}{4\pi} \int_{\mathbb{R}^2} f(\xi, \eta) \ln\left[(x-\xi)^2 + (y-\eta)^2\right] d\xi \, d\eta. \tag{6.95}$$

$\square$

Example 6.4.19. We find the fundamental solution of the Helmholtz equation

$$(\nabla^2 + k^2)u = \frac{1}{r^2} \frac{\partial}{\partial r}\left(r^2 \frac{\partial u}{\partial r}\right) + k^2 u = 0, \quad 0 < r < \infty, \tag{6.96}$$

with the radiation condition

$$\lim_{r \to \infty} r(u_r + iku) = 0. \tag{6.97}$$

In this case, the Green's function must satisfy

$$\nabla^2 G + k^2 G = \frac{\delta(r)}{4\pi r^2}. \tag{6.98}$$

Clearly, G satisfies

$$G_{rr} + \frac{2}{r} G_r + k^2 G = 0 \quad \text{for } r > 0$$

or

$$(rG)_{rr} + k^2(rG) = 0.$$

This equation admits a solution of the form

$$rG = Ae^{ikr} + Be^{-ikr}$$

or

$$G = A\frac{e^{ikr}}{r} + B\frac{e^{-ikr}}{r}.$$

In order for G to satisfy the radiation condition, we need $A = 0$ and thus

$$G = B\frac{e^{-ikr}}{r}. \tag{6.99}$$

To determine B, we find

$$\lim_{\varepsilon \to 0} \int_{S_\varepsilon} \frac{\partial G}{\partial n} ds = -\lim_{\varepsilon \to 0} \int_{S_\varepsilon} \frac{B}{r} e^{-ikr} \left(\frac{1}{r} + ik \right) ds = 1,$$

from which we obtain $B = -1/(4\pi)$. Consequently, the fundamental solution $G(r)$ for (6.96) is

$$G(r) = -\frac{e^{-ikr}}{4\pi r}. \tag{6.100}$$

Physically, this represents outgoing spherical waves radiating away from the source at the origin. With a point source at a point ξ, the Green's function G has the representation

$$G(\mathbf{x}, \boldsymbol{\xi}) = -\frac{e^{-i|\mathbf{x} - \boldsymbol{\xi}|k}}{4\pi |\mathbf{x} - \boldsymbol{\xi}|}, \tag{6.101}$$

where $\mathbf{x}$ and $\boldsymbol{\xi}$ are position vectors in $\mathbb{R}^3$.

Finally, when $k = 0$, these results reduce exactly to the known solution for the Poisson equation. □

Example 6.4.20. The acoustic (or light) waves are incident from $z = -\infty$ onto a solid screen at $z = 0$ with a tiny hole at the origin. Describe the behavior of the waves after they pass through the hole.

We seek solution of the wave equation

$$u_{tt} = c^2 \nabla^2 u$$

in the three-dimensional half-space $z > 0$. We assume that for $z < 0$, waves are propagating in from $z = -\infty$, and hence, we have the representation

$$u = \frac{u_0}{ik} e^{-i(kz - \omega t)}, \quad k = \omega/c.$$

At the solid screen $(z = 0)$, $\partial u/\partial z = 0$, and at the hole, we take

$$\frac{\partial u}{\partial z} = u_0 e^{-i\omega t}.$$

To determine u for $z > 0$, we seek solutions in the form $u = \Phi(\mathbf{x}) e^{-i\omega t}$ so that Φ satisfies the Helmholtz equation

$$\left(\nabla^2 + k^2\right)\Phi = 0$$

with the boundary conditions

$$\frac{\partial \Phi}{\partial z} = 0 \quad \text{at the solid screen } (z = 0),$$

$$\frac{\partial \Phi}{\partial z} = u_0 \quad \text{at the hole.}$$

Now the problem can be solved using Example 6.4.19. □

Example 6.4.21. (Green's function of the one-dimensional wave equation) We consider the one-dimensional inhomogeneous wave equation

$$u_{tt} - c^2 u_{xx} = p(x, t), \quad x \in \mathbb{R}, t > 0, \tag{6.102}$$

with the initial and boundary conditions

$$u(x, 0) = 0, \quad u_t(x, t) = 0 \quad \text{for } x \in \mathbb{R}, \tag{6.103}$$

$$u(x, t) \to 0 \quad \text{as } |x| \to \infty. \tag{6.104}$$

In this case, the Green's function G must satisfy the equation

$$G_{tt} - c^2 G_{xx} = \delta(x)\delta(t), \quad x \in \mathbb{R}, t > 0 \tag{6.105}$$

and the initial and boundary conditions (6.103) and (6.104). We apply the joint Laplace and Fourier transform defined by

$$\widetilde{\overline{G}}(k, s) = \frac{1}{\sqrt{2\pi}} \int_{-\infty}^{\infty} e^{-ikx} dx \int_0^{\infty} e^{-st} G(x, t) dt. \tag{6.106}$$

The solution of the transformed problem is

$$\widetilde{\overline{G}}(k, s) = \frac{1}{\sqrt{2\pi}} \frac{1}{(s^2 + c^2 k^2)}. \tag{6.107}$$

The inverse Laplace and Fourier transform gives the solution

$$G(x, t) = \frac{1}{2c} H(ct - |x|), \tag{6.108}$$

where H is the Heaviside unit step function.

With a source at (ξ, τ), the Green's function takes the form

$$G(x, t; \xi, \tau) = \frac{1}{2c} H(c(t - \tau), |x - \xi|). \tag{6.109}$$

This function is also called the *Riemann function* for the wave equation. It follows from (6.109) that $G = 0$ unless the point (x, t) lies within the characteristic cone defined by the inequality $c(t - \tau) > |x - \xi|$.

Finally, the solution of the inhomogeneous wave equation (6.102) is given by

$$
\begin{aligned}
u(x, t) &= \int_0^t d\tau \int_{-\infty}^{\infty} G(x, t; \xi, \tau) p(\xi, \tau) d\xi \\
&= \frac{1}{2c} \int_0^t d\tau \int_{-\infty}^{\infty} H(c(t - \tau), |x - \xi|) p(\xi, \tau) d\xi.
\end{aligned}
$$

Since $H = 1$ for $x - c(t - \tau) < \xi < x + c(t - \tau)$ and zero outside, the solution is

$$u(x, t) = \frac{1}{2c} \int_0^t d\tau \int_{x-c(t-\tau)}^{x+c(t-\tau)} p(\xi, \tau) d\xi. \tag{6.110}$$

□

Example 6.4.22. (Green's function of the one-dimensional diffusion equation) We consider the inhomogeneous one-dimensional diffusion equation

$$u_t - K u_{xx} = q(x, t), \quad x \in \mathbb{R}, t > 0 \tag{6.111}$$

with the initial and boundary conditions

$$u(x, 0) = 0 \quad \text{for } x \in \mathbb{R} \quad \text{and} \quad u(x, t) \to 0 \quad \text{as } |x| \to \infty, t > 0. \tag{6.112}$$

The Green's function must satisfy the equation

$$G_t - KG_{xx} = \delta(x)\delta(t), \quad x \in \mathbb{R}, t > 0, \tag{6.113}$$

with the initial and boundary conditions (6.112). Using the joint Laplace and Fourier transform defined by (6.106), we obtain the solution of the transformed problem as

$$\overline{\tilde{G}}(k, s) = \frac{1}{\sqrt{2\pi}} \frac{1}{(s + Kk^2)}. \tag{6.114}$$

The joint inverse transform gives the solution

$$G(x, t) = \frac{1}{\sqrt{4\pi Kt}} \exp\left(-\frac{x^2}{4Kt}\right). \tag{6.115}$$

Note that $G(x, t)$ is an even function of x for all $t > 0$. The amplitude (or peak height) of G decreases inversely with $\sqrt{Kt}$, whereas the width of the peak increases with $\sqrt{Kt}$. The spatial distribution of $G(x, t)$ is Gaussian, and, hence, it can easily be drawn against x for different values of $2\sqrt{Kt}$.

If the source is located at (ξ, τ) instead of $(0, 0)$, the corresponding Green's function assumes the form

$$G(x, t; \xi, \tau) = \frac{1}{\sqrt{4K\pi(t - \tau)}} \exp\left(-\frac{(x - \xi)^2}{4K(t - \tau)}\right). \tag{6.116}$$

Thus, the solution of (6.111) is given by

$$u(x, t) = \int_0^t d\tau \int_{-\infty}^{\infty} q(\xi, \tau)G(x, t; \xi, \tau)d\xi. \tag{6.117}$$

Finally, the solution of the inhomogeneous diffusion equation (6.111) with inhomogeneous initial data

$$u(x, 0) = f(x), \quad x \in \mathbb{R} \tag{6.118}$$

can be expressed as

$$u(x, t) = \int_{-\infty}^{\infty} f(\xi)G(x, t; \xi, 0)d\xi + \int_0^t d\tau \int_{-\infty}^{\infty} q(\xi, \tau)G(x, t; \xi, \tau)d\xi, \tag{6.119}$$

where $G(x, t; \xi, \tau)$ is given by (6.116). $\square$

Example 6.4.23. (Green's function of the one-dimensional Klein–Gordon equation) We consider the inhomogeneous Klein–Gordon equation

$$u_{tt} - c^2 u_{xx} + d^2 u = p(x, t), \quad x \in \mathbb{R}, t > 0, \tag{6.120}$$

with the initial and boundary conditions

$$u(x, 0) = 0 = u_t(x, 0) \quad \text{for } x \in \mathbb{R}, \tag{6.121}$$

$$u(x, t) \to 0 \quad \text{as } |x| \to \infty, \ t > 0, \tag{6.122}$$

where c and d are constants.

The Green's function G associated with this problem satisfies the equation

$$G_{tt} - c^2 G_{xx} + d^2 G = \delta(x)\delta(t), \quad x \in \mathbb{R}, \ t > 0, \tag{6.123}$$

with the same initial and boundary data (6.121) and (6.122). Application of the joint Laplace and Fourier transform yields the solution for the transformed problem as

$$\overline{\tilde{G}}(k, s) = \frac{1}{\sqrt{2\pi}} \frac{1}{(s^2 + \alpha^2)}, \tag{6.124}$$

where $\alpha^2 = c^2 k^2 + d^2$.

The joint inverse transform leads to the solution

$$\begin{aligned}
G(x, t) &= \frac{1}{\sqrt{2\pi}} \mathcal{F}^{-1} \left\{ \frac{\sin \alpha t}{\alpha} \right\} \\
&= \frac{1}{2\pi c} \int_{-\infty}^{\infty} \frac{\sin\left(ct\sqrt{k^2 + \frac{d^2}{c^2}}\right)}{\sqrt{k^2 + \frac{d^2}{c^2}}} e^{ikx} \, dk \\
&= \frac{1}{2c} J_0 \left[\frac{d}{c} \sqrt{c^2 t^2 - x^2} \right] H(ct - |x|) \\
&= \begin{cases} \frac{1}{2c} J_0 \left[\frac{d}{c} \sqrt{c^2 t^2 - x^2} \right] & \text{if } |x| < ct, \\ 0 & \text{if } |x| > ct. \end{cases}
\end{aligned}$$

In the limit as $d \to 0$, the Green's function of the Klein–Gordon equation reduces to that of the standard wave equation (6.102).

If the source is located at (ξ, τ), the corresponding Green's function takes the form

$$G(x, t; \xi, \tau) = \begin{cases} \frac{1}{2c} J_0 \left[\frac{d}{c} \sqrt{c^2 (t - \tau)^2 - (x - \xi)^2} \right] & \text{if } |x - \xi| < c(t - \tau), \\ 0 & \text{if } |x - \xi| > c(t - \tau). \end{cases} \tag{6.125}$$

Finally, the solution of (6.120) with the initial data

$$u(x, 0) = f(x) \quad \text{and} \quad u_t(x, 0) = g(x), \quad x \in \mathbb{R}, \tag{6.126}$$

can be written as

$$u(x, t) = \int_0^t d\tau \int_{-\infty}^{\infty} p(\xi, \tau) G(x, t; \xi, \tau) d\xi$$

$$+ \int_{-\infty}^{\infty} \left[g(\xi) G(x, t; \xi, 0) - f(\xi) G_\tau(x, t; \xi, 0) \right] d\xi, \quad (6.127)$$

where $G = 0$ for $\tau > t$, and, hence, the integration with respect to τ extends only to t. It also follows from (6.125) that G is nonzero when $|x - \xi| < c(t - \tau)$, which is equivalent to $x - c(t - \tau) < \xi < x + c(t - \tau)$. Consequently, the double integral in (6.127) becomes

$$\int_0^t d\tau \int_{-\infty}^{\infty} p(\xi, \tau) G(x, t; \xi, \tau) d\xi$$

$$= \int_0^t d\tau \int_{x-c(t-\tau)}^{x+c(t-\tau)} p(\xi, \tau) J_0 \left[\frac{d}{c} \sqrt{c^2 (t - \tau)^2 - (x - \xi)^2} \right] d\xi. \quad (6.128)$$

Note that $G(x, t; \xi, 0)$ is nonzero in $(x - ct, x + ct)$ and vanishes outside of this interval. Consequently,

$$\int_{-\infty}^{\infty} g(\xi) G(x, t; \xi, 0) d\xi = \frac{1}{2c} \int_{x-ct}^{x+ct} g(\xi) J_0 \left[\frac{d}{c} \sqrt{c^2 t^2 - (x - \xi)^2} \right] d\xi. \quad (6.129)$$

In terms of the Heaviside function, the Green's function (6.125) can be written as

$$G(x, t; \xi, \tau) = \frac{1}{2c} J_0 \left[\frac{d}{c} \sqrt{c^2 (t - \tau)^2 - (x - \xi)^2} \right] H\big(x + ct - (\xi + c\tau)\big)$$

$$\times H\big(\xi + ct - (x + c\tau)\big). \quad (6.130)$$

This gives

$$\left[G(x, t; \xi, \tau) \right]_{\tau=0} = \frac{td}{2} \sqrt{c^2 t^2 - (x - \xi)^2} J_0' \left[\frac{d}{c} \sqrt{c^2 t^2 - (x - \xi)^2} \right]$$

$$\times H(x - \xi + ct) H(\xi - x + ct)$$

$$+ \frac{1}{2} J_0 \left[\frac{d}{c} \sqrt{c^2 t^2 - (x - \xi)^2} \right]$$

$$\times \left[\delta(x - \xi + ct) H(\xi - x + ct) \right.$$

$$\left. + \delta(\xi - x + ct) H(x - \xi + ct) \right].$$

In view of this result combined with the property of the delta function with $H(2ct) = 1$, we obtain

$$-\int_{-\infty}^{\infty} f(x)G_\tau(x, t; \xi, 0)d\xi$$

$$= -\frac{td}{2}\int_{x-ct}^{x+ct} \frac{J_1\left[\frac{d}{c}\sqrt{c^2t^2 - (x-\xi)^2}\right]}{\sqrt{c^2t^2 - (x-\xi)^2}}f(\xi)d\xi$$

$$+ \frac{1}{2}\int_{-\infty}^{\infty} J_0\left[\frac{d}{c}\sqrt{c^2t^2 - (x-\xi)^2}\right]$$

$$\times\left[\delta(x - \xi + ct)H(\xi - x + ct) + \delta(\xi - x + ct)H(x - \xi + ct)\right]f(\xi)d\xi$$

$$= -\frac{td}{2}\int_{x-ct}^{x+ct} \frac{J_1\left[\frac{d}{c}\sqrt{c^2t^2 - (x-\xi)^2}\right]}{\sqrt{c^2t^2 - (x-\xi)^2}}f(\xi)d\xi + \frac{1}{2}[f(x - ct) + f(x + ct)].$$

$$(6.131)$$

Combining (6.128), (6.129), and (6.131), solution (6.127) takes the final form

$$u(x, t) = \frac{1}{2}[f(x - ct) + f(x + ct)]$$

$$+ \frac{1}{2c}\int_{x-ct}^{x+ct} J_0\left[\frac{d}{c}\sqrt{c^2t^2 - (x-\xi)^2}\right]g(\xi)d\xi$$

$$- \frac{td}{2}\int_{x-ct}^{x+ct} \frac{J_1\left[\frac{d}{c}\sqrt{c^2t^2 - (x-\xi)^2}\right]}{\sqrt{c^2t^2 - (x-\xi)^2}}f(\xi)d\xi$$

$$+ \frac{1}{2}\int_0^t d\tau \int_{x-c(t-\tau)}^{x+c(t-\tau)} J_0\left[\frac{d}{c}\sqrt{c^2(t-\tau)^2 - (x-\xi)^2}\right]p(\xi, \tau)d\xi.$$

$$(6.132)$$

If $d = 0$, this solution reduces to that of the Cauchy problem for the inhomogeneous wave equation. □

6.5 Weak Solutions of Elliptic Boundary Value Problems

We consider the Dirichlet problem for the second-order elliptic operator

$$-\nabla^2 u = f \quad \text{in } \Omega, \qquad u = 0 \quad \text{on } \partial\Omega, \tag{6.133}$$

where $\Omega \subset \mathbb{R}^N$ is a bounded open set, $\partial\Omega$ is the boundary of Ω, and $f \in C(\Omega)$ is a given function.

By definition, a classical solution u of this problem is a function $u \in C^2(\mathrm{cl}\,\Omega)$, which satisfies (6.133) at every point. We assume that u is a classical solution and multiply Equation (6.133) by $\varphi \in \mathcal{D}(\Omega)$, and then we integrate to obtain

$$-\int_\Omega \varphi \nabla^2 u d\tau = \int_\Omega \varphi f d\tau. \tag{6.134}$$

Since $\varphi = 0$ on $\partial\Omega$, an application of the Green's first identity to (6.134) yields

$$\int_\Omega \nabla u \cdot \nabla \varphi d\tau = \int_\Omega f\varphi d\tau \tag{6.135}$$

for every $\varphi \in \mathcal{D}(\Omega)$. This does not require any information on the second derivatives of u.

On the other hand, if $f \notin C(\Omega)$ the problem (6.133) does not have a classical solution. It is then necessary to generalize the solution in an appropriate manner. If $f \in L^2(\Omega)$, Equation (6.135) makes sense if $\nabla u \in L^2(\Omega)$. If $u \in H^1_0(\Omega)$, where $H^1_0(\Omega)$ is the subspace of $H^1(\Omega)$ consisting of functions vanishing on $\partial\Omega$, and if the derivatives $\partial u/\partial x_k$ are considered in the generalized sense, then it follows from the definition of the Sobolev space that $\partial u/\partial x_k \in L^2(\Omega)$. Then if $u \in H^1_0(\Omega)$ and u satisfies (6.135), then it is a weak solution of (6.133).

Since $H^1_0(\Omega)$ is the closure of $\mathcal{D}(\Omega)$, it is a dense subspace of $H^1_0(\Omega)$. Therefore, solving Equation (6.135) is equivalent to finding $u \in H^1_0(\Omega)$ such that

$$\langle \nabla u, \nabla \varphi \rangle = \langle f, \varphi \rangle \quad \text{for all } \varphi \in \mathcal{D}(\Omega), \tag{6.136}$$

where $\langle \cdot, \cdot \rangle$ is the inner product in $L^2(\Omega)$: $\langle \varphi, \psi \rangle = \int_\Omega \varphi\psi$. Equation (6.136) is known as the *variational* or *weak* formulation of the problem (6.133).

Theorem 6.5.1. *Let Ω be a bounded open subset of $\mathbb{R}^N$, and let $f \in L^2(\Omega)$. Then there exists a unique weak solution $u \in H^1_0(\Omega)$ satisfying (6.136). Furthermore, $u \in H^1_0(\Omega)$ is a solution of (6.136) if and only if*

$$J(u) = \min_{v \in H^1_0(\Omega)} J(v), \tag{6.137}$$

where

$$J(v) = \frac{1}{2} \int_\Omega \nabla v \cdot \nabla v d\tau - \int_\Omega f v d\tau. \tag{6.138}$$

Proof: In order to apply the Lax–Milgram theorem 4.3.16, we set $H = H^1_0(\Omega)$ and, for $u, v \in H^1_0(\Omega)$,

$$a(u, v) = \int_\Omega \nabla u \cdot \nabla v d\tau. \tag{6.139}$$

We first show that $a(\cdot, \cdot)$ is coercive, that is, there exists a positive constant K such that

$$\int_\Omega |\nabla u|^2 d\tau \geq K\|u\|_1^2 = K\left[\int_\Omega |\nabla u|^2 d\tau + \int_\Omega u^2 d\tau\right] \quad \text{for all } u \in H.$$

This readily follows from *Friedrichs' first inequality*

$$\int_\Omega |\nabla u|^2 d\tau \geq \alpha \int_\Omega u^2 d\tau, \quad u \in H, \tag{6.140}$$

where α is a positive constant. Thus,

$$\int_\Omega |\nabla u|^2 d\tau = \frac{1}{2}\int_\Omega |\nabla u|^2 d\tau + \frac{1}{2}\int_\Omega |\nabla u|^2 d\tau$$

$$\geq \frac{1}{2}\int_\Omega |\nabla u|^2 d\tau + \frac{\alpha}{2}\int_\Omega u^2 d\tau \geq K\|u\|_1^2, \tag{6.141}$$

where $K = \min\{1/2, \alpha/2\}$ and $u \in H$.

To prove the boundedness of $a(\cdot, \cdot)$, we note that

$$a(u, u) = \int_\Omega |\nabla u|^2 d\tau \leq \int_\Omega \left(|\nabla u|^2 + u^2\right)d\tau = \|u\|_1^2. \tag{6.142}$$

Thus, $a(\cdot, \cdot)$ is bounded, symmetric, and coercive. So, by the Lax–Milgram theorem 4.3.16, there exists a unique weak solution of Equation (6.136).

We next consider the Neumann boundary value problem

$$-\nabla^2 u + bu = f \quad \text{in } \Omega, \tag{6.143a}$$

$$\frac{\partial u}{\partial n} = 0 \quad \text{on } \partial\Omega, \tag{6.143b}$$

where $\Omega \subset \mathbb{R}^N$ is a bounded open set, n is the exterior unit normal to $\partial\Omega$, and b is a non-negative constant.

According to Green's first identity (6.55), if u is a classical solution, then $u \in H_0^1(\Omega)$ and it satisfies the equation

$$\int_\Omega \nabla u \cdot \nabla v d\tau + \int_\Omega v\nabla^2 u d\tau = \int_{\partial\Omega} v\frac{\partial u}{\partial n}ds.$$

Or, equivalently, by (6.143a) and (6.143b),

$$\int_\Omega \nabla u \cdot \nabla v d\tau + \int_\Omega buv d\tau = \int_\Omega fv d\tau \tag{6.144}$$

for every $v \in H_0^1(\Omega)$.

If $f \in L^2(\Omega)$, then we define a weak solution of (6.143a) and (6.143b) as $u \in H_0^1(\Omega)$ satisfying (6.144).

Consider the bilinear from associated with the operator $A = -\nabla^2 + b$:

$$a(u, v) = \int_\Omega \left[(\nabla u \cdot \nabla v) + buv \right] d\tau. \tag{6.145}$$

Clearly, a is a bilinear form on $H^1(\Omega)$ and

$$a(u, v) = \int_\Omega (\nabla u \cdot \nabla v + buv) d\tau$$

$$\leq \max(1, b_1) \int_\Omega (\nabla v \cdot \nabla v + uv) d\tau$$

$$= M(u, v) \leq M\|u\|\|v\|,$$

where $0 < b \leq b_1$, $M = \max(1, b_1)$, and a is continuous.

On the other hand,

$$a(u, u) = \int_\Omega \left(\nabla u \cdot \nabla u + bu^2 \right) d\tau \geq \min(1, b_0)\|u\|^2,$$

where $0 < b_0 \leq b$. Therefore, a is a continuous and coercive bilinear form. Then, by the Lax–Milgram theorem, there exists a unique solution $u \in H^1(\Omega)$ such that

$$a(u, v) = \langle f, v \rangle \tag{6.146}$$

for all $v \in H^1(\Omega)$. This u is called the *weak solution* of the equation $Au = f$, that is, u is the unique solution of the Neumann boundary value problem (6.143a) and (6.143b). Furthermore, the solution minimizes the functional

$$J(v) = \frac{1}{2} \int_\Omega \left(\nabla v \cdot \nabla v + bv^2 \right) d\tau - \int_\Omega f v d\tau. \tag{6.147}$$

□

Example 6.5.2. Consider the boundary value problem

$$-\nabla^2 u + a_0 u = f \quad \text{in } \Omega \subset \mathbb{R}^2, \tag{6.148a}$$

$$u = 0 \qquad \text{on } \partial\Omega, \tag{6.148b}$$

where a_0 is a positive constant. Set $Tu = -\nabla^2 u + a_0 u$.

Define an inner product in $H_0^1(\Omega)$

$$\langle u, v \rangle = \int_\Omega (u_x v_x + u_y v_y + uv) dx dy, \tag{6.149}$$

a bilinear form in $H_0^1(\Omega)$

$$a(u, v) = \langle v, Tu \rangle = \int_\Omega v\big(-\nabla^2 u + a_0 u\big)dxdy, \qquad (6.150)$$

and a functional on $H_0^1(\Omega)$

$$I(v) = \int_\Omega fvdxdy. \qquad (6.151)$$

A quadratic form for this problem can be defined in $H_0^1(\Omega)$ by

$$I(u) = \frac{1}{2}a(u, u) - I(u) = \int_\Omega \left[\frac{1}{2}\{(u_x^2 + u_y^2) + a_0 u^2\} - fu\right]dxdy.$$

The bilinear form a is symmetric, bounded, and positive definite. The boundedness follows from the Schwarz inequality

$$|a(u, v)| \leq \sqrt{\int_\Omega (|u_x|^2 + |u_y|^2)dxdy}\sqrt{\int_\Omega (|v_x|^2 + |v_y|^2)dxdy}$$

$$+ a_0\sqrt{\int_\Omega |u|^2 dxdy}\sqrt{\int_\Omega |v|^2 dxdy}$$

$$\leq K\|u\|\|v\|,$$

where $K = \max(1, a_0)$.

The positive definiteness follows from (6.150) by setting $u = v$:

$$a(u, u) = \int_\Omega \big(|\nabla u|^2 + a_0 u^2\big)dxdy \geq a_0 \int_\Omega \big(|\nabla u|^2 + |u|^2\big)dxdy = \alpha\|u\|^2,$$

where $\alpha = \min(1, a_0)$.

Note that $I(v)$ is bounded. Hence, it follows from the Lax–Milgram theorem that the problem $a(u, v) = I(v)$ has a unique solution in $H_0^1(\Omega)$. $\quad\square$

We can generalize the preceding result to cover the case of *second-order elliptic equations* defined on an open bounded set $\Omega \subset \mathbb{R}^N$ with smooth boundary $\partial\Omega$. We now consider the boundary value problem

$$Tu = f \quad \text{in } \Omega \subset \mathbb{R}^N, \qquad (6.152a)$$

$$u = 0 \quad \text{on } \partial\Omega, \qquad (6.152b)$$

where

$$Tu = -\sum_{i,j=1}^N \frac{\partial}{\partial x_i}\left[a_{ij}\frac{\partial u}{\partial x_j}\right] + a_0 u,$$

$a_{ij} \in C^1(\text{cl}\,\Omega)$, $1 \le i, j \le N$, $a_0 \in C^1(\text{cl}\,\Omega)$, $x = (x_1, \ldots, _N) \in \mathbb{R}^N$. The differential operator T is said to be in *divergence form*. It is called *uniformly elliptic* if the ellipticity condition

$$\sum_{i,j=1}^{N} a_{ij}(x)\xi_i\xi_j \ge K|\xi|^2 = K\left(\xi_1^2 + \cdots + \xi_N^2\right) \tag{6.153}$$

is satisfied for all $\xi \in \mathbb{R}^N$, $x \in \Omega$, and K is positive and independent of x and ξ.

If $f \in L^2(\Omega)$, a weak solution of (6.152a) and (6.152b) is given by

$$\int_\Omega \sum_{i,j=1}^{N} a_{ij} \frac{\partial u}{\partial x_i} \frac{\partial v}{\partial x_j}\, d\tau + \int_\Omega a_0 uv\, d\tau = \int_\Omega fv\, d\tau \tag{6.154}$$

for all $v \in H_0^1(\Omega)$.

It can readily be verified that every classical solution is a weak solution. Conversely, every sufficiently smooth weak solution is a classical solution.

We next define a bilinear form in $H_0^1(\Omega)$ by

$$a(u, v) = \int_\Omega \sum_{i,j=1}^{N} a_{ij} \frac{\partial u}{\partial x_i} \frac{\partial v}{\partial x_j}\, d\tau + \int_\Omega a_0 uv\, d\tau \tag{6.155}$$

and the norm

$$\|u\| = \sqrt{\int_\Omega \sum_{n=1}^{N} \left|\frac{\partial u}{\partial x_n}\right|^2 d\tau}. \tag{6.156}$$

If $a_0(x) \ge 0$ for all $x \in \Omega$, then, in view of the ellipticity condition (6.153),

$$a(u, u) = \int_\Omega \sum_{i,j=1}^{N} a_{ij} \frac{\partial u}{\partial x_i} \frac{\partial u}{\partial x_j}\, d\tau + \int_\Omega a_0 u^2\, d\tau$$

$$\ge \int_\Omega \sum_{i,j=1}^{N} a_{ij} \frac{\partial u}{\partial x_i} \frac{\partial u}{\partial x_j}\, d\tau$$

$$\ge K \sum_{n=1}^{N} \left(\frac{\partial u}{\partial x_n}\right)^2 = K\|u\|^2.$$

It can be checked that the form $a(u, v)$ is bounded in $H_0^1(\Omega)$, that is,

$$|a(u, v)| \le M\|u\|\|v\| \tag{6.157}$$

for some constant M and all $u, v \in H_0^1(\Omega)$.

If a is symmetric, that is, $a_{ij} = a_{ji}$ for all $i, j \in \mathbb{N}$, then by the Lax–Milgram theorem, there exists a unique solution $u \in H_0^1(\Omega)$ such that

$$a(u, v) = \langle f, v \rangle \tag{6.158}$$

for all $v \in H_0^1(\Omega)$. Consequently, u satisfies Equation (6.154). In other words, the unique solution u minimizes the functional

$$J(v) = \frac{1}{2} \int_\Omega \sum_{i,j=1}^N a_{ij} \frac{\partial v}{\partial x_i} \frac{\partial v}{\partial x_j} d\tau + \frac{1}{2} \int_\Omega a_0 v^2 d\tau - \int_\Omega f v d\tau \tag{6.159}$$

on $H_0^1(\Omega)$.

To define a weak solution through (6.154), it suffices to assume a_{ij}, a_0 are bounded on Ω. Hence, u is the weak solution of the equation $Tu = f$, that is, u is the unique weak solution of the elliptic boundary value problem (6.152a) and (6.152b).

More generally, we consider the following second-order elliptic boundary value problem:

$$Tu = f \quad \text{in } \Omega \subset \mathbb{R}^N, \qquad u = 0 \quad \text{on } \partial\Omega, \tag{6.160}$$

where

$$Tu = -\sum_{i,j=1}^N \frac{\partial}{\partial x_i} \left[a_{ij} \frac{\partial u}{\partial x_j} \right] + \sum_{i=1}^N a_i \frac{\partial u}{\partial x_i} + a_0 u,$$

where the a_{ij}'s satisfy the ellipticity condition (6.153) and $a_i \in \mathcal{C}(\text{cl}\,\Omega)$, $1 \le i \le N$.

A weak solution is a $u \in H_0^1(\Omega)$ satisfying

$$a(u, v) = \langle f, v \rangle \tag{6.161}$$

for every $v \in H_0^1(\Omega)$, where

$$a(u, v) = \int_\Omega \sum_{i,j=1}^N a_{ij} \frac{\partial u}{\partial x_i} \frac{\partial v}{\partial x_j} d\tau + \int_\Omega \sum_{i=1}^N a_i \frac{\partial u}{\partial x_i} v d\tau + \int_\Omega a_0 u v d\tau. \tag{6.162}$$

This bilinear form is not always symmetric. If it is symmetric, bounded, and coercive, then there exists a unique solution by the Lax–Milgram theorem.

6.6 Examples of Applications of the Fourier Transform to Partial Differential Equations

Example 6.6.1. (One-dimensional diffusion equation with no sources or sinks) Consider the initial value problem for the one-dimensional diffusion equation with no sources or sinks:

$$u_t = Ku_{xx}, \quad -\infty < x < \infty, t > 0, \tag{6.163}$$

where K is a constant, with the initial data

$$u(x, 0) = f(x). \tag{6.164}$$

This kind of problem can often be solved by the use of the Fourier transform

$$\tilde{u}(k, t) = \frac{1}{\sqrt{2\pi}} \int_{-\infty}^{\infty} e^{-ikx} u(x, t)dx.$$

When the Fourier transform is applied to (6.163) and (6.164), we obtain

$$\tilde{u}_t = -Kk^2\tilde{u},$$

$$\tilde{u}(k, 0) = \tilde{f}(k) = \frac{1}{\sqrt{2\pi}} \int_{-\infty}^{\infty} e^{-ikx} f(x)dx.$$

The solution of the transformed system is

$$\tilde{u}(k, t) = \tilde{f}(k)e^{-Kk^2t}. \tag{6.165}$$

The inverse Fourier transform gives the solution

$$u(x, t) = \frac{1}{\sqrt{2\pi}} \int_{-\infty}^{\infty} \tilde{f}(k)e^{ikx - Kk^2t} dk,$$

which is, by the convolution Theorem 5.11.11,

$$u(x, t) = \frac{1}{\sqrt{2\pi}} \int_{-\infty}^{\infty} f(\xi)g(x - \xi)d\xi, \tag{6.166}$$

where

$$g(x) = \mathcal{F}^{-1}\left\{e^{-Ktk^2}\right\} = \frac{1}{\sqrt{2\pi}} \int_{-\infty}^{\infty} e^{ikx - Ktk^2} dk$$

$$= \frac{1}{\sqrt{2\pi}} \int_{-\infty}^{\infty} \exp\left[-Kt\left(k - \frac{ix}{2Kt}\right)^2 - \frac{x^2}{4Kt}\right] dk$$

$$= \frac{1}{\sqrt{2\pi}}\left(\frac{\pi}{Kt}\right)^{1/2} \exp\left(-\frac{x^2}{4Kt}\right) = \frac{1}{\sqrt{2Kt}} \exp\left(-\frac{x^2}{4Kt}\right).$$

Thus, the solution (6.166) becomes

$$u(x, t) = \frac{1}{\sqrt{4\pi Kt}} \int_{-\infty}^{\infty} f(\xi) \exp\left[-\frac{(x - \xi)^2}{4Kt}\right] d\xi. \tag{6.167}$$

The integrand involved in the integral solution consists of the initial data $f(x)$ and the Green's function $G(x, t)$:

$$G(x, t) = \frac{1}{\sqrt{4\pi Kt}} \exp\left[-\frac{(x - \xi)^2}{4Kt}\right]. \qquad (6.168)$$

Since

$$\lim_{t \to 0+} \frac{1}{\sqrt{4\pi Kt}} \exp\left[-\frac{(x - \xi)^2}{4Kt}\right] = \delta(x - \xi), \qquad (6.169)$$

if we let $t \to 0+$, the solution becomes

$$u(x, 0) = f(x).$$

Consider now the initial value problem

$$u_t = u_{xx} + u_{yy}, \quad -\infty < x, y < \infty, \ t > 0, \qquad (6.170)$$

$$u(x, y, 0) = f(x, y). \qquad (6.171)$$

The function

$$G(x, y, t) = \frac{1}{4\pi t} \exp\left[-\frac{x^2 + y^2}{4t}\right] \qquad (6.172)$$

satisfies Equation (6.170). From this we can construct the formal solution

$$u(x, y, t) = \frac{1}{4\pi t} \int_{\mathbb{R}^2} f(\xi, \eta) \exp\left[-\frac{(x - \xi)^2 + (y - \eta)^2}{4t}\right] d\xi\, d\eta. \qquad (6.173)$$

Similarly, a formal solution of the initial value problem for the three-dimensional diffusion equation

$$u_t = \nabla^2 u, \quad -\infty < x, y, z < \infty, \ t > 0, \qquad (6.174)$$

$$u(x, y, z, 0) = f(x, y, z), \qquad (6.175)$$

is

$$u(x, y, z, t) = \frac{1}{8(\pi t)^{3/2}} \int_{\mathbb{R}^3} f(\xi, \eta, \zeta) e^{-r^2/4t} d\xi\, d\eta\, d\zeta, \qquad (6.176)$$

where

$$r^2 = (x - \xi)^2 + (y - \eta)^2 + (z - \zeta)^2. \qquad \square$$

Example 6.6.2. (One-dimensional wave equation) We obtain the *d'Alembert solution* of the Cauchy problem for a one-dimensional wave equation

$$u_{tt} = c^2 u_{xx}, \quad -\infty < x < \infty, \ t > 0, \qquad (6.177)$$

$$u(x, 0) = f(x), \quad u_t(x, 0) = g(x). \qquad (6.178)$$

We apply the joint Fourier and Laplace transform defined by

$$\tilde{u}(k, s) = \frac{1}{\sqrt{2\pi}} \int_{-\infty}^{\infty} e^{-ikx} dx \int_0^{\infty} e^{-st} u(x, t) dt. \tag{6.179}$$

The transformed Cauchy problem has the solution in the form

$$\tilde{u}(k, s) - \frac{s\tilde{f}(k) + \tilde{g}(k)}{s^2 + c^2 k^2}. \tag{6.180}$$

The joint inverse transformation gives the solution

$$u(x, t) = \frac{1}{\sqrt{2\pi}} \int_{-\infty}^{\infty} e^{ikx} \mathcal{L}^{-1} \left\{ \frac{s\tilde{f}(k) + \tilde{g}(k)}{s^2 + c^2 k^2} \right\} dk, \tag{6.181}$$

where $\mathcal{L}^{-1}$ is the inverse Laplace transform operator. Finally, we obtain

$$
\begin{aligned}
u(x, t) &= \frac{1}{\sqrt{2\pi}} \int_{-\infty}^{\infty} e^{ikx} \left[\tilde{f}(k) \cos ckt + \frac{\tilde{g}(k)}{ck} \sin ckt \right] dk \\
&= \frac{1}{\sqrt{2\pi}} \int_{-\infty}^{\infty} \frac{1}{2} e^{ikx} \left[e^{ickt} + e^{-ickt} \right] \tilde{f}(k) dk \\
&\quad + \frac{1}{\sqrt{2\pi}} \frac{1}{2ic} \int_{-\infty}^{\infty} \frac{1}{k} e^{ikx} \left[e^{ickt} - e^{-ickt} \right] \tilde{g}(k) dk \\
&= \frac{1}{2} [f(x + ct) + f(x - ct)] + \frac{1}{\sqrt{2\pi}} \frac{1}{2c} \int_{-\infty}^{\infty} \tilde{g}(k) dk \int_{x-ct}^{x+ct} e^{ik\zeta} d\zeta \\
&= \frac{1}{2} [f(x + ct) + f(x - ct)] + \frac{1}{2c} \int_{x-ct}^{x+ct} g(\zeta) d\zeta. \tag{6.182}
\end{aligned}
$$

This is the classical d'Alembert solution (Jean Le Rond d'Alembert (1717–1783)). It can be shown, by direct substitution, that it is the unique solution of the wave equation provided f is twice continuously differentiable and g is once differentiable. This essentially proves the existence and uniqueness of the d'Alembert solution. It can also be shown, by direct substitution, that the solution (6.182) is uniquely determined by the initial data. It is important to point out that the solution u depends only on the initial values at points between $x - ct$ and $x + ct$ and not at all on initial values outside this interval on the line $t = 0$. This interval is called the *domain of dependence* of the variables (x, t). Moreover, it can be proved that the solution depends continuously on the initial data, that is, the problem is well posed.

In particular, if $g(x) = 0$, the d'Alembert solution (6.182) reduces to

$$u(x, t) = \frac{1}{2} [f(x + ct) + f(x - ct)]. \tag{6.183}$$

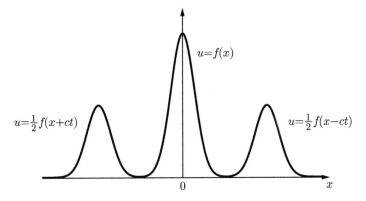

Figure 6.3   Splitting of initial data with equal waves.

Physically, this solution shows that the initial data $u(x, 0) = f(x)$ are split into two identical traveling waves of half amplitude moving in opposite directions with constant velocity c as shown in Figure 6.3.

In general, the physical significance of the d'Alembert solution can be investigated by rewriting the solution (6.182) in the form

$$u(x, t) = \frac{1}{2}f(x - ct) - \frac{1}{2c}\int_0^{x-ct} g(\zeta)d\zeta + \frac{1}{2}f(x + ct) + \frac{1}{2c}\int_0^{x+ct} g(\zeta)d\zeta$$

$$= \Phi(x - ct) + \Psi(x + ct), \qquad (6.184)$$

where

$$\Phi(\xi) = \frac{1}{2}f(\xi) - \frac{1}{2c}\int_0^{\xi} g(\zeta)d\zeta, \qquad \Psi(\eta) = \frac{1}{2}f(\eta) + \frac{1}{2c}\int_0^{\eta} g(\zeta)d\zeta.$$

Physically, $\Phi(x - ct)$ represents a wave traveling in the positive x-direction with constant speed c without change of shape. Similarly, $\Psi(x + ct)$ also represents a wave propagating in the negative x-direction with the same speed c without change of shape.

Finally, the solution of the inhomogeneous wave equation (6.102) with the initial data (6.178) can be obtained by adding (6.110) to the d'Alembert solution (6.182) in the form

$$u(x, t) = \frac{1}{2}\left[f(x + ct) + f(x - ct)\right] + \frac{1}{2c}\int_{x-ct}^{x+ct} g(\zeta)d\zeta$$

$$+ \frac{1}{2c}\int_0^t d\tau \int_{x-c(t-\tau)}^{x+c(t-\tau)} p(\zeta, \tau)d\zeta. \qquad (6.185)$$

□

Example 6.6.3. (Laplace's equation in a half-plane)   We consider the Dirichlet problem consisting of the Laplace equation

$$u_{xx} + u_{yy} = 0, \quad -\infty < x < \infty, \; y \geq 0 \tag{6.186}$$

with the boundary conditions

$$u(x, 0) = f(x), \tag{6.187}$$

$$u(x, y) \to 0 \quad \text{as } r = \sqrt{x^2 + y^2} \to \infty. \tag{6.188}$$

We introduce the Fourier transform

$$\tilde{u}(k, y) = \frac{1}{\sqrt{2\pi}} \int_{-\infty}^{\infty} e^{-ikx} u(x, y) dx$$

so that from (6.186) to (6.188), we find

$$\frac{d^2 \tilde{u}}{dy^2} - k^2 \tilde{u} = 0,$$

$$\tilde{u}(k, 0) = \tilde{f}(k), \qquad \tilde{u}(k, y) \to 0 \quad \text{as } y \to \infty.$$

Thus, the solution of the transformed system is

$$\tilde{u}(k, y) = \tilde{f}(k) e^{-|k|y}. \tag{6.189}$$

From the convolution Theorem 5.11.11, we obtain the solution

$$u(x, y) = \frac{1}{\sqrt{2\pi}} \int_{-\infty}^{\infty} f(\xi) g(x - \xi) d\xi, \tag{6.190}$$

where

$$g(x) = \mathcal{F}^{-1}\left\{ e^{-|k|y} \right\} = \sqrt{\frac{2}{\pi}} \frac{y}{x^2 + y^2}. \tag{6.191}$$

Consequently, the solution (6.190) becomes

$$u(x, y) = \frac{y}{\pi} \int_{-\infty}^{\infty} \frac{f(\xi) d\xi}{(x - \xi)^2 + y^2}, \quad y > 0. \tag{6.192}$$

This is the well-known *Poisson integral* formula in the half plane. Note that

$$\lim_{y \to 0^+} u(x, y) = \int_{-\infty}^{\infty} f(\xi) \left[\lim_{y \to 0^+} \frac{y}{\pi} \frac{1}{(x - \xi)^2 + y^2} \right] d\xi = \int_{-\infty}^{\infty} f(\xi) \delta(x - \xi) d\xi,$$

where Cauchy's definition of the delta function is used, that is,

$$\delta(x - \xi) = \lim_{y \to 0^+} \frac{y}{\pi} \frac{1}{(x - \xi)^2 + y^2}.$$

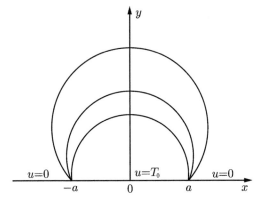

Figure 6.4 Isothermal curves representing a family of circular curves.

This may be recognized as a solution of the Laplace equation for a dipole source at $(x, y) = (\xi, 0)$.

In particular, when

$$f(x) = T_0 H(a - |x|),$$

the solution (6.192) reduces to

$$u(x, y) = \frac{yT_0}{\pi} \int_{-a}^{a} \frac{d\xi}{(\xi - x)^2 + y^2} \tag{6.193}$$

$$= \frac{T_0}{\pi} \left[\tan^{-1}\left(\frac{x+a}{y}\right) - \tan^{-1}\left(\frac{x-a}{y}\right) \right]$$

$$= \frac{T_0}{\pi} \tan^{-1}\left(\frac{2ay}{x^2 + y^2 - a^2}\right). \tag{6.194}$$

The curves in the upper half-plane for which the steady-state temperature is constant are known as *isothermal curves*. In this case, these curves represent a family of circular curves

$$x^2 + y^2 - \alpha y = a^2 \tag{6.195}$$

with centers on the y-axis and fixed end points on the x-axis at $x = \pm a$ as shown in Figure 6.4.

Similarly, we can solve the Dirichlet problem for the three-dimensional Laplace equation in the half-space:

$$u_{xx} + u_{yy} + u_{zz} = 0, \quad -\infty < x, y < \infty, \ z \geq 0, \tag{6.196}$$

with the initial and boundary conditions

$$u(x, y, 0) = f(x, y), \tag{6.197}$$

$$u(x, y, z) \to 0 \quad \text{as } r = \sqrt{x^2 + y^2 + z^2} \to \infty. \tag{6.198}$$

Application of the double Fourier transform gives the solution

$$u(x, y, z) = \frac{z}{2\pi} \int_{\mathbb{R}^2} \frac{f(\xi, \eta)d\xi\,d\eta}{[(x - \xi)^2 + (y - \eta)^2 + z^2]^{3/2}}. \tag{6.199}$$

□

Example 6.6.4. We find the potential for the irrotational two-dimensional in-viscid fluid flow filling the half-space.

For the irrotational motion, the curl of the velocity vector **v** is zero every-where, so that **v** can be expressed in terms of the velocity potential φ by $\mathbf{v} = -\nabla\varphi$. The continuity equation, $\operatorname{div}\mathbf{u} = 0$, reduces to the Laplace equa-tion for φ:

$$\nabla^2\varphi = \varphi_{xx} + \varphi_{yy} = 0, \quad -\infty < x < \infty, y \geq 0. \tag{6.200}$$

The fluid is introduced normally to the half-space through the strip $|x| \leq a$ of the plane $y = 0$. This gives the boundary conditions

$$\varphi_y = -\frac{U}{2}f(x) \qquad \text{on } y = 0, \tag{6.201}$$

$$(u, v) = -(\varphi_x, \varphi_y) \to (0, 0) \quad \text{as } y \to \infty, \tag{6.202}$$

where U is a constant and f is a given function.

Application of the Fourier transform

$$\tilde{\varphi}(k, y) = \frac{1}{\sqrt{2\pi}} \int_{-\infty}^{\infty} e^{-ikx}\varphi(x, y)dx \tag{6.203}$$

enables us to solve the problem. The formal solution has the form

$$\varphi(x, y) = \frac{U}{\sqrt{2\pi}} \int_{-\infty}^{\infty} \frac{\tilde{f}(k)}{|k|} e^{ikx - |k|y}dk. \tag{6.204}$$

In particular, if $f(x) = H(a - |x|)$, then $\tilde{f}(k) = (2/\sqrt{2\pi})(\sin ka)/k$ so that the solution becomes

$$\varphi(x, y) = \frac{U}{2\pi} \int_{-\infty}^{\infty} \frac{\sin ka}{k|k|} e^{ikx - |k|y}dk. \tag{6.205}$$

Thus, the velocity component in the y direction is given by

$$v = -\varphi_y = \frac{U}{2\pi} \operatorname{Re} \int_{-\infty}^{\infty} \frac{\sin ka}{k} e^{ikx - |k|y}dk$$

$$= \frac{U}{2\pi} \int_{-\infty}^{\infty} \frac{1}{k} \sin ka \cos kx e^{-|k|y} dk$$

$$= \frac{U}{4\pi} \int_{-\infty}^{\infty} \{\sin k(x+a) - \sin k(x-a)\} \frac{e^{-|k|y}}{k} dk, \qquad (6.206)$$

where Re stands for the real part.

Using the result

$$\int_0^{\infty} \sin \alpha k \frac{e^{-ky}}{k} dk = \frac{\pi}{2} - \tan^{-1} \frac{y}{\alpha},$$

Solution (6.206) for v becomes

$$v = \frac{U}{2\pi} \left[\tan^{-1} \frac{y}{x-a} - \tan^{-1} \frac{y}{x+a} \right]. \qquad (6.207)$$

Similarly, for the x-component of the velocity, we obtain

$$u = -\varphi_x = -\frac{iU}{2\pi} \int_{-\infty}^{\infty} \frac{\sin ka}{k} e^{ikx - |k|y} dk = \frac{U}{2\pi} \ln \frac{r_2}{r_1}, \qquad (6.208)$$

where $r_1^2 = (x-a)^2 + y^2$ and $r_2^2 = (x+a)^2 + y^2$.

Introducing a complex potential $w = \varphi + i\psi$, we obtain

$$\frac{dw}{dz} = \frac{\partial \varphi}{\partial x} - i \frac{\partial \varphi}{\partial y} = -u + iv, \qquad (6.209)$$

which can be written, by (6.206) to (6.208), in the form

$$\frac{dw}{dz} = \frac{U}{2\pi} \left[\ln \frac{r_1}{r_2} + i(\theta_1 - \theta_2) \right] = \frac{U}{2\pi} \ln \frac{z-a}{z+a}, \qquad (6.210)$$

where $\tan \theta_1 = y/(x-a)$ and $\tan \theta_2 = y/(x+a)$.

Integrating (6.210) with respect to z gives the complex potential

$$w = \frac{U}{2\pi} \left[2a + (z-a) \ln(z-a) - (z+a) \ln(z+a) \right]. \qquad (6.211)$$

□

Example 6.6.5. (The Navier–Stokes equation)   The Navier–Stokes equation (Claude Louis Marie Henri Navier (1785–1836), George Gabriel Stokes (1819–1903)) in a viscous fluid of constant density ρ and constant kinematic viscosity ν with no external forces is

$$\frac{D\mathbf{u}}{Dt} = -\frac{1}{\rho} \nabla p + \nu \nabla^2 \mathbf{u}, \qquad (6.212)$$

where $\mathbf{u} = (u, v, w)$ is the local Eulerian fluid velocity at a point $\mathbf{x} = (x, y, z)$ and at time t, $p(\mathbf{x}, t)$ is the pressure, and the total derivative following the motion is

$$\frac{D}{Dt} = \frac{\partial}{\partial t} + \mathbf{u} \cdot \nabla, \tag{6.213}$$

which consists of an unsteady term and a convective term.

We next introduce the vorticity vector $\boldsymbol{\omega} = (\xi, \eta, \zeta)$ in rectangular Cartesian coordinates

$$\xi = w_y - v_z, \tag{6.214a}$$

$$\eta = u_z - w_x, \tag{6.214b}$$

$$\zeta = v_x - u_y. \tag{6.214c}$$

Using the vector identity

$$\mathbf{u} \times \operatorname{curl} \mathbf{u} = \frac{1}{2} \nabla (\mathbf{u} \cdot \mathbf{u}) - \mathbf{u} \cdot \nabla \mathbf{u} \tag{6.215}$$

with $q^2 = \mathbf{u} \cdot \mathbf{u}$, Equation (6.212) assumes the form

$$\frac{\partial \mathbf{u}}{\partial t} + \mathbf{u} \cdot \nabla \mathbf{u} = -\nabla \left(\frac{p}{\rho} + \frac{1}{2} q^2 \right) + \nu \nabla^2 \mathbf{u}. \tag{6.216}$$

Taking the curl of both sides of this equation, the pressure term disappears and hence we get

$$\frac{\partial \boldsymbol{\omega}}{\partial t} = \operatorname{curl}(\mathbf{u} \times \boldsymbol{\omega}) + \nu \nabla^2 \boldsymbol{\omega} \tag{6.217}$$

$$= -\mathbf{u} \cdot \nabla \boldsymbol{\omega} + \boldsymbol{\omega} \cdot \nabla \mathbf{u} + \nu \nabla^2 \boldsymbol{\omega} \tag{6.218}$$

in which the continuity equations, $\nabla \cdot \mathbf{u} = 0$ and $\nabla \cdot \boldsymbol{\omega} = 0$, are used. Equation (6.218) can be also written in the form

$$\frac{D \boldsymbol{\omega}}{Dt} = \boldsymbol{\omega} \cdot \nabla \mathbf{u} + \nu \nabla^2 \boldsymbol{\omega}. \tag{6.219}$$

This equation (or its equivalent form (6.218)) is called the *vorticity transport equation* and represents the rate of change in vorticity $\boldsymbol{\omega}$, which is described by three terms on the right-hand side of (6.218). The first term, $\mathbf{u} \cdot \nabla \boldsymbol{\omega}$, is the familiar rate of change due to convection of fluid in which the vorticity is nonuniform past a given point. The second term, $\boldsymbol{\omega} \cdot \nabla \mathbf{u}$, describes the stretching of vortex lines, and the last term, $\nu \nabla^2 \boldsymbol{\omega}$, represents the rate of change of $\boldsymbol{\omega}$ due to molecular diffusion of vorticity in exactly the way that $\nu \nabla^2 \mathbf{u}$ represents the contribution to the acceleration from the diffusion of velocity (or momentum).

In the case of two-dimensional flow, $\boldsymbol{\omega}$ is everywhere normal to the plane of flow, and $\boldsymbol{\omega} \cdot \nabla \mathbf{u} = 0$. Equation (6.219) then reduces to the scalar equation

$$\frac{D\omega}{Dt} = \nu \nabla^2 \omega, \tag{6.220}$$

so that only convection and viscous conduction occur. In terms of the stream function ψ, where $u = \psi_y$, $v = -\psi_x$ (and $\omega = -\nabla^2 \psi$) satisfy the continuity equation identically, Equation (6.218) assumes the form

$$\left(\frac{\partial}{\partial t} + \frac{\partial \psi}{\partial y}\frac{\partial}{\partial x} - \frac{\partial \psi}{\partial x}\frac{\partial}{\partial y}\right)\nabla^2 \psi = \nu \nabla^4 \psi. \tag{6.221}$$

In the steady-state $\partial/\partial t = 0$, and if the velocity of the fluid is very small and the viscosity is very large, all terms on the left-hand side of (6.221) can be neglected in the first approximation. Consequently, (6.221) reduces to the biharmonic equation

$$\nabla^4 \psi = 0. \tag{6.222}$$

We solve this biharmonic equation for the viscous fluid bounded by the plane $y = 0$ with the fluid introduced through a strip $|x| < a$ of this plane with a given velocity. Thus, the boundary conditions are

$$\psi_y = 0 \quad \text{on } y = 0, \tag{6.223a}$$

$$\psi_x = f(x)H\big(a - |x|\big). \tag{6.223b}$$

Furthermore, we consider only solutions which tend to zero at large distance from the strip.

Application of the Fourier transform to $\psi(x, y)$ with respect to x reduces (6.222) and (6.223a) to (6.223b) to

$$\left(\frac{d^2}{dy^2} - k^2\right)^2 \tilde{\psi}(k, y) = 0, \tag{6.224}$$

$$\tilde{\psi}_y = 0, \qquad y = 0, \tag{6.225a}$$

$$ik\tilde{\psi} = \tilde{f}(k), \quad y = 0, \tag{6.225b}$$

where

$$\tilde{f}(k) = \frac{1}{\sqrt{2\pi}} \int_{-a}^{a} e^{-ikx} f(x)\,dx.$$

Thus, the solution of the system (6.224) and (6.225a)–(6.225b) is

$$\tilde{\psi}(k, y) = -\frac{i}{k}\big[1 + |k|y\big]\tilde{f}(k)e^{-|k|y}. \tag{6.226}$$

By the means of the inverse Fourier transform combined with the convolution theorem, we obtain

$$\psi(x, y) = \frac{1}{\sqrt{2\pi}} \int_{-\infty}^{\infty} f(x')g(x - x')dx', \tag{6.227}$$

where

$$g(x) = \frac{1}{\sqrt{2\pi}} \int_{-\infty}^{\infty} \left[-\frac{i}{k}(1 + |k|y)e^{-|k|y + ikx} \right] dk$$

$$= \sqrt{\frac{2}{\pi}} \left[\tan^{-1}\frac{x}{y} + \frac{xy}{x^2 + y^2} \right]. \tag{6.228}$$

Finally, the solution of the boundary value problem is

$$\psi(x, y) = \frac{1}{\pi} \int_{-\infty}^{\infty} \left[\tan^{-1}\left(\frac{x - x'}{y}\right) + \frac{(x - x')y}{(x - x')^2 + y^2} \right] f(x')dx'. \tag{6.229}$$

In particular, if $f(x) = \delta(x)$, the solution (6.229) becomes

$$\psi(x, y) = \frac{1}{\pi} \left[\tan^{-1}\left(\frac{x}{y}\right) + \frac{xy}{x^2 + y^2} \right]. \qquad \square$$

Example 6.6.6. In wave propagation problems in applied mathematics, a typical initial boundary problem is

$$u_{tt} = -Lu \qquad \text{in } \Omega, \tag{6.230}$$

$$u = 0 \qquad \text{on } \partial\Omega, \tag{6.231}$$

$$u = f(x, y) \qquad \text{at } t = 0, \tag{6.232a}$$

$$u_t = g(x, y) \qquad \text{at } t = 0, \tag{6.232b}$$

where Ω is a bounded domain in $\mathbb{R}^2$ with a smooth boundary $\partial\Omega$, and L is the differential operator

$$Lu = \frac{1}{r(x, y)} \left[-\frac{\partial}{\partial x} p_1(x, y)\frac{\partial u}{\partial x} - \frac{\partial}{\partial y} p_2(x, y)\frac{\partial u}{\partial y} + q(x, y)u \right], \tag{6.233}$$

r, p_1, p_2, and q are continuous positive functions on the closure of Ω, and p_1 and p_2 have continuous derivatives.

We solve the problem in the Hilbert space $L^2(\Omega)$, which consists of functions $f(x, y)$ with the square of the norm

$$\|f\|^2 = \int_{\Omega} r(x, y)|f(x, y)|^2 dxdy < \infty. \tag{6.234}$$

We seek a separable solution of (6.230) in the form

$$u(x, y, t) = \sum_{n=1}^{\infty} a_n(t)\varphi_n(x, y), \tag{6.235}$$

where φ_n are the eigenfunctions of L with eigenvalues λ_n, so that $L\varphi_n = \lambda_n\varphi_n$. We also assume that

$$f(x, y) = \sum_{n=1}^{\infty} f_n\varphi_n(x, y), \tag{6.236a}$$

$$g(x, y) = \sum_{n=1}^{\infty} g_n\varphi_n(x, y). \tag{6.236b}$$

A simple substitution of (6.235) into (6.230) gives

$$\sum_{n=1}^{\infty} a_n''(t)\varphi_n(x, y) = -\sum_{n=1}^{\infty} \lambda_n a_n(t)\varphi_n(x, y).$$

Thus, the functions a_n satisfy the equation

$$a_n''(t) + \lambda_n a_n(t) = 0 \tag{6.237}$$

so that

$$a_n(t) = A_n \frac{\sin\sqrt{\lambda_n}t}{\sqrt{\lambda_n}} + B_n \cos\sqrt{\lambda_n}t.$$

To satisfy the initial conditions, we set $A_n = g_n$ and $B_n = f_n$, and the final solution becomes

$$u(x, y, t) = \sum_{n=1}^{\infty} \left[f_n \cos\sqrt{\lambda_n}t + g_n\frac{\sin\sqrt{\lambda_n}t}{\sqrt{\lambda_n}} \right]\varphi_n(x, y). \tag{6.238}$$

This clearly satisfies the given equation, as well as boundary and initial conditions. □

Example 6.6.7. In a general diffusion process, a typical initial boundary problem is

$$u_t = -Lu \qquad \text{in } \Omega \subset \mathbb{R}^2, \tag{6.239}$$

$$u = 0 \qquad \text{on } \partial\Omega, \tag{6.240}$$

$$u(x, y, t) = f(x, y) \quad \text{at } t = 0, \tag{6.241}$$

where L is given by (6.233).

Following the argument similar to that employed in Example 6.6.6, we write the solution in the form

$$u = \sum_{n=1}^{\infty} a_n(t)\varphi_n(x, y),$$

and we assume that

$$f(x, y) = \sum_{n=1}^{\infty} f_n \varphi_n(x, y).$$

Consequently, the functions a_n satisfy the equation

$$a'_n(t) = -\lambda_n a_n(t)$$

so that

$$a_n(t) = A_n e^{-\lambda_n t}.$$

We choose $A_n = f_n$ to obtain the solution

$$u(x, y, t) = \sum_{n=1}^{\infty} f_n e^{-\lambda_n t} \varphi_n(x, y). \tag{6.242}$$

A simple check reveals that (6.242) satisfies the differential equation and the boundary and initial conditions.   □

Example 6.6.8. (The linearized Korteweg–de Vries equation)   The linearized Korteweg–de Vries (KdV) (Diederik Johannes Korteweg (1848–1941)) equation for the free surface evolution $\eta(x, t)$ in an inviscid water of constant depth h is

$$\eta_t + c\eta_x + \frac{ch^2}{6}\eta_{xxx} = 0, \quad -\infty < x < \infty, \ t > 0, \tag{6.243}$$

where $c = \sqrt{gh}$ is the shallow water speed.

We solve Equation (6.243) with the initial condition

$$\eta(x, 0) = f(x), \quad -\infty < x < \infty. \tag{6.244}$$

Application of the Fourier transform, with respect to x, to the KdV system gives the solution in the form

$$E(k, t) = \hat{f}(k) \exp\left[ikct\left(\frac{k^2 h^2}{6} - 1\right)\right],$$

where $E(k, t) = \mathcal{F}\{\eta(x, t)\}$. The inverse Fourier transform gives

$$\eta(x, t) = \frac{1}{\sqrt{2\pi}} \int_{-\infty}^{\infty} \hat{f}(k) \exp\left[ik\left\{(x - ct) + \left(\frac{cth^2}{6}\right)k^2\right\}\right] dk. \tag{6.245}$$

In particular, if $f(x) = \delta(x)$, then (6.245) reduces to the *Airy integral*

$$\eta(x, t) = \frac{1}{2\pi} \int_0^\infty \cos\left[k(x - ct) + \left(\frac{cth^2}{6} \right) k^3 \right] dk, \tag{6.246}$$

which is, in terms of the Airy function,

$$\eta(x, t) = \left(\frac{cth^2}{6} \right)^{-1/3} Ai\left[\left(\frac{cth^2}{6} \right)^{-1/3} (x - ct) \right], \tag{6.247}$$

where the *Airy function Ai(z)* is defined by

$$Ai(z) = \frac{1}{\pi} \int_0^\infty \cos\left(kz + \frac{1}{3} k^3 \right) dk. \tag{6.248}$$

$\square$

6.7 Exercises

1. Let f and g be continuous functions on $\mathbb{R}^N$. Show that if

$$\int_{\mathbb{R}^N} f(x)\varphi(x)dx = \int_{\mathbb{R}^N} g(x)\varphi(x)dx$$

for every $\varphi \in C^\infty(\mathbb{R}^N)$ with compact support, then

$$f(x) = g(x) \quad \text{for every } x \in \mathbb{R}^N.$$

2. Show that a test function ψ is of the form $\psi(x) = (x\varphi(x))'$, where φ is a test function if and only if

$$\int_{-\infty}^0 \psi(x)dx = 0 \quad \text{and} \quad \int_0^\infty \psi(x)dx = 0.$$

3. Show that $\mathcal{D}$ is a vector space.

4. Show that if $\varphi, \psi \in \mathcal{D}$, then

   (a) $f\varphi \in \mathcal{D}$ for every smooth function f.

   (b) $\{\varphi(Ax)\} \in \mathcal{D}$ for every affine transformation A of $\mathbb{R}^N$ onto $\mathbb{R}^N$.

   (c) $\varphi * \psi \in \mathcal{D}$.

5. Construct a test function φ such that $\varphi(x) = 1$ for $|x| \leq 1$, and $\varphi(x) = 0$ for $|x| \geq 2$.

6. Which of the following expressions define a distribution?

(a) $\langle f, \varphi \rangle = \sum_{n=1}^{m} \varphi^{(n)}(0)$.

(b) $\langle f, \varphi \rangle = \sum_{n=1}^{m} \varphi(x_n), x_1, \ldots, x_m \in \mathbb{R}$ are fixed.

(c) $\langle f, \varphi \rangle = \sum_{n=1}^{\infty} \varphi^{(n)}(0)$.

(d) $\langle f, \varphi \rangle = \sum_{n=1}^{\infty} \varphi(x_n), x_1, x_2, \ldots \in \mathbb{R}$ are fixed.

(e) $\langle f, \varphi \rangle = \sum_{n=1}^{m} \varphi^{(n)}(x_n), x_1, \ldots, x_m \in \mathbb{R}$ are fixed.

(f) $\langle f, \varphi \rangle = (\varphi(0))^2$.

(g) $\langle f, \varphi \rangle = \sup \varphi$.

(h) $\langle f, \varphi \rangle = \int_{-\infty}^{\infty} |\varphi(t)| dt$.

(i) $\langle f, \varphi \rangle = \int_{a}^{b} \varphi(t) dt$.

(j) $\langle f, \varphi \rangle = \sum_{n=1}^{\infty} \varphi(x_n)$, where $\lim_{n \to \infty} x_n = 0$.

7. Let $\varphi_n \xrightarrow{D} \varphi$ and $\psi_n \xrightarrow{D} \psi$. Prove the following:

(a) $a\varphi_n + b\psi_n \xrightarrow{D} a\varphi + b\psi$ for any scalars a, b.

(b) $f\varphi_n \xrightarrow{D} f\varphi$ for any smooth function f defined on $\mathbb{R}^N$.

(c) $\varphi_n \circ A \xrightarrow{D} \varphi \circ A$ for any affine transformation A of $\mathbb{R}^N$ onto $\mathbb{R}^N$.

(d) $D^\alpha \varphi_n \xrightarrow{D} D^\alpha \varphi$ for any multi-index α.

8. Is the convergence in $\mathcal{D}$ metrizable?

9. Let f be a locally integrable function on $\mathbb{R}^N$. Prove that the functional F on $\mathcal{D}$ defined by

$$\langle F, \varphi \rangle = \int_{\mathbb{R}^N} f\varphi$$

is a distribution.

10. Find the nth distributional derivative of $f(x) = |x|$.

11. Let $f_n(x) = \sin nx$. Show that $f_n \to 0$ in the distributional sense.

12. Let $\{f_n\}$ be the sequence of functions on $\mathbb{R}$ defined by

$$f_n(x) = \begin{cases} 0, & \text{if } x < -1/2n; \\ n, & \text{if } -1/2n \le x \le 1/2n; \\ 0, & \text{if } x > 1/2n. \end{cases}$$

Show that the sequence converges to the Dirac delta distribution.

13. Show that the sequence of Gaussian functions on $\mathbb{R}$ defined by

$$f_n(x) = \sqrt{\frac{n}{\pi}} e^{-n^2 x^2}, \quad n = 1, 2, \ldots,$$

converges to the Dirac delta distribution.

14. Show that the sequence of functions on $\mathbb{R}$ defined by

$$f_n(x) = \frac{\sin nx}{\pi x}, \quad n = 1, 2, \ldots,$$

converges to the Dirac delta distribution.

15. Let $\varphi_0 \in \mathcal{D}(\mathbb{R})$ be a fixed test function such that $\int_{-\infty}^{\infty} \varphi_0(x) dx = 1$. Show that every test function $\varphi \in \mathcal{D}(\mathbb{R})$ can be represented in the form

$$\varphi = K\varphi_0 + \varphi_1,$$

where K is a constant and φ_1 is a test function such that $\int_{-\infty}^{\infty} \varphi_1(x) dx = 0$. Moreover, the representation is unique.

16. Let Ω be an open set in $\mathbb{R}^N$. Show that if $F \in \mathcal{D}'(\Omega)$ and g is a smooth function on Ω, then

$$\frac{\partial}{\partial x_j}(gF) = \left(\frac{\partial}{\partial x_j} g\right) F + g\left(\frac{\partial}{\partial x_j} F\right),$$

for $j = 1, \ldots, N$.

17. Prove the Leibniz formula for the product of smooth function and a distribution (Theorem 6.2.27).

18. Consider the two-dimensional diffusion equation

$$u_t - K\nabla^2 u = f(x, y)\delta(t), \quad (x, y) \in \mathbb{R}^2, t > 0$$

with the initial and boundary conditions

$$u(x, y, 0) = 0 \quad \text{for all } (x, y) \in \mathbb{R}^2,$$
$$u(x, y, t) \to 0 \quad \text{as } r = \sqrt{x^2 + y^2} \to \infty.$$

Show that the Green's function is

$$G\big((x, y), (\xi, \eta)\big) = \frac{1}{4\pi Kt} \exp\left[-\frac{(x - \xi)^2 + (y - \eta)^2}{4Kt}\right].$$

Generalize the result for the n-dimensional diffusion equation.

19. Consider the three-dimensional inhomogeneous wave equation

$$u_{tt} - c^2 \nabla^2 u = f(x, y, z, t), \quad (x, y, z) \in \mathbb{R}^3, t > 0$$

with the initial and boundary conditions

$$u(x, y, z, 0) = 0 = u_t(x, y, z, 0),$$
$$u(x, y, z, t) \to 0 \quad \text{as } \sqrt{x^2 + y^2 + z^2} \to \infty.$$

Show that the Green's function of this problem is

$$G(x, y, z, t) = \frac{1}{4\pi c^2 \sqrt{x^2 + y^2 + z^2}} \delta\left(t - \frac{1}{c}\sqrt{x^2 + y^2 + z^2}\right).$$

20. Find the Green's function $G(x, y, z, t)$ for the three-dimensional Klein–Gordon equation where $G(x, y, z, t)$ satisfies the equation

$$G_{tt} - c^2 \nabla^2 G + d^2 G = \delta(x)\delta(y)\delta(z)\delta(t), \quad (x, y, z) \in \mathbb{R}^3, \ t > 0$$

with the initial and boundary conditions

$$G(x, y, z, 0) = 0 = G_t(x, y, z, 0) \quad \text{for all } (x, y, z) \in \mathbb{R}^3$$

and

$$G(x, y, z, t) \to 0 \quad \text{as } \sqrt{x^2 + y^2 + z^2} \to \infty, \ t > 0.$$

21. Use the joint Laplace and Fourier transform to solve the inhomogeneous Cauchy problem

$$u_{tt} - c^2 u_{xx} = p(x, t), \quad x \in \mathbb{R}, \ t > 0,$$
$$u(x, 0) = f(x) \quad \text{and} \quad u_t(x, 0) = g(x) \quad \text{for all } x \in \mathbb{R}.$$

22. Apply the joint Fourier and Laplace transform to obtain the Green's function for the wave equation

$$G_{tt} - c^2 G_{xx} = \delta(x)\delta(t), \quad -\infty < x < \infty, \ t > 0,$$
$$G(x, 0) = G_t(x, 0) = 0.$$

23. (a) Show that the fundamental solution $G(x, \xi, t)$ for the Cauchy problem

$$G_{tt} = c^2 G_{xx}, \quad -\infty < x < \infty, \ t > 0,$$
$$G(x, 0) = 0, \qquad G_t(x, 0) = \delta(x - \xi),$$

is

$$G(x, \xi, t) = \frac{1}{2c}\big[H(x - \xi + ct) - H(x - \xi - ct)\big].$$

(b) Use this fundamental solution to solve a more general wave problem

$$u_{tt} = c^2 u_{xx}, \quad -\infty < x < \infty, \ t > 0,$$
$$u(x, 0) = 0, \quad u_t(x, 0) = g(x).$$

24. Prove the existence of the weak solution of the Dirichlet boundary value problem

$$-\nabla^2 u + cu = f \quad \text{in } \Omega \subset \mathbb{R}^2, \qquad u = 0 \quad \text{on } \partial\Omega,$$

where c is a positive function of x and y. Show that the weak solution is given by

$$\int_\Omega v\big(-\nabla^2 u + cu\big)d\tau = \int_\Omega fv d\tau,$$

where $u, v \in H_0^1(\Omega)$.

25. Show that the Dirichlet problem for the biharmonic operator

$$\nabla^4 u = f \qquad \text{in } \Omega, \quad f \in L^2(\Omega),$$
$$u = \frac{\partial u}{\partial n} = 0 \quad \text{on } \partial\Omega,$$

where $\Omega \subset \mathbb{R}^N$, has a weak solution $u \in H_0^2(\Omega)$ given by

$$\int_\Omega \Delta u \Delta v d\tau = \int_\Omega fv d\tau \quad \text{for every } v \in H_0^2(\Omega).$$

26. Show that the boundary value problem

$$-\Delta u + u = f \quad \text{in } \mathbb{R}^N, \quad f \in L^2(\mathbb{R}^N),$$
$$u \to 0 \quad \text{as } |x| \to \infty$$

has a unique solution $u \in H^1(\mathbb{R}^N)$ such that

$$\int_{\mathbb{R}^N} \nabla u \cdot \nabla v d\tau + \int_{\mathbb{R}^N} uv d\tau = \int_{\mathbb{R}^N} fv d\tau$$

for all $v \in H^1(\mathbb{R}^N)$.

27. Let $\Omega \subset \mathbb{R}^N$ be a bounded open set. Consider the Robin boundary value problem

$$-\Delta u + u = f \quad \text{in } \Omega, \quad f \in L^2(\Omega),$$

$$\frac{\partial u}{\partial n} + \alpha u = 0 \quad \text{on } \partial\Omega, \quad \alpha > 0.$$

Show that there exists a unique solution $u \in H_0^1(\Omega)$ such that

$$a(u, v) = \langle f, v \rangle \quad \text{for every } v \in H_0^1(\Omega),$$

where

$$a(u, v) = \int_\Omega \nabla u \cdot \nabla v d\tau + \int_\Omega uv d\tau + \alpha \int_{\partial\Omega} uv d\tau \quad \text{and} \quad u, v \in H_0^1(\Omega).$$

28. Use the Fourier transform method to show that the solution of the telegrapher's problem

$$u_{tt} + au_t + bu = c^2 u_{xx}, \quad -\infty < x, t < \infty,$$
$$u(0, t) = f(t), \qquad u_x(0, t) = g(t)$$

is

$$u(x, t) = \frac{1}{2\pi} \int_{\infty}^{\infty} \left[\tilde{f}(k) \cos\{x\alpha(k)\} + \frac{\tilde{g}(k)}{\alpha(k)} \sin\{x\alpha(k)\} \right] e^{ikx} dk,$$

where

$$\alpha(k) = \frac{b + ika - k^2}{c^2},$$

and $\tilde{f}$ and $\tilde{g}$ are the Fourier transforms of f and g, respectively.

29. Find the solution of the Neumann problem in the half-plane

$$u_{xx} + u_{yy} = 0, \qquad -\infty < x < \infty, \ y > 0,$$
$$u_y(x, 0) = g(x), \qquad -\infty < x < \infty,$$
$$u \text{ is bounded as } y \to \infty, \quad u, u_x \text{ vanish as } |x| \to \infty.$$

30. Find the solution of the system

$$u_{xx} + u_{yy} = 0, \qquad -\infty < x < \infty, \ 0 \le y \le a,$$
$$u(x, 0) = f(x), \qquad u(x, a) = g(x).$$

31. Solve the boundary value problem

$$u_{xx} + u_{yy} = 0, \quad -\infty < x < \infty, \ 0 \leq y \leq a,$$
$$u(x, 0) = f(x), \quad u_y(x, a) = 0.$$

32. Show that the solution of the slow motion of viscous fluid through a slit governed by the biharmonic Equation (6.222) with the boundary conditions

$$\psi_y = g(x)H(a - |x|), \quad \psi_x = 0 \quad \text{on } y = 0$$

is

$$\psi(x, y) = \frac{y^2}{\pi} \int_{-\infty}^{\infty} \frac{g(x')dx'}{(x - x')^2 + y^2}.$$

33. Use the analysis of Example 6.6.4 to find the solution of the two-dimensional steady flow of an inviscid liquid through a slit in a plane rigid boundary $y = 0$. The problem is to find the velocity potential $\varphi(x, y)$ satisfying the Laplace equation with the boundary conditions

$$\varphi = H(a - |x|), \quad v = -\varphi_y = (a^2 - x^2)^{1/2} H(a - |x|) \quad \text{on } y = 0.$$

34. If $E(u, v)$ is a bilinear form defined by the Dirichlet integral (6.67) of a self-adjoint operator L, prove the following:

(a) $E(\alpha u + \beta v, \alpha u + \beta v) = \alpha^2 E(u, v) + 2\alpha\beta E(u, v) + \beta^2 E(v, v)$, where α and β are constants.

(b) $(E(u, v))^2 \leq E(u, u)E(v, v)$, if $b \leq 0$ and L is an elliptic operator.

35. (a) Show that the solution of the one-dimensional Schrödinger equation for a free particle of mass m

$$i\hbar \frac{\partial \psi}{\partial t} = -\frac{\hbar^2}{2m} \frac{\partial^2 \psi}{\partial x^2}$$

is

$$\psi(x, t) = \frac{N}{b} \exp\left(-\frac{x^2}{2b^2}\right), \quad b = \left(a^2 + \frac{i\hbar t}{m}\right)^{1/2},$$

where N and b are constants.

(b) Show that the Gaussian probability density is

$$|\psi|^2 = \frac{|N|^2}{ac} \exp\left(-\frac{x^2}{c^2}\right)$$

and its mean width is

$$\delta = \frac{c}{\sqrt{2}}, \quad \text{where } c = \left(a^2 + \frac{\hbar^2 t^2}{m^2 a^2}\right)^{1/2}.$$

36. (a) Show that the solution of the one-dimensional Schrödinger Equation (6.50) with the finite square well potential, $V(x) = V_0 H(|x| - a)$, is

   (i) $\psi_1(x) = A \exp(\kappa x),\ x \leq -a$, where $\kappa^2 = \frac{2m}{\hbar^2}(V_0 - E) > 0$,

   (ii) $\psi_2(x) = B \sin kx + C \cos kx,\ |x| < a$, where $k^2 = \frac{2mE}{\hbar}$,

   (iii) $\psi_3(x) = D \exp(-\kappa x),\ x \geq a$,

   where A, B, C, and D are arbitrary constants.

   (b) Use the matching conditions at $x = \pm a$ to show that the eigenvalue equations are $k \cot ak = -\kappa$ and $k \tan ak = \kappa$,

   (c) Discuss the even and odd solutions of the Schrödinger equation.

37. If $g_t(x) = \frac{1}{\sqrt{2kt}} \exp(-\frac{x^2}{4kt})$, show that $g_t(x) * g_s(x) = g_{t+s}(x)$.

Mathematical Foundations of Quantum Mechanics

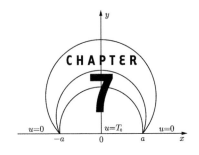

CHAPTER 7

"The tool which serves as intermediary between theory and practice, between thought and observation, is mathematics; it is mathematics which builds the linking bridges and gives the ever more reliable forms. From this it has come about that our entire contemporary culture, in as much as it is based on the intellectual penetration and the exploitation of nature, has its foundations in mathematics. Already Galileo said: one can understand nature only when one has learned the language is mathematics and these signs are mathematical figures. ... Without mathematics, the astronomy and physics of today would be impossible; these sciences, in their theoretical branches, virtually dissolve into mathematics."

David Hilbert

"The research worker, in his efforts to express the fundamental laws of Nature in mathematical form, should strive mainly for mathematical beauty. He should take simplicity into consideration in a subordinate way to beauty. ... It often happens that the requirements of simplicity and beauty are the same, but where they clash the latter must take precedence."

Paul Dirac

7.1 Introduction

This chapter is devoted to the Hilbert space formalism of quantum mechanics in general terms without aiming at mathematical rigor or generality. The theory of Hilbert spaces and the so-called Hermitian operators are essential for the mathematical foundation of quantum mechanics. Although physical concepts and their interpretations underlying quantum mechanics are radically different from those of classical physics, the relation between these two subjects is of special importance. Our aim is to discuss the basic concepts and postulates of quantum mechanics in terms of the results of the theory of operators in a Hilbert space.

7.2 Basic Concepts and Equations of Classical Mechanics

In classical mechanics, matter is usually made up of point particles. We assume that a macroscopic system is made up of a number of electrically neutral particles, of which the ith particle has mass m_i and coordinates (x_i, y_i, z_i). They move under the action of mutually interacting forces according to Newton's laws of motion (Sir Isaac Newton (1643–1727)). With the forces $\mathbf{F} = (F_{x_i}, F_{y_i}, F_{z_i})$ acting on the ith particle, Newton's Second Law of motion has the form

$$m_i \frac{d^2}{dt^2}(x_i, y_i, z_i) = (F_{x_i}, F_{y_i}, F_{z_i}). \tag{7.1}$$

In vector notation, they are written as

$$m_i \frac{d^2 \mathbf{r}_i}{dt^2} = \mathbf{F}_i, \tag{7.2}$$

where $\mathbf{r}_i = (x_i, y_i, z_i)$ and $\mathbf{F}_i = (F_{x_i}, F_{y_i}, F_{z_i})$.

In general, the particles are assumed to be nonrelativistic, that is, their speed is much less than that of light. The total force acting on the particles is assumed to be irrotational ($\nabla \times \mathbf{F} = 0$) and hence, can be derived from a potential energy function, $V = V(\mathbf{r}, t)$ using

$$\mathbf{F}_i = -\nabla_i V, \tag{7.3}$$

where $\nabla_i = (\partial/\partial x_i, \partial/\partial y_i, \partial/\partial z_i)$ and the potential V is a scalar function of all coordinates (three for each particle) and time t.

A field of force is often called *conservative*, if it is irrotational and independent of time t. When a particle is under the action of conservative forces, it can be shown that the total work done by such forces in carrying a system about a closed path is zero.

In view of (7.3), Newton's equations (7.2) can be written in a more general form

$$-\nabla_i V = \mathbf{F}_i = \frac{dp_i}{dt}, \tag{7.4}$$

where $p_i = m\mathbf{v}_i = m\dot{\mathbf{r}}_i = (m\dot{x}_i, m\dot{y}_i, m\dot{z}_i)$ is the *linear momentum* of the particle.

The main problem of classical dynamics is to describe the general motion of a particle (or a system of particles) when its initial position and velocity are prescribed and the total force acting on the particle (or system) is given.

The kinetic energy T_i of the ith particle is a function of three time-derivatives of position or velocities $\dot{\mathbf{r}} \equiv d\mathbf{r}/dt$ and is defined by

$$T_i = \frac{1}{2}m_i(\dot{x}_i^2 + \dot{y}_i^2 + \dot{z}_i^2) = \frac{1}{2}m_i\left(\frac{d\mathbf{r}_i}{dt}\right) \cdot \left(\frac{d\mathbf{r}_i}{dt}\right). \tag{7.5}$$

Hence, the kinetic energy of the system of particles is

$$T = \sum_i T_i = \sum_i \frac{1}{2}m_i\left(\frac{d\mathbf{r}_i}{dt}\right) \cdot \left(\frac{d\mathbf{r}_i}{dt}\right). \tag{7.6}$$

Using (7.5), Newton's Equations (7.4) can be expressed as a set of equations in terms of kinetic and potential energies for each particle:

$$\frac{d}{dt}\left(\frac{\partial T}{\partial \dot{r}_i}\right) + \nabla_i V = 0. \tag{7.7}$$

Or, more explicitly,

$$\left[\frac{d}{dt}\left(\frac{\partial T}{\partial \dot{x}_i}\right), \frac{d}{dt}\left(\frac{\partial T}{\partial \dot{y}_i}\right), \frac{d}{dt}\left(\frac{\partial T}{\partial \dot{z}_i}\right)\right] + \left[\frac{\partial V}{\partial x_i}, \frac{\partial V}{\partial y_i}, \frac{\partial V}{\partial z_i}\right] = 0. \tag{7.8}$$

The total energy E of a particle is defined as the sum of its kinetic and potential energies and, in general, is a function of $\mathbf{r}$, $\dot{\mathbf{r}}$, and t, that is,

$$E = T + V(\mathbf{r}, t). \tag{7.9}$$

A dynamical function is called a *constant of motion* if it does not change in time as the particle moves along its trajectory. Such a function is said to be *conserved*. When a particle is under the action of conservative forces, its energy E is always a constant of the motion, that is, energy is conserved. From (7.9) and the assumption that V is independent of t, we obtain

$$\frac{dE}{dt} = m\mathbf{v} \cdot \frac{d\mathbf{v}}{dt} + \mathbf{v} \cdot \nabla V = \mathbf{v} \cdot \left(\frac{d\mathbf{p}}{dt} + \nabla V\right) = 0, \tag{7.10}$$

by (7.4). Hence E is a constant of motion.

The *Lagrangian function L* of the system is defined as the difference of the kinetic and potential energies. More precisely,

$$L = T - V = L(x_i, y_i, z_i, \dot{x}_i, \dot{y}_i, \dot{z}_i), \tag{7.11}$$

where the set of six arguments of L stands for one set of six arguments for each particle. Consequently,

$$\frac{\partial L}{\partial \dot{x}_i} = \frac{\partial T}{\partial \dot{x}_i} \tag{7.12a}$$

and

$$\frac{\partial L}{\partial x_i} = -\frac{\partial V}{\partial x_i} \tag{7.12b}$$

so that Equation (7.8) can be expressed in terms of the Lagrangian function L as

$$\frac{d}{dt}\left(\frac{\partial L}{\partial \dot{x}_i}\right) - \frac{\partial L}{\partial x_i} = 0, \tag{7.13a}$$

$$\frac{d}{dt}\left(\frac{\partial L}{\partial \dot{y}_i}\right) - \frac{\partial L}{\partial y_i} = 0, \tag{7.13b}$$

$$\frac{d}{dt}\left(\frac{\partial L}{\partial \dot{z}_i}\right) - \frac{\partial L}{\partial z_i} = 0. \tag{7.13c}$$

These are clearly the transformed version of Newton's equations and are known as the *Lagrange equations of motion* for each particle.

There is no reason to restrict Newton's equations to Cartesian coordinates. In fact, many dynamical problems that cannot be solved in Cartesian coordinates can be solved using other coordinate systems. So it would be more general and more convenient to develop a formalism equivalent to Newton's (or Lagrange's) equations which is independent of any particular system of coordinates. Thus, the transformation of Newton's equations leads to an important idea of coordinate transformation which can be introduced by defining the Cartesian coordinates (x_i, y_i, z_i) in terms of new coordinates q_j and time t where $j = 1, 2, 3, \ldots, 3n$ for n particles in a system. Since the configuration of the system at time t is specified by values of q_j and t,

$$x_i = x_i(q_1, q_2, \ldots, q_{3n}, t), \tag{7.14a}$$

$$y_i = y_i(q_1, q_2, \ldots, q_{3n}, t), \tag{7.14b}$$

$$z_i = z_i(q_1, q_2, \ldots, q_{3n}, t), \tag{7.14c}$$

where q_j are called the *generalized coordinates*, which are independent so that there can be no functional relation between them. The only requirement is that the generalized coordinates be independent and that they uniquely determine the position of a particle. Conversely, a set of generalized coordinates may be defined in terms of the Cartesian variables by a transformation of the form $q_j = q_j(\mathbf{r}, t)$ or simply $q_j = q_j(\mathbf{r})$. There can be no integrable linear relation connecting their increments δq_j. The name *generalized* is justified because the nature of the coordinates q_j are not specified. They may be distances, angles, areas, or any other set of numbers such that when their values are known, the configuration of the system is uniquely prescribed.

A dynamical system is called *holonomic* if there are no nonintegrable relations between δq_j's. Otherwise, it is *nonholonomic*. Thus, in a holonomic system, not only are the coordinates q_j independent, but so also are the infinitesimals δq_j. In such a system, it is therefore possible to effect a displacement such that q_k becomes $q_k + \delta q_k$ while the remaining ones $q_1, q_2, \ldots, q_{k-1}, q_{k+1}, \ldots, q_{3n}$ retain constant values.

It can be shown that Lagrange's equations (7.13a) to (7.13c) have the same form in terms of q_j for a holonomic system. To prove this result, we first compute $\dot{x}_i, \dot{y}_i, \dot{z}_i$ from (7.14a) to (7.14c) to obtain

$$\dot{x}_i = \frac{\partial x_i}{\partial q_1}\dot{q}_1 + \cdots + \frac{\partial x_i}{\partial q_{3n}}\dot{q}_{3n} + \frac{\partial x_i}{\partial t},$$

$$\dot{y}_i = \frac{\partial y_i}{\partial q_1}\dot{q}_1 + \cdots + \frac{\partial y_i}{\partial q_{3n}}\dot{q}_{3n} + \frac{\partial y_i}{\partial t},$$

$$\dot{z}_i = \frac{\partial z_i}{\partial q_1}\dot{q}_1 + \cdots + \frac{\partial z_i}{\partial q_{3n}}\dot{q}_{3n} + \frac{\partial z_i}{\partial t}.$$

Evidently,

$$\frac{\partial \dot{x}_i}{\partial \dot{q}_j} = \frac{\partial x_i}{\partial q_j}, \tag{7.15a}$$

$$\frac{\partial \dot{y}_i}{\partial \dot{q}_j} = \frac{\partial y_i}{\partial q_j}, \tag{7.15b}$$

$$\frac{\partial \dot{z}_i}{\partial \dot{q}_j} = \frac{\partial z_i}{\partial q_j}. \tag{7.15c}$$

The kinetic energy of the system is

$$T = \frac{1}{2}\sum m_i\left(\dot{x}_i^2 + \dot{y}_i^2 + \dot{z}_i^2\right)$$

$$= \frac{1}{2}\left[a_{11}\dot{q}_1^2 + a_{22}\dot{q}_2^2 + \cdots + 2a_{12}\dot{q}_1\dot{q}_2 + \cdots + 2a_1\dot{q}_1 + \cdots + a\right]$$

$$= \frac{1}{2}\sum_{j=1}^{3n}\sum_{k=1}^{3n} a_{jk}\dot{q}_j\dot{q}_k, \tag{7.16}$$

where

$$a_{jk} = \sum m_i\left(\frac{\partial x_i}{\partial q_j}\frac{\partial x_i}{\partial q_k} + \frac{\partial y_i}{\partial q_j}\frac{\partial y_i}{\partial q_k} + \frac{\partial z_i}{\partial q_j}\frac{\partial z_i}{\partial q_k}\right) = a_{kj}, \tag{7.17a}$$

$$a_j = \sum m_i \left(\frac{\partial x_i}{\partial q_j} \frac{\partial x_i}{\partial t} + \frac{\partial y_i}{\partial q_j} \frac{\partial y_i}{\partial t} + \frac{\partial z_i}{\partial q_j} \frac{\partial z_i}{\partial t} \right), \tag{7.17b}$$

$$a = \sum m_i \left[\left(\frac{\partial x_i}{\partial t} \right)^2 + \left(\frac{\partial y_i}{\partial t} \right)^2 + \left(\frac{\partial z_i}{\partial t} \right)^2 \right]. \tag{7.17c}$$

It also follows from (7.15a) that

$$\ddot{x}_i \frac{\partial x_i}{\partial q_j} = \ddot{x}_i \frac{\partial \dot{x}_i}{\partial \dot{q}_j} = \frac{d}{dt} \left(\dot{x}_i \frac{\partial \dot{x}_i}{\partial \dot{q}_j} \right) - \dot{x}_i \frac{d}{dt} \left(\frac{\partial x_i}{\partial q_j} \right)$$

$$= \frac{d}{dt} \left[\frac{\partial}{\partial \dot{q}_j} \left(\frac{1}{2} \dot{x}_i^2 \right) \right] - \frac{\partial}{\partial q_j} \left(\frac{1}{2} \dot{x}_i^2 \right).$$

Thus, it turns out that

$$\frac{d}{dt} \left(\frac{\partial T}{\partial \dot{q}_j} \right) - \frac{\partial T}{\partial q_j} = \sum m_i \left(\ddot{x}_i \frac{\partial x_i}{\partial q_j} + \ddot{y}_i \frac{\partial y_i}{\partial q_j} + \ddot{z} \frac{\partial z_i}{\partial q_j} \right)$$

$$= \sum \left(X_i \frac{\partial x_i}{\partial q_j} + Y_i \frac{\partial y_i}{\partial q_j} + Z_i \frac{\partial z_i}{\partial q_j} \right), \tag{7.18}$$

where X_i, Y_i, Z_i are the components of the total force (internal and external) acting on the ith particle.

In view of the fact that the system is holonomic, a displacement can be made such that q_j becomes $q_j + \delta q_j$, while all other q's remain constant. If $Q_j \delta q_j$ is the work done in this displacement, then

$$Q_j \delta q_j = \sum (X_i \delta x_i + Y_i \delta y_i + Z_i \delta z_i)$$

$$= \sum \left(X_i \frac{\partial x_i}{\partial q_j} + Y_i \frac{\partial y_i}{\partial q_j} + Z_i \frac{\partial z_i}{\partial q_j} \right) \delta q_j.$$

Substituting this result into (7.18), we obtain

$$\frac{d}{dt} \left(\frac{\partial T}{\partial \dot{q}_j} \right) - \frac{\partial T}{\partial q_j} = Q_j. \tag{7.19}$$

These are called *Lagrange's equations of motion* for a holonomic dynamical system. A great advantage of these equations for the solution of dynamical problems is that forces which do no work, such as reactions at smooth hinges, do not appear. However, these are to be included in the usual equations of motion.

An important modification of (7.19) can be made for conservative dynamical systems. In such cases,

$$Q_j = -\frac{\partial V}{\partial q_j}, \tag{7.20}$$

where V is the potential energy of the system and is a function of the generalized coordinates $q_1, q_2, \ldots, q_{3n}$ and possibly of the time t. Since V does not involve the velocities $\dot{q}_j$, $\partial V / \partial \dot{q}_j = 0$. Thus, Lagrange's equation (7.19) can be written in terms of L as

$$\frac{d}{dt}\left(\frac{\partial L}{\partial \dot{q}_j}\right) - \frac{\partial L}{\partial q_j} = 0, \quad j = 1, 2, \ldots, 3n. \tag{7.21}$$

If the kinetic energy of the system T is a homogeneous quadratic function of velocities $\dot{q}_j$, then

$$2T = \sum \dot{q}_j \frac{\partial T}{\partial \dot{q}_j} = \sum \dot{q}_j \frac{\partial L}{\partial \dot{q}_j}. \tag{7.22}$$

Assuming that V does not involve the time t explicitly, it follows from (7.21) that

$$\frac{d}{dt}(T + V) = \frac{d}{dt}(2T - L) = \sum \left[\frac{d}{dt}\left(\dot{q}_j \frac{\partial L}{\partial \dot{q}_j}\right) - \ddot{q}_j \frac{\partial L}{\partial \dot{q}_j} - \dot{q}_j \frac{\partial L}{\partial q_j}\right]$$

$$= \sum \left[\dot{q}_j \frac{d}{dt}\left(\frac{\partial L}{\partial \dot{q}_j}\right) - \dot{q}_j \frac{\partial L}{\partial q_j}\right] = 0. \tag{7.23}$$

This shows that $T + V$ is constant for a conservative system.

To discuss the so-called *Hamilton equations of motion* (Sir William Rowan Hamilton (1805–1865)), we introduce the concepts of *generalized momentum* p_j and *generalized force* F_j by

$$p_j = \frac{\partial L}{\partial \dot{q}_j} \tag{7.24a}$$

and

$$F_j = \frac{\partial L}{\partial q_j}. \tag{7.24b}$$

Consequently, the Lagrange equations (7.21) become

$$\frac{\partial L}{\partial q_j} = \frac{d}{dt}p_j = \dot{p}_j, \tag{7.25}$$

where p_j and q_j are usually called *conjugate variables*.

In Cartesian coordinates, this result reduces to the familiar equation for p_{x_j}:

$$p_{x_j} = \frac{\partial L}{\partial \dot{x}} = \frac{\partial T}{\partial \dot{x}} = m\dot{x}_j. \tag{7.26}$$

Any equations that will hold in any coordinate system are of the most interest and useful for applications. To develop such equations of motion, we now introduce a new function, called the *Hamiltonian* (or *Hamilton's function*) H, which is defined in terms of the Lagrangian L and two conjugate variables p_j and q_j by

$$H(p, q) = \sum_j p_j \dot{q}_j - L, \tag{7.27}$$

where $L = L(q_j, \dot{q}_j, t)$ is, in general, a function of q_j, $\dot{q}_j$, t, and $\dot{q}_j$ enters through the kinetic energy as a quadratic term. Equation (7.24a) will give p_j as a linear function of $\dot{q}_j$. This system of linear equations involving p_j and $\dot{q}_j$ can be solved to determine $\dot{q}_j$ in terms of p_j, and then the $\dot{q}_j$'s can in principle be eliminated from (7.27). This essentially means that H can always be expressed as a function of p_j, q_j, and t so that $H \equiv H(p_j, q_j, t)$. Thus,

$$dH = \sum \frac{\partial H}{\partial p_j} dp_j + \sum \frac{\partial H}{\partial q_j} dq_j + \frac{\partial H}{\partial t} dt. \tag{7.28}$$

On the other hand, differentiation of H in (7.27) with respect to t gives

$$\frac{dH}{dt} = \sum p_j \frac{d}{dt} \dot{q}_j + \sum \dot{q}_j \frac{d}{dt} p_j - \sum \frac{\partial L}{\partial q_j} \frac{d}{dt} q_j - \sum \frac{\partial L}{\partial \dot{q}_j} \frac{d}{dt} \dot{q}_j - \frac{\partial L}{\partial t}, \tag{7.29}$$

or

$$dH = \sum p_j d\dot{q}_j + \sum \dot{q}_j dp_j - \sum \frac{\partial L}{\partial q_j} dq_j - \sum \frac{\partial L}{\partial \dot{q}_j} d\dot{q}_j - \frac{\partial L}{\partial t} dt. \tag{7.30}$$

In view of (7.24a), this equation becomes

$$dH = \sum \dot{q}_j dp_j - \sum \frac{\partial L}{\partial q_j} dq_j - \frac{\partial L}{\partial t} dt. \tag{7.31}$$

Evidently, the two expressions of dH given in (7.28) and (7.31) must be equal so that the coefficients of the corresponding differentials can be equated to obtain

$$\dot{q}_j = \frac{\partial H}{\partial p_j}, \tag{7.32a}$$

$$-\frac{\partial L}{\partial q_j} = \frac{\partial H}{\partial q_j}, \tag{7.32b}$$

$$-\frac{\partial L}{\partial t} = \frac{\partial H}{\partial t}. \tag{7.32c}$$

Using the Lagrange equations (7.21) and (7.24a), the first two of the preceding equations become, for each j,

$$\frac{dq_j}{dt} = \frac{\partial H}{\partial p_j}, \tag{7.33a}$$

$$\frac{dp_j}{dt} = -\frac{\partial H}{\partial q_j}, \quad j = 1, 2, \ldots, 3n. \tag{7.33b}$$

These are known as *Hamilton's canonical equations* of motion. They constitute a set of $6n$ coupled first-order equations that reflects the symmetry except for a negative sign. Thus, the Hamilton equations are completely equivalent to Lagrange's Equations (7.13a) to (7.13c), which represent a set of $3n$ coupled second-order differential equations. These equations possess a unique solution if the initial data are prescribed at some time $t = t_0$. In other words, Hamilton's equations completely determine the position and momentum at all times provided the initial data are given. This shows that the fundamental laws of classical mechanics are completely deterministic.

The Lagrange–Hamilton theory can be employed to deduce (a) the law of conservation of energy and (b) that H is equal to the total energy.

To derive (a), we assume that L, and therefore, H in (7.27) do not involve the time t explicitly. Consequently,

$$\frac{dH}{dt} = \frac{d}{dt}\left[\sum p_j \dot{q}_j - L(q_j, \dot{q}_j)\right]$$

$$= \sum\left(p_j \ddot{q}_j + \dot{p}_j \dot{q}_j - \frac{\partial L}{\partial q_j}\dot{q}_j - \frac{\partial L}{\partial \dot{q}_j}\ddot{q}_j\right)$$

$$= \sum(p_j \ddot{q}_j + \dot{p}_j \dot{q}_j - \dot{p}_j \dot{q}_j - p_j \ddot{q}_j) = 0.$$

This shows that H is constant.

To prove the second property, we assume that the coordinate transformations (7.14a) to (7.14c) do not depend explicitly on time t. We note that T in $L = T - V$ is given by (7.16) where the coefficients a_{jk} are symmetric functions of the generalized coordinates q_j. On the other hand, V is, in general, independent of $\dot{q}_j$ and hence,

$$p_j = \frac{\partial L}{\partial \dot{q}_j} = \frac{\partial T}{\partial \dot{q}_j} = \sum_{k=1}^{3n} a_{jk}\dot{q}_k.$$

Thus, the Hamiltonian H becomes

$$H = \sum_{j=1}^{3n} p_j \dot{q}_j - L = \sum_{j=1}^{3n} \dot{q}_j\left(\sum_{j=1}^{3n} a_{jk}\dot{q}_k\right) - L$$

$$= 2T - L = T + V.$$

Thus, H is equal to the total energy. Since H was proved to be a constant, the sum of the kinetic and potential energies is constant. This is the celebrated *law of conservation of energy.*

We consider a conservative holonomic system that is described by the generalized coordinates $q_1, q_2, \ldots, q_n$. For any complete set of specified initial conditions, each of the coordinates q_j must be a single-valued function of time t. Thus, we assume that these functions are known and have the form $q_1 = q_1(t), q_2 = q_2(t), \ldots, q_n = q_n(t)$. Then these equations may be regarded as the parametric equations of a path in n-dimensional Euclidean space, and the motion of the system can be related to that of a point which moves along this path. Given the initial state, the motion in subsequent times is uniquely determined by Newton's laws of motion. Therefore, there exists a unique path in the n-dimensional Euclidean space for a given set of initial data. It is of interest to compare this path with another one in the n-dimensional space, the two paths having the same end points and as such that they are traversed in the same time τ. Assuming that at any instant of time the difference between the positions of two points which trace out the two paths is infinitesimally small, we denote the variation between the two paths at any instant by δ. Both paths have the same end points so that $\delta q_1(0) = \delta q_1(\tau) = \cdots = \delta q_n(0) = \delta q_n(\tau) = 0$. However, the variations $\delta q_1(t), \delta q_2(t), \ldots, \delta q_n(t)$ do not vanish at any time t between 0 and τ. Then it follows that

$$\delta L = \sum \left[\frac{\partial L}{\partial \dot{q}_j} \delta \dot{q}_j + \frac{\partial L}{\partial q_j} \delta q_j \right]$$

$$= \sum \left[\frac{\partial L}{\partial \dot{q}_j} \frac{d}{dt} (\delta q_j) + \frac{d}{dt} \left(\frac{\partial L}{\partial \dot{q}_j} \right) \delta q_j \right]$$

$$= \frac{d}{dt} \sum \left(\frac{\partial L}{\partial \dot{q}_j} \delta q_j \right).$$

Thus,

$$\delta \int_0^\tau L \, dt = \int_0^\tau \delta L \, dt = \int_0^\tau \frac{d}{dt} \left(\sum \left(\frac{\partial L}{\partial \dot{q}_j} \delta q_j \right) \right) dt$$

$$= \int_0^\tau d \left(\sum \frac{\partial L}{\partial \dot{q}_j} \delta q_j \right)$$

$$= \left[\sum \frac{\partial L}{\partial \dot{q}_j} \delta q_j \right]_{t=\tau} - \left[\sum \frac{\partial L}{\partial \dot{q}_j} \delta q_j \right]_{t=0} = 0,$$

since both the paths have the same end points.

Evidently,

$$\delta \int_0^\tau L \, dt = 0. \tag{7.34}$$

This result is well known as *Hamilton's variational principle* provided the Lagrangian $L = L(q_j, \dot{q}_j)$. This principle was obtained from Newton's law of motion.

Conversely, it can be shown that Newton's laws can be derived from Hamilton's principle if $L = L(q_j, \dot{q}_j)$. Clearly,

$$\delta L = \sum \left[\frac{\partial L}{\partial \dot{q}_j} \delta \dot{q}_j + \frac{\partial L}{\partial q_j} \delta q_j \right].$$

By Hamilton's principle

$$0 = \delta \int_0^\tau L dt = \int_0^\tau \delta L dt = \sum \int_0^\tau \left(\frac{\partial L}{\partial \dot{q}_j} \delta \dot{q}_j + \frac{\partial L}{\partial q_j} \delta q_j \right) dt$$

$$= \sum \int_0^\tau \left[-\frac{d}{dt} \left(\frac{\partial L}{\partial \dot{q}_j} \right) \delta q_j + \frac{\partial L}{\partial q_j} \delta q_j \right] dt + \left[\sum \frac{\partial L}{\partial \dot{q}_j} \delta q_j \right]_0^\tau,$$

where the last result is obtained by integration by parts.

Since $\delta q_j = 0$ at $t = 0$ and $t = \tau$, the last term vanishes and the preceding expression gives

$$\sum \int_0^\tau \left[\frac{\partial L}{\partial q_j} - \frac{d}{dt} \left(\frac{\partial L}{\partial \dot{q}_j} \right) \right] \delta q_j dt = 0$$

for all δq_j and all τ. Thus, the integrand must vanish, which yields Lagrange's equation

$$\frac{d}{dt} \left(\frac{\partial L}{\partial \dot{q}_j} \right) - \frac{\partial L}{\partial q_j} = 0, \tag{7.35}$$

which is the transformed version of Newton's equations. Hence, Newton's laws can be derived from these equations.

Hamilton's principle shows that motion according to Newton's laws is distinguished from all other kinds of motion by having the property that the integral $\int L dt$ for any given time interval has a stationary value. Hence, it is regarded as a fundamental principle of classical mechanics from which everything else can be derived.

Poisson's Brackets in Mechanics

The equations of motion for any canonical function $F(p_i, q_i, t)$ can be expressed, using Hamilton's Equations (7.33a) and (7.33b), as

$$\frac{dF}{dt} = \sum_{i=1}^{3n} \left(\frac{\partial F}{\partial q_i} \dot{q}_i + \frac{\partial F}{\partial p_i} \dot{p}_i \right) + \frac{\partial F}{\partial t}$$

$$= \sum_{i=1}^{3n} \left(\frac{\partial F}{\partial q_i} \frac{\partial H}{\partial p_i} - \frac{\partial F}{\partial p_i} \frac{\partial H}{\partial q_i} \right) + \frac{\partial F}{\partial t}$$

$$= \{F, H\} + \frac{\partial F}{\partial t}, \qquad (7.36)$$

where $\{F, H\}$ is called the *Poisson bracket* of two functions F and H.

If the canonical function F does not explicitly depend on time t, then $\partial F/\partial t = 0$ so that (7.36) becomes

$$\frac{dF}{dt} - \{F, H\}. \qquad (7.37)$$

In addition, if $\{F, H\} = 0$, then F is a constant of the motion. In fact, (7.37) really includes the Hamilton equations, which can be verified by setting $F = p_i$, $F = q_i$, or $F = H$.

It readily follows the definition of the Poisson bracket that

$$\{q_i, p_j\} = \delta_{ij}, \qquad (7.38a)$$

$$\{q_i, q_j\} = \{p_i, p_j\} = 0, \qquad (7.38b)$$

where δ_{ij} is the Kronecker delta notation.

These are the fundamental Poisson brackets for the canonically conjugate variables p_i and q_i.

Any relation involving Poisson's brackets must be invariant under a canonical transformation. This is often used as an alternative definition of a canonical transformation.

It can also be verified that the components of the *angular momentum* $\mathbf{L} = \mathbf{r} \times \mathbf{p}$ satisfy

$$\{L_i, L_j\} = L_k, \quad i, j, k = x, y, z \text{ in cyclic order} \qquad (7.39)$$

and

$$\left\{L_i, L^2\right\} = 0, \qquad (7.40)$$

where $L^2 = L_x^2 + L_y^2 + L_z^2$.

It also follows from the definition of the Poisson bracket that the derivative of a canonical function with respect to generalized coordinates q_j is equal to the Poisson bracket of that function with the canonically conjugate momentum p_j, that is,

$$\frac{\partial F}{\partial q_j} = \{F, p_j\}. \qquad (7.41)$$

In particular, we obtain

$$\frac{\partial F}{\partial x} = \{F, p_x\}, \qquad (7.42a)$$

$$\frac{\partial F}{\partial y} = \{F, p_y\}, \qquad (7.42b)$$

$$\frac{\partial F}{\partial z} = \{F, p_z\}, \tag{7.42c}$$

or equivalently,

$$F(x + dx, y, z) = F(x, y, z) + \{F, p_x\}dx, \tag{7.43a}$$

$$F(x, y + dy, z) = F(x, y, z) + \{F, p_y\}dy, \tag{7.43b}$$

$$F(x, y, z + dz) = F(x, y, z) + \{F, p_z\}dz. \tag{7.43c}$$

Thus, the canonical momenta p_x, p_y, and p_z are called the *generators* of infinitesimal translations along the x, y, and z directions, respectively.

In general, a mechanical description of a physical system requires the concepts of (a) variables or observables, (b) states, (c) equations of motion. Physically, measurable quantities are called *observables*. In classical mechanics, examples of variables or observables are position, momentum, angular momentum, and energy, which are the characteristics of a physical system. They can be measured experimentally and are represented by dynamical variables, which are well-defined functions of two canonically conjugate variables (generalized coordinates and generalized momenta). So the observables in classical mechanics are completely deterministic. There are *states* which describe values of the observables at given times. The state of a physical system at a time $t = t_0 > 0$ is uniquely determined by the appropriate physical law and the initial state at $t = 0$. For example, the state of a system of n interacting particles is determined by assigning $3n$ position coordinates and $3n$ velocity coordinates. Finally, there are *equations of motion* which determine how the values of the observables change in time. As mentioned earlier, Newton's equations, Lagrange's equations, or Hamilton's equations are well-known examples of equations of motion.

7.3 Basic Concepts and Postulates of Quantum Mechanics

Classical physics breaks down at the levels of atoms and molecules. Historically, the first indication of a breakdown of classical ideas occurred in the rather complex phenomenon of the so-called black body radiation, which essentially deals with electromagnetic radiation in a container in equilibrium with its surroundings. In other words, the black body radiation is concerned with the thermodynamics of the exchange of energy between radiation and matter. According to principles of classical physics, this exchange of energy is assumed to be continuous in the sense that light of frequency v can give up any amount of energy on absorption, the exact amount in any particular case depending on the energy intensity of the light beam. Specifically, in 1900 Max

Planck (Max Karl Ernst Ludwig Planck (1858–1947)) first postulated that the vibrating particles of matter are regarded to act as harmonic oscillators and do not emit or absorb light continuously, but instead only in discrete quantities.

Mathematically, the radiation of frequency v can only exchange energy with matter in units of hv, where h is the *Planck constant* of numerical value

$$h = 2\pi\hbar = 6.625 \times 10^{-27} \text{ erg sec} = 4.14 \times 10^{-21} \text{ MeV sec} \qquad (7.44)$$

and $\hbar$ is called the *universal constant*. Clearly, h has dimension (energy × time) of *action*, which is a dynamical quantity in classical mechanics.

Equivalently, Planck's quantum postulate can be stated by saying that radiation of frequency v behaves like a stream of photons of energy

$$E = hv = \hbar\omega, \qquad (7.45)$$

which may be emitted or absorbed by matter where $\omega = 2\pi v$ is the angular frequency. Clearly, Planck's constant h measures the degree of discreteness that was required to explain the energy distribution of the black body radiation. Thus, the concept of discreteness is fundamental in quantum mechanics, but it is totally unacceptable in classical physics. Finally, it is important to point out that the Planck equation (7.45) is fairly general so that it can be applied to any quantum system as a fundamental relation between its energy E and the frequency of v of an oscillation associated with the system.

Also, the failure of classical concepts when applied to the motion of electrons appeared most clearly in connection with the hydrogen atom. According to the Rutherford (Ernest Rutherford (1871–1937)) classical model, an atom can be considered as a negatively charged electron orbiting around a relatively massive, positively charged nucleus. With the neglect of radiation, this system is exactly similar to the motion of a planet round the sun, with gravitational attraction between the masses being replaced by the Coulomb attraction between the charges. The potential energy of the Coulomb attraction between the fixed nucleus charge $+Ze$ and the electron of charge $-e$ is $V(r) = -Ze^2/r$. The hydrogen atom consists of two particles: the nucleus, a proton of mass m_p and charge $+e$ ($Z = 1$), and an electron of mass m_e and charge $-e$. The nucleus is small and heavy ($m_p/m_e \sim 2000$) and the radius of the proton $\sim 10^{-3}$ times the atomic radius. According to the classical atomic theory of Rutherford, the attractive potential would cause the electron to orbit around the nucleus, and the orbiting electron constitutes a rapidly accelerating charge, which according to Maxwell's theory acts as a source of radiant energy (James Clerk Maxwell (1831–1879)). Thus, the accelerated charged electron would continuously radiate energy, and in a matter of 10^{-10} seconds the electron should coalesce with the nucleus, causing the atom to collapse.

On the other hand, the frequency of the emitted radiation is related to that of the electron in its orbit. As the electron radiates energy, this frequency, according to classical theory, must change rapidly but continuously, thus giving rise to radiation with a continuous range of frequencies. Thus, the Rutherford classical atomic model has two important qualitative weaknesses:

(a) The atom should be very unstable.

(b) It should radiate energy over a continuous range of frequencies.

Both of these results are totally contradicted by experiments.

The original problem of quantum mechanics was to investigate the stability of atoms and molecules, as well as to explain the discrete frequency spectra of the emitted radiation by excited atoms. The remarkable success in predicting observed atomic and molecular spectra is one of the major triumphs of quantum mechanics.

In this chapter, we present the basic principles of quantum mechanics as postulates that will then be used to discuss various consequences. No attempt will be made to derive or justify these postulates. Both the number and content of the basic postulates are to some extent a matter of individual choice. The postulates together with their consequences form a basic but limited theory of quantum mechanics.

It has been mentioned in previous sections that classical mechanics identifies the *state* of a physical system with the values of certain *observables* (for example, the position x and the momentum p) of the system. On the other hand, quantum mechanics makes a very clear distinction between states and observables. So we begin with the first postulate concerning the *state* of a quantum system.

Postulate I. (The state vector) *Every possible state of a given system in quantum mechanics corresponds to a separable Hilbert space over the complex number field. A state of the system is represented by a nonzero vector in the space, and every nonzero scalar multiple of a state vector represents the same state. Conversely, every nonzero vector in the Hilbert space and its nonzero scalar multiples represent the same physical state of the system. The particular state vector to which the state of the system corresponds at time t is denoted by $\Psi(x, t)$ and is called the time-dependent state vector of the system. The state of a physical system is completely described by this state vector $\Psi(x, t)$ in the sense that almost all information about the system at time t can be obtained from the vector $\Psi(x, t)$.*

Usually, a state vector is denoted by $\psi(x)$. In the Dirac notation, any general state vector $\psi(x)$ is written as

$$\psi(x) = \langle x | \psi \rangle \tag{7.46}$$

and its complex conjugate as

$$\overline{\psi(x)} = \langle \psi | x \rangle. \tag{7.47}$$

This postulate makes several assertions. First, all physical properties of a given system are unchanged if it is multiplied by a nonzero scalar. We can remove this arbitrariness by imposing the normalizing condition

$$\int \overline{\psi(x)}\psi(x)dx = \int |\psi(x)|^2 dx = 1, \tag{7.48}$$

where the integral is taken over all admissible values of x. Or, equivalently,

$$\int \langle \psi|x\rangle\langle x|\psi\rangle dx = \int |\langle \psi|x\rangle|^2 dx = 1, \tag{7.49}$$

or, simply as an abbreviated form,

$$\langle \psi|\psi\rangle = 1. \tag{7.50}$$

Clearly, the norm of the state vector is unity. In other words, if c is any scalar of modulus unity satisfying $|c|^2 = 1$, then the two state vectors $\psi(x)$ and $\varphi(x) = c\psi(x)$ are the same in the sense that they correspond to the same physical state.

Clearly, for any two state vectors $|\varphi\rangle$ and $|\psi\rangle$, the complex number $\langle \varphi|\psi\rangle$ represents the inner product in the space and is given by

$$\langle \varphi|\psi\rangle = \int \overline{\varphi(x)}\psi(x)dx = \int \langle \varphi|x\rangle\langle x|\psi\rangle dx. \tag{7.51}$$

This result combined with (7.46) and (7.47) immediately gives

$$\langle \varphi|\psi\rangle = \overline{\langle \psi|\varphi\rangle}. \tag{7.52}$$

Two state vectors $|\varphi\rangle$ and $|\psi\rangle$ are called *orthogonal* if

$$\langle \varphi|\psi\rangle = 0. \tag{7.53}$$

A set of state vectors $|\psi_1\rangle, |\psi_2\rangle, \ldots$ is orthonormal if $\langle \psi_i|\psi_j\rangle = \delta_{ij}$. An arbitrary vector $|\psi\rangle$ in the state space can be expressed as a sum of a complete set of orthonormal vectors $|\psi_i\rangle$ in the form

$$|\psi\rangle = \sum_i c_i|\psi_i\rangle, \tag{7.54}$$

where the coefficients c_i are given by

$$c_i = \langle \psi_i|\psi\rangle. \tag{7.55}$$

Obviously, this postulate implies the principle of superposition. The number of elements in an orthonormal set of state vectors is called the *dimension* of the state space. If the dimension is finite, then (7.54) is a finite series. On the other hand, if the dimension is infinite, (7.54) is an infinite series and hence the question of convergence will arise.

Second, Postulate I asserts that all that can possibly be known about the state of the quantum system at time t can be obtained from its state vector $\Psi(x, t)$. On the contrary, in classical mechanics, the state of a system at time t is completely described if and only if the state variables $q(t)$ and $p(t)$ are given as definite real numbers. However, the x in $\Psi(x, t)$ does not depend in any way on the time variable t. The third assertion of Postulate I is that the functional dependence of Ψ on t is essentially different from its dependence on x. Indeed, as a Hilbert space vector, $\Psi(x, t)$ is a function of x alone, and t serves as a parameter. More precisely, $\Psi(x, t_1)$ and $\Psi(x, t_2)$ are to be considered as two different vectors in the space, that is, two different functions of x, which specify the state of the system at two different times t_1 and t_2. The state vector $\Psi(x, t)$ is often referred to as the *state function* or the *wave function* of the quantum system. Naturally, the time evolution of the state function is another important matter, which will be considered later on.

In classical mechanics, *observables* are simply dynamical variables such as position, momentum, and functions of position and momentum. Examples of observables also include angular momentum and energy, which are well-defined functions of two conjugate variables: generalized momentum and generalized coordinates. The *measurement* (or *observation*) of a classical observable is a well-defined physical operation that, when performed on the system, yields a single real number, the so-called *value* of the observable. In the study of classical systems, it is tacitly assumed that the operations of measurement do not significantly disturb their motion. More precisely, any disturbances caused by measurement are negligible. So, in classical mechanics, there is no real difference between the *mathematical representation* of an observable and the *values* of the observables. These lead us to formulate the second postulate of quantum mechanics.

Postulate II. (Observable operators and their values)

(a) *To every physical observable in quantum mechanics, there corresponds in the Hilbert space a linear Hermitian operator $\hat{A}$, which has a complete set of orthonormal eigenvectors $\{\psi_n\}$ with the corresponding eigenvalues $\{\lambda_n\}$ such that*

$$\hat{A}\psi_n = \lambda_n \psi_n, \quad n = 1, 2, 3, \dots . \tag{7.56}$$

Conversely, to each such operator in the Hilbert space there corresponds some physical observable.

(b) *The only possible values of a physical observable are the various eigenvalues.*

According to this postulate, a quantum observable is mathematically represented by a linear Hermitian operator on an infinite dimensional separable

Hilbert space and there is one such operator for each quantum observable such as the position, the momentum, the energy, and so on. Such operators in quantum mechanics are called *observable operators*, which usually satisfy the rules of the theory of operators in the Hilbert space. So the quantum observables are operators in a Hilbert space. However, an observable in classical mechanics is simply a real function of conjugate variables in Euclidean space.

This postulate has several remarkable mathematical consequences. The first consequence is that the Hermitian property of an observable operator $\hat{A}$ ensures that its eigenvalues $\{\lambda_n\}$ are real. The real eigenvalues correspond to possible measurable random quantities of the observable operator $\hat{A}$. The second consequence of Postulate II is that the nature of the representation of the eigenvalues $\{\lambda_n\}$ seems to suggest that they form a discrete set, which leads to the possibility for allowing certain observable operators to be "quantized." The discreteness is one of the inherent features of quantum phenomena. However, the eigenvalues of an observable operator can be discrete or continuous or a combination of the two depending entirely on the particular observable operator.

The most profound difference between these rules of two operators and the rules for manipulating classical observables (or dynamical variables) is that two classical observables A, B always commute ($AB = BA$), but two observable operators $\hat{A}$ and $\hat{B}$ do not commute in general. The *commutator* of these operators is defined by

$$\left[\hat{A}, \hat{B}\right] = \hat{A}\hat{B} - \hat{B}\hat{A}. \tag{7.57}$$

This is essentially the difference between operating first with $\hat{B}$ and then $\hat{A}$, and first with $\hat{A}$ and then $\hat{B}$. So, in general, the commutator of two observable operators is nonzero. Physically, the nonzero value must be associated with the magnitude of the unavoidable disturbances between the two measurements of $\hat{A}$ and $\hat{B}$ in two different orders. It is appropriate to mention here what we mean by a measurement from a simple physical point of view. A *measurement* is simply a well-defined physical operation which, when performed on a physical system, yields a single real number without any error or ambiguity in the sense that there is no experimental uncertainty associated with the number obtained. If $\hat{A}$ and $\hat{B}$ represent the measurements of two particular quantum observables (for example, $\hat{A}$ might mean measurement of position and $\hat{B}$ that of momentum), then $\hat{A}\hat{B} - \hat{B}\hat{A} \neq 0$. This means that measurement in quantum theory is fundamentally different from that in classical mechanics. A measurement of quantum observables does affect the system, and so it may be expected to appear explicitly in the theory. In other words, it may not be possible to measure two different observable operators of a system simultaneously, and measurement of one may change the value of the other. Physically, observation (or measurement) of something always involves shining light (photons) on it, which means striking it with photons. If the position is to be determined accurately, the wavelengths of the corresponding waves must be very small, their frequency correspondingly very high, and the momentum of the photons consequently above a certain limit. A blow with such a stream of photons may significantly disturb the observed system if it is small enough.

In particular, the position observable, $\hat{x}$ and the corresponding momentum observable, $\hat{p} = -i\hbar\partial/\partial x$ satisfy the celebrated Heisenberg commutation relation

$$[\hat{x}, \hat{p}] = i\hbar. \qquad (7.58)$$

This follows immediately from the fact that

$$[\hat{x}, \hat{p}]\psi = (\hat{x}\hat{p} - \hat{p}\hat{x})\psi = -i\hbar\hat{x}\frac{\partial\psi}{\partial x} + i\hbar\hat{x}\frac{\partial\psi}{\partial x} + i\hbar\psi = i\hbar\psi.$$

Clearly, the concept of the commutator of two observable operators is of fundamental physical importance and plays a crucial role in the mathematical development of quantum mechanics.

Two observables are called *complementary* if the corresponding operators associated with them do not commute. In one-dimensional problems, the position and momentum are examples of complementary observables. So the simultaneous eigenstates of $\hat{x}$ and $\hat{p}$ do not exist. On the other hand, two observables are said to be *compatible* if measurements of one do not affect the value of the other. So the corresponding operators associated with them must commute. Any two classical observables always commute. In quantum mechanics, energy and momentum are examples of compatible observables. More precisely, if

$$\hat{A} = \hat{p} = -i\hbar\frac{\partial}{\partial x} \quad \text{and} \quad \hat{B} = \hat{T} = \frac{\hat{p}^2}{2m} = -\frac{\hbar^2}{2m}\frac{\partial^2}{\partial x^2},$$

then

$$[\hat{p}, \hat{T}]\psi = \left[-i\hbar\frac{\partial}{\partial x}, -\frac{\hbar^2}{2m}\frac{\partial^2}{\partial x^2}\right]\psi = \frac{i\hbar^3}{2m}\left(\frac{\partial}{\partial x}\frac{\partial^2}{\partial x^2} - \frac{\partial^2}{\partial x^2}\frac{\partial}{\partial x}\right)\psi = 0$$

for any state vector $\psi(x)$. This means that energy and momentum of a free particle can be determined precisely and simultaneously. However, this is not true if the particle moves in a potential field $V(x)$. In this case,

$$\hat{p} = -i\hbar\frac{\partial}{\partial x} \quad \text{and} \quad \hat{H} = \frac{\hat{p}^2}{2m} + \hat{V}(\hat{x}) = \hat{T} + \hat{V}(\hat{x})$$

so that

$$[\hat{p}, \hat{H}]\psi = [\hat{p}, \hat{T} + \hat{V}]\psi = [\hat{p}, \hat{T}]\psi + [\hat{p}, \hat{V}]\psi = -i\hbar\psi\frac{\partial\hat{V}}{\partial x} \neq 0.$$

This implies that the momentum and energy of a particle moving in a potential field cannot be determined precisely and simultaneously.

The third consequence of Postulate II is that the eigenvectors belonging to different eigenvalues of an observable operator form a complete orthonormal set, that is, for all n and m,

$$\langle \psi_n | \psi_m \rangle = \int_{-\infty}^{\infty} \overline{\psi_n}(x) \psi_m(x) dx = \delta_{mm} \tag{7.59}$$

and any state vector $\psi(x)$ can be expanded in terms of the eigenvectors $\{\psi_n\}$ as

$$\psi(x) = \sum_{n=1}^{\infty} \langle \psi_n | \psi \rangle \psi_n(x). \tag{7.60}$$

In other words, the set of such eigenvectors $\{\psi_n(x)\}$ forms an orthonormal basis in the associated Hilbert space. This set $\{\psi_n\}$ is often called the *eigenbasis* of the observable operator $\hat{A}$.

Finally, in classical mechanics, both the state and the observable depend on time t. However, no such characterization holds true for the state and the observables in quantum theory. It follows from Postulates I and II that the state (or wave) function $\Psi(x, t)$ can change with time t, but the observable operator $\hat{A}$ along with its eigenbasis $\{\psi_n(x)\}$ and its real eigenvalues $\{\lambda_n\}$ are all independent of time t. If we express the wave function $\Psi(x, t)$ in terms of the eigenvectors $\{\psi_n(x)\}$ as in (7.60), that is,

$$\Psi(x, t) = \sum_{n=1}^{\infty} \langle \psi_n | \Psi \rangle \psi_n(x), \tag{7.61}$$

then the coefficients of this expansion of $\Psi(x, t)$ relative to the eigenbasis $\psi_n(x)$ of $\hat{A}$ are time-dependent scalars of the form

$$\langle \psi_n | \Psi \rangle = \int_{-\infty}^{\infty} \overline{\psi_n}(x) \Psi(x, t) dx, \quad n = 1, 2, 3, \ldots. \tag{7.62}$$

Such an expansion of the wave function $\Psi(x, t)$ in terms of the eigenbasis of the observable operator $\hat{A}$ plays a fundamental role in the development of quantum mechanics. However, the fact that $\hat{A}$, $\psi_n(x)$, and λ_n are all independent of time does not necessarily mean that the measured values of the observable operator $\hat{A}$ will be constant in time. At any rate, it remains to be specified how the result of a particular measurement depends on the state of the system at the time of the measurement.

It is then necessary to establish some sort of logical connection between physical observables in classical mechanics and in quantum mechanics. This

Observables	Classical quantities	Quantum operators
Position	x or $\mathbf{r}$	$\hat{x}$ or $\hat{r}$
Linear momentum	$p_x = m\dot{x}$	$\hat{p}_x = -i\hbar\frac{\partial}{\partial x}$
Linear momentum	$\mathbf{p} = (p_x, p_y, p_z)$	$\hat{\mathbf{p}} = -i\hbar\nabla = -i\hbar\left(\frac{\partial}{\partial x}, \frac{\partial}{\partial y}, \frac{\partial}{\partial z}\right)$
Potential energy	$V(\mathbf{r}, t)$	$\hat{V}(\hat{r}, t)$
Kinetic energy	$T = \frac{p^2}{2m}$	$\hat{T} = -\frac{\hbar^2}{2m}\nabla^2$ $= -\frac{\hbar^2}{2m}\left(\frac{\partial^2}{\partial x^2} + \frac{\partial^2}{\partial y^2} + \frac{\partial^2}{\partial z^2}\right)$
Hamiltonian	$H(x, p) = \frac{p^2}{2m} + V(x)$	$\hat{H}(\hat{x}, \hat{p}) = -\frac{\hbar^2}{2m}\nabla^2 + \hat{V}(\hat{x})$
Angular momentum	$L_x = yp_z - xp_y$	$\hat{L}_x = \frac{\hbar}{i}\left(y\frac{\partial}{\partial z} - z\frac{\partial}{\partial y}\right)$
$\mathbf{L} = \mathbf{r} \times \mathbf{p}$	$L_y = zp_x - xp_z$	$\hat{L}_y = \frac{\hbar}{i}\left(z\frac{\partial}{\partial x} - x\frac{\partial}{\partial z}\right)$
	$L_z = xp_y - yp_x$	$\hat{L}_z = \frac{\hbar}{i}\left(x\frac{\partial}{\partial y} - y\frac{\partial}{\partial x}\right)$

Table 7.1 Physical quantities in classical mechanics and the corresponding quantum mechanical operators

connection can be made through the celebrated *Correspondence Principle*, which is the topic of the next postulate.

Postulate III. (Correspondence Principle) *A quantum observable operator corresponding to a dynamical variable is obtained by replacing the canonical variable in classical mechanics by the corresponding quantum mechanical operator.*

This postulate gives a method to formulate the quantum mechanical operators from the classical dynamical variables which are made of the canonical conjugate variables p and q. Like so many other things in quantum mechanics, the Correspondence Principle, first proposed by Niels Bohr (Niels Henrik David Bohr (1885–1962)), is a very deep subject and has many remarkable ramifications.

Thus, we have a list of correspondences between physical quantities in classical mechanics and the quantum mechanical operators, as shown in Table 7.1.

In general, any observable in classical mechanics is some well-behaved function of position and momentum, $f(x, p)$ and is represented in quantum mechanics by the operator $f(\hat{x}, \hat{p})$. In one dimension, $\hat{x} = x$ and $\hat{p} = -i\hbar(d/dx)$ so that

$$f(\hat{x}, \hat{p}) = f\left(\hat{x}, -i\hbar\frac{d}{dx}\right).$$

The position, momentum, angular momentum, and energy are not the only observables in quantum mechanics; there may be several other observables, some of them having no analogues in classical mechanics. For example, quantum observables such as spin and isospin, which characterize some other attributes of the quantum system, have no classical analogue.

However, this postulate does not say anything about how to specify the quantum observables corresponding to the basic canonical variables. An answer to this point is provided by the following postulate.

Postulate IV. (Quantization)   *Every pair of canonically conjugate observable operators satisfies the following Heisenberg (Werner Karl Heisenberg (1901–1976)) commutation relations:*

$$[\hat{q}_m, \hat{q}_n] = 0 = [\hat{p}_m, \hat{p}_n], \tag{7.63}$$

$$[\hat{q}_m, \hat{p}_n] = i\hbar\hat{\delta}_{mn}, \tag{7.64}$$

where $\hat{q}_m$ is the observable operator corresponding to the generalized coordinates, and $\hat{p}_m$ is the momentum operator corresponding to the generalized momentum.

This postulate is often called the *principle of quantization*. Clearly, $i[\hat{p}_n, \hat{q}_n] = -\hbar$.

In Cartesian coordinates, canonical operators $\hat{x}_m$ and $\hat{p}_m$ satisfy

$$[\hat{x}_m, \hat{x}_n] = 0 = [\hat{p}_m, \hat{p}_n], \qquad [\hat{x}_m, \hat{p}_n] = i\hbar\delta_{mn},$$

where the subscripts m, n take the values 1, 2, 3 for the x, y, z components of $\hat{x}_m$ and $\hat{p}_m$.

Obviously, in one dimension,

$$[\hat{p}, \hat{x}]\psi = (\hat{p}\hat{x} - \hat{x}\hat{p})\psi = -i\hbar\frac{\partial}{\partial x}(x\psi) + i\hbar x\frac{\partial\psi}{\partial x} = -i\hbar\psi$$

so that

$$[\hat{x}, \hat{p}] = \hat{x}\hat{p} - \hat{p}\hat{x} = i\hbar. \tag{7.65}$$

Any observable operator that has a classical analogue can always be expressed as a function of q_m and p_m, and the quantization rules of such an observable will follow from (7.63) and (7.64). For example, the angular momentum operator $\hat{L} = (\hat{L}_x, \hat{L}_y, \hat{L}_z)$ satisfies the following properties:

$$\left[\hat{L}_x, \hat{L}_y\right] = i\hbar\hat{L}_z, \quad \left[\hat{L}_y, \hat{L}_z\right] = i\hbar\hat{L}_x, \quad \text{and} \quad \left[\hat{L}_z, \hat{L}_x\right] = i\hbar\hat{L}_y. \tag{7.66}$$

It is now necessary to establish some sort of mathematical relationship between the state vector and the observable operator of a quantum system. This mathematical relationship is formulated through the concept of measurement. The quantum theory of measurement thus forms the keystone of the mathematical structure of quantum mechanics. According to Postulate II, the only real values that can ever be predicted for an observable are the eigenvalues of the corresponding operator of a given physical system. As to which one of these eigenvalues will be obtained in any given situation, the answer to this question is expected to be determined by the particular form of a state vector of the physical system at the time of measurement. But the operations of measurement generally affect the system. Therefore, it is almost impossible to predict a single measurement with absolute certainty. Naturally, this difficulty can somehow be resolved by dealing with a large number of measurements under identical conditions and then introducing the average of all results so obtained. This average, called the *expectation value* of an observable operator, is defined as follows:

Definition 7.3.1. (Expectation value)
The expectation value $\langle \hat{A} \rangle$ of an observable operator $\hat{A}$ in the state $\psi(x)$ of a physical system is defined by

$$\langle \hat{A} \rangle = \frac{\langle \psi | \hat{A} \psi \rangle}{\langle \psi | \psi \rangle}. \tag{7.67}$$

If the state ψ is normalized, then the expectation value is

$$\langle \hat{A} \rangle = \langle \psi | \hat{A} \psi \rangle. \tag{7.68}$$

In terms of the expectation value of $\hat{A}$, we define the root-mean-square deviation, $(\nabla \hat{A})$, which measures the dispersion around the mean value $\langle \hat{A} \rangle$.

Definition 7.3.2. (Root-mean-square deviation)
The root-mean-square deviation $(\nabla \hat{A})$ is defined by the square root of the expectation value of $(\hat{A} - \langle \hat{A} \rangle)^2$ in the state ψ in which $\langle \hat{A} \rangle$ is computed.

It follows from the definition that

$$\left(\nabla \hat{A}\right)^2 = \left\langle \left(\hat{A} - \langle \hat{A} \rangle\right)^2 \right\rangle = \left\langle \psi \middle| \left(\hat{A} - \langle \hat{A} \rangle\right)^2 \psi \right\rangle$$
$$= \left\langle \left(\hat{A} - \langle \hat{A} \rangle\right)\psi \middle| \left(\hat{A} - \langle \hat{A} \rangle\right)\psi \right\rangle = \left\| \left(\hat{A} - \langle \hat{A} \rangle\right)\psi \right\|^2 \tag{7.69}$$

because $\hat{A}$ is Hermitian and $\langle \hat{A} \rangle$ is real.

Theorem 7.3.3.

$$\text{(a)} \quad \left(\nabla \hat{A}\right)^2 = \langle \hat{A}^2 \rangle - \langle \hat{A} \rangle^2, \tag{7.70}$$

$$\text{(b)} \quad \langle \hat{A}^2 \rangle = \left\| \hat{A}\psi \right\|^2. \tag{7.71}$$

Proof: (a)

$$\left(\nabla\hat{A}\right)^2 = \langle(\hat{A} - \langle\hat{A}\rangle)^2\rangle = \langle\psi|(\hat{A} - \langle\hat{A}\rangle)^2\psi\rangle$$
$$= \langle\psi|(\hat{A}^2 - 2\hat{A}\langle\hat{A}\rangle + \langle\hat{A}\rangle^2)\psi\rangle$$
$$= \langle\psi|\hat{A}^2\psi\rangle - 2\langle\psi|\hat{A}\psi\rangle\langle\hat{A}\rangle + \langle\hat{A}\rangle^2\langle\psi|\psi\rangle$$
$$= \langle\hat{A}^2\rangle - 2\langle\hat{A}\rangle^2 + \langle\hat{A}\rangle^2 \quad \text{(since } \langle\psi|\psi\rangle = 1\text{)}$$
$$= \langle\hat{A}^2\rangle - \langle\hat{A}\rangle^2.$$

(b)

$$\langle\hat{A}^2\rangle = \langle\psi|\hat{A}^2\psi\rangle = \langle\hat{A}\psi|\hat{A}\psi\rangle = \|\hat{A}\psi\|^2. \qquad \square$$

Theorem 7.3.4. *A necessary and sufficient condition for a physical system to be in an eigenstate of an observable $\hat{A}$ is $\nabla\hat{A} = 0$.*

Proof: The condition is necessary. Suppose the system is an eigenstate ψ of $\hat{A}$ with a real eigenvalue λ so that $\hat{A}\psi = \lambda\psi$. Clearly,

$$\langle\hat{A}\rangle = \langle\psi|\hat{A}\psi\rangle = \lambda,$$
$$\langle\hat{A}^2\rangle = \|\hat{A}\psi\|^2 = \lambda^2.$$

Hence, by Theorem 7.3.3(a), $(\nabla\hat{A}) = 0$.

The condition $\nabla\hat{A} = 0$ is sufficient. Obviously,

$$0 = \left(\nabla\hat{A}\right)^2 = \|(\hat{A} - \langle\hat{A}\rangle)\psi\|^2$$

implies that

$$(\hat{A} - \langle\hat{A}\rangle)\psi = 0$$

or

$$\hat{A}\psi = \langle\hat{A}\rangle\psi.$$

Thus, ψ is an eigenvector of $\hat{A}$ with an eigenvalue $\langle\hat{A}\rangle$. $\qquad \square$

If $\psi_n(x)$ is an eigenvector of $\hat{A}$ with the corresponding eigenvalue λ_n so that $\hat{A}\psi_n = \lambda_n\psi_n$ is satisfied for all n, then the expectation value in the orthonormal eigenstate $\psi_n(x)$ is given by

$$\langle\hat{A}\rangle = \frac{\langle\psi_n|\hat{A}\psi_n\rangle}{\langle\psi_n|\psi_n\rangle} = \frac{\langle\psi_n|\lambda_n\psi_n\rangle}{\langle\psi_n|\psi_n\rangle} = \lambda_n. \tag{7.72}$$

On the other hand, if $\psi(x)$ is not an eigenvector of $\hat{A}$, then it can be expanded in terms of a complete orthonormal set of eigenvectors $\{\psi_n(x)\}$. Consequently,

$$\psi(x) = \sum_{n=1}^{\infty} c_n \psi_n(x) = \sum_{n=1}^{\infty} \langle \psi_n | \psi \rangle \psi_n(x) \tag{7.73}$$

so that

$$\|\psi\|^2 = \sum_{n=1}^{\infty} |c_n|^2 = \sum_{n=1}^{\infty} |\langle \psi_n | \psi \rangle|^2. \tag{7.74}$$

Then the expectation value given by (7.67) assumes the form

$$\langle \hat{A} \rangle = \frac{1}{\|\psi\|^2} \left\langle \sum_{m=1}^{\infty} c_m \psi_m \,\middle|\, A \sum_{n=1}^{\infty} c_n \psi_n \right\rangle$$

$$= \frac{1}{\|\psi\|^2} \left\langle \sum_{m=1}^{\infty} c_m \psi_m \,\middle|\, \sum_{n=1}^{\infty} c_n \lambda_n \psi_n \right\rangle$$

$$= \frac{1}{\|\psi\|^2} \sum_{m,n} c_m \bar{c}_n \lambda_n \langle \psi_m | \psi_n \rangle$$

$$= \frac{1}{\|\psi\|^2} \sum_{m,n} c_m \bar{c}_n \lambda_n \delta_{mn}$$

$$= \frac{\sum_n |c_n|^2 \lambda_n}{\|\psi\|^2} = \frac{\sum_{n=1}^{\infty} |c_n|^2 \lambda_n}{\sum_{n=1}^{\infty} |c_n|^2} \tag{7.75}$$

$$= \sum_{n=1}^{\infty} \left(\frac{|c_n|^2}{\sum_{n=1}^{\infty} |c_n|^2} \right) \lambda_n = \sum_{n=1}^{\infty} a_n \lambda_n, \tag{7.76}$$

where

$$a_n = \frac{|c_n|^2}{\sum_{n=1}^{\infty} |c_n|^2} = \frac{|\langle \psi_n | \psi \rangle|^2}{\sum_{n=1}^{\infty} |\langle \psi_n | \psi \rangle|^2}. \tag{7.77}$$

The result (7.76) can be interpreted as the outcome of a large number of measurements under the identical conditions. In other words, $\langle \hat{A} \rangle$ is the *weighted average* of a large number of measurements of the system. According to Postulate II, each measurement yields one of the eigenvalues of $\hat{a}$ belonging to the set $\{\lambda_n\}$. The numerator of (7.77) represents the frequency with which the eigenvalues $\{\lambda_n\}$ occur in the measurement, while the denominator of (7.77) is the total number of measurements. Hence, the ratio of the frequency, $|\langle \psi_n | \psi \rangle|^2$, and the total number of measurements, $\sum_{n=1}^{\infty} |\langle \psi_n | \psi \rangle|^2$, represents the weight a_n of the eigenvalues in the measurement. This clearly leads to a statistical interpretation of a_n as the probability that a measurement will yield the eigenvalue λ_n for the operator $\hat{A}$ in the state $\psi(x)$ of the physical system. Furthermore, it follows from (7.77) that the total probability $\sum_{n=1}^{\infty} a_n = 1$ reconfirms the statistical interpretation, since a_n is the probability that a single measurement yields an eigenvalue of $\hat{A}$.

For a normalized state vector ψ,

$$\|\psi\|^2 = \sum_{n=1}^{\infty} |c_n|^2 = \left| \langle \psi_n | \psi \rangle \right|^2 = 1, \tag{7.78}$$

$$a_n = |c_n|^2 = \left| \langle \psi_n | \psi \rangle \right|^2. \tag{7.79}$$

Consequently, (7.76) reduces to

$$\langle \hat{A} \rangle = \sum_{n=1}^{\infty} \left| \langle \psi_n | \psi \rangle \right|^2 \lambda_n. \tag{7.80}$$

Thus, the quantity $|c_n|^2 = |\langle \psi_n | \psi \rangle|^2$ can be recognized as the *probability* that the measurement will yield the eigenvalue λ_n of $\hat{A}$ in the normalized state $\psi(x)$. Clearly, $0 \leq |\langle \psi_n | \psi \rangle|^2 \leq 1$.

All these results lead us to Postulate V, which is essentially concerned with the outcome of a measurement of a quantum observable of a physical system in a given state.

Postulate V. (Outcome of quantum measurement) *If an observable operator $\hat{A}$ has eigenbasis $\{\psi_n\}$ with the corresponding eigenvalues $\{\lambda_n\}$, then the probability that the measurement will yield the eigenvalue λ_n of $\hat{A}$ of the system in the normalized state $\psi(x)$ is*

$$P(\lambda_n) = \left| \langle \psi_n | \psi \rangle \right|^2. \tag{7.81}$$

This postulate makes several radical assertions. First, it is impossible to predict with absolute certainty the outcome of a measurement that is made on a quantum system in a completely defined state. This is because of the fact that any measurement entails an interaction between the measuring equipment and the measured system. Thus, the prediction of a quantum measurement is of statistical nature. This is perhaps the most radical feature of the quantum theory of measurement and completely contrary to the spirit of classical mechanics, which is totally based on the deterministic principle. Second, when a quantum system is in a state $\psi(x)$, the value obtained in measurement of an observable operator $\hat{A}$ is a random variable with a probability distribution whose mean value is associated with a large number of measurements in this particular state. This leads to a statistical interpretation of quantum measurements. Third, if two repeated measurements of some observable are performed on a system in a given state, the results of these two measurements will not necessarily be the same, since the operations of measurements affect the system in general. In classical mechanics, two or more repeated measurements of a classical observable will always produce identical results. So this postulate makes the most remarkable difference between quantum and classical mechanics because of the special role played by the process of measurement in quantum mechanics.

In the quantum theory of measurement, it is important to determine the degree to which a random measurement deviates from the expectation value. The *standard deviation*, which is a measure of the dispersion around the expectation value, is usually called the *uncertainty* $\Delta \hat{A}$ (or *root-mean-square deviation*) introduced in Definition 7.3.2.

7.4 The Heisenberg Uncertainty Principle

If $\hat{A}$ and $\hat{B}$ are Hermitian observables, the uncertainties $\Delta \hat{A}$ and $\Delta \hat{B}$ in the measurement of $\hat{A}$ and $\hat{B}$ in the given state $\psi(x)$ are given by

$$\Delta \hat{A} = \left(\langle (\hat{A} - \langle \hat{A} \rangle)^2 \rangle \right)^{1/2}$$
$$= \langle \psi | (\hat{A} - \langle \hat{A} \rangle)^2 \psi \rangle^{1/2}$$
$$= \langle (\hat{A} - \langle \hat{A} \rangle) \psi | (\hat{A} - \langle \hat{A} \rangle) \psi \rangle^{1/2}$$
$$= \left\| (\hat{A} - \langle \hat{A} \rangle) \psi \right\| = \| \psi_1 \|, \tag{7.82}$$

where

$$\psi_1 = (\hat{A} - \langle \hat{A} \rangle) \psi. \tag{7.83}$$

Similarly, we write

$$\Delta \hat{B} = \| \psi_2 \|, \tag{7.84a}$$

where

$$\psi_2 = (\hat{B} - \langle \hat{B} \rangle) \psi. \tag{7.84b}$$

Since the state vector ψ is an element of a Hilbert space, there exists a correlation between the uncertainties. The correlation is well known as the *Generalized Uncertainty Principle*.

Theorem 7.4.1. (The Uncertainty Principle) *If $\hat{A}$ and $\hat{B}$ are Hermitian operators, then for any state vector*

$$\Delta \hat{A} \Delta \hat{B} > \frac{1}{2} \left| \frac{1}{i} \langle [\hat{A}, \hat{B}] \rangle \right|. \tag{7.85}$$

Proof: We have

$$\Delta \hat{A} = \|\psi_1\|,$$
$$\Delta \hat{B} = \|\psi_2\|, \tag{7.86}$$

$$\psi_1 = \left(\hat{A} - \langle\hat{A}\rangle\right)\psi,$$
$$\psi_2 = \left(\hat{B} - \langle\hat{B}\rangle\right)\psi. \tag{7.87}$$

According to Schwarz's inequality, we have, for any two vectors ψ_1 and ψ_2,

$$\|\psi_1\|\|\psi_2\| \geq \left|\langle\psi_1|\psi_2\rangle\right|$$
$$\geq \left|\mathrm{Im}\langle\psi_1|\psi_2\rangle\right|$$
$$\geq \left|\frac{1}{2i}\left\{\langle\psi_1|\psi_2\rangle - \overline{\langle\psi_1|\psi_2\rangle}\right\}\right|$$
$$\geq \left|\frac{1}{2i}\left\{\langle\psi_1|\psi_2\rangle - \langle\psi_2|\psi_1\rangle\right\}\right|.$$

Evidently,

$$\Delta\hat{A}\Delta\hat{B} \geq \left|\frac{1}{2i}\left\{\langle(\hat{A} - \langle\hat{A}\rangle)\psi|(\hat{B} - \langle\hat{B}\rangle)\psi\rangle - \langle(\hat{B} - \langle\hat{B}\rangle)\psi|(\hat{A} - \langle\hat{A}\rangle)\psi\rangle\right\}\right|$$
$$\geq \left|\frac{1}{2i}\langle\psi|(\hat{A}\hat{B} - \hat{B}\hat{A})\psi\rangle\right|$$
$$\geq \frac{1}{2}\left|\frac{1}{i}\langle[\hat{A}, \hat{B}]\rangle\right|. \qquad \square$$

Corollary 7.4.2. *For conjugate operators q_j and p_k, the following holds*

$$\Delta q_j \Delta p_k \geq \frac{\hbar}{2}\delta_{jk}, \tag{7.88a}$$

$$\Delta x_j \Delta p_j \geq \frac{\hbar}{2}, \tag{7.88b}$$

$$\Delta x_j \Delta v_j \geq \frac{\hbar}{2m}, \tag{7.88c}$$

where v_j is the velocity vector and m is the mass.

Proof: Substituting $\hat{A} = q_j$ and $\hat{B} = p_k$ in (7.85), we obtain (7.88a). In Cartesian coordinates, (7.88a) reduces to (7.88b). Finally, with the velocity vector v_j and mass m, (7.88b) becomes (7.88c). $\qquad \square$

It is important to point out that (7.88b) is the famous uncertainty relation between the position and momentum first discovered by Heisenberg. According to the Heisenberg uncertainty relation, the position and momentum of a particle cannot be determined exactly and simultaneously. In other words, it is impossible to design an experiment that can measure the precise values of both the position and momentum of a particle. Quantum mechanics is essentially built upon the Heisenberg Uncertainty Principle, and hence this principle is perhaps the most fundamental result in quantum theory.

Finally, since $\hbar$ is very small, $\hbar/2m$ is extremely small for any macroscopic system, and hence (7.88c) reveals why the Uncertainty Principle is of no importance in classical mechanics. So when the system is large, $\hbar$ is negligible and all observable operators commute with each other. Consequently, all observables can then be represented by ordinary algebraic variables and can be measured without any mutual disturbance. According to the Corresponding Principle, the definitive relations between these variables, such as position, energy and momentum, are the same as in classical mechanics in the limit $\hbar \to 0$. So the distinction between quantum and classical systems is not made on the sole basis of spatial extension, but in units of Planck's constant h. Evidently, the size of a physical system becomes important and must be considered. Thus, for example, for an electron in an atom the typical "action" (length $\times$ momentum or time $\times$ energy) is the product of the Bohr radius 10^{-8} cm, and the momentum in the Bohr orbit 10^{-19} so that their product is exactly of the order of h, so quantum effects are absolutely essential. Thus, classical mechanics can be considered as the limit $\hbar \to 0$ of quantum mechanics.

7.5 The Schrödinger Equation of Motion

In our review of classical mechanics in Section 7.2, the problem of the time evolution of the classical state has been discussed in some detail. It follows from the review that there is a very close connection between energy and time as established in the Hamiltonian formulation of classical mechanics. In fact, it is the energy written as the Hamiltonian function $H(\mathbf{x}, p)$ that really governs the time evolution of the classical systems. On the other hand, our previous discussion reveals that there is a remarkable distinction between the state of a system and the physical observables in quantum mechanics. With a reasonably good understanding of these concepts, it is not too difficult to comprehend the time-evolution problem in quantum mechanics. Before we actually embark on the problem, it is necessary to preview some other related ideas of mechanics.

In general, any dynamical quantity that varies in space and time constitutes a wave that can typically be represented by the wave function

$$\Psi(\mathbf{r}, t) = A(\mathbf{r}) \exp\left[-i\left(2\pi \nu t + \varphi(\mathbf{r})\right)\right], \tag{7.89}$$

where A is the amplitude, ν is the frequency, and φ is the phase of the wave. This can also be expressed in the form

$$\Psi(\mathbf{r}, t) = \psi(\mathbf{r})e^{-2\pi i\nu t}, \tag{7.90}$$

where

$$\psi(\mathbf{r}) = A(\mathbf{r})\exp[-i\varphi(\mathbf{r})]. \tag{7.91}$$

According to the Planck equation ($E = h\nu$), if a quantum system has energy E, its state vector $|\Psi(\mathbf{r}, t)\rangle$ at time t should contain a factor $\exp(-2\pi i\nu t) = \exp[-(iE/\hbar)t]$ so that

$$\left|\Psi(\mathbf{r}, t)\right\rangle = e^{-iEt/\hbar}\left|\Psi(\mathbf{r}, 0)\right\rangle. \tag{7.92}$$

Clearly, energy is an observable, and hence, for a system to have a definite energy E, it must be in an eigenstate of this observable. If this is the case, Equation (7.92) expresses the fact that the state vector at time t is different from that at $t = 0$ only by a scalar factor, and so it describes the same physical state. For this reason an eigenstate of energy is called a *stationary state* of the system.

The following postulate of quantum mechanics is concerned with the existence of the energy operator and the time development of a quantum system.

Postulate VI.

(a) (Hamiltonian operator) *For every physical system there exists a linear Hermitian operator $\hat{H}$, the so-called Hamiltonian operator which represents the observable operator corresponding to the total energy of the state vector of the system.*

(b) (Schrödinger's equation) *If a physical system is not disturbed by any experiment, the Hamiltonian operator $\hat{H}$ determines the time development of the state vector of the system $\Psi(\mathbf{r}, t)$ through the partial differential equation*

$$i\hbar\frac{\partial\Psi}{\partial t} = \hat{H}\Psi(\mathbf{r}, t). \tag{7.93}$$

This is called the *time-dependent Schrödinger equation*, and represents the fundamental equation of motion in quantum mechanics first discovered by Erwin Schrödinger (1887–1961). Its ultimate justification is that it leads to predictions which are in remarkable agreement with experimental findings. As in equations of motion in classical mechanics, Equation (7.93) is completely deterministic in the sense that, given the state $\Psi(\mathbf{r}, t)$ at some time $t = t_0$, Equation (7.93) will uniquely determine the state $\Psi(\mathbf{r}, t)$ at some other time t. The Hamiltonian operator $\hat{H}$ corresponds to the force in classical mechanics because the total energy $\hat{H}$ includes the potential energy which gives the force in classical mechanics. So $\hat{H}$ is equivalent to the force field.

For a single particle of mass m moving in space, the classical state is described by the position and momentum vectors $(\mathbf{r}, \mathbf{p})$. If the particle is under the action of a force $\mathbf{F}(\mathbf{r})$ which is derived from a potential $V(\mathbf{r})$ so that $\mathbf{F} = -\nabla V$, the Hamiltonian function H is

$$H(\mathbf{r}, \mathbf{p}) = T + V = \frac{p^2}{2m} + V(\mathbf{r}). \tag{7.94}$$

The Hamiltonian operator $\hat{H}$ corresponding to this classical function H in quantum mechanics is derived by replacing $\mathbf{r}$ and $\mathbf{p}$ with the operators $\hat{\mathbf{r}}$ and $\hat{\mathbf{p}} = -i\hbar\nabla$, respectively. Consequently, the Schrödinger equation (7.93) assumes the form

$$i\hbar\frac{\partial\Psi}{\partial t} = \left[-\frac{\hbar^2}{2m}\nabla^2 + V(\hat{\mathbf{r}})\right]\Psi = \hat{H}\Psi, \tag{7.95}$$

where

$$\hat{H} = -\frac{\hbar^2}{2m}\nabla^2 + V(\hat{\mathbf{r}}) \tag{7.96}$$

and $\Psi(\mathbf{r}, t)$ belongs to the Hilbert space $L^2(\mathbf{R}^3)$.

This postulate has three main consequences. First, it asserts that (7.92) can be derived as a solution of the Schrödinger equation (7.93). We assume that $\hat{H}$ has a purely discrete spectrum of eigenvalues so that it has a complete set of eigenstates $|\psi_n(\mathbf{r})\rangle$ with the corresponding eigenvalues E_n. Then $|\Psi(\mathbf{r}, t)\rangle$ can be expanded in terms of the complete eigenstates as

$$|\Psi(\mathbf{r}, t)\rangle = \sum_n a_n(t)|\psi_n(\mathbf{r})\rangle. \tag{7.97}$$

Substituting this into (7.93) gives

$$i\hbar\sum\frac{da_n}{dt}|\psi_n\rangle = \sum_n E_n a_n(t)|\psi_n\rangle,$$

which, equating the coefficients of $|\psi_n\rangle$, yields

$$i\hbar\frac{da_n}{dt} = E_n a_n(t). \tag{7.98}$$

Hence, the solution of this equation is

$$a_n(t) = a_n(0)\exp\left(-\frac{iE_n t}{\hbar}\right), \tag{7.99}$$

where $a_n(0)$ is the initial value of $a_n(t)$ at $t = 0$. Thus, the time-dependent solution is

$$\big|\Psi(\mathbf{r}, t)\big\rangle = \sum_n a_n(0) \exp\left(-\frac{iE_n t}{\hbar}\right)\big|\psi_n(\mathbf{r})\big\rangle. \tag{7.100}$$

The second consequence of (7.93) is the de Broglie wave (Louis Victor Pierre Raymond duc de Broglie (1892–1987)) for free particles of definite momentum $p = \hbar k$, which is required to explain electron diffraction. The one-dimensional de Broglie wave is

$$\langle x|p_a\rangle = a e^{ikx} = a e^{ipx/\hbar}, \tag{7.101}$$

where a is a constant.

If it is normalized by using the orthonormality condition for continuous eigenvalues $\langle \alpha | \alpha' \rangle = \delta(\alpha - \alpha')$, where the simplest representation of the Dirac delta function is

$$\delta(\alpha) = \frac{1}{2\pi} \int_{-\infty}^{\infty} e^{i\alpha x} dx,$$

then

$$\langle p|p'\rangle = \int \langle p|x\rangle \langle x|p'\rangle dx$$

$$= |a|^2 \int_{-\infty}^{\infty} \exp[-i(p - p')x/\hbar] dx$$

$$= \delta(p - p') \tag{7.102}$$

so that $|a|^2 = (2\pi\hbar)^{-1}$.

In the case of a free particle of definite momentum,

$$\langle x|\psi\rangle = \langle x|p\rangle = \frac{1}{\sqrt{2\pi\hbar}} e^{ipx/\hbar}.$$

According to (7.101), the time-dependent solution of (7.93) is

$$\big|\Psi(x, t)\big\rangle = \frac{1}{\sqrt{2\pi\hbar}} \exp[i(px - Et)/\hbar], \tag{7.103}$$

where $E = (p^2/2m)$.

This is the complete de Broglie wave. It appears now for the first time as a consequence of the general quantum mechanical formalism and is direct evidence for the correctness of the postulated Schrödinger equation of motion.

The third consequence of the Schrödinger equation is the conservation law of total probability in space and time. The Schrödinger equation (7.95) is satisfied by the complex wave function of real variables so that its real and imaginary parts separately satisfy (7.95). The complex conjugate equation is obtained by changing the sign of the imaginary part everywhere so that

$$-i\hbar\frac{\partial\overline{\Psi}}{\partial t} = -\frac{\hbar^2}{2m}\nabla^2\overline{\Psi} + V(\mathbf{r})\overline{\Psi}. \tag{7.104}$$

If the general solution of (7.95) has the form

$$\Psi(\mathbf{r}, t) = \sum_n a_n\psi_n(\mathbf{r})\exp\left[-\frac{iE_n t}{\hbar}\right], \tag{7.105}$$

then its complex conjugate satisfies (7.104) and has the representation

$$\overline{\Psi}(\mathbf{r}, t) = \sum_n \overline{a}_n\overline{\psi}_n(\mathbf{r})\exp\left[\frac{iE_n t}{\hbar}\right]. \tag{7.106}$$

Then the time rate of change of the *probability density* $P(\mathbf{r}, t) = \Psi\overline{\Psi} = |\Psi|^2$ at any point is

$$\frac{\partial}{\partial t}(\overline{\Psi}\Psi) = \overline{\Psi}\frac{\partial\Psi}{\partial t} + \Psi\frac{\partial\overline{\Psi}}{\partial t},$$

which is, by (7.95) and (7.104),

$$\frac{\partial}{\partial t}(\overline{\Psi}\Psi) = -\frac{i\hbar}{2m}\left[\Psi(\nabla^2\overline{\Psi}) - \overline{\Psi}(\nabla^2\Psi)\right]$$

$$= -\frac{i\hbar}{2m}\,\text{div}\left[\Psi(\nabla\overline{\Psi}) - \overline{\Psi}(\nabla\Psi)\right]$$

$$= -\,\text{div}\,\mathbf{J},$$

or, equivalently,

$$\frac{\partial P}{\partial t} + \text{div}\,\mathbf{J} = 0, \tag{7.107}$$

where the vector $\mathbf{J}$ may be called the *probability current density* or the *probability flux* defined by

$$\mathbf{J} = \frac{i\hbar}{2m}\left[\Psi(\nabla\overline{\Psi}) - \overline{\Psi}(\nabla\Psi)\right]. \tag{7.108}$$

Equation (7.107) represents the conservation of total probability in space and time, that is, it neither created nor destroyed. Equivalently, any net increase (or decrease) of probability within a region in space is associated with a net inflow (or outflow) or probability. If the system is in a stationary state, then $\partial P/\partial t = 0$ and hence $\text{div}\,\mathbf{J} = 0$, that is, the probability flow is solenoidal. It is important to point out that the derivation of the *equation of continuity for probability flow* (7.107) has been based upon the real potential function. The complex potential energy function is physically unrealistic because it leads to complex energy eigenvalues. However, it has been utilized extensively in modeling nuclear scattering interactions.

Equation (7.107) is identical with the *equation of continuity*, which represents the conservation of mass in classical hydrodynamics and the conservation of charge in classical electrodynamics provided there are no sources or sinks in the system.

The interpretation of the probability flux **J** as the probability current density leads to the representation of momentum by the operator $-i\hbar\nabla$ and velocity by the operator $(\hbar/im)\nabla$. It is evident from the definition (7.108) of **J** that

$$\mathbf{J}(\mathbf{r}, t) = \mathrm{Re}\left(\overline{\Psi}\frac{\hbar}{im}\nabla\Psi\right), \qquad (7.109a)$$

or

$$\mathbf{J}(\mathbf{r}, t) = \mathrm{Im}\left(\overline{\Psi}\frac{\hbar}{m}\nabla\Psi\right). \qquad (7.109b)$$

From the definition, the probability density $P(\mathbf{r}, t) = |\Psi(\mathbf{r}, t)|^2$ is determined by modulus squared of the wave function $\Psi(\mathbf{r}, t)$. This means that $P(\mathbf{r}, t)d\mathbf{r}$ is the probability of finding the particle in its volume element $d\mathbf{r} = dxdydz$ about its position $\mathbf{r}$ at time t, when a large number of precise measurements are made on independent particles each of which is described by the single particle wave function $\Psi(\mathbf{r}, t)$. The normalizing condition ensures that the total probability of finding the particle somewhere in the space must be unity, so that for each fixed t,

$$\langle\Psi|\Psi\rangle = \iiint |\Psi(\mathbf{r}, t)|^2 d\mathbf{r} = 1, \qquad (7.110)$$

where the integral is taken over the entire region of the space. A general statement of the boundary condition on the wave function is needed so that it can be normalized in principle. This probability interpretation of the wave function was first introduced by Max Born (1882–1970) in 1926 and plays a fundamental role in the development of quantum mechanics.

The fourth consequence of the postulate is that the time evolution of the state vector as described by Equation (7.93) is consistent with the requirement that the state vector always has a unit norm. If $\Psi(x, 0) = \Psi_0$ and $\langle\Psi_0|\Psi_0\rangle = 1$, it can be proved that, for all t,

$$\frac{d}{dt}\langle\Psi|\Psi\rangle = 0. \qquad (7.111)$$

We have

$$\frac{d}{dt}\langle\Psi|\Psi\rangle = \left\langle\frac{\partial\Psi}{\partial t}\bigg|\Psi\right\rangle + \left\langle\Psi\bigg|\frac{\partial\Psi}{\partial t}\right\rangle$$

$$= \frac{1}{i\hbar} \langle \hat{H}\Psi | \Psi \rangle \frac{1}{i\hbar} \langle \Psi | \hat{H}\Psi \rangle \quad \text{by (7.93)}$$

$$= \frac{1}{i\hbar} [\langle \Psi | \hat{H}\Psi \rangle \langle \Psi | \hat{H}\Psi \rangle] = 0, \tag{7.112}$$

since $\hat{H}$ is Hermitian. This means that $\|\Psi\| = $ constant. But $\|\Psi_0\| = 1$, and hence the constant is equal to 1.

The fifth consequence of Postulate VI(b) is that (7.93) leads to the analogue of the Hamilton–Jacobi equation in classical mechanics. Equation (7.93) can also be rewritten as

$$\frac{\partial}{\partial t}\left(\frac{\hbar}{i} \ln \Psi\right) + \hat{H} = 0, \tag{7.113}$$

or, equivalently,

$$\frac{\partial S}{\partial t} + \hat{H} = 0, \tag{7.114}$$

where $S = (\hbar/i) \ln \Psi$ is called *Hamilton's principal function* and may be interpreted as the *action* so that $\Psi = \exp((i/\hbar)S)$. So (7.114) is the analogue of the Hamilton–Jacobi equation.

A careful interpretation of this postulate further reveals how the states described by the wave functions and the observables represented by the Hermitian operators are directly related to actual measurements. Conversely, the question is how can information obtained about a physical system be incorporated in the wave functions. If an observable $\hat{A}$ of a quantum system is measured in a state $|\Psi(\mathbf{r}, t)\rangle$, then Postulate II determines the value λ as an eigenvalue of $\hat{A}$. If the same measurement of $\hat{A}$ is repeated in a sufficiently short time interval, then the same value of λ is expected. However, if the time interval between the two measurements is long enough, then the system is expected to change appreciably due to the time variation according to the Schrödinger equation. Consequently, the eigenvalue λ of $\hat{A}$ will be different. However, if the observable $\hat{A}$ is a constant of motion, it will retain the same value λ obtained by the first measurement. Thus, this case is very much similar to that in classical mechanics where the time evolution of a classical observable is uniquely determined by the classical equations of motion. This discussion also leads to the question of simultaneous measurement of all observables in a quantum system. In general, quantum observables are mutually incompatible in the sense that measurements of one of them partially or completely destroy any knowledge or information about the others. So it is not possible to determine the simultaneous values of all observables of a quantum system.

Example 7.5.1. We shall discuss the plane wave solution of the time-dependent Schrödinger equation for a particle under the action of a constant potential. Obtain the solution of a free particle with the initial condition $\Psi(\mathbf{r}, 0) = \Psi_0(\mathbf{r})$.

The time-dependent Schrödinger equation for a particle under the potential $V(\mathbf{r})$ is

$$i\hbar \Psi_t = \left[V(\mathbf{r}) - \frac{\hbar^2}{2m} \nabla^2 \right] \Psi. \tag{7.115}$$

With a constant potential $V(\mathbf{r}) = \text{constant} = V$, a plane wave solution of the form

$$\Psi(\mathbf{r}, t) = A \exp\left[i(\boldsymbol{\kappa} \cdot \mathbf{r} - \omega t) \right]$$

is possible provided the dispersion relation

$$\omega \hbar = V + \frac{\hbar^2 \kappa^2}{2m} \tag{7.116}$$

is satisfied by frequency ω and wavenumber vector $\boldsymbol{\kappa}$. The group velocity of the wave is given by

$$\mathbf{C}_g = \nabla_\kappa \omega = \frac{\hbar \boldsymbol{\kappa}}{m}. \tag{7.117}$$

Through the Correspondence Principle, $\hbar\omega$ is to be interpreted as the to-
tal energy, $\hbar^2 \kappa^2 / 2m$ as the kinetic energy, $\hbar\boldsymbol{\kappa}$ as the particle momentum, and
hence the group velocity as the classical velocity.

We next solve the Schrödinger equation for a free particle ($V \equiv 0$) with the initial condition

$$\Psi(\mathbf{r}, 0) = \Psi_0(\mathbf{r}), \tag{7.118}$$

where $\Psi_0(\mathbf{r})$ has a compact support in $\mathbb{R}^3$.

Apply the Fourier transform $\tilde{\Psi}(\boldsymbol{\kappa}, t)$ of $\Psi(\mathbf{r}, t)$ defined by

$$\tilde{\Psi}(\boldsymbol{\kappa}, t) = \frac{1}{(2\pi)^{3/2}} \int\!\!\!\int\!\!\!\int_{-\infty}^{\infty} e^{-i\boldsymbol{\kappa} \cdot \mathbf{r}} \Psi(\mathbf{r}, t) d\mathbf{r},$$

where $\boldsymbol{\kappa} = (k, l, m)$.

Then the Schrödinger equation and the initial condition become

$$i\hbar \tilde{\Psi}_t = \frac{\hbar^2 \kappa^2}{2m} \tilde{\Psi}, \qquad \tilde{\Psi}(\boldsymbol{\kappa}, 0) = \tilde{\Psi}_0(\boldsymbol{\kappa}).$$

Hence, the transformed solution of this system is

$$\tilde{\Psi}(\boldsymbol{\kappa}, t) = \tilde{\Psi}_0(\boldsymbol{\kappa}) \exp\left[-\frac{i\hbar}{2m} \kappa^2 t \right]. \tag{7.119}$$

The inverse Fourier transformation gives

$$\Psi(\mathbf{r}, t) = \frac{1}{(2\pi)^{3/2}} \iiint\limits_{-\infty}^{\infty} \tilde{\Psi}_0(\boldsymbol{\kappa}) \exp\left[i\left(\boldsymbol{\kappa} \cdot \mathbf{r} - \frac{\hbar \kappa^2 t}{2m}\right)\right] d\boldsymbol{\kappa}$$

$$= \frac{1}{(2\pi)^{3/2}} \iiint\limits_{-\infty}^{\infty} \tilde{\Psi}_0(\boldsymbol{\kappa}) \exp\left[i\kappa\left(\frac{\boldsymbol{\kappa} \cdot \mathbf{r}}{\kappa} - \frac{\hbar \kappa t}{2m}\right)\right] d\boldsymbol{\kappa}. \quad (7.120)$$

Thus, the solution $\Psi(\mathbf{r}, t)$ is a continuous superposition of functions of the form

$$\exp\left[i\kappa\left(\frac{\boldsymbol{\kappa} \cdot \mathbf{r}}{\kappa} - \frac{\hbar \kappa}{2m} t\right)\right]. \quad (7.121)$$

Such a function represents a plane wave traveling with velocity $\hbar \kappa / 2m$ in a direction normal to the plane $\boldsymbol{\kappa} \cdot \mathbf{r} = 0$ and having a wavelength $2\pi / \kappa$. If $\Psi_0(\boldsymbol{\kappa})$ is zero outside a small region about a fixed point $\boldsymbol{\kappa}_0 = (k_0, l_0, m_0)$, then the x, y, and z components of momentum $\mathbf{p}_0$ are certain to have values very near to $\hbar k_0$, $\hbar l_0$, and $\hbar m_0$, respectively, and $\Psi(\mathbf{r}, t)$ is a superposition of plane waves with wavenumbers $\kappa_0 = [(k_0^2 + l_0^2 + m_0^2)]^{1/2} = p_0/\hbar$, where $p_0 = [(\hbar k_0)^2 + (\hbar l_0)^2 + (\hbar m_0)^2]^{1/2}$ is the total momentum. More precisely, in a state where the momentum is certainly very close to p_0, the state function is a superposition of plane waves with wavenumbers close to the value $p_0/\hbar$. In this sense, a free particle of momentum p is associated with a wave of wavenumber $\kappa_0 = p_0/\hbar$ (or of wavelength $\lambda_0 = 2\pi \hbar / p_0$).

In the one-dimensional case, the solution takes the form

$$\Psi(x, t) = \frac{1}{\sqrt{2\pi}} \int_{-\infty}^{\infty} \tilde{\Psi}_0(k) \exp\left[i\left(kx - \gamma t k^2\right)\right] dk \quad \left(\gamma = \frac{\hbar}{2m}\right)$$

$$= \frac{1}{2\pi} \int_{-\infty}^{\infty} \psi_0(\alpha) d\alpha \int_{-\infty}^{\infty} \exp\left[i\left(kx - k\alpha - \gamma k^2 t\right)\right] dk$$

$$= \frac{1}{2\pi} \int_{-\infty}^{\infty} \psi_0(\alpha) d\alpha \int_{-\infty}^{\infty} \exp\left[-i\gamma t\left(k - \frac{x - \alpha}{2\gamma t}\right)^2 + \frac{i(x - \alpha)^2}{4\gamma t}\right] dk$$

$$= \frac{1 - i}{2\sqrt{2\pi \gamma t}} \int_{-\infty}^{\infty} \psi_0(\alpha) \exp\left[\frac{i(x - \alpha)^2}{4\gamma t}\right] d\alpha, \quad (7.122)$$

where the last equality follows from the equality $\int_{-\infty}^{\infty} e^{-i\alpha x^2} d\alpha = \sqrt{\pi/2\alpha} \times (1 - i)$. The integral in (7.122) would be convergent for a suitably chosen function $\psi_0(x)$. $\qquad \square$

Example 7.5.2. (The free particle de Broglie wave and the wave–particle duality) The interference phenomena, of which diffraction is typical, require the wave picture for a satisfactory explanation. The typical wave is given by

$$\Psi(\mathbf{r}, t) = a \exp\left[i(\boldsymbol{\kappa} \cdot \mathbf{r} - \omega t)\right], \tag{7.123}$$

where a is the amplitude, $\boldsymbol{\kappa}$ is the wavenumber vector, and ω is the frequency representing the fundamental physical characteristics of the wave. The wave propagates with a constant velocity of light $c = \omega/|\boldsymbol{\kappa}|$.

According to the famous Planck hypothesis, the radiation of frequency ω behaves like a stream of photons of energy $E = \hbar\omega$ which may be emitted or absorbed by matter. Since they travel with the velocity of light, according to the theory of special relativity, their rest mass must be zero. The relativistic equation for energy E and momentum p is

$$\left(E^2/c^2\right) = p^2 + m^2 c^2. \tag{7.124}$$

Thus, for photons ($m = 0$), $p = E/c$. A simple elimination of c gives

$$E = \hbar\omega, \tag{7.125a}$$

$$\mathbf{p} = \hbar\boldsymbol{\kappa}. \tag{7.125b}$$

This shows clearly the basic relation between the particle parameters $(E, \mathbf{p})$ of the photons, and the wave parameters $(\omega, \boldsymbol{\kappa})$ of the corresponding wave.

It is well known from diffraction experiments with electrons that a beam of electrons traveling with a definite momentum of magnitude p in the x-direction is directly associated with the wavelength of the corresponding wave so that $p = \hbar\kappa$. According to Max Born's physical interpretation of $\Psi(\mathbf{r}, t)$, $|\Psi(\mathbf{r}, t)|^2$ represents the density of electrons, that is, the number of electrons per unit volume at $(\mathbf{r}, t)$. De Broglie made a remarkable hypothesis that the result (7.125a) and (7.125b) which relates the wave and the particle aspects of radiation should apply also to electrons. Consequently, an electron of given energy and momentum must be associated with the de Broglie wave

$$\Psi(\mathbf{r}, t) = a \exp\left[\frac{i}{\hbar}(\mathbf{p} \cdot \mathbf{r} - Et)\right]. \tag{7.126}$$

So we can identify the wave function of a free particle of mass m in the form

$$\psi(\mathbf{r}) = a e^{i\boldsymbol{\kappa} \cdot \mathbf{r}}, \tag{7.127}$$

where the wavenumber $\boldsymbol{\kappa}$ and the momentum $\mathbf{p}$ are related by (7.125b). Clearly

$$-i\hbar \nabla \psi = \mathbf{p}\psi, \tag{7.128}$$

$$-\frac{\hbar^2}{2m} \nabla^2 \psi = \left(\frac{p^2}{2m}\right)\psi. \tag{7.129}$$

These results show that the wave function is a simultaneous eigenfunction of momentum and energy with eigenvalues $\mathbf{p}$ and $p^2/2m$, respectively, for all values of $\mathbf{p}$. This is due to the fact that for a free particle the total energy H is a function of the momentum only ($H = p^2/2m$) as in classical mechanics. The eigenfunctions are mutually orthogonal which follows from the Dirac formalism for eigenvalue problems with continuous spectra:

$$\left(ae^{i\boldsymbol{\kappa}'\cdot\mathbf{r}}, ae^{i\boldsymbol{\kappa}\cdot\mathbf{r}}\right) = \int |a|^2 e^{i(\boldsymbol{\kappa}'-\boldsymbol{\kappa})\cdot\mathbf{r}} d\mathbf{r} = (2\pi)^3 |a|^2 \delta(\boldsymbol{\kappa} - \boldsymbol{\kappa}'). \tag{7.130}$$

$\square$

Example 7.5.3. (Momentum wave function $\boldsymbol{\psi}(\mathbf{p})$ as the three-dimensional Fourier transform of $\boldsymbol{\psi}(\mathbf{r})$) The wave function $\psi(\mathbf{r}) = \psi(x, y, z)$ is the solution of the time-independent Schrödinger equation

$$\nabla^2 \psi + \frac{2m}{\hbar^2}(E - V)\psi = 0, \tag{7.131}$$

where E is the total energy and $V(x, y, z)$ is the potential of the field in which a particle moves. To avoid the frequent repetition of the constant $2m/\hbar^2$ in the equation, *atomic units* are used. If we use the electronic charge e, the electronic mass m, and the length $a = (\hbar^2/me^2)$ as the units of charge, mass, and length, respectively, the Schrödinger equation (7.131) becomes

$$\nabla^2 \psi + 2(E - V)\psi = 0, \tag{7.132}$$

and the units of energy and velocity are now e^2/a and $e^2/\hbar = c/137$, respectively, where $\hbar c/e^2 = 137$ and c is the speed of light.

According to the Dirac transformation theory, $\tilde{\psi}(\mathbf{p})$ is the three-dimensional Fourier transform of the wave function $\psi(\mathbf{r})$ provided atomic units are used:

$$\tilde{\psi}(\mathbf{p}) = \frac{1}{(2\pi)^{3/2}} \int\int\int_{-\infty}^{\infty} e^{-i\mathbf{p}\cdot\mathbf{r}} \psi(\mathbf{r}) d\mathbf{r}.$$

It follows from the Parseval relation of Fourier transforms that

$$\int\int\int_{-\infty}^{\infty} |\psi(\mathbf{r})|^2 d\mathbf{r} = 1 = \int\int\int_{-\infty}^{\infty} |\tilde{\psi}(\mathbf{p})|^2 d\mathbf{p},$$

where $\psi(\mathbf{r})$ is normalized to unity and hence $\tilde{\psi}(\mathbf{p})$ is automatically normalized to unity.

The momentum density function $I(p)$ is defined by

$$I(p) = \int_0^\infty |\tilde{\psi}(\mathbf{p})|^2 p^2 d\Omega, \tag{7.133}$$

where $d\Omega$ is the element of solid angle for $\mathbf{p}$. Physically, $I(p)dp$ can be interpreted as the probability that the magnitude of the momentum of the electron lies between p and $p + dp$ so that the total probability is unity, that is,

$$\int_0^\infty I(p)dp = 1. \qquad (7.134)$$

In the Schrödinger interpretation, the momentum $\mathbf{p} = -i\nabla\psi$ in the sense that the mean value of the x-component of $\mathbf{p}$ is given by the equation

$$\langle p_x \rangle = -i \iiint_{-\infty}^{\infty} \psi^*(\mathbf{r})\frac{\partial\psi(\mathbf{r})}{\partial x}d\mathbf{r}. \qquad (7.135)$$

Since $\mathcal{F}\{-i\partial\psi/\partial x\} = p_x\tilde{\psi}(\mathbf{p})$,

$$\langle p_x \rangle = \iiint_{-\infty}^{\infty} p_x\tilde{\psi}^*(\mathbf{p})\tilde{\psi}(\mathbf{p})d\mathbf{p}. \qquad (7.136)$$

Similarly, the mean momentum is given by

$$\langle p \rangle = \int_0^\infty I(p)dp. \qquad (7.137)$$

For a single electron in a hydrogen-like atom of nuclear charge Z, we choose the Coulomb potential

$$V(r) = -\frac{Z}{r},$$

where $r^2 = x^2 + y^2 + z^2$, and for the hydrogen atom $Z = 1$, for ionized helium $Z = 2$, and for doubly ionized lithium $Z = 3$. It can be readily verified by direct substitution in equation (7.132) that

$$\psi(\mathbf{r}) = \left(\frac{Z^3}{\pi}\right)^{1/2} e^{-Zr} \qquad (7.138)$$

is a solution of (7.132) with $E = -\frac{1}{2}Z^2$. The constant involved in the solution for $\psi(\mathbf{r})$ is chosen so that $\psi(\mathbf{r})$ is normalized to unity. The momentum wave function associated with this wave function is

$$\tilde{\psi}(\mathbf{p}) = \left(\frac{Z^3}{\pi}\right)^{1/2} \frac{1}{(2\pi)^{3/2}} \iiint_{-\infty}^{\infty} e^{-Zr - i(\mathbf{p}\cdot\mathbf{r})} d\mathbf{r}$$

$$= \frac{(8Z^5)^{1/2}}{\pi(p^2 + Z^2)^2}, \qquad (7.139)$$

so that

$$I(p) = 4\pi p^2 |\psi(\mathbf{p})|^2 = \frac{32 p^2 Z^5}{\pi (p^2 + Z^2)^4}. \tag{7.140}$$

Hence, the mean momentum is

$$\langle p \rangle = \int_0^\infty p I(p) dp$$

$$= \frac{32 Z^5}{\pi} \left[\int_0^\infty \frac{p dp}{(p^2 + Z^2)^3} - Z^2 \int_0^\infty \frac{p dp}{(p^2 + Z^2)^4} \right] = \frac{8Z}{3\pi}. \tag{7.141}$$

In terms of the unit of momentum $mc/137$, the mean momentum has therefore the value $8mcZ/411\pi$. In the nomenclature of spectroscopy, the wave function $\psi(r) = (1/\sqrt{\pi}) \exp(-r)$ is the 1s wave function of a hydrogen atom. So, in a 1s orbit, the mean momentum of the electron is $8mZe^2/3\pi h$ because the fine structure constant is $137 = \hbar c/e^2$. $\qquad \square$

Example 7.5.4. (Motion of a wave-packet in a field of force) We consider a three-dimensional wave-packet $\Psi(\mathbf{r}, t)$ and calculate the x-component of its velocity given by

$$\frac{d}{dt}\langle x \rangle = \frac{d}{dt} \int \Psi^* x \Psi d\tau = \int \frac{\partial \Psi^*}{\partial t} x \Psi d\tau + \int \Psi^* x \frac{\partial \Psi}{\partial t} d\tau,$$

where the integration is taken over the whole space. We next substitute for $\partial \Psi/\partial t$ and $\partial \Psi^*/\partial t$ from the Schrödinger equation (7.95) and its complex conjugate to obtain

$$\frac{d}{dt}\langle x \rangle = -\frac{i\hbar}{2m} \left[\int (\nabla^2 \Psi^*) x \Psi d\tau - \int \Psi^* x (\nabla^2 \Psi) d\tau \right]. \tag{7.142}$$

Then we apply Green's theorem

$$\int [(\nabla^2 \Psi^*) x \Psi - \Psi^* (\nabla^2 x \Psi)] d\tau = \int [(\nabla \Psi^*) x \Psi - \Psi^* (\nabla x \Psi)] ds,$$

where the right-hand side represents the surface integral which is taken over an infinitely distant bounding surface. Since the wave-packet represents a localized particle, it must vanish all over the above surface. Consequently, the surface integral vanishes and hence

$$\int (\nabla^2 \Psi^*) x \Psi d\tau = \int \Psi^* (\nabla^2 x \Psi) d\tau.$$

We substitute this result into (7.142) to obtain

$$\frac{d}{dt}\langle x \rangle = -\frac{i\hbar}{2m} \int [\Psi^* (\nabla^2 x \Psi) - \Psi^* x \nabla^2 \Psi] d\tau$$

$$= -\frac{i\hbar}{2m} \int \Psi^* \left[x\nabla^2\Psi + 2\frac{\partial\Psi}{\partial x} - x\nabla^2\Psi \right] d\tau$$

$$= \frac{1}{m} \int \Psi^* \left[\frac{\hbar}{i} \frac{\partial\Psi}{\partial x} \right] d\tau = \frac{1}{m} \langle p_x \rangle. \qquad (7.143)$$

Similarly,

$$\frac{d}{dt} \langle p_x \rangle = \frac{\hbar}{i} \frac{d}{dt} \int \Psi^* \frac{\partial\Psi}{\partial x} d\tau$$

$$= \frac{\hbar}{i} \left[\int \frac{\partial\Psi^*}{\partial t} \frac{\partial\Psi}{\partial x} d\tau + \int \Psi^* \frac{\partial}{\partial x} \left(\frac{\partial\Psi}{\partial t} \right) d\tau \right]$$

$$= \int \left(-\frac{\hbar^2}{2m}\nabla^2\Psi^* + V\Psi^* \right) \frac{\partial\Psi}{\partial x} d\tau + \int \Psi^* \frac{\partial}{\partial x} \left(\frac{\hbar^2}{2m}\nabla^2\Psi - V\Psi \right) d\tau$$

$$= \frac{\hbar^2}{2m} \int \left(\Psi^*\nabla^2\Psi_x - \Psi_x\nabla^2\Psi^* \right) d\tau + \int \left[V\Psi^*\Psi_x - \Psi^* \frac{\partial}{\partial x}(V\Psi) \right] d\tau.$$

The first integral can be transformed into a surface integral by Green's theorem. It is assumed that this integral vanishes when taken over a very large surface. Consequently,

$$\frac{d}{dt} \langle p_x \rangle = \int \Psi^* \left[V\frac{\partial\Psi}{\partial x} - \frac{\partial}{\partial x}(V\Psi) \right] d\tau$$

$$= \int \Psi^* \left(-\frac{\partial V}{\partial x} \right) \Psi d\tau = \left\langle -\frac{\partial V}{\partial x} \right\rangle. \qquad (7.144)$$

This result is called *Ehrenfest's theorem* (Paul Ehrenfest (1880–1933)).   □

Example 7.5.5. (The three-dimensional time-independent Schrödinger equation in a spherically symmetric potential) We solve the Schrödinger equation (7.131) for a spherically symmetric potential $V(\mathbf{r}) = V(r)$. This equation may be written in spherical polar coordinates (r, θ, φ) in the form:

$$\frac{1}{r^2}\frac{\partial}{\partial r}\left(r^2\frac{\partial\psi}{\partial r} \right) + \frac{1}{r^2\sin\theta}\frac{\partial}{\partial\theta}\left(\sin\theta\frac{\partial\psi}{\partial\theta} \right) + \frac{1}{r^2\sin^2\theta}\frac{\partial^2\psi}{\partial\varphi^2} + K\big[E - V(r)\big]\psi = 0,$$
$$(7.145)$$

where $\psi = \psi(r, \theta, \varphi)$, $K = 2m/\hbar^2$, $0 \le r < \infty$, $0 \le \theta \le \pi$, and $0 \le \varphi \le 2\pi$.

We seek a separable solution of the form

$$\psi = R(r)Y(\theta, \varphi)$$

and then substitute into (7.145) to obtain two equations

$$\frac{d}{dr}\left[r^2\frac{dR}{dr} \right] + \big[K(E - V)r^2 - \lambda \big]R = 0, \qquad (7.146)$$

$$\left[\frac{1}{\sin\theta} \frac{\partial}{\partial\theta} \left(\sin\theta \frac{\partial}{\partial\theta} \right) + \frac{1}{\sin^2\theta} \frac{\partial^2}{\partial\varphi^2} \right] Y + \lambda Y = 0, \qquad (7.147)$$

where λ is a separation constant.

We first solve (7.147) by separation of variables through $Y(\theta, \varphi) = \Theta(\theta) \times \Phi(\varphi)$ so that this equation becomes

$$\sin\theta \frac{d}{d\theta} \left(\sin\theta \frac{d\Theta}{d\theta} \right) + \left(\lambda \sin^2\theta - m^2 \right) \Theta = 0 \qquad (7.148)$$

and

$$\frac{d^2\Phi}{d\varphi^2} + m^2 \Phi = 0, \qquad (7.149)$$

where m^2 is the second separation constant.

The general solution of (7.149) is

$$\Phi = Ae^{im\varphi} + Be^{-im\varphi},$$

where the arbitrary constants A and B are to be determined from the boundary conditions on $\psi(r, \theta, \varphi) = R(r)\Theta(\theta)\Phi(\varphi)$.

According to the basic postulate of quantum mechanics, the wave function for a particle without spin must have a definite value at every point in space. In particular, ψ must have the same value whether the azimuthal coordinate φ is φ or $\varphi + 2\pi$, that is, $\Phi(\varphi) = \Phi(\varphi + 2\pi)$. Consequently, the solution for Φ has the form

$$\Phi = Ce^{im\varphi}, \quad m = 0, \pm 1, \pm 2, \ldots, \qquad (7.150)$$

where C is an arbitrary constant.

To solve equation (7.148), it is convenient to change the variable $x = \cos\theta$, $\Theta(\theta) = u(x)$, $-1 \leq x \leq 1$, so that it becomes

$$\frac{d}{dx} \left[(1 - x^2) \frac{du}{dx} \right] + \left(\lambda - \frac{m^2}{1 - x^2} \right) u = 0. \qquad (7.151)$$

When $m = 0$, this is the Legendre equation, which gives the Legendre polynomials $P_l(x)$ of degree l as solutions provided $\lambda = l(l+1)$, where l is a positive integer or zero.

When $m \neq 0$, (7.151) with $\lambda = l(l+1)$ admits solutions which are well known as the associated Legendre functions $P_l^m(x)$ of degree l and order m defined by

$$P_l^m(x) = (1 - x^2)^{m/2} \frac{d^m}{dx^m} P_l(x), \quad x = \cos\theta. \qquad (7.152)$$

Clearly, $P_l^m(x)$ vanishes when $m > l$. As for negative integral values of m, it can be readily shown that

$$P_l^{-m}(x) = (-1)^m \frac{(l-m)!}{(l+m)!} P_l^m(x). \tag{7.153}$$

Hence, the functions $P_l^{-m}(x)$ differ from $P_l^m(x)$ by a constant factor, and as a consequence, m can take only non-negative integral values. Thus, the associated Legendre functions $P_l^m(x)$ with $|m| \le l$ are the only nonsingular and physically acceptable solutions of (7.151). Since $|m| \le l$, when $l = 0$, we have $m = 0$; when $l = 1$, we have $m = -1, 0, +1$; when $l = 2$, we have $m = -2, -1, 0, 1, 2$; and so on. This means that, given l, there are exactly $(2l + 1)$ different values of $m = -l, \dots, -1, 0, 1, \dots, l$. The numbers l and m are called the *orbital quantum number* and the *magnetic quantum number*, respectively.

It is convenient to write down the solutions of (7.147) as functions which are normalized with respect to an integration over the whole solid angle. They are called *spherical harmonics* and are given by

$$Y_l^m(\theta, \varphi) = \left[\frac{(2l+1)}{4\pi} \frac{(l-m)!}{(l+m)!} \right]^{1/2} (-1)^m e^{im\varphi} P_l^m(\cos\theta), \quad m \ge 0, \tag{7.154}$$

where

$$P_l^m(\cos\theta) = \sin^{|m|}(\theta) \frac{d^{|m|}}{dx^{|m|}} P_l(x), \quad x = \cos\theta, \tag{7.155}$$

and

$$P_l(x) = \left(2^l l!\right)^{-1} \frac{d^l}{dx^l} (x^2 - 1)^l. \tag{7.156}$$

The identity (7.156) is known as the Rodrigues formula for the Legendre polynomial $P_l(x)$ of degree l.

To obtain the complete solution of (7.145), we combine the spherical harmonic solutions (7.154) with the solutions of the radial equation (7.146). For a given $\lambda = l(l + 1)$, (7.146) possesses radial eigenfunctions $R_{n,l}(r)$ in the normalized form corresponding to energy eigenvalues $E_{n,l}$ of the system. These energy levels may form a discrete spectrum or a continuous spectrum depending on the nature of the given potential $V(r)$. Finally, the solutions of the eigenvalue problem (7.145) are of the form

$$\psi_{n,l,m}(r, \theta, \varphi) = R_{n,l}(r) Y_l^m(\theta, \varphi), \tag{7.157}$$

where n is usually called the *principal quantum number*, which can have the values $1, 2, 3, \dots$. □

7.6 The Schrödinger Picture

State vectors in quantum mechanics describe values or probabilities for physical observables at a given time. As in classical mechanics, equations of motion in quantum mechanics determine how these values or probabilities change in time. It is sufficient for a quantum state to specify expectation values of observable operators, and hence the equations of motion are required to describe the time evolution of the expectation values. The expectation value of an observable operator in the state represented by the normalized state vector $|\Psi(\mathbf{r}, t)\rangle$ is given by

$$\langle \hat{A} \rangle = \langle \hat{A} \rangle_\Psi = \langle \Psi | \hat{A} \Psi \rangle. \tag{7.158}$$

The time development of the expectation value can therefore be described in at least three different ways:

(a) The state vector $\Psi(\mathbf{r}, t)$, which will be simply written as $\Psi(t)$, changes with time t, and each observable quantity is represented by an operator $\hat{A}$, which is independent of time t. Then the expectation value has the time dependence

$$\langle \hat{A} \rangle(t) = \langle \Psi(t) | \hat{A} \Psi(t) \rangle. \tag{7.159}$$

This is called the *Schrödinger picture*.

(b) Each observable operator $\hat{A}$ changes with time t and the state vector Ψ is independent of time. The expectation value has the time dependence

$$\langle \hat{A}(t) \rangle = \langle \Psi | \hat{A}(t) \Psi \rangle. \tag{7.160}$$

This is known as the *Heisenberg picture*.

(c) Both the state vector $\Psi(t)$ and observable operator $\hat{A}(t)$ change with time t. This is known as the *interaction picture*, which is a combination of the Schrödinger and the Heisenberg pictures.

To describe the Schrödinger picture, it is convenient to use the time-dependent Schrödinger equation (7.93) and to rewrite it in terms of a time-dependent state vector $\Psi(t)$ and the Hamiltonian H:

$$i\hbar \frac{d}{dt} \Psi(t) = \hat{H} \Psi(t), \tag{7.161}$$

where the total time derivative is used here since the dependence of Ψ on the space coordinates or other variables do not appear *explicitly*. The solution of

(7.161) is given by

$$\Psi(t) = \exp\left[-\frac{i}{\hbar}\int_0^t \hat{H}dt\right]\Psi_0 = \hat{U}(t)\Psi_0, \qquad (7.162)$$

where $\Psi_0 = \Psi(0)$, or more explicitly, $\Psi_0 = \Psi(\mathbf{r}, 0)$, and the operator $\hat{U}(t)$ is given by

$$\hat{U}(t) = \exp\left[-\frac{i}{\hbar}\int_0^t \hat{H}dt\right] = \exp\left[if(\hat{H})\right], \qquad (7.163)$$

and

$$f(\hat{H}) = -\frac{1}{\hbar}\int_0^t \hat{H}dt \qquad (7.164)$$

is a real function of the Hamiltonian operator $\hat{H}$. Since the adjoint of $\hat{U}(t)$ is

$$\hat{U}^*(t) = \exp\left[-if(\hat{H})\right], \qquad (7.165a)$$

then

$$\hat{U}^*\hat{U} = \hat{U}\hat{U}^* = \mathcal{I}. \qquad (7.165b)$$

Clearly, $\hat{U}(t)$ is a unitary operator. Consequently, for any Φ and Ψ, we obtain

$$\left(\hat{U}\Phi|\hat{U}\Psi\right) = \left(\Phi|\hat{U}^*\hat{U}\Psi\right) = \left(\Phi|\Psi\right). \qquad (7.166)$$

In particular, if $\hat{H}$ does not depend on time t explicitly ($d\hat{H}/dt = 0$), then the solution (7.162) becomes

$$\Psi(t) = \exp\left(-\frac{i\hat{H}}{\hbar}t\right)\Psi(0) = \hat{U}(t)\Psi(0), \qquad (7.167)$$

where the unitary operator $\hat{U}(t)$ has the form

$$\hat{U}(t) = \exp\left(-\frac{i\hat{H}t}{\hbar}\right). \qquad (7.168)$$

More generally, the solution (7.167) has the form

$$\Psi(t) = \exp\left[-\frac{i}{\hbar}\hat{H}(t - t_0)\right]\Psi(t_0) = \hat{U}(t, t_0)\Psi(t_0). \qquad (7.169)$$

It follows from the properties of unitary operators that $\hat{U}(t)$ has an inverse $\hat{U}^{-1}(t)$, $\hat{U}^{-1} = \hat{U}^*$ and $\|\hat{U}\Psi_0\| = \|\Psi_0\|$. So $\hat{U}(t)$ is isometric.

The upshot of this analysis is that if a physical system is not disturbed by an experiment, there exists a unitary operator $\hat{U}(t)$, defined on the infinite dimensional Hilbert space of the system, which describes the time evolution of the state vector from a given time to any time t:

$$\Psi(t) = \hat{U}(t, t_0)\Psi(t_0), \tag{7.170}$$

where $\hat{U}$ is called the *time-evolution* (or *time-development*) *operator*. The operator $\hat{U}(t, t_0)$ does not depend on $\Psi(t_0)$. The immediate consequence is that

$$\Psi(t_2) = \hat{U}(t_2, t_1)\Psi(t_1) = \hat{U}(t_2, t_1)\hat{U}(t_1, t_0)\Psi(t_0) = \hat{U}(t_2, t_0)\Psi(t_0).$$

Hence, the time-evolution development operator has the property

$$\hat{U}(t_2, t_0) = \hat{U}(t_2, t_1)\hat{U}(t_1, t_0). \tag{7.171}$$

It follows from (7.170) that

$$\hat{U}(t, t) = \mathcal{I}. \tag{7.172}$$

Thus,

$$\hat{U}(t, t_0)\hat{U}(t_0, t) = \hat{U}(t_0, t)\hat{U}(t, t_0) = \mathcal{I}.$$

In other words,

$$\hat{U}^{-1}(t, t_0) = \hat{U}(t_0, t). \tag{7.173}$$

Writing $t_2 = t$ and $t_1 = t - \delta t$ where δt is infinitesimal in (7.171), we obtain

$$\hat{U}(t, t_0) = \hat{U}(t, t - \delta t)\hat{U}(t - \delta t, t_0). \tag{7.174}$$

Now, $\hat{U}(t, t - \delta t)$ is an infinitesimal unitary operator, which may be written in the form

$$\hat{U}(t, t - \delta t) = 1 - \frac{i}{\hbar}\delta t \hat{H}(t), \tag{7.175}$$

where the factor $i/\hbar$ is introduced for convenience of interpretation, and $\hat{H}(t)$ is a Hermitian operator, and is usually referred to as the *generator* of the infinitesimal unitary transformation. We next substitute (7.175) into (7.174) to obtain

$$\hat{U}(t, t_0) = \hat{U}(t - \delta t, t_0) - \frac{i}{\hbar}\delta t \hat{H}(t)\hat{U}(t - \delta t, t_0)$$

or

$$\lim_{\delta t \to 0}\frac{1}{\delta t}\left[\hat{U}(t, t_0) - \hat{U}(t - \delta t, t_0)\right] = \frac{1}{i\hbar}\hat{H}(t)\hat{U}(t, t_0).$$

This leads to the evolution equation

$$i\hbar\frac{\partial}{\partial t}\hat{U}(t, t_0) = \hat{H}(t)\hat{U}(t, t_0). \tag{7.176}$$

Obviously, its complex conjugate equation is

$$-i\hbar\frac{\partial}{\partial t}\hat{U}^*(t, t_0) = \hat{U}^*(t, t_0)\hat{H}(t). \tag{7.177}$$

It follows from (7.176) that this equation determines $\hat{U}(t, t_0)$ for any time t with the initial condition $\hat{U}(t_0, t_0) = 1$. So an application of the operator equation (7.176) to the state vector $\Psi(t_0)$ gives

$$i\hbar\frac{\partial}{\partial t}\hat{U}(t, t_0)\Psi(t_0) = \hat{H}(t)\hat{U}(t, t_0)\Psi(t_0)$$

which is, by (7.170),

$$i\hbar\frac{\partial\Psi}{\partial t} = \hat{H}(t)\Psi(t). \tag{7.178}$$

This is the time-dependent Schrödinger equation for the state vector $\Psi(t)$.

The time rate of change of the expectation value of an observable $\hat{A}$ in the state $\Psi(t)$ in the Schrödinger picture is given by

$$\frac{d}{dt}\langle\hat{A}\rangle = \frac{d}{dt}\langle\Psi|\hat{A}\Psi\rangle. \tag{7.179}$$

By direct differentiation of the right-hand side, we find

$$\frac{d}{dt}\langle\hat{A}\rangle = \left\langle\frac{\partial\Psi}{\partial t}\bigg|\hat{A}\Psi\right\rangle + \left\langle\Psi\bigg|\hat{A}\frac{\partial\Psi}{\partial t}\right\rangle. \tag{7.180}$$

This can be expressed in terms of the Hamiltonian operator $\hat{H}$ by using (7.93) and its Hermitian conjugate

$$-i\hbar\frac{\partial}{\partial t}\langle\Psi| = \langle\Psi(t)|\hat{H}, \tag{7.181}$$

in the form

$$\begin{aligned}
\frac{d}{dt}\langle\hat{A}\rangle &= \frac{i}{\hbar}\langle\hat{H}\Psi|\hat{A}\Psi\rangle - \frac{i}{\hbar}\langle\Psi|\hat{A}\hat{H}\Psi\rangle \\
&= \left\langle\Psi\bigg|\frac{i}{\hbar}(\hat{H}\hat{A} - \hat{A}\hat{H})\Psi\right\rangle \\
&= \left\langle\Psi\bigg|\frac{i}{\hbar}[\hat{H}, \hat{A}]\Psi\right\rangle \\
&= \left\langle\frac{i}{\hbar}[\hat{H}, \hat{A}]\right\rangle.
\end{aligned} \tag{7.182}$$

This is the basic time-evolution equation for the expectation value of an observable operator. In particular, if $\hat{H}$ commutes with $\hat{A}$, then $[\hat{H}, \hat{A}] = 0$, and the expectation value of $\hat{A}$ is a constant of motion, that is, $\langle\hat{A}\rangle$ is conserved (or invariant with respect to time t). Moreover, if $[\hat{H}, \hat{A}^2] = 0$, then $\langle\hat{A}\rangle$ is a constant of time, and $\Delta\hat{A}^2 = \langle\hat{A}^2\rangle - \langle\hat{A}\rangle^2$ is also a constant of time. In fact, any function of an observable $\hat{A}$ has a conserved expectation value. The observable is then said to be *conserved*. In particular, a particle which is initially in an eigenstate of $\hat{A}$ belonging to an eigenvalue λ remains in that condition at all times. Finally, since $\hat{H}$ always commutes with itself, then $\langle\hat{H}\rangle$ and $\Delta\hat{H}$ are always conserved, which leads to the result in classical mechanics that the energy is a constant of the motion.

If the quantum system is in one of the stationary states given by

$$\psi = \psi_n = \varphi_n(r) \exp\left[-\left(\frac{i}{\hbar}\right)E_n t\right],$$

then

$$\langle\psi_n|[\hat{A}, \hat{H}]\psi_n\rangle = \langle\varphi_n|[\hat{A}, \hat{H}]\varphi_n\rangle$$
$$= (\bar{E}_n - E_n)\langle\varphi_n|\hat{A}\varphi_n\rangle, \quad \text{since } \hat{H}\varphi_n = E_n\varphi_n$$
$$= 0, \quad \text{since } \bar{E}_n = E_n.$$

It follows from (7.182) that

$$\frac{d}{dt}\langle\hat{A}\rangle = 0$$

for the stationary state.

In the presence of an external field, $\hat{A}$ may depend on time t; then an additional term $\langle\Psi|\frac{\partial\hat{A}}{\partial t}\Psi\rangle$ must be added to the right-hand side of (7.180) so that (7.182) becomes

$$\frac{d}{dt}\langle\hat{A}\rangle = \left\langle\Psi\left|\left(\frac{\partial\hat{A}}{\partial t} + \frac{i}{\hbar}[\hat{H}, \hat{A}]\right)\Psi\right.\right\rangle.$$

The right-hand side is clearly equal to the expectation value of the operator in the brackets so that

$$\frac{d}{dt}\langle\hat{A}\rangle = \left\langle\left(\frac{\partial\hat{A}}{\partial t} + \frac{i}{\hbar}[\hat{H}, \hat{A}]\right)\right\rangle$$
$$= \left\langle\frac{\partial\hat{A}}{\partial t}\right\rangle + \frac{i}{\hbar}\langle[\hat{H}, \hat{A}]\rangle, \tag{7.183}$$

or, equivalently,

$$i\hbar\frac{d}{dt}\langle\hat{A}\rangle = i\hbar\left\langle\frac{\partial\hat{A}}{\partial t}\right\rangle + \langle[\hat{A}, \hat{H}]\rangle. \tag{7.184}$$

It is then possible to define an operator $d\hat{A}/dt$ and call it the *total time derivative* of $\hat{A}$, by requiring the equality of the expectation values

$$\left\langle \frac{d\hat{A}}{dt} \right\rangle = \frac{d}{dt}\langle\hat{A}\rangle \tag{7.185}$$

for every state. In view of (7.184), the dynamical equation (7.183) assumes the operator form

$$\frac{d\hat{A}}{dt} = \frac{\partial\hat{A}}{\partial t} + \frac{1}{i\hbar}[\hat{A}, \hat{H}]. \tag{7.186}$$

We shall now give several examples of concrete applications of (7.182) and (7.183):

(a)

$$\hat{A} = \hat{\mathbf{p}} = -i\hbar\nabla.$$

Since $\hat{\mathbf{p}}$ does not involve time explicitly, $\partial\hat{\mathbf{p}}/\partial t = 0$. The commutator

$$[\hat{H}, \hat{\mathbf{p}}] = \hat{H}\hat{\mathbf{p}} - \hat{\mathbf{p}}\hat{H} = \left(\frac{\hat{\mathbf{p}}^2}{2m} + \hat{V}\right)\hat{\mathbf{p}} - \hat{\mathbf{p}}\left(\frac{\hat{\mathbf{p}}^2}{2m} + \hat{V}\right) = \hat{V}\hat{\mathbf{p}} - \hat{\mathbf{p}}\hat{V}$$

gives the result

$$[\hat{H}, \hat{\mathbf{p}}]\psi = -i\hbar\hat{V}\nabla\psi + i\hbar\nabla(\hat{V}\psi) = i\hbar\psi\nabla\hat{V}.$$

Substituting this result into (7.183), we obtain

$$\frac{d}{dt}\langle\hat{\mathbf{p}}\rangle = -\langle\nabla\hat{V}\rangle = \langle\mathbf{F}\rangle, \tag{7.187}$$

where $\mathbf{F}$ is the classical force field. This result is known as *Ehrenfest's theorem*.

For a free particle, $\hat{V} \equiv 0$; then (7.187) becomes

$$\frac{d}{dt}\langle\hat{\mathbf{p}}\rangle = 0. \tag{7.188}$$

This corresponds to the classical conservation law of linear momentum for a free particle.

(b)

$$\hat{A} = \hat{\mathbf{r}}.$$

Then $\partial\hat{\mathbf{r}}/\partial t = 0$, and

$$[\hat{H}, \hat{\mathbf{r}}] = \hat{H}\hat{\mathbf{r}} - \hat{\mathbf{r}}\hat{H} = \left(\frac{\hat{\mathbf{p}}^2}{2m} + \hat{V}\right)\hat{\mathbf{r}} - \hat{\mathbf{r}}\left(\frac{\hat{\mathbf{p}}^2}{2m} + \hat{V}\right)$$

$$= \frac{1}{2m}[\hat{\mathbf{p}}^2, \hat{\mathbf{r}}], \tag{7.189}$$

since $\hat{V}$ commutes with $\hat{\mathbf{r}}$, both being multiplicative operators. The nth component of the commutator on the right-hand side of (7.189) is

$$[\hat{\mathbf{p}}^2, \hat{r}_n] = [\hat{p}_n^2, \hat{r}_n] = [\hat{p}_n, \hat{r}_n]\hat{p}_n + \hat{p}_n[\hat{p}_n, \hat{r}_n] = -2i\hbar\hat{p}_n$$

and so, in general,

$$[\hat{\mathbf{p}}^2, \hat{\mathbf{r}}] = -2i\hbar\hat{\mathbf{p}}. \tag{7.190}$$

Substituting this result into (7.189), Equation (7.183) gives

$$\frac{d}{dt}\langle\hat{\mathbf{r}}\rangle = \frac{1}{m}\langle\hat{\mathbf{p}}\rangle. \tag{7.191}$$

This corresponds to the classical result

$$m\frac{d\mathbf{r}}{dt} = \mathbf{p}. \tag{7.192}$$

(c) $\hat{A} = \hat{H} = \hat{T} + \hat{V}$ so that the total energy is such that the potential $\hat{V}$ does not depend on time t.

Consequently,

$$\frac{\partial\hat{H}}{\partial t} = 0 \quad \text{and} \quad [\hat{H}, \hat{H}] = 0.$$

Equation (7.183) implies that

$$\frac{d}{dt}\langle\hat{H}\rangle = 0. \tag{7.193}$$

This result corresponds to the classical conservation law of total energy. Similarly, all the conservation laws of classical mechanics can be reproduced in this way. Finally, all three applications (a) to (c) are remarkable examples of the Correspondence Principle in the sense that the expectation values of observables in quantum mechanics behave in the same way as the observables themselves do in classical mechanics.

7.7 The Heisenberg Picture and the Heisenberg Equation of Motion

We denote a state vector and an observable operator in the Schrödinger picture by Ψ_S and $\hat{A}_S$, and in the Heisenberg picture by Ψ_H and $\hat{A}_H$. According to the

Heisenberg picture, $\hat{A}_H$ depends on time t, but Ψ_H does not depend on t. If we define $\Psi_H(t)$ by

$$\Psi_H(t) = \hat{U}^{-1}(t, t_0)\Psi_S(t), \tag{7.194}$$

where $\Psi_H(t)$ is independent of t, and $\hat{U}(t, t_0)$ is a unitary operator, then (7.194) represents a unitary operator in the vector space. In view of (7.170), the result becomes

$$\Psi_H(t) = \hat{U}^{-1}(t, t_0)\hat{U}(t, t_0)\Psi_S(t_0) = \Psi_S(t_0). \tag{7.195}$$

So it turns out that a Heisenberg operator $\hat{A}_H$ is related to a Schrödinger operator $\hat{A}_S$ by the unitary transformation law

$$\begin{aligned} \hat{A}_H &= \hat{U}^{-1}(t, t_0)\hat{A}_S\hat{U}(t, t_0) \\ &= \hat{U}^*(t, t_0)\hat{A}_S\hat{U}(t, t_0). \end{aligned} \tag{7.196}$$

Differentiating this result with respect to t gives

$$\frac{d\hat{A}_H}{dt} = \frac{\partial \hat{U}^*}{\partial t}\hat{A}_S\hat{U} + \hat{U}^*\hat{A}_S\frac{\partial \hat{U}}{\partial t}. \tag{7.197}$$

In view of the differential equations (7.176) and (7.177), Equation (7.197) takes the form

$$\begin{aligned} \frac{d\hat{A}_H}{dt} &= \frac{1}{i\hbar}\left(\hat{U}^*\hat{A}_S\hat{H}_S\hat{U} - \hat{U}^*\hat{H}_S\hat{A}_S\hat{U}\right) \\ &= \frac{1}{i\hbar}\left(\hat{U}^*\hat{A}_S\hat{U}\hat{U}^*\hat{H}_S\hat{U} - \hat{U}^*\hat{H}_S\hat{U}\hat{U}^*\hat{A}_S\hat{U}\right) \\ &= \frac{1}{i\hbar}\left(\hat{A}_H\hat{H}_H - \hat{H}_H\hat{A}_H\right) \\ &= \frac{1}{i\hbar}\left[\hat{A}_H, \hat{H}_H\right]. \end{aligned} \tag{7.198}$$

If $\hat{A}_H$ depends on time t explicitly, then (7.198) can be generalized to

$$\frac{d\hat{A}_H}{dt} = \frac{\partial \hat{A}_H}{\partial t} + \frac{1}{i\hbar}\left[\hat{A}_H, \hat{H}_H\right], \tag{7.199}$$

where the first term on the right-hand side of (7.199) is defined by

$$\frac{\partial \hat{A}_H}{\partial t} = \hat{U}\frac{\partial \hat{A}_S}{\partial t}\hat{U}^*. \tag{7.200}$$

Equations (7.198) and (7.199) are the celebrated *Heisenberg equations of motion* for the operator $\hat{A}_H$. They are similar in form to the Hamilton equations of motion in classical mechanics. The strong resemblance between (7.199) and (7.36) suggests that quantum analogues of the classical equations of motion can be obtained in general by substituting the commutator bracket divided by $i\hbar$ for the Poisson bracket

$$\{A, B\} \to \frac{1}{i\hbar}[\hat{A}, \hat{B}]. \tag{7.201}$$

Note that the Schrödinger Equation (7.93) is most suitable for quantum mechanic calculations. The Heisenberg equation (7.198) or (7.199) is more closely related to classical theory and can be shown to imply that classical mechanics is indeed the limit of quantum mechanics as $\hbar \to 0$.

For the basic canonical operators $\hat{q}_j$ and $\hat{p}_j$, Equation (7.198) assumes the form

$$\frac{d\hat{q}_j}{dt} = \frac{1}{i\hbar}[\hat{q}_j, \hat{H}] = \frac{1}{i\hbar}\left(i\hbar\frac{\partial \hat{H}}{\partial \hat{p}_j}\right) = \frac{\partial \hat{H}}{\partial \hat{p}_j},$$

$$\frac{d\hat{p}_j}{dt} = \frac{1}{i\hbar}[\hat{p}_j, \hat{H}] = \frac{1}{i\hbar}\left(-i\hbar\frac{\partial \hat{H}}{\partial \hat{q}_j}\right) = -\frac{\partial \hat{H}}{\partial \hat{q}_j}.$$

These are identical with the Hamilton equations of motion in classical mechanics.

Theorem 7.7.1. *The equations of motion for the expectation values in the Heisenberg picture and in the Schrödinger picture are the same.*

Proof: We have

$$\frac{d}{dt}\langle \hat{A}_H \rangle = \frac{d}{dt}\langle \Psi_H | \hat{A}_H(t)\Psi_H \rangle$$

$$= \left\langle \Psi_H \left| \frac{d\hat{A}}{dt} \Psi_H \right.\right\rangle$$

$$= \left\langle \frac{1}{i\hbar}[\hat{A}_H, \hat{H}_H] \right\rangle, \tag{7.202}$$

where $\hat{A}_H$ does not depend on time explicitly, and Ψ_H is independent of time.

However, if $\hat{A}_H$ depends on time t explicitly, then

$$\frac{d}{dt}\langle \hat{A}_H \rangle = \left\langle \frac{\partial \hat{A}_H}{\partial t} \right\rangle + \left\langle \frac{1}{i\hbar}[\hat{A}_H, \hat{H}_H] \right\rangle. \tag{7.203}$$

Thus, the equations of motion (7.202) and (7.203) for $\langle \hat{A}_H \rangle$ are identically the same for $\langle \hat{A}_S \rangle$, which satisfies (7.182) and (7.183). □

It follows that the Heisenberg picture is more akin to classical dynamics than the Schrödinger picture. The latter does, in fact, emphasize the Hamilton–Jacobi formalism of classical mechanics, and the operator $d\hat{A}/dt$ in the Schrödinger picture is defined by (7.186).

In view of the fact that the Heisenberg Equation (7.198) refers to operators, it is not of much practical importance in particular quantum mechanic problems. This leads to a dependence on all the expectation values of the operator for any given state Ψ. However, if any observable operator $\hat{C}(t)$ commutes with $\hat{H}$, such as $\hat{L}^2$ or $\hat{m}_z$, then the following conclusions are evident. It follows from (7.198) that

$$\frac{d\hat{C}}{dt} = \frac{1}{i\hbar}\left[\hat{C}(t), \hat{H}\right] = 0. \qquad (7.204)$$

Taking the expectation value for any state, it follows from (7.202) that

$$\frac{d}{dt}\langle\hat{C}(t)\rangle = 0. \qquad (7.205)$$

This shows that $\langle\hat{C}(t)\rangle$ does not change with time. If the system is in an eigenstate of $\hat{C}$ at $t = 0$, the state will remain in the eigenstate at any subsequent time because the operator does not change with time. All such operators $\hat{C}(t)$ satisfying the commutator relation

$$\left[\hat{C}(t), \hat{H}\right] = 0 \qquad (7.206)$$

are also called *constants* of the motion. They are the generalizations of the conserved quantities of classical mechanics. For example, the total momentum operator for a free particle is a constant of motion. For a particle in any central potential field, the total angular momentum and each separate component all commute with the Hamiltonian $\hat{H}$ and, hence, according to the preceding argument, they are also constants of the motion.

Example 7.7.2. We apply the Heisenberg equation to $\hat{A}_H(t) = \hat{x}(t)$ for the one-dimensional motion of a particle moving in a potential $V(\hat{x})$ so that the Hamiltonian operator $\hat{H}$ is given by

$$\hat{H} = \frac{\hat{p}^2}{2m} + V(\hat{x}).$$

Then

$$i\hbar\frac{d\hat{x}(t)}{dt} = \left[\hat{x}(t), \hat{H}(t)\right]$$

$$= \hat{x}\hat{H} - \hat{H}\hat{x}$$

$$= \hat{x}\left(\frac{\hat{p}^2}{2m} + V(\hat{x})\right) - \left(\frac{\hat{p}^2}{2m} + V(\hat{x})\right)\hat{x}$$

$$= \frac{1}{2m}\left(\hat{x}\hat{p}^2 - \hat{p}^2\hat{x}\right)$$

$$= \frac{1}{2m}\left([\hat{x}, \hat{p}]\hat{p} + \hat{p}[\hat{x}, \hat{p}]\right)$$

$$= \frac{2i\hbar\hat{p}}{2m} = \frac{i\hbar\hat{p}}{m}.$$

Therefore,

$$\frac{d\hat{x}(t)}{dt} = \frac{\hat{p}(t)}{m}. \tag{7.207}$$

Similarly, if $\hat{A}(t) = \hat{p}(t)$, then

$$i\hbar\frac{d\hat{p}(t)}{dt} = \left[\hat{p}(t), \hat{H}\right]$$

$$= \hat{p}(t)\left(\frac{\hat{p}^2}{2m} + V\right) - \left(\frac{\hat{p}^2}{2m} + V\right)\hat{p}(t)$$

$$= \left(-i\hbar\frac{\partial}{\partial x}\right)V + i\hbar V\frac{\partial}{\partial x}1$$

$$= -i\hbar\frac{\partial V(x)}{\partial x}.$$

Hence,

$$\frac{d}{dt}\hat{p}(t) = -\frac{\partial V(x)}{\partial x}. \tag{7.208}$$

Equations (7.207) and (7.208) may readily be generalized to three dimensions. Equation (7.208) is the direct generalization to operators of Newton's second law of motion. This shows that the Schrödinger Equation (7.93) or equivalently the Heisenberg Equation (7.198) implies that the time-dependent observable-operators, defined by (7.196), satisfy exactly the same equations as the corresponding classical variables.   □

7.8 The Interaction Picture

In this picture, the total Hamiltonian operator $\hat{H}(t)$ is expressed as the sum of two terms

$$\hat{H}(t) = \hat{H}^{(0)} + \hat{H}^{(1)}(t), \tag{7.209}$$

where $\hat{H}^{(0)}$ is the time-independent term representing the Hamiltonian of the system in the absence of an external field, and $\hat{H}^{(1)}(t)$ represents the time-dependent term that arises in the presence of an external field. In the absence of the latter term, the time-evolution operator is obtained from (7.168) in the form

$$\hat{U}_0(t, t_0) = \exp\left[-\frac{i}{\hbar}(t - t_0)\hat{H}^{(0)}\right]. \tag{7.210}$$

Both the state vector $\Psi_I(t)$ and the operator $\hat{A}_I(t)$ depend on time t and are defined by

$$\Psi_I(t) = \hat{U}_0^{-1}(t, t_0)\Psi_S(t) = \exp\left[\frac{i}{\hbar}(t - t_0)\hat{H}^{(0)}\right]\Psi_S(t), \tag{7.211}$$

$$\hat{A}_I(t) = \hat{U}_0^{-1}(t, t_0)\hat{A}_S\hat{U}_0(t, t_0)$$

$$= \exp\left[\frac{i}{\hbar}(t - t_0)\hat{H}^{(0)}\right]\hat{A}_S \exp\left[-\frac{i}{\hbar}(t - t_0)\hat{H}^{(0)}\right], \tag{7.212}$$

where $\Psi_S(t)$ is the state vector and $\hat{A}_S$ is the operator in the Schrödinger picture so that

$$i\hbar\frac{\partial \Psi_S}{\partial t} = \hat{H}(t)\Psi_S(t) = \left[\hat{H}^{(0)} + \hat{H}^{(1)}(t)\right]\Psi_S(t), \tag{7.213}$$

$$\frac{d\hat{A}_S}{dt} = \frac{\partial \hat{A}_S}{\partial t}. \tag{7.214}$$

It follows from (7.211) and (7.213) that

$$i\hbar\frac{\partial \Psi_I}{\partial t} = i\hbar\frac{\partial}{\partial t}\left\{\exp\left[\frac{i}{\hbar}(t - t_0)\hat{H}^{(0)}\right]\Psi_S(t)\right\}$$

$$= -\hat{H}^{(0)}\Psi_I + \exp\left[\frac{i}{\hbar}(t - t_0)\hat{H}^{(0)}\right]\left(i\hbar\frac{\partial \Psi_S}{\partial t}\right)$$

$$= -\hat{H}^{(0)}\Psi_I + \exp\left[\frac{i}{\hbar}(t - t_0)\hat{H}^{(0)}\right]\left[\hat{H}^{(0)} + \hat{H}^{(1)}(t)\right]\Psi_S(t)$$

$$= -\hat{H}^{(0)}\Psi_I + \hat{H}^{(0)}\Psi_I + \hat{U}_0^{-1}(t, t_0)\hat{H}^{(1)}(t)\hat{U}_0(t, t_0)\Psi_I(t)$$

$$= \hat{H}_I^{(1)}(t)\Psi_I(t), \tag{7.215}$$

where

$$\hat{H}_I^{(1)}(t) = \hat{U}_0^{-1}(t, t_0)\hat{H}^{(1)}\hat{U}_0(t, t_0). \tag{7.216}$$

On the other hand, it also follows from (7.212) and (7.214) that

$$\frac{d\hat{A}_I}{dt} = \frac{\partial \hat{A}_I}{\partial t} + \frac{1}{i\hbar}[\hat{A}_I, \hat{H}_I^{(0)}], \tag{7.217}$$

where

$$\hat{H}_I^{(0)} = \hat{U}_0^{-1}(t, t_0)\hat{H}^{(0)}\hat{U}_0(t, t_0) = \hat{H}^{(0)}. \tag{7.218}$$

These results show that the state vector $\Psi_I(t)$ in the interaction picture satisfies the Schrödinger equation (7.215) with the Hamiltonian $\hat{H}_I^{(1)}$, while the operator $\hat{A}_I(t)$ obeys the Heisenberg equation with the time-independent Hamiltonian $\hat{H}^{(0)}$.

7.9 The Linear Harmonic Oscillator

According to classical mechanics, a harmonic oscillator is a particle of mass m moving under the action of a force $F = -m\omega^2 x$. The equation of motion is then

$$\frac{d^2x}{dt^2} + \omega^2 x = 0. \tag{7.219}$$

The solution of this equation with the initial conditions, $x(0) = a$, $\dot{x}(0) = 0$ is

$$x = a\cos\omega t. \tag{7.220}$$

This represents an oscillatory motion of angular frequency ω and amplitude a. The potential is related to the force by $F = -\partial V/\partial x$ so that $V(x) = \frac{1}{2}m\omega^2 x^2$. The energy of the oscillatory motion is the potential energy when the particle is at the extreme position. Therefore, the total energy is

$$E = \frac{1}{2}ma^2\omega^2. \tag{7.221}$$

Since the amplitude a can have any non-negative value, the total energy E can have any value greater than or equal to zero. In other words, the energy forms a continuous spectrum.

We next consider the quantum theory of the harmonic oscillator. The total energy of the system is represented by the Hamiltonian operator

$$\hat{H} = \hat{T} + \hat{V} = \frac{\hat{p}^2}{2m} + \frac{1}{2}m\omega^2\hat{x}^2. \tag{7.222}$$

It is convenient to introduce two dimensionless operators $\hat{a}$ and $\hat{a}^*$ by

$$\hat{a} = \sqrt{\frac{m}{2}}\omega\hat{x} + \frac{i}{\sqrt{2m}}\hat{p}, \tag{7.223}$$

$$\hat{a}^* = \sqrt{\frac{m}{2}}\omega\hat{x} - \frac{i}{\sqrt{2m}}\hat{p}. \tag{7.224}$$

Since $\hat{x}$ and $\hat{p}$ are Hermitian operators, it follows that

$$\langle \psi_1 | \hat{a}\psi_2 \rangle = \langle \hat{a}^*\psi_1 | \psi_2 \rangle \quad \text{and} \quad \langle \psi_1 | \hat{a}^*\psi_2 \rangle = \langle \hat{a}\psi_1 | \psi_2 \rangle$$

for any two wave functions ψ_1 and ψ_2. Thus, the operators $\hat{a}$ and $\hat{a}^*$ are not Hermitian and hence they do not represent physical observables. However, $\hat{a}\hat{a}^*$ and $\hat{a}^*\hat{a}$ are Hermitian operators, because they can be represented as real functions of $\hat{H}$:

$$\hat{a}\hat{a}^* = \frac{\hat{p}^2}{2m} + \frac{m\omega^2}{2}\hat{x} - \frac{i\omega}{2}[\hat{x},\hat{p}] = \hat{H} + \frac{1}{2}\hbar\omega,$$

$$\hat{a}^*\hat{a} = \frac{\hat{p}^2}{2m} + \frac{m\omega^2}{2}\hat{x} + \frac{i\omega}{2}[\hat{x},\hat{p}] = \hat{H} - \frac{1}{2}\hbar\omega,$$

and hence $\hat{H}$ can be written in terms of $\hat{a}$ and $\hat{a}^*$ as

$$\hat{H} = \hat{a}^*\hat{a} + \frac{1}{2}\hbar\omega = \hat{a}\hat{a}^* - \frac{1}{2}\hbar\omega \tag{7.225}$$

so that

$$[\hat{a},\hat{a}^*] = \hbar\omega, \qquad [\hat{H},\hat{a}^*] = [\hat{a},\hat{a}^*]\hat{a}^* = \hbar\omega\hat{a}^*. \tag{7.226}$$

The eigenstate of energy E_n is $|E_n\rangle$ and

$$\hat{H}|E_n\rangle = E_n|E_n\rangle. \tag{7.227}$$

Using (7.225), we rewrite (7.227) either as

$$\hat{a}^*\hat{a}|E_n\rangle = \left(E_n - \frac{1}{2}\hbar\omega\right)|E_n\rangle \tag{7.228}$$

or

$$\hat{a}\hat{a}^*|E_n\rangle = \left(E_n + \frac{1}{2}\hbar\omega\right)|E_n\rangle. \tag{7.229}$$

Multiplying (7.228) by $\hat{a}$, we obtain

$$\hat{a}\hat{a}^*\hat{a}|E_n\rangle = \left(E_n - \frac{1}{2}\hbar\omega\right)\hat{a}|E_n\rangle. \tag{7.230}$$

Then, either

$$\hat{a}|E_n\rangle = 0 \tag{7.231}$$

or

$$\hat{a}|E_n\rangle = |E_{n-1}\rangle. \tag{7.232}$$

This result is used to rewrite (7.230) as

$$\hat{a}\hat{a}^*|E_{n-1}\rangle = \left(E_n - \hbar\omega + \frac{1}{2}\hbar\omega \right)|E_{n-1}\rangle. \tag{7.233}$$

This is identical with (7.229) for E_{n-1}, provided

$$E_{n-1} = E_n - \hbar\omega. \tag{7.234}$$

Thus, given any eigenvector $|E_n\rangle$, it is possible to generate a new eigenvector $|E_{n-1}\rangle$, by (7.232), unless $|E_n\rangle$ is the lowest state $|E_0\rangle$. In this case, (7.231) is satisfied. It follows from (7.228) for $n = 0$ that

$$E_0 - \frac{1}{2}\hbar\omega = 0. \tag{7.235}$$

This determines the *lowest* (or *ground*) *state energy*. Clearly, it follows from (7.232) that $\hat{a}$ is the operator which annihilates energy in the system in quantum units of $\hbar\omega$, and $\hat{a}$ is called the *annihilation operator*.

Similarly, multiplication of (7.229) by $\hat{a}^*$ gives

$$\hat{a}^*\hat{a}\hat{a}^*|E_n\rangle = \left(E_n + \frac{1}{2}\hbar\omega \right)\hat{a}^*|E_n\rangle. \tag{7.236}$$

Then, either

$$\hat{a}^*|E_n\rangle = 0 \tag{7.237}$$

or

$$\hat{a}^*|E_n\rangle = |E_{n+1}\rangle. \tag{7.238}$$

This result is used to rewrite (7.236) as

$$\hat{a}^*\hat{a}|E_{n+1}\rangle = \left(E_n + \hbar\omega - \frac{1}{2}\hbar\omega \right)|E_{n+1}\rangle. \tag{7.239}$$

This is identical with equation (7.228) for E_{n+1} provided

$$E_{n+1} = E_n + \hbar\omega. \tag{7.240}$$

It follows that, given any eigenstate $|E_n\rangle$, it is also possible to generate a new eigenvector $|E_{n+1}\rangle$, by (7.238), with the eigenvalues given by (7.240), unless E_n is the highest energy level, in which case (7.237) is satisfied. But the potential is an increasing function of x and hence there is no highest level and the creation of higher energy levels is always possible. Thus, the operator $\hat{a}^*$ generates energy in the system in quantum units of $\hbar\omega$ and is called the *creation operator*. It then follows from (7.235) and (7.240) that the general energy level is

$$E_n = \left(n + \frac{1}{2} \right)\hbar\omega, \quad n = 0, 1, 2, \ldots. \tag{7.241}$$

This obviously represents a discrete set of energies. Thus, in quantum mechanics, a stationary state of the harmonic oscillator can assume only one of the values from the set E_n. The energy is thus quantized and forms a discrete spectrum. According to classical mechanics, the energy forms a continuous spectrum, that is, all non-negative numbers are allowed for the energy of a simple harmonic oscillator. This shows a remarkable contrast between the results of the classical and quantum theory.

The non-negative integer n, which characterizes the energy eigenvalues (and hence eigenfunctions), is called the *quantum number*. The value of $n = 0$ corresponds to the minimum value of the quantum number with energy

$$E_0 = \frac{1}{2}\hbar\omega. \tag{7.242}$$

This is called the *lowest* (or *ground*) *state energy*, which never vanishes, as the lowest possible classical energy would. The ground state energy E_0 is proportional to $\hbar$, representing a quantum phenomenon. The discrete energy spectrum is in perfect agreement with the quantization rules of the quantum theory.

To determine the energy eigenfunctions ψ_n belonging to E_n, it is convenient to write the annihilation and creation operators as $\hat{A} \equiv \hat{a}/\sqrt{\hbar\omega}$ and $\hat{A}^* \equiv \hat{a}^*\sqrt{\hbar\omega}$ and replace $\hat{p}$ by $-i\hbar(\partial/\partial x)$ so that

$$\hat{A} = \frac{1}{\sqrt{2}}\left[(\hbar\,m\omega)^{1/2}\frac{\partial}{\partial x} + (m\omega/\hbar)^{1/2}\hat{x}\right] = \frac{1}{\sqrt{2}}\left(\frac{\partial}{\partial\eta} + \hat{\eta}\right), \tag{7.243}$$

$$\hat{A}^* = \frac{1}{\sqrt{2}}\left[-(\hbar\,m\omega)^{1/2}\frac{\partial}{\partial x} + (m\omega/\hbar)^{1/2}\hat{x}\right] = \frac{1}{\sqrt{2}}\left(-\frac{\partial}{\partial\eta} + \hat{\eta}\right), \tag{7.244}$$

where $\hat{\eta} = (m\omega/\hbar)^{1/2}\hat{x}$. Consequently,

$$\hat{A}\hat{A}^* = \frac{\hat{H}}{\hbar\omega} + \frac{1}{2}, \tag{7.245a}$$

$$\hat{A}^*\hat{A} = \frac{\hat{H}}{\hbar\omega} - \frac{1}{2}. \tag{7.245b}$$

Since ψ_0 is the eigenfunction corresponding to the lowest energy, E_0, $\hat{A}\psi_0 = 0$ or

$$\frac{d\psi_0}{d\eta} + \eta\psi_0 = 0. \tag{7.246}$$

Its normalized solution can be written as

$$\psi_0(\eta) = \left(\frac{m\omega}{\pi\hbar}\right)^{1/4} e^{-\eta^2/2}. \tag{7.247}$$

All other eigenfunctions ψ_n can be calculated from ψ_0 by successive applications of the creation operator $\hat{A}^*$, and thus ψ_n is proportional to $(\hat{A}^*)^n \psi_0$. We also note that

$$\langle \hat{A}^* \psi_n | \hat{A}^* \psi_n \rangle = \langle \psi_n | \hat{A} \hat{A}^* \psi_n \rangle = \left\langle \psi_n \left| \left(\frac{\hat{H}}{\hbar \omega} + \frac{1}{2} \right) \psi_n \right\rangle \right. = (n+1) \langle \psi_n | \psi_n \rangle \tag{7.248}$$

so that if ψ_n is normalized, so is $\psi_{n+1} = (n+1)^{-1/2} \hat{A}^* \psi_n$. Thus, it turns out that

$$\psi_n = (n!)^{-1/2} \left(\hat{A}^* \right)^n \psi_0$$

$$= (2^n n!)^{-1/2} \left(-\frac{d}{d\eta} + \eta \right)^n \exp\left(-\frac{\eta^2}{2} \right). \tag{7.249}$$

This result can be simplified by using the operator identities

$$\left(-\frac{d}{d\eta} + \eta \right) = -e^{\eta^2/2} \frac{d}{d\eta} e^{-\eta^2/2}, \tag{7.250a}$$

$$\left(-\frac{d}{d\eta} + \eta \right)^n = (-1)^n e^{\eta^2/2} \frac{d^n}{d\eta^n} e^{-\eta^2/2}, \tag{7.250b}$$

so that the final form of ψ_n is

$$\psi_n(\eta) = (2^n n!)^{-1/2} \left(\frac{m\omega}{\pi \hbar} \right)^{1/4} e^{-\eta^2/2} \left[(-1)^n e^{\eta^2} \frac{d^n}{d\eta^n} e^{-\eta^2} \right] \tag{7.251}$$

$$= (2^n n!)^{-1/2} \left(\frac{m\omega}{\pi \hbar} \right)^{1/4} e^{-\eta^2/2} H_n(\eta), \quad n = 0, 1, 2, \ldots, \tag{7.252}$$

where the result in the square brackets in (7.251) defines $H_n(\eta)$, the Hermite polynomials of degree n.

Example 7.9.1. (The Schrödinger equation treatment of Planck's simple harmonic oscillator) The quantum mechanical motion of the Planck oscillator is described by the one-dimensional Schrödinger equation

$$\frac{d^2 \psi}{dx^2} + \frac{2M}{\hbar^2} \left(E - \frac{1}{2} M\omega^2 x^2 \right) \psi = 0. \tag{7.253}$$

In terms of the constants

$$\beta = \frac{2ME}{\hbar^2}, \tag{7.254a}$$

$$\alpha = \frac{M\omega}{\hbar} > 0, \tag{7.254b}$$

and an independent variable $x' = x\sqrt{\alpha}$, Equation (7.253) becomes, dropping the prime,

$$\frac{d^2\psi}{dx^2} + \left(\frac{\beta}{\alpha} - x\right)\psi = 0. \tag{7.255}$$

The eigenfunctions of this equation are the Hermite orthogonal functions

$$\psi_n(x) - A_n II_n\left(x\sqrt{\alpha}\right)\exp\left(-\frac{\alpha x^2}{2}\right) \tag{7.256}$$

with the corresponding eigenvalues

$$\frac{\beta}{\alpha} = (2n + 1), \tag{7.257}$$

where $H_n(x)$ is the Hermite polynomial of degree n. Substituting the values of α and β, it turns out that

$$E \equiv E_n = \left(\frac{2n+1}{2}\right)\omega\hbar, \quad n = 0, 1, 2, \ldots. \tag{7.258}$$

The so-called half-integral multiples of the energy quanta are the characteristics of the oscillator, that is, the *odd multiples* of $\frac{1}{2}\omega\hbar$. This result is remarkably the same as in the Heisenberg theory. In view of the following properties of the Hermite polynomials

$$H_0(x) = 1, \quad H_1(x) = 2x, \quad \text{and} \quad H_2(x) = 4x^2 - 2,$$

it follows that the first eigenfunction $\psi_0(x)$ represents a Gaussian distribution curve, and the second eigenfunction $\psi_1(x)$ vanishes at the origin and corresponds to a Maxwellian distribution curve for positive x, which is continued toward negative values of x so that it is an odd function of x. The third eigenfunction $\psi_2(x)$ is negative at the origin and has two symmetric zeros $\pm1/\sqrt{2}$ and so on. Thus, the geometrical shape of these eigenfunctions can easily be determined. It is also important that the roots of successive polynomials separate one another. □

7.10 Angular Momentum Operators

The orbital angular momentum operators $\hat{L}_x$, $\hat{L}_y$, and $\hat{L}_z$ were introduced in Section 7.3. It has been shown that they obey the commutation relations (7.66). Using the spherical polar coordinates (r, θ, φ), which are related to rectangular Cartesian coordinates (x, y, z) by

$$x = r\sin\theta\cos\varphi, \tag{7.259a}$$

$$y = r \sin \theta \sin \varphi, \tag{7.259b}$$

$$z = r \cos \theta, \tag{7.259c}$$

combined with the chain rule for differentiation

$$\frac{\partial}{\partial x} = \frac{\partial r}{\partial x}\frac{\partial}{\partial r} + \frac{\partial \theta}{\partial x}\frac{\partial}{\partial \theta} + \frac{\partial \varphi}{\partial x}\frac{\partial}{\partial \varphi}$$

and similar results for $\partial/\partial y$ and $\partial/\partial z$, the angular momentum operators can be expressed in angular variables

$$\hat{L}_x = i\hbar \left(\sin \varphi \frac{\partial}{\partial \theta} + \cot \theta \cos \varphi \frac{\partial}{\partial \varphi} \right), \tag{7.260a}$$

$$\hat{L}_y = i\hbar \left(-\cos \varphi \frac{\partial}{\partial \theta} + \cot \theta \sin \varphi \frac{\partial}{\partial \varphi} \right), \tag{7.260b}$$

$$\hat{L}_z = -i\hbar \frac{\partial}{\partial \varphi}, \tag{7.260c}$$

$$\hat{L}^2 = \hat{L}_x^2 + \hat{L}_y^2 + \hat{L}_z^2 = -\hbar^2 \left[\frac{1}{\sin \theta}\frac{\partial}{\partial \theta}\left(\sin \theta \frac{\partial}{\partial \theta} \right) + \frac{1}{\sin^2 \theta}\frac{\partial^2}{\partial \varphi^2} \right]. \tag{7.261}$$

From (7.259c) and (7.261) it is easy to verify that

$$\left[\hat{L}^2, \hat{L}_z \right] = 0. \tag{7.262}$$

It also follows from (7.259c) and (7.157) that

$$\hat{L}_z Y_l^m(\theta, \varphi) = (\hbar m) Y_l^m(\theta, \varphi). \tag{7.263}$$

For any given value of l, the possible eigenvalues of the z-component of the angular momentum, $\hat{L}_z$ are

$$L_z \equiv \hbar m, \quad m = 0, \pm 1, \pm 2, \ldots, \pm l, \tag{7.264}$$

giving $(2l + 1)$ admissible values.

On the other hand, it can easily be verified with the aid of (7.261), (7.147), and (7.157) that, with $\lambda = l(l + 1)$,

$$\hat{L}^2 Y_l^m(\theta, \varphi) = \left[\hbar^2 l(l + 1) \right] Y_l^m(\theta, \varphi), \tag{7.265}$$

where $|m| \leq l$, and $l = 0, 1, 2, \ldots$.

This shows that the eigenvalues of $\hat{L}^2$ are

$$L^2 \equiv \hbar^2 l(l + 1), \quad l = 0, 1, 2, 3, \ldots. \tag{7.266}$$

Evidently, the spherical harmonics $Y_l^m(\theta, \varphi)$ are the simultaneous eigenfunctions of $\hat{L}^2$ and $\hat{L}_z$. The eigenvalues of the total angular momentum $\hat{L}^2$ are $\hbar^2 l(l+1)$, $l = 0, 1, 2, \ldots$, and those of $\hat{L}_z$ are $m\hbar$, $m = 0, \pm 1, \ldots, \pm l$. Thus, a measurement of L^2 can yield as its result only the values $0, 2\hbar^2, 6\hbar^2, 12\hbar^2, \ldots$. The total angular momentum states with l values $0, 1, 2, 3, 4$ are known, for historical reasons, as S, P, D, F, G states, respectively. Similarly, the measured values of $\hat{L}_z$ are only $0, \pm\hbar, \pm 2\hbar, \ldots$. Hence, both $\hat{L}^2$ and $\hat{L}_z$ are quantized and can only reveal one of the specified discrete values upon measurement.

It is convenient to define two operators $\hat{L}_+$ and $\hat{L}_-$ by

$$\hat{L}_+ = \hat{L}_x + i\hat{L}_y, \tag{7.267a}$$

$$\hat{L}_- = \hat{L}_x - i\hat{L}_y. \tag{7.267b}$$

Theorem 7.10.1.   (a) $\hat{L}_+$ and $\hat{L}_-$ are non-Hermitian operators, (b) $\hat{L}_+\hat{L}_-$ and $\hat{L}_-\hat{L}_+$ are Hermitian.

Proof: Since $\hat{L}_x$ and $\hat{L}_y$ are Hermitian,

$$\langle \hat{L}_+\psi_1 | \psi_2 \rangle = \langle \psi_1 | \hat{L}_-\psi_2 \rangle, \qquad \langle \hat{L}_-\psi_1 | \psi_2 \rangle = \langle \psi_1 | \hat{L}_+\psi_2 \rangle$$

for any two wave functions ψ_1 and ψ_2. Thus, $\hat{L}_+$ and $\hat{L}_-$ are not Hermitian operators, and hence they do not represent observables.

On the other hand,

$$\hat{L}_+\hat{L}_- = \left(\hat{L}_x + i\hat{L}_y\right)\left(\hat{L}_x - i\hat{L}_y\right) = \hat{L}_x^2 + \hat{L}_y^2 - i\left[\hat{L}_x, \hat{L}_y\right]$$

$$= \hat{L}_x^2 + \hat{L}_y^2 + \hbar\hat{L}_z = \hat{L}^2 - \hat{L}_z\left(\hat{L}_z - \hbar\right). \tag{7.268}$$

Similarly,

$$\hat{L}_-\hat{L}_+ = \hat{L}_x^2 + \hat{L}_y^2 - \hbar\hat{L}_z = \hat{L}^2 - \hat{L}_z\left(\hat{L}_z + \hbar\right). \tag{7.269}$$

Thus, both $\hat{L}_+\hat{L}_-$ and $\hat{L}_-\hat{L}_+$ are expressed as real functions of $\hat{L}^2$ and $\hat{L}_z$. Hence, they are Hermitian operators.   □

Since the orbital angular momentum can only take on integer values, this result indicates the necessity for some generalization of this formalism. It is necessary to introduce matrix operators of size $n \times n$ defined by

$$\hat{A} = \begin{pmatrix} \langle 1|\hat{A}1\rangle & \langle 1|\hat{A}2\rangle & \cdots & \langle 1|\hat{A}n\rangle \\ \langle 2|\hat{A}1\rangle & \langle 2|\hat{A}2\rangle & \cdots & \langle 2|\hat{A}n\rangle \\ \vdots & \vdots & \ddots & \vdots \\ \langle n|\hat{A}1\rangle & \langle n|\hat{A}2\rangle & \cdots & \langle n|\hat{A}n\rangle \end{pmatrix},$$

where $\langle i|\hat{A}j\rangle$, $i = 1, 2, \ldots, n$, $j = 1, 2, 3, \ldots, n$ are the elements of the matrix.

An eigenvalue for the matrix operator, $\hat{A}$ is defined by

$$\hat{A}|u_n\rangle = a_n|u_n\rangle, \tag{7.270}$$

where a_n and $|u_n\rangle$ are the eigenvalues and eigenvectors of $\hat{A}$, respectively.

If the matrix $\hat{A}$ is of the diagonal form

$$\hat{A} = \begin{pmatrix} a_1 & 0 \\ 0 & a_2 \end{pmatrix}, \tag{7.271}$$

it is very easy to check by direct substitution into (7.270) that the eigenvalues are a_1 and a_2 with eigenvectors

$$|u_1\rangle = \begin{pmatrix} 1 \\ 0 \end{pmatrix}, \tag{7.272a}$$

$$|u_2\rangle = \begin{pmatrix} 0 \\ 1 \end{pmatrix}. \tag{7.272b}$$

We next assume the definition of the orbital angular momentum operators given in Section 7.3 and derive the commutation relations (7.66). We then take these commutation relations as the definition of the general angular momentum operators $\hat{\mathbf{M}} = (\hat{M}_x, \hat{M}_y, \hat{M}_z)$.

Definition 7.10.2.   (The Pauli spin matrices)

The *Pauli spin matrices* are defined by

$$\hat{\sigma}_x = \begin{pmatrix} 0 & 1 \\ 1 & 0 \end{pmatrix}, \quad \hat{\sigma}_y = \begin{pmatrix} 0 & -i \\ i & 0 \end{pmatrix}, \quad \hat{\sigma}_z = \begin{pmatrix} 1 & 0 \\ 0 & -1 \end{pmatrix}. \tag{7.273}$$

The Pauli spin matrices were defined by Wolfgang Ernst Pauli (1900–1958) in 1927. These matrices satisfy the following relations

$$\hat{\sigma}_x\hat{\sigma}_z = -\hat{\sigma}_y\hat{\sigma}_x = i\hat{\sigma}_z, \tag{7.274a}$$

$$\hat{\sigma}_y\hat{\sigma}_z = -\hat{\sigma}_z\hat{\sigma}_y = i\hat{\sigma}_x, \tag{7.274b}$$

$$\hat{\sigma}_z\hat{\sigma}_x = -\hat{\sigma}_x\hat{\sigma}_z = i\hat{\sigma}_y. \tag{7.274c}$$

It is very simple to check that the matrices

$$\hat{M}_x = \frac{1}{2}\hbar\hat{\sigma}_x, \tag{7.275a}$$

$$\hat{M}_y = \frac{1}{2}\hbar\hat{\sigma}_y, \tag{7.275b}$$

$$\hat{M}_z = \frac{1}{2}\hbar\hat{\sigma}_z \tag{7.275c}$$

satisfy the commutation relations (7.66).

Also it follows directly that

$$\hat{M}^2 = \hat{M}_x^2 + \hat{M}_y^2 + \hat{M}_z^2 = \frac{1}{2}\left(\frac{1}{2}+1\right)\hbar^2\begin{pmatrix}1 & 0 \\ 0 & 1\end{pmatrix}, \tag{7.276}$$

so that

$$\left[\hat{M}^2, \hat{M}_x\right] = \left[\hat{M}^2, \hat{M}_y\right] = \left[\hat{M}^2, \hat{M}_z\right] = 0. \tag{7.277}$$

According to (7.271), (7.272a), and (7.272b), the eigenvalues of $\hat{M}_z$ are

$$M_z = \pm\frac{1}{2}\hbar \tag{7.278}$$

with the corresponding eigenvectors

$$|u_{+1/2}\rangle = \begin{pmatrix}1 \\ 0\end{pmatrix}, \tag{7.279a}$$

$$|u_{-1/2}\rangle = \begin{pmatrix}0 \\ 1\end{pmatrix}. \tag{7.279b}$$

The same vectors are also eigenvectors of $\hat{M}^2$, each corresponding to the eigenvalue

$$M^2 = \frac{1}{2}\left(\frac{1}{2}+1\right)\hbar^2. \tag{7.280}$$

The results (7.278) and (7.280) are precisely the same as (7.264) and (7.266), which were established for the orbital angular momentum, except that in orbital angular momentum operators the possible values of l and m were restricted to integers. With the commutation relations (7.66) as the defining property for the angular momentum operators $\hat{M}_x$, $\hat{M}_y$, $\hat{M}_z$, and allowing matrix representation of these operators, it clearly indicates the possibility of non-integral values for $l = \frac{1}{2}$ and $m = \pm\frac{1}{2}$. Since this is not associated with orbital motion, it must be the *intrinsic angular momentum* or *spin* of the particle itself. The so-called electron spin thus represents an example of an angular momentum $\frac{1}{2}$.

Physically, an electron as an electrically charged particle interacts with an external magnetic field. The effect of the interaction is a shifting in atomic energy levels which results in a change in spectral lines. The observed spectrum of an atom in the presence of a magnetic field is known as the *anomalous Zeeman effect*. Experimental facts in spectroscopy including the Zeeman effect led to the discovery of the electron spin. Many other spectroscopic phenomena such as the multiplicity of numerous lines cannot be explained without the introduction of this new hypothesis. So the new hypothesis has not only resolved many important difficulties, but is essential for an understanding of many atomic and nuclear phenomena.

To establish the possibility of both integer and half-integer values of the angular momentum, the operator techniques can be applied to the problem of the angular momentum.

We take the commutation relations as the defining property of the angular momentum operators **M**:

$$\left[\hat{M}_x, \hat{M}_y\right] = i\hbar\hat{M}_z, \tag{7.281a}$$

$$\left[\hat{M}_y, \hat{M}_z\right] = i\hbar\hat{M}_x, \tag{7.281b}$$

$$\left[\hat{M}_z, \hat{M}_x\right] = i\hbar\hat{M}_y. \tag{7.281c}$$

It is convenient to introduce two new operators

$$\hat{M}_+ \equiv \hat{M}_x + i\hat{M}_y, \tag{7.282a}$$

$$\hat{M}_- \equiv \hat{M}_x - i\hat{M}_y. \tag{7.282b}$$

By direct substitution, it turns out from (7.281a) to (7.281c) that

$$\left[\hat{M}_z, \hat{M}_+\right] = \hbar\hat{M}_+, \tag{7.283a}$$

$$\left[\hat{M}_z, \hat{M}_-\right] = -\hbar\hat{M}_-, \tag{7.283b}$$

$$\left[\hat{M}_+, \hat{M}_-\right] = 2\hbar\hat{M}_z. \tag{7.283c}$$

In view of the fact that $\hat{M}^2$ and $\hat{M}_z$ commute, we can introduce a state $|\lambda, m\rangle$, which is simultaneously an eigenstate of both operators. Hence,

$$\hat{M}^2|\lambda, m\rangle = \hbar^2\lambda|\lambda, m\rangle, \tag{7.284}$$

$$\hat{M}_z|\lambda, m\rangle = \hbar m|\lambda, m\rangle. \tag{7.285}$$

So the main problem is to determine the possible values of λ and m implied by the relations (7.281a) to (7.281c). For a given value of the magnitude of the angular momentum λ, from a physical point of view, the possible values of $\hat{M}_z$, determined by m, must lie in the finite domain with values bounded by $m_{\max}$ and $m_{\min}$. This observation is important in the subsequent discussion. It now follows from (7.285) that

$$\hat{M}_+\hat{M}_z|\lambda, m\rangle = \hbar m\hat{M}_+|\lambda, m\rangle. \tag{7.286}$$

In view of (7.283a), we obtain

$$\hat{M}_+\hat{M}_z = \hat{M}_z\hat{M}_+ - \left[\hat{M}_z, \hat{M}_+\right] = \hat{M}_z\hat{M}_+ - \hbar M_+, \tag{7.287}$$

which is substituted into (7.286) to obtain

$$\hat{M}_z\hat{M}_+|\lambda, m\rangle = \hbar(m + 1)\hat{M}_+|\lambda, m\rangle. \tag{7.288}$$

Clearly, either

$$\hat{M}_+|\lambda, m\rangle = 0 \tag{7.289}$$

or

$$\hat{M}_+|\lambda, m\rangle = |\lambda, m + 1\rangle. \tag{7.290}$$

Consequently, (7.288) becomes

$$\hat{M}_z|\lambda, m + 1\rangle = \hbar(m + 1)|\lambda, m + 1\rangle. \tag{7.291}$$

Thus, it follows from (7.285) and (7.291) that given an eigenstate $|\lambda, m\rangle$, a new eigenstate $|\lambda, m + 1\rangle$ can be generated by (7.290) with eigenvalue $\hbar(m + 1)$, unless $m = m_{\max}$, in which case (7.289) holds. The admissible values of m differ by integers.

An argument similar to the preceding is applied to $\hat{M}_-\hat{M}_z$ to show that either

$$\hat{M}_-|\lambda, m\rangle = 0, \tag{7.292}$$

or

$$\hat{M}_-|\lambda, m\rangle = |\lambda, m - 1\rangle, \tag{7.293}$$

so that

$$\hat{M}_z|\lambda, m - 1\rangle = \hbar(m - 1)|\lambda, m - 1\rangle. \tag{7.294}$$

So, given any eigenstate $|\lambda, m\rangle$, a new eigenstate can be generated by (7.293) with eigenvalues $\hbar(m - 1)$, unless $m = m_{\min}$, in which case (7.292) applies.

We have

$$\hat{M}_-\hat{M}_+ = \left(\hat{M}_x - i\hat{M}_y\right)\left(\hat{M}_x + i\hat{M}_y\right)$$
$$= \hat{M}_x^2 + \hat{M}_y^2 + i\left[\hat{M}_x, \hat{M}_y\right] = \hat{M}^2 - \hat{M}_z^2 - \hbar\hat{M}_z. \tag{7.295}$$

If the operator $\hat{M}_-\hat{M}_+$ acts on $|\lambda, m_{\max}\rangle$, it follows by using (7.289) that

$$\left(\hat{M}^2 - \hat{M}_z^2 - \hbar\hat{M}_z\right)|\lambda, m_{\max}\rangle = \hat{M}_-\hat{M}_+|\lambda, m_{\max}\rangle = 0. \tag{7.296}$$

Therefore, by (7.284) and (7.285),

$$\left(\lambda - m_{\max}^2 - m_{\max}\right)|\lambda, m_{\max}\rangle = 0,$$

or

$$\lambda = m_{\max}^2 + m_{\max}. \tag{7.297}$$

Similarly,

$$\hat{M}_+\hat{M}_- = \hat{M}^2 - \hat{M}_z^2 + \hbar\hat{M}_z. \tag{7.298}$$

If this operator acts on $|\lambda, m_{min}\rangle$ and (7.292) is used, we obtain

$$\lambda - m_{min}^2 + m_{min} = 0. \tag{7.299}$$

If we equate the two results for λ from (7.297) and (7.299), it turns out that

$$(m_{max} + m_{min})(m_{min} - m_{max} - 1) = 0. \tag{7.300}$$

Thus,

$$m_{max} = -m_{min}. \tag{7.301}$$

Therefore, the admissible values of m lie symmetrically about the origin. Since the extreme values differ by an integer, it follows that

$$m_{max} - m_{min} = 2l, \tag{7.302}$$

where

$$l = 0, \frac{1}{2}, 1, \frac{3}{2}, \ldots. \tag{7.303}$$

These results combined with (7.301) show that

$$-l \leq m \leq l \quad (2l + 1 \text{ values}). \tag{7.304}$$

Finally, it follows from (7.297) and (7.302) that

$$\lambda = l(l + 1), \quad l = 0, \frac{1}{2}, 1, \frac{3}{2}, \ldots. \tag{7.305}$$

This is a definite proof for integer and half-integer eigenvalues for the angular momentum. Particles with integral spin are called the *Bosons* after S. N. Bose (1894–1974), those with half-integral spins are known as *Fermions* after Enrico Fermi (1901–1954).

The two different kinds of angular momentum operators can be combined to define the *total angular momentum*

$$\hat{\mathbf{J}} = \hat{\mathbf{L}} + \hat{\mathbf{M}} \tag{7.306}$$

with the components $\hat{J}_x = \hat{L}_x + \hat{M}_x, \hat{J}_y = \hat{L}_y + \hat{M}_y, \hat{J}_z = \hat{L}_z + \hat{M}_z$.

It follows from the properties of $\hat{L}$ and $\hat{M}$ that $\hat{J}$ satisfies the usual commutation relations

$$[\hat{J}_x, \hat{J}_y] = i\hbar\hat{J}_z, \tag{7.307a}$$

$$[\hat{J}_y, \hat{J}_z] = i\hbar\hat{J}_x, \tag{7.307b}$$

$$[\hat{J}_z, \hat{J}_x] = i\hbar\hat{J}_y, \tag{7.307c}$$

and hence,

$$\left[\hat{J}_x, \hat{\mathbf{J}}^2\right] = \left[\hat{J}_y, \hat{\mathbf{J}}^2\right] = \left[\hat{J}_z, \hat{\mathbf{J}}^2\right] = 0, \tag{7.308}$$

where

$$\hat{\mathbf{J}}^2 = \hat{J}_x^2 + \hat{J}_y^2 + \hat{J}_z^2. \tag{7.309}$$

It can readily be shown that

$$\hat{\mathbf{J}}^2 |l, m\rangle = l(l+1)\hbar^2 |l, m\rangle, \tag{7.310}$$

$$\hat{J}_z |l, m\rangle = \hbar m |l, m\rangle. \tag{7.311}$$

This means that the eigenvalues of $\hat{\mathbf{J}}^2$ and $\hat{J}_z$ are $l(l+1)\hbar^2$ and $\hbar m$, respectively, where $|m| \le l$ and the quantum numbers may be either integers or half-integers.

Finally, it follows that

$$\left[\hat{J}^2, \hat{L}_z\right] = 2\hat{M}_x\left[\hat{L}_x, \hat{L}_z\right] + 2\hat{M}_y\left[\hat{L}_y, \hat{L}_z\right] = \hbar\left[\hat{L}_+, \hat{L}_-\right], \tag{7.312}$$

$$\left[\hat{J}^2, \hat{M}_z\right] = -\hbar\left[\hat{L}_+, \hat{L}_-\right]. \tag{7.313}$$

7.11 The Dirac Relativistic Wave Equation

The Schrödinger wave equation, which forms the basis of quantum mechanics, does not satisfy the requirements of the special theory of relativity, namely, invariance under the Lorenz transformations. Dirac first formulated a relativistic version of the Schrödinger equation, which is universally known as the *Dirac equation*. His famous generalization of the Schrödinger equation to a set of four first-order partial differential equations met the requirements of relativity and incorporated spin properties of particles without abandoning the general physical aspects of the wave-mechanical description. Historically, the Dirac equation is considered as fundamental in the sense that it led to the discovery of antiparticles, to the basic understanding of electron's magnetic moment, and to more precise prediction of the spectra in atoms. With the standard Hamiltonian form, the Dirac equation is

$$i\hbar\frac{\partial\psi}{\partial t} = \hat{H}\psi \tag{7.314}$$

where the *Dirac Hamiltonian* is

$$\hat{H} = c\boldsymbol{\alpha} \cdot \hat{\mathbf{p}} + \beta mc^2, \tag{7.315}$$

where $\boldsymbol{\alpha} = (\alpha_1, \alpha_2, \alpha_3)$ and β are assumed to be dimensionless constants (independent of space and time) determined by the requirement that

$$H^2 = (pc)^2 + \left(mc^2\right)^2. \tag{7.316}$$

Keeping in mind the operator nature of α and β, we use the condition (7.316) so that

$$\left(c\boldsymbol{\alpha} \cdot \mathbf{p} + \beta mc^2\right) = (pc)^2 + \left(mc^2\right)^2,$$

or, preserving the order of relevant factors, we obtain

$$\sum_i \alpha_i^2 p_i^2 c^2 + \frac{1}{2}\sum_{i \neq j}(\alpha_i\alpha_j + \alpha_j\alpha_i)p_ip_jc^2 + \sum_i(\alpha_i\beta + \beta\alpha_i)mc^2p_i^2 + \beta^2\left(mc^2\right)^2$$

$$= (pc)^2 + \left(mc^2\right)^2.$$

Comparing terms, it turns out that

$$\alpha_i^2 = \beta^2 = 1,$$

or

$$\alpha_1^2 = \alpha_2^2 = \alpha_3^2 = \beta^2 = 1, \tag{7.317}$$

and

$$\alpha_1\alpha_2 + \alpha_2\alpha_1 = \alpha_2\alpha_3 + \alpha_3\alpha_2 = \alpha_3\alpha_1 + \alpha_1\alpha_3$$

$$= \alpha_1\beta + \beta\alpha_1 = \alpha_2\beta + \beta\alpha_2 = \alpha_3\beta + \beta\alpha_3 = 0. \tag{7.318}$$

The four matrices mutually anticommute and their squares are unity. So the problem is to find these matrices with the properties (7.317) and (7.318). These matrices are called the *Dirac matrices* whose algebra is seen to be identical to that of Pauli matrices except that there are four of them. Since the Pauli matrices (along with the unit matrix) exhaust the independent 2×2 matrices, the four Dirac matrices cannot be represented by 2×2 matrices. It turns out that 3×3 matrices do not suffice, and the smallest matrices that do are 4×4. These are not uniquely defined by the commutation relations, but the conventional choice is

$$\beta = \begin{pmatrix} 1 & 0 & 0 & 0 \\ 0 & 1 & 0 & 0 \\ 0 & 0 & -1 & 0 \\ 0 & 0 & 0 & -1 \end{pmatrix}, \qquad \alpha_1 = \begin{pmatrix} 0 & 0 & 0 & 1 \\ 0 & 0 & 1 & 0 \\ 0 & 1 & 0 & 0 \\ 1 & 0 & 0 & 0 \end{pmatrix},$$

$$\alpha_2 = \begin{pmatrix} 0 & 0 & 0 & -i \\ 0 & 0 & i & 0 \\ 0 & -i & 0 & 0 \\ i & 0 & 0 & 0 \end{pmatrix}, \qquad \alpha_3 = \begin{pmatrix} 0 & 0 & 1 & 0 \\ 0 & 0 & 0 & -1 \\ 1 & 0 & 0 & 0 \\ 0 & -1 & 0 & 0 \end{pmatrix}.$$

These matrices can also be expressed in terms of the 2×2 Pauli matrices σ_j as

$$\beta = \begin{pmatrix} I & 0 \\ 0 & -I \end{pmatrix}, \qquad \alpha_j = \begin{pmatrix} 0 & \sigma_j \\ \sigma_j & 0 \end{pmatrix}, \qquad j = 1, 2, 3, \tag{7.319}$$

where every element of these 2×2 matrices is itself to be understood as 2×2 matrices.

Since the Dirac matrices are 4×4, the Dirac equation makes sense if the Dirac wave function is a column matrix with four rows as

$$\psi = \begin{pmatrix} \psi_1 \\ \psi_2 \\ \psi_3 \\ \psi_4 \end{pmatrix}.$$

The Dirac equation can be put in a more appealing form by introducing a new set of 4×4 matrices as

$$\gamma^0 = \beta, \quad \gamma^1 = \beta\alpha_1, \quad \gamma^2 = \beta\alpha_2, \quad \gamma^3 = \beta\alpha_3, \tag{7.320}$$

and by replacing the momentum operator $\mathbf{p}$ with $-i\hbar\nabla$ in (7.314) and (7.315). It turns out that the Dirac equation assumes the form

$$i\hbar\frac{\partial\psi}{\partial t} = \left(-i\hbar c\hat{\alpha} \cdot \nabla + \beta mc^2\right)\psi, \tag{7.321}$$

or, equivalently,

$$\frac{1}{c}\gamma^0\frac{\partial\psi}{\partial t} + \left(\gamma^j \cdot \nabla\right)\psi + i\kappa\psi = 0, \quad \kappa = \left(\frac{mc}{\hbar}\right), \tag{7.322}$$

where the explicit form of the Dirac γ matrices are

$$\gamma^0 = \begin{pmatrix} I & 0 \\ 0 & I \end{pmatrix}, \quad \gamma^j = \begin{pmatrix} 0 & \sigma_j \\ -\sigma_j & 0 \end{pmatrix}, \quad j = 1, 2, 3. \tag{7.323}$$

These matrices satisfy the following properties

$$\left(\gamma^0\right)^* = \gamma^0, \quad \left(\gamma^j\right)^* = \gamma^j \quad \text{for } j = 1, 2, 3, \tag{7.324}$$

where $*$ represents the conjugate transpose

$$\left(\gamma^0\right)^2 = I, \quad \left(\gamma^j\right)^2 = -I, \quad \text{for } j = 1, 2, 3, \tag{7.325}$$

and

$$\gamma^k\gamma^\ell + \gamma^\ell\gamma^k = 0 \quad \text{for } k \neq \ell, \; k, \ell = 0, 1, 2, 3. \tag{7.326}$$

Equation (7.322) is a conventional form of the Dirac equation and its solution ψ is called the *spinor wave function* or simply, *spinor*.

The Dirac equation (7.322) admits plane wave solutions of the form

$$\psi_j(\mathbf{r}, t) = a_j \exp\left[i(\mathbf{k} \cdot \mathbf{r} - \omega t)\right] = a_j \exp\left[-i\hbar(\mathbf{p} \cdot \mathbf{r} - Et)\right], \quad j = 1, 2, 3, 4, \tag{7.327}$$

where a_j are scalars. These solutions represent the eigenfunctions of the energy and momentum operators with eigenvalues $\hbar\omega$ and $\hbar\mathbf{k}$, respectively. Explicit solutions can be found for momentum $\mathbf{p}$ by choosing a sign for the energy $E_+ = H_+ = +(c^2\mathbf{p}^2 + m^2c^4)^{1/2}$. Then there are two sets of linearly independent solutions for a_j. Similarly, there are two sets of solutions corresponding to the negative square root of energy, $H_- = E_- = -(c^2\mathbf{p}^2 + m^2c^4)^{1/2}$. The former solutions correspond to positive energy and the latter to negative energy. In the nonrelativistic limit, in which $H_+ = -H_-$ is close to mc^2 and large compared to $c\mathbf{p}$. The physical distinction between the two solutions for each sign of energy can be seen by defining three *new* 4×4 *spin* matrices $\tilde{\gamma}^j$ as

$$\tilde{\gamma}^j = \begin{pmatrix} \sigma_j & 0 \\ 0 & \sigma_j \end{pmatrix}, \quad j = 1, 2, 3, \tag{7.328}$$

where σ_j are the 2×2 Pauli matrices.

The Dirac equation implies the Klein–Gordon equation. Indeed, the former is a square root of the latter in the sense

$$\left[\frac{1}{c}\gamma^0\frac{\partial}{\partial t} + (\gamma^j \cdot \nabla) + i\kappa\right]^2 \psi = \left(\frac{1}{c^2}\frac{\partial^2}{\partial t^2} - \nabla^2 + \kappa^2\right)\psi \tag{7.329}$$

as operators.

The equations of quantum electrodynamics form a system that combines the Dirac and Maxwell equations with nonlinear coupling terms. Physically, they describe interaction between electrons governed by the Dirac equation and photons described by the Maxwell equations. The predictions of the Dirac theory are in remarkable agreement with experimental observations, and hence it is the most accurate theory in all of physical sciences.

7.12 Exercises

1. (a) Use the Lagrangian, $L = \frac{1}{2}m(\dot{x}^2 + \dot{y}^2 + \dot{z}^2) - \frac{1}{2}k(x^2 + y^2 + z^2)$ for the three-dimensional isotropic harmonic oscillator, and Lagrange's equations of motion to show that the total energy is constant where k is a constant.

   (b) Show that the Lagrangian for the oscillator in spherical polar coordinates (r, θ, φ) is

   $$L = T - V = \frac{1}{2}m\left(\dot{r}^2 + r^2\dot{\theta}^2 + r^2\sin^2\varphi\dot{\varphi}^2\right) - \frac{k^2}{r},$$

   where $k = 4\pi^2 m\omega^2$.

   Hence, write down the Lagrange equations of motion.

2. Consider a single particle of mass m moving in a plane under a conservative force with potential $V(r)$, where r is the distance from the origin of coordinates. With r and θ as generalized coordinates describing the motion of the particle, show that the corresponding momenta are

$$p_r = \frac{\partial L}{\partial \dot{r}} = m\dot{r}, \qquad p_\theta = \frac{\partial L}{\partial \dot{\theta}} = mr^2\dot{\theta},$$

where $L = T - V = \frac{1}{2}m(\dot{r}^2 + r^2\dot{\theta}^2) - V(r)$. Hence show that

$$H = \frac{p_r^2}{2m} + \frac{p_\theta^2}{2mr^2} + V(r), \qquad mr^2\dot{\theta} = \text{constant}, \qquad m(\ddot{r} - r\dot{\theta}^2) = -\frac{\partial V}{\partial r}.$$

Give an interpretation of each of these results.

3. If A is a complex dynamical function of q and p, A^* is its complex conjugate, and if the Poisson bracket $\{A, A^*\} = i$, compute $\{A, AA^*\}$, $\{A, A^*A\}$, $\{A^*, AA^*\}$, and $\{A^*, A^*A\}$.

4. Find the Hamiltonian and Hamilton's equations of motion for

   (a) The simple harmonic oscillator with $T = \frac{1}{2}m\dot{x}^2$ and $V = \frac{1}{2}kx^2$ and

   (b) the planetary motions with $T = \frac{1}{2}m(\dot{r}^2 + r^2\dot{\theta}^2)$, and $V = m\mu(\frac{1}{2}a - 1/r)$. In this case, derive the differential equations for the central orbit.

5. Establish the following results for the Poisson brackets:

   (a) $\{A, B\} = -\{B, A\}$.

   (b) $\{(A + B), C\} = \{A, C\} + \{B, C\}$.

   (c) $\{AB, C\} = \{A, C\}B + A\{B, C\}$.

   (d) $\{A, \alpha\} = 0$.

   (e) $\{A, \{B, C\}\} + \{B, \{C, A\}\} + \{C, \{A, B\}\} = 0$ (Jacobi's identity),

   where A, B, and C are canonical functions and α is a scalar.

6. Show that the following results hold for commutators:

   (a) $[\hat{A}, \hat{B}] = -[\hat{B}, \hat{A}]$.

   (b) $[\hat{A} + \hat{B}, \hat{C}] = [\hat{A}, \hat{C}] + [\hat{B}, \hat{C}]$.

   (c) $[\hat{A}, \hat{B} + \hat{C}] = [\hat{A}, \hat{B}] + [\hat{A}, \hat{C}]$.

   (d) $[\hat{A}\hat{B}, \hat{C}] = [\hat{A}, \hat{C}]\hat{B} + \hat{A}[\hat{B}, \hat{C}]$.

   (e) $[\hat{A}, \hat{B}\hat{C}] = [\hat{A}, \hat{B}]\hat{C} + \hat{B}[\hat{A}, \hat{C}]$.

   (f) $[\hat{A}, [\hat{B}, \hat{C}]] + [\hat{B}, [\hat{C}, \hat{A}]] + [\hat{C}, [\hat{A}, \hat{B}]] = 0$ (Jacobi's identity).

(g) $[\hat{A}^2, \hat{B}] = \hat{A}[\hat{A}, \hat{B}] + [\hat{A}, \hat{B}]\hat{A}$,

(h) $[\hat{A}, \alpha] = 0$, where α is a scalar.

7. For the three-dimensional position and momentum operators of a particle, prove that

$$[\hat{r}_i, \hat{p}_j] = i\hbar\delta_{ij},$$

where the suffixes i, j take the values 1, 2, 3 for the x, y, z components of $\hat{\mathbf{r}}$ and $\hat{\mathbf{p}}$, respectively.

8. By direct evaluation for canonically conjugate variables q and p, show that

(a) $[p^2, q^2] = 2\hbar^2 - 4i\hbar pq$.

(b) $[p, q^2] = -2i\hbar q$.

(c) $[\hat{x}^2, \hat{p}_x^2] = 2\hbar^2 - 4i\hat{x}\hat{p}_x$.

(d) $[\hat{p}_x, \hat{x}^2] = -2i\hbar\hat{x}$.

9. If A and B are two operators which both commute with their commutator $[\hat{A}, \hat{B}]$, prove that

$$[\hat{A}, \hat{B}^n] = n\hat{B}^{n-1}[\hat{A}, \hat{B}],$$
$$[\hat{A}^n, \hat{B}] = n\hat{A}^{n-1}[\hat{A}, \hat{B}].$$

10. Establish the following commutator relations:

$$[\hat{L}^2, \hat{L}_x] = [\hat{L}^2, \hat{L}_y] = [\hat{L}^2, \hat{L}_z] = 0.$$

11. Show that

$$[\hat{L}_+, \hat{L}_-] = \hbar\hat{L}_z,$$
$$[\hat{L}_+, \hat{L}_z] = -\hbar\hat{L}_+,$$
$$[\hat{L}_-, \hat{L}_z] = \hbar\hat{L}_-,$$
$$[\hat{L}_x, \hat{L}_+] = \hbar\hat{L}_-,$$
$$[\hat{L}^2, \hat{L}_+] = 0.$$

12. Prove that

$$\hat{J}^2 = \hat{L}^2 + \hat{M}^2 + 2\hat{L} \cdot \hat{M} = \hat{L}^2 + \hat{M}^2 + 2\hat{L}_z\hat{M}_z + \hat{L}_+\hat{L}_- + \hat{L}_-\hat{L}_+,$$
$$2\hat{L} \cdot \hat{M} = \hat{J}^2 - \hat{L}^2 - \hat{M}^2.$$

13. Show that the probability for a position measurement on the state $\Psi(x, t)$ to yield a value somewhere between x_1 and x_2 is

$$P(x_1, x_2, t) = \int_{x_1}^{x_2} \overline{\Psi} \Psi \, dx = \int_{x_1}^{x_2} |\Psi(x, t)|^2 dx.$$

Using the Schrödinger equations, derive the result

$$\frac{d}{dt} P(x_1, x_2, t) = J(x_1, t) - J(x_2, t),$$

where

$$J(x, t) = \frac{i\hbar}{2m} \left[\Psi \frac{\partial \overline{\Psi}}{\partial x} - \overline{\Psi} \frac{\partial \Psi}{\partial x} \right].$$

14. Use the inner product

$$\langle \varphi, \psi \rangle = \int_{-\infty}^{\infty} \overline{\varphi} \psi \, dx,$$

and the property $\langle \varphi, \psi \rangle \to (0, 0)$ as $|x| \to \infty$, to show that the position operator $\hat{x} = x$, the momentum operator $\hat{p} = -i\hbar \partial/\partial x$, and the energy operator $\hat{H} = (\hat{p}^2/2m) + \hat{V}(\hat{x})$ are Hermitian operators.

15. Establish the following commutation relations for the orbital angular momentum operators:

$$[\hat{L}_x, \hat{x}] = 0,$$
$$[\hat{L}_x, \hat{y}] = i\hbar \hat{z},$$
$$[\hat{L}_x, \hat{z}] = -i\hbar \hat{y},$$
$$[\hat{L}_x, \hat{p}_x] = 0,$$
$$[\hat{L}_x, \hat{p}_y] = i\hbar \hat{p}_z,$$
$$[\hat{L}_x, \hat{p}_z] = -i\hbar \hat{p}_y.$$

16. Prove the Heisenberg uncertainty relation for the harmonic oscillator

$$\Delta x \Delta p \geq \frac{1}{2} \hbar.$$

17. If $\hat{A}$ and $\hat{B}$ are constants of motion, show that the commutator $i[\hat{A}, \hat{B}]$ is also a constant of motion.

18. Show that, for the linear harmonic oscillator,

$$[\hat{H}, \hat{A}] = (-\hbar\omega)\hat{A}, \qquad [\hat{H}, \hat{A}^*] = (\hbar\omega)\hat{A}^*,$$

where

$$\hat{A} = \hat{a}/\sqrt{\hbar\omega} \quad \text{and} \quad \hat{A}^* = \hat{a}^*/\sqrt{\hbar\omega}.$$

19. For the three-dimensional anisotropic Planck's oscillator, the Hamiltonian is given by

$$H_r = \frac{1}{2m}p_r^2 + \frac{1}{2}m\omega_r^2 x_r^2, \quad r = 1, 2, 3,$$

so that the Hamiltonian $H = H_1 + H_2 + H_3$ and the total energy $E = E_1 + E_2 + E_3$, where E_1, E_2, E_3 are energies of each of the independent degrees of freedom. Show that

$$E = \left(n_1 + \frac{1}{2}\right)\hbar\omega_1 + \left(n_1 + \frac{1}{2}\right)\hbar\omega_2 + \left(n_3 + \frac{1}{2}\right)\hbar\omega_3.$$

In the case of an isotropic oscillator, $\omega_1 = \omega_2 = \omega_3 = \omega$, derive the result

$$E_N = \left(N + \frac{3}{2}\right)\hbar\omega, \quad N = (n_1 + n_2 + n_3) = 0, 1, 2, 3, \ldots.$$

20. Prove the compatibility theorem which states that any one of the following conditions implies the other two:

(a) $\hat{A}$ and $\hat{B}$ are compatible,

(b) $\hat{A}$ and $\hat{B}$ possess a common eigenbasis,

(c) $\hat{A}$ and $\hat{B}$ commute,

where A and B are two observables with corresponding operators $\hat{A}$ and $\hat{B}$.

21. If the eigenvectors $\{\psi_n(x)\}$ form an orthonormal basis in a Hilbert space, show that any state vector $\psi(x)$ satisfies the result

$$(\psi, \psi) = \sum_{n=1}^{\infty} |(\psi_n, \psi)|^2.$$

22. If $\hat{A}' \equiv \hat{A} - \langle\hat{A}\rangle$ and $\hat{B}' \equiv \hat{B} - \langle\hat{B}\rangle$, prove the following results:

(a) $\hat{A}'$ and $\hat{B}'$ are Hermitian operators.

(b) $[\hat{A}', \hat{B}'] = [\hat{A}, \hat{B}].$

(c) $\langle \hat{A}'\psi, \hat{A}'\psi \rangle = (\Delta \hat{A})^2.$

Use these results to establish the generalized uncertainty relation.

23. Using $\langle \hat{A} \rangle = \int_{-\infty}^{\infty} \Psi^*(x)[\hat{A}\Psi(x)]dx$, prove that the expectation values of position and momentum in the state $\Psi(x, t)$ are

$$\langle \hat{x} \rangle = \int_{-\infty}^{\infty} x|\Psi(x, t)|^2 dx, \qquad \langle \hat{p} \rangle = -i\hbar \int_{\infty}^{\infty} \Psi^* \frac{\partial}{\partial x}\Psi(x, t)dx.$$

Also show that

$$\langle \hat{x}^2 \rangle = \int_{\infty}^{\infty} x^2|\Psi(x, t)|^2 dx, \qquad \langle \hat{p}^2 \rangle = \hbar^2 \int_{-\infty}^{\infty} \left|\frac{\partial}{\partial x}\Psi(x, t)\right|^2 dx.$$

24. Apply the basic commutation relations $[\hat{x}_i, \hat{p}_j] = i\hbar\delta_{ij}$ and rules of commutator algebra to show that

$$[\hat{x}\hat{p}_x, \hat{H}] = \frac{i\hbar}{m}\hat{p}_x^2 + \hat{x}[\hat{p}_x, \hat{V}], \qquad [\hat{y}\hat{p}_y, \hat{H}] = \frac{i\hbar}{m}\hat{p}_y^2 + \hat{y}[\hat{p}_y, \hat{V}],$$

$$[\hat{z}\hat{p}_z, \hat{H}] = \frac{i\hbar}{m}\hat{p}_z^2 + \hat{z}[\hat{p}_z, \hat{V}].$$

Combine these results to obtain the Heisenberg equation of motion for the operator $\mathbf{r} \cdot \mathbf{p}$

$$\frac{d}{dt}\langle \mathbf{r} \cdot \mathbf{p} \rangle = \left\langle \frac{\mathbf{p}^2}{m} \right\rangle - \langle \mathbf{r} \cdot \nabla V \rangle.$$

Prove the *Virial Theorem* for the stationary states:

$$2\langle T \rangle = \langle \mathbf{r} \cdot \nabla V \rangle.$$

25. Use the results in Exercise 9 for $\hat{A} = \hat{x}$ and $\hat{B} = \hat{p}_x$ to prove that for any Hamiltonian of the form

$$\hat{H} = \frac{\hat{p}_x^2}{2m} + \alpha\hat{x}^n,$$

the following relation holds:

$$[\hat{x}\hat{p}_x, \hat{H}] = i\hbar\left(\frac{\hat{p}_x^2}{m} - \alpha n\hat{x}^n\right) = i\hbar(2\hat{T} - n\hat{V}).$$

26. Use the Hamiltonian operator for the one-dimensional simple harmonic oscillator in the form

$$\hat{H} = \frac{1}{2m}\hat{p}^2 + \frac{1}{2}m\omega^2\hat{x}^2,$$

and then introduce the nondimensional variables

$$\hat{X} = \left(\frac{m\omega}{2\hbar}\right)^{1/2} \hat{x}, \qquad \hat{P} = \frac{1}{(2m\hbar\omega)^{1/2}} \hat{p}.$$

(a) Show that

   (i) $\hat{X}$ and $\hat{P}$ are Hermitian operators.

   (ii) $\hat{H} = \hbar\omega(\hat{P}^2 + \hat{X}^2)$.

   (iii) $[\hat{X}, \hat{P}] = \frac{1}{2}i$.

(b) If $\hat{Q} = \hat{X} + i\hat{P}$ and $\hat{Q}^* = \hat{X} - i\hat{P}$, show that

$$\hat{Q}\hat{Q}^* = \hat{X}^2 + \hat{P}^2 + \frac{1}{2}, \qquad \hat{Q}^*\hat{Q} = \hat{X}^2 + \hat{P}^2 - \frac{1}{2},$$

$$\hat{H} = \hbar\omega\left(\hat{Q}^*\hat{Q} + \frac{1}{2}\right),$$

where the algebra of the operators $\hat{Q}$ and $\hat{Q}^*$ is defined by the commutation relation

$$[\hat{Q}, \hat{Q}^*] = 1.$$

27. If $\hat{\mathbf{A}}$ and $\hat{\mathbf{B}}$ are two vector operators that commute with the Pauli spin matrices but do not commute between themselves, prove the Dirac identity

$$(\hat{\boldsymbol{\sigma}} \cdot \hat{\mathbf{A}})(\hat{\boldsymbol{\sigma}} \cdot \hat{\mathbf{B}}) = (\hat{\mathbf{A}} \cdot \hat{\mathbf{B}}) + i(\hat{\mathbf{A}} \times \hat{\mathbf{B}}) \cdot \hat{\boldsymbol{\sigma}},$$

where $\hat{\boldsymbol{\sigma}} = (\hat{\sigma}_x, \hat{\sigma}_y, \hat{\sigma}_z)$.

28. For the linear harmonic oscillator, establish that

$$[\hat{H}, \hat{a}] = -\hbar\omega\hat{a}.$$

Show that this implies that $\hat{a}|\psi\rangle$ is the eigenstate of $\hat{H}$ with eigenvalues $(E - \hbar\omega)$. Hence, deduce that (a) $E \geq \frac{1}{2}\hbar\omega$, and (b) $E = \frac{1}{2}\hbar\omega$ if and only if $a|\psi\rangle = 0$.

29. For the free Dirac particles, show that the Dirac Hamiltonian operator $\hat{H} = c\sum_k \hat{\alpha}_k \hat{p}_k + \hat{\beta}mc^2$ does not commute with angular momentum operators $\hat{L}_k$ where $k = 1, 2, 3$.

30. Show that the plane wave solutions

$$\psi_j(\mathbf{r}, t) = a_j \exp\left[-i\hbar(\mathbf{p} \cdot \mathbf{r} - Et)\right], \quad j = 1, 2, 3, 4$$

of the Dirac equation (7.321) lead to the relativistic energy relations

$$E_\pm = \pm\left(c^2\mathbf{p}^2 + m^2c^4\right)^{1/2}.$$

31. Consider a one-dimensional simple harmonic oscillator system consisting of a single particle of mass m and a spring whose constant is k. With the canonical coordinates (q, p), $p = \frac{\partial S}{\partial q}$, the Hamiltonian of this conservative system is

$$H = \frac{p^2}{2m} + \frac{kq^2}{2} = \text{total energy} = E.$$

Show that

$$q = \sqrt{\frac{2E}{k}}\cos\omega(t - \tau), \qquad p = -\sqrt{2Em}\sin\omega(t - \tau),$$

where $\omega = \sqrt{k/m}$ is the natural frequency of this mass-spring system.

32. (a) Consider an n-dimensional conservative system with $H = E$ so that the Hamilton principle function S and the Hamilton characteristic function W are related by

$$S(\mathbf{q}, \mathbf{p}, t) = W(\mathbf{q}, \mathbf{p}) - Et,$$

where $\mathbf{q}$ and $\mathbf{p}$ are n-dimensional vectors. Show that the phase velocity c_p of the wave is given by

$$c_p = \frac{ds}{dt} = \frac{E}{|\nabla W|},$$

where ds is the element of the arc length normal to the surface of constant S or W.

(b) For a single particle without constraints, show that the phase velocity is $c_p = E/p$ and the momentum is $\mathbf{p} = \nabla W$.

33. For an infinite square well with the potential $V(x) = V_0 H(|x| - a)$, the solution of the one-dimensional time-independent Schrödinger equation is

(a) $\psi_1(x) = Ae^{\kappa x}$ $(x \leq -a)$, $\kappa = \sqrt{\frac{2m}{\hbar^2}(V_0 - E)}$ $(V_0 > E)$,

(b) $\psi_2(x) = 0$, $x \geq a$,

(c) $\psi_3(x) = A\sin kx + B\cos kx$, for $|x| \leq a$, where A, B are arbitrary constants and $k^2 = \frac{2mE}{\hbar^2}$.

Hence, find the energy levels for an even solution ($\cos ak = 0$) and for an odd solution ($\sin ak = 0$).

34. (a) For the nonrelativistic Dirac equation, if a particle moves along the z-axis, show that there are four degenerate eigenvalues E_+, E_+, E_-, and E_- where $E_\pm = \pm\sqrt{c^2 p_z^2 + m^2 c^4}$.

(b) Show that the corresponding eigensolutions for the wave functions are

$$
\begin{array}{cccc}
\psi_1 = 1, & \psi_2 = 0, & \psi_3 = a, & \psi_4 = 0; \\
\psi_1 = 0, & \psi_2 = 1, & \psi_3 = 0, & \psi_4 = -a; \\
\psi_1 = -b, & \psi_2 = 0, & \psi_3 = 1, & \psi_4 = 0; \\
\psi_1 = 0, & \psi_2 = b, & \psi_3 = 0, & \psi_4 = 1;
\end{array}
$$

where

$$
a = \frac{cp_z}{mc^2 + E_+} \quad \text{and} \quad b = \frac{cp_z}{mc^2 - E_-}.
$$

35. (a) Show that the simplest relativistic wave equation is

$$
p_\mu p^\mu \psi = m^2 c^2 \psi,
$$

where $p_\mu = (\frac{E}{c}, \mathbf{p})$ and $p^\mu = (\frac{E}{c}, -\mathbf{p})$.

(b) Show also that the wave function ψ satisfies the Klein–Gordon equation

$$
\left(\Box + \frac{m^2 c^2}{\hbar^2}\right)\psi = 0,
$$

where

$$
\Box \equiv \partial_\mu \partial^\mu \equiv \left(\frac{1}{c^2}\frac{\partial^2}{\partial t^2} - \nabla^2\right).
$$

(c) Show that the Klein–Gordon equation has plane wave solutions with positive as well as negative energy.

(d) Derive

$$
\frac{\partial}{\partial t}\rho(\mathbf{r}, t) = -\nabla \cdot \mathbf{J}(\mathbf{r}, t),
$$

where

$$
\rho(\mathbf{r}, t) = \frac{i\hbar}{2mc^2}\left(\bar{\psi}\frac{\partial \psi}{\partial t} - \psi\frac{\partial \bar{\psi}}{\partial t}\right)
$$

and $\mathbf{J}(\mathbf{r}, t)$ is defined by (7.108).

Wavelets and Wavelet Transforms

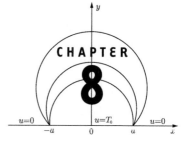

"Wavelets are without doubt an exciting and intuitive concept. The concept brings with it a new way of thinking, which is absolutely essential and was entirely missing in previously existing algorithms."

Yves Meyer

"Multiresolution analysis provides a natural framework for the understanding of wavelet bases and for the construction of new examples. The history of the formulation of multiresolution analysis is a beautiful example of applications stimulating theoretical development."

Ingrid Daubechies

8.1 Brief Historical Remarks

The concept of *wavelets* or *ondelettes* started to appear in the literature only in the early 1980s. This new concept can be viewed as a synthesis of various ideas which originated from different disciplines including mathematics (Calderón–Zygmund operators and Littlewood–Paley theory), physics (coherent states formalism in quantum mechanics and renormalization group), and engineering (quadratic mirror filters, sideband coding in signal processing, and pyramidal algorithms in image processing). In 1982, Jean Morlet, a French geophysical engineer, first introduced the idea of wavelet transform as a new mathematical tool for seismic signal analysis. It was Alex Grossmann, a French theoretical physicist, who quickly recognized the importance of the Morlet wavelet transform, which is similar to coherent states formalism in quantum mechanics, and developed an exact inversion formula for the wavelet transform. In 1984, the joint venture of Morlet and Grossmann led to a detailed mathematical study of the continuous wavelet transforms and their various applications. It has become clear from their work that, analogous to the Fourier expansions, the wavelet theory has provided a new method for decomposing a function or a signal.

In many applications, especially in the time-frequency analysis of a signal, the standard Fourier transform analysis is not adequate because the Fourier transform of the signal does not contain any local information. This is probably the major weakness of the Fourier transform analysis, since it neglects the idea of frequencies changing with time or, equivalently, the notion of finding the frequency spectrum of a signal locally in time. To eliminate this weakness, in 1946 Dennis Gabor, a Hungarian–British physicist and engineer, first introduced the *window-Fourier transform* (or the *short-time Fourier transform*) or more appropriately the *Gabor transform* by using a Gaussian distribution function as the window function. The idea is to use a window function to localize the Fourier transform, then shift the window to another position, and so on. The remarkable difference of the Gabor transform is the local aspect of the Fourier transform analysis (with resolution in time equal to the size of the window). In fact, it deals with a discrete set of coefficients, which allows efficient numerical computation of those coefficients. However, the Gabor wavelets suffered from some serious algorithmic difficulties which were successfully resolved by Henrique Malvar in 1987. Malvar introduced new wavelets, which are now known as the *Malvar wavelets* and fall within the general framework of window Fourier analysis. From an algorithmic point of view, the Malvar wavelets are much more effective and superior to other wavelets including Gabor wavelets and Morlet–Grossmann wavelets.

In 1985, Yves Meyer, a French pure mathematician, recognized immediately the deep connection between the Calderón formula in harmonic analysis and the new algorithm discovered by Morlet and Grossmann. Using the knowledge of the Calderón–Zygmund operators and the Littlewood–Paley theory, Meyer was able to give a mathematical foundation for wavelet theory. The first major achievement of wavelet analysis was Daubechies, Grossmann, and Meyer's (1986) construction of a "painless" nonorthogonal wavelet expansion. During 1985 and 1986, further work of Meyer and Lemarié on the first construction of a smooth orthonormal wavelet basis on $\mathbb{R}$ and $\mathbb{R}^N$ marked the beginning of their famous contributions to the wavelet theory. At the same time, Stéphans Mallat recognized that some quadratic mirror filters play an important role for the construction of orthogonal wavelet bases generalizing the Haar system. Meyer (1986) and Mallat (1988) realized that the orthogonal wavelet bases could be constructed systematically from a general formalism. Their collaboration culminated with the remarkable discovery by Mallat (1989a,b) of a new formalism, which is the so-called *multiresolution analysis*. It was also Mallat who constructed the wavelet decomposition and reconstruction algorithms using the multiresolution analysis. Mallat's brilliant work was the major source of many new developments in wavelets. A few months later, G. Battle (1987) and Lamarié (1988) independently proposed the construction of spline orthogonal wavelets with exponential decay. While reviewing two books on wavelets in 1993, Meyer states: "Wavelets are without doubt an exciting and intuitive concept. The concept brings with it a new way of thinking, which is absolutely essential and was entirely missing in previously existing algorithms."

Mathematically, the fundamental idea of multiresolution analysis is to represent a function (or signal) f as a limit of successive approximations, each of which is a finer version of the function f. These successive approximations correspond to different levels of resolutions. Thus, multiresolution analysis is a formal approach to constructing orthogonal wavelet bases using a definite set of rules and procedures. The key feature of this analysis is to describe mathematically the process of studying signals or images at different scales. The basic principle of the MRA deals with the decomposition of the whole function space into individual subspaces $V_n \subset V_{n+1}$ so that the space V_{n+1} consists of all rescaled functions in V_n. This essentially means a decomposition of each function (or signal) into components of different scales (or frequencies) so that an individual component of the original function f occurs in each subspace. These components can describe finer and finer versions of the original function f. For example, a function is resolved at scales $\Delta t = 2^0, 2^{-1}, \ldots, 2^{-n}$. In audio signals, these scales are basically *octaves*, which represent higher and higher frequency components. For images and, indeed, for all signals, the simultaneous existence of a multiscale may also be referred to as *multiresolution*. From the point of view of practical application, MRA is really an effective mathematical framework for hierarchical decomposition of an image (or signal) into components of different scales (or frequencies).

Inspired by the work of Meyer, Ingrid Daubechies (1988) made a new remarkable contribution to wavelet theory by constructing families of compactly supported orthonormal wavelets with some degree of smoothness. Her 1988 paper had a tremendous positive impact on the study of wavelets and their diverse applications. This work significantly explained the connection between the continuous wavelets on $\mathbb{R}$, and the discrete wavelets on $\mathbb{Z}$ or $\mathbb{Z}_N$, where the latter has become useful for digital signal analysis. The idea of frames was introduced by Duffin and Schaeffer (1952) and subsequently studied in some detail by Daubechies (1990, 1992). In spite of tremendous success, experts in wavelet theory recognized that it is difficult to construct wavelets that are symmetric, orthogonal and compactly supported. In order to overcome this difficulty, Cohen et al. (1992a,b) studied bi-orthogonal wavelets in some detail. Chui and Wang (1991, 1992) introduced compactly supported spline wavelets and semi-orthogonal wavelet analysis. On the other hand, Beylkin, Coifman, and Rokhlin (1991) and Beylkin (1992) have successfully applied the multiresolution analysis generated by a completely orthogonal scaling function to study a wide variety of integral operators on $L^2(\mathbb{R})$ by a matrix in a wavelet basis. This work culminated with the remarkable discovery of new algorithms in numerical analysis. Consequently, some significant progress has been made in boundary element methods, finite element methods, and numerical solutions of partial differential equations using wavelet analysis.

As a natural extension of wavelet analysis, Coifman (1992a,b) in collaboration with Meyer and Wickerhauser discovered wavelet packets, which can be used to design efficient schemes for the representation and compression of acoustic signals and images. This led them to the construction of a library of orthogonal bases by extending the method of multiresolution decomposition and using the quadratic mirror filters. Recently, there have also been significant applications of wavelet analysis to a wide variety of problems in many diverse fields including mathematics, physics, medicine, computer science, and engineering.

We close this historical introduction by citing some of the applications, which include addressing problems in signal processing, computer vision, seismology, turbulence, computer graphics, image processing, structures of the galaxies in the Universe, digital communication, pattern recognition, approximation theory, quantum optics, biomedical engineering, sampling theory, matrix theory, operator theory, differential equations, numerical analysis, statistics and multiscale segmentation of well logs, natural scenes, and mammalian visual systems. Wavelets allow complex information such as music, speech, images, patterns, and so on to be decomposed into elementary form, called building blocks (wavelets). The information is subsequently reconstructed with high precision. To describe the present state of wavelet research, Meyer (1993, page 31) wrote as follows:

> Today the boundaries between mathematics and signal and image processing have faded, and mathematics has benefited from the rediscovery of wavelets by experts from other disciplines. The detour through signal and image processing was the most direct path leading from Haar basis to Daubechies's wavelets.

8.2 Continuous Wavelet Transforms

An integral transform is an operator T on a space of functions on some $\Omega \subset \mathbb{R}^N$ which is defined by

$$(Tf)(y) = \int_\Omega K(x, y) f(x) dx.$$

The properties of the transform depend on the function K, which is called the *kernel* of the transform. For example, in the case of the Fourier transform $K(x, y) = e^{-ixy}$. Note that y can be interpreted as a scaling factor. We take the exponential function $\varphi(x) = e^{-ix}$ and then generate a one-parameter family of functions by taking scaled copies of φ, that is $\varphi_\alpha(x) = e^{-i\alpha x}$, for all $\alpha \in \mathbb{R}$. The continuous wavelet transform is similar to the Fourier transform in the sense that it is based on a single function ψ and that this function is scaled. But unlike the Fourier transform, we also shift the function, thus generating a two-parameter family of functions $\psi_{a,b}$. It is convenient to define $\psi_{a,b}$ as follows:

$$\psi_{a,b}(x) = |a|^{-1/2} \psi\left(\frac{x-b}{a}\right).$$

Then the continuous wavelet transform is defined by

$$(W_\psi f)(a,b) = \int_{-\infty}^{\infty} f(t)\overline{\psi_{a,b}(t)}dt = |a|^{-1/2}\int_{-\infty}^{\infty} f(t)\overline{\psi\left(\frac{t-b}{a}\right)}dt.$$

The continuous wavelet transform is not a single transform like the Fourier transform, but any transform obtained in this way. Properties of a particular transform will depend on the choice of ψ. One of the first properties we expect of any integral transform is that the original function can be reconstructed from the transform. We will prove a theorem, which gives conditions on ψ which guarantee invertibility of the transform. First we need to define the object of our study more precisely.

Definition 8.2.1. (Wavelet)
By a *wavelet* we mean a function $\psi \in L^2(\mathbb{R})$ satisfying the *admissibility condition*

$$\int_{-\infty}^{\infty} \frac{|\widehat{\psi}(\omega)|^2}{|\omega|}d\omega < \infty, \tag{8.1}$$

where $\widehat{\psi}$ is the Fourier transform ψ, that is,

$$\widehat{\psi}(\omega) = \frac{1}{\sqrt{2\pi}}\int_{-\infty}^{\infty} e^{-i\omega x}\psi(x)dx.$$

If $\psi \in L^2(\mathbb{R})$, then $\psi_{a,b}(x) \in L^2(\mathbb{R})$ for all a, b. Indeed,

$$\left\|\psi_{a,b}(t)\right\|^2 = |a|^{-1}\int_{-\infty}^{\infty}\left|\psi\left(\frac{x-b}{a}\right)\right|^2 dt = \int_{-\infty}^{\infty}|\psi(t)|^2 dt = \|\psi\|^2. \tag{8.2}$$

The Fourier transform of $\psi_{a,b}(x)$ is given by

$$\widehat{\psi}_{a,b}(\omega) = |a|^{-1/2}\frac{1}{\sqrt{2\pi}}\int_{-\infty}^{\infty} e^{-i\omega x}\psi\left(\frac{x-b}{a}\right)dx = \sqrt{|a|}e^{-ib\omega}\widehat{\psi}(a\omega). \tag{8.3}$$

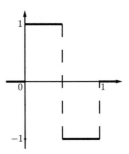

Figure 8.1   The Haar wavelet.

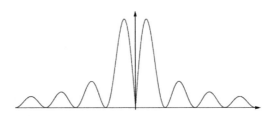

Figure 8.2   The absolute value of the Fourier transform of the Haar wavelet.

Example 8.2.2. (Haar wavelet)   Let

$$\psi(x) = \begin{cases} 1, & 0 \le x < \frac{1}{2}, \\ -1, & \frac{1}{2} \le x < 1, \\ 0, & \text{otherwise} \end{cases}$$

(see Figure 8.1). Then

$$\widehat{\psi}(\omega) = \frac{1}{\sqrt{(2\pi)}} \frac{(\sin \frac{\omega}{4})^2}{\frac{\omega}{4}} e^{-i(\omega - \pi)/2}$$

and

$$\int_{-\infty}^{\infty} \frac{|\widehat{\psi}(\omega)|^2}{|\omega|} d\omega = \frac{8}{\pi} \int_{-\infty}^{\infty} \frac{|\sin \frac{\omega}{4}|^4}{|\omega|^3} d\omega < \infty.$$

The Haar wavelet is one of the classic examples. It is well localized in the time domain, but it is not continuous. The absolute value of the Fourier transform of the Haar wavelet, $|\widehat{\psi}(\omega)|$, is plotted in Figure 8.2. This figure clearly indicates that the Haar wavelet has poor frequency localization, since it does not have compact support in the frequency domain. The function $|\widehat{\psi}(\omega)|$ is even and attains its maximum at the frequency $\omega_0 \sim 4.662$. The rate of decay as $\omega \to \infty$ is as ω^{-1}. The reason for the slow decay is discontinuity of ψ. Its discontinuous nature is a serious weakness in many applications. However, the Haar wavelet is one of the most fundamental examples that illustrate major features of the general wavelet theory.   □

Theorem 8.2.3. *Let ψ be a wavelet and let φ be a bounded integrable function. Then the function $\psi * \varphi$ is a wavelet.*

Proof: Since

$$\int_{-\infty}^{\infty} |\psi * \varphi(x)|^2 dx = \int_{-\infty}^{\infty} \left| \int_{-\infty}^{\infty} \psi(x-u)\varphi(u)du \right|^2 dx$$

$$\leq \int_{-\infty}^{\infty} \left(\int_{-\infty}^{\infty} |\psi(x-u)||\varphi(u)|du \right)^2 dx$$

$$= \int_{-\infty}^{\infty} \left(\int_{-\infty}^{\infty} |\psi(x-u)||\varphi(u)|^{1/2}|\varphi(u)|^{1/2}du \right)^2 dx$$

$$\leq \int_{-\infty}^{\infty} \left(\int_{-\infty}^{\infty} |\psi(x-u)|^2|\varphi(u)|du \int_{-\infty}^{\infty} |\varphi(u)|du \right) dx$$

$$\leq \int_{-\infty}^{\infty} |\varphi(u)|du \int_{-\infty}^{\infty}\int_{-\infty}^{\infty} |\psi(x-u)|^2|\varphi(u)|dx\,du$$

$$= \left(\int_{-\infty}^{\infty} |\varphi(u)|du \right)^2 \int_{-\infty}^{\infty} |\psi(x)|^2 dx < \infty,$$

we have $\psi * \varphi \in L^2(\mathbb{R})$. Moreover,

$$\int_{-\infty}^{\infty} \frac{|\widehat{\psi * \varphi}(\omega)|^2}{|\omega|} d\omega = \int_{-\infty}^{\infty} \frac{|\widehat{\psi}(\omega)\widehat{\varphi}(\omega)|^2}{|\omega|} d\omega$$

$$= \int_{-\infty}^{\infty} \frac{|\widehat{\psi}(\omega)|^2}{|\omega|} |\widehat{\varphi}(\omega)|^2 d\omega$$

$$\leq \sup|\widehat{\varphi}(\omega)|^2 \int_{-\infty}^{\infty} \frac{|\widehat{\psi}(\omega)|^2}{|\omega|} d\omega < \infty.$$

Thus, the function $\psi * \varphi$ is a wavelet.   □

Example 8.2.4. Theorem 8.2.3 can be used to generate examples of wavelets. For example, if we take the Haar wavelet and convolve it with the following function

$$\varphi(x) = \begin{cases} 0, & x < 0, \\ 1, & 0 \leq x \leq 1, \\ 0, & x \geq 1, \end{cases}$$

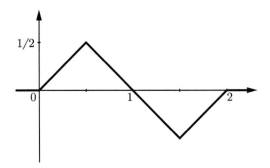

Figure 8.3   A continuous wavelet.

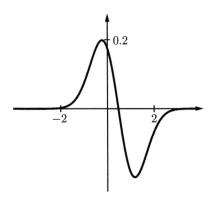

Figure 8.4   A smooth wavelet.

then we obtain a simple continuous function, see Figure 8.3. If we convolve the Haar wavelet with $\varphi(x) = e^{-x^2}$, then the obtained wavelet is smooth, see Figure 8.4.   □

Definition 8.2.5.   (Continuous wavelet transform)
Let $\psi \in L^2(\mathbb{R})$ and let, for $a, b \in \mathbb{R}, a \neq 0$,

$$\psi_{a,b}(x) = |a|^{-1/2} \psi\left(\frac{x-b}{a}\right).$$

The integral transform W_ψ defined on $L^2(\mathbb{R})$ by

$$(W_\psi f)(a, b) = \int_{-\infty}^{\infty} f(t)\overline{\psi_{a,b}(t)}dt = \langle f, \psi_{a,b} \rangle \tag{8.4}$$

is called a *continuous wavelet transform*.

The function ψ is often called the *mother wavelet* or the *analyzing wavelet*. The parameter b can be interpreted as the time translation and a is a scaling parameter which measures the degree of compression or scale.

Lemma 8.2.6. *For any $f \in L^2(\mathbb{R})$, we have*

$$\mathcal{F}\{(W_\psi f)(a, b)\} = \sqrt{2\pi |a|}\, \hat{f}(\omega)\overline{\hat{\psi}(a\omega)}. \tag{8.5}$$

Proof: Using the Parseval formula for the Fourier transform, it follows from (8.4) that

$$(W_\psi f)(a, b) = \langle f, \psi_{a,b} \rangle = \langle \hat{f}, \hat{\psi}_{a,b} \rangle$$

$$= \frac{1}{\sqrt{2\pi}} \int_{-\infty}^{\infty} \left\{ \sqrt{2\pi |a|}\, \hat{f}(\omega)\overline{\hat{\psi}(a\omega)} \right\} e^{ib\omega}\, d\omega. \tag{8.6}$$

This means that

$$\mathcal{F}\{(W_\psi f)(a, b)\} = \frac{1}{\sqrt{2\pi}} \int_{-\infty}^{\infty} e^{-ib\omega} (W_\psi f)(a, b)\, db$$

$$= \sqrt{2\pi |a|}\, \hat{f}(\omega)\overline{\hat{\psi}(a\omega)}. \tag{8.7}$$

□

Theorem 8.2.7. (Parseval's relation for wavelet transforms) *Let $\psi \in L^2(\mathbb{R})$ satisfy*

$$C_\psi = 2\pi \int_{-\infty}^{\infty} \frac{|\hat{\psi}(\omega)|^2}{|\omega|}\, d\omega < \infty. \tag{8.8}$$

Then, for any $f, g \in L^2(\mathbb{R})$, we have

$$\int_{-\infty}^{\infty}\int_{-\infty}^{\infty} (W_\psi f)(a, b)\overline{(W_\psi g)(a, b)} \frac{db\, da}{a^2} = C_\psi \langle f, g \rangle. \tag{8.9}$$

Proof: From (8.6) we get

$$(W_\psi f)(a, b) = \sqrt{|a|} \int_{-\infty}^{\infty} \hat{f}(\omega) e^{ib\omega} \overline{\hat{\psi}(a\omega)}\, d\omega \tag{8.10}$$

and

$$\overline{(W_\psi g)(a, b)} = \sqrt{|a|} \int_{-\infty}^{\infty} \overline{\hat{g}(\sigma)} e^{-ib\sigma} \hat{\psi}(a\sigma)\, d\sigma. \tag{8.11}$$

Substituting (8.10) and (8.11) in the left side of (8.9) gives

$$\int_{-\infty}^{\infty}\int_{-\infty}^{\infty} (W_\psi f)(a, b)\overline{(W_\psi g)(a, b)} \frac{db\, da}{a^2}$$

$$= \int_{-\infty}^{\infty} \int_{-\infty}^{\infty} \frac{db\,da}{a^2} \int_{-\infty}^{\infty} \int_{-\infty}^{\infty} |a|\hat{f}(\omega)\hat{g}(\sigma)\,\overline{\hat{\psi}(a\omega)}\hat{\psi}(a\sigma)e^{ib(\omega-\sigma)}\,d\omega\,d\sigma$$

which is, by interchanging the order of integration,

$$= 2\pi \int_{-\infty}^{\infty} \frac{da}{|a|} \int_{-\infty}^{\infty} \int_{-\infty}^{\infty} \hat{f}(\omega)\hat{g}(\sigma)\,\overline{\hat{\psi}(a\omega)}\hat{\psi}(a\sigma)\,d\omega\,d\sigma \frac{1}{2\pi}\int_{-\infty}^{\infty} e^{ib(\omega-\sigma)}\,db$$

$$= 2\pi \int_{-\infty}^{\infty} \frac{da}{|a|} \int_{-\infty}^{\infty} \int_{-\infty}^{\infty} \hat{f}(\omega)\hat{g}(\sigma)\,\overline{\hat{\psi}(a\omega)}\hat{\psi}(a\sigma)\delta(\sigma-\omega)\,d\omega\,d\sigma$$

$$= 2\pi \int_{-\infty}^{\infty} \frac{da}{|a|} \int_{-\infty}^{\infty} \hat{f}(\omega)\overline{\hat{g}(\omega)}|\hat{\psi}(a\omega)|^2\,d\omega$$

and finally, again interchanging the order of integration and putting $a\omega = x$,

$$= 2\pi \int_{-\infty}^{\infty} \hat{f}(\omega)\overline{\hat{g}(\omega)}d\omega \int_{-\infty}^{\infty} \frac{|\hat{\psi}(x)|^2}{|x|}dx = C_{\psi}\langle\hat{f},\hat{g}\rangle = C_{\psi}\langle f,g\rangle. \quad (8.12)$$

$\square$

If $f = g$, then (8.9) takes the form

$$\int_{-\infty}^{\infty} \int_{-\infty}^{\infty} |(W_{\psi}f)(a,b)|^2 \frac{db\,da}{a^2} = C_{\psi}\|f\|^2. \quad (8.13)$$

This shows that, except for the factor C_{ψ}, the wavelet transform is an isometry from $L^2(\mathbb{R})$ to $L^2(\mathbb{R})$.

Theorem 8.2.8. (Inversion formula) *If $f \in L^2(\mathbb{R})$, then*

$$f(x) = \frac{1}{C_{\psi}} \int_{-\infty}^{\infty} \int_{-\infty}^{\infty} (W_{\psi}f)(a,b)\psi_{a,b}(x) \frac{db\,da}{a^2}, \quad (8.14)$$

where the equality holds almost everywhere.

Proof: For any $g \in L^2(\mathbb{R})$, we have

$$C_{\psi}\langle f,g\rangle = \int_{-\infty}^{\infty} \int_{-\infty}^{\infty} (W_{\psi}f)(a,b)\overline{(W_{\psi}g)(a,b)} \frac{db\,da}{a^2}$$

$$= \int_{-\infty}^{\infty} \int_{-\infty}^{\infty} (W_{\psi}f)(a,b) \overline{\int_{-\infty}^{\infty} g(t)\overline{\psi_{a,b}(t)}dt} \frac{db\,da}{a^2}$$

$$= \int_{-\infty}^{\infty} \int_{-\infty}^{\infty} \int_{-\infty}^{\infty} (W_{\psi}f)(a,b)\psi_{a,b}(t) \frac{db\,da}{a^2}\overline{g(t)}dt$$

$$= \left\langle \int_{-\infty}^{\infty} \int_{-\infty}^{\infty} (W_{\psi}f)(a,b)\psi_{a,b} \frac{db\,da}{a^2}, g\right\rangle.$$

Since g is an arbitrary element of $L^2(\mathbb{R})$, the inversion formula (8.14) follows. □

The following theorem summarizes some elementary properties of the continuous wavelet transform. The straightforward proofs are left as exercises.

Theorem 8.2.9. *Let ψ and φ be wavelets and let $f, g \in L^2(\mathbb{R})$.*

(a) $(W_\psi(\alpha f + \beta g))(a, b) = \alpha(W_\psi f)(a, b) + \beta(W_\psi g)(a, b)$ *for any $\alpha, \beta \in \mathbb{C}$,*

(b) $(W_\psi(T_c f))(a, b) = (W_\psi f)(a, b - c)$, *where T_c is the translation operator defined by $T_c f(t) = f(t - c)$,*

(c) $(W_\psi(D_c f))(a, b) = \frac{1}{\sqrt{c}}(W_\psi f)(\frac{a}{c}, \frac{b}{c})$, *where c is a positive number and D_c is the dilation operator defined by $D_c f(t) = \frac{1}{c} f(\frac{t}{c})$,*

(d) $(W_\psi \varphi)(a, b) = \overline{(W_\varphi \psi)(\frac{1}{a}, -\frac{b}{a})}$, $a \neq 0$,

(e) $(W_{\alpha\psi + \beta\varphi} f)(a, b) = \overline{\alpha}(W_\psi f)(a, b) + \overline{\beta}(W_\varphi f)(a, b)$ *for any $\alpha, \beta \in \mathbb{C}$,*

(f) $(W_{P\psi} Pf)(a, b) = (W_\psi f)(a, -b)$, *where P is the parity operator defined by $Pf(t) = f(-t)$,*

(g) $(W_{T_c \psi} f)(a, b) = (W_\psi f)(a, b + ca)$,

(h) $(W_{D_c \psi} f)(a, b) = \frac{1}{\sqrt{c}}(W_\psi f)(ac, b)$, $c > 0$.

For the wavelets to be useful analyzing functions, the mother wavelet must have certain properties. One such property is defined by the condition (8.1) which guarantees existence of the inversion formula for the continuous wavelet transform. If $\psi \in L^1(\mathbb{R})$, then its Fourier transform $\widehat{\psi}$ is continuous. If $\widehat{\psi}$ is continuous, C_ψ can be finite only if $\widehat{\psi}(0) = 0$, or equivalently, $\int_{-\infty}^{\infty} \psi(t) dt = 0$. This means that ψ must be an oscillatory function with zero mean. Condition (8.1) also imposes a restriction on the rate of decay of $|\widehat{\psi}(\omega)|^2$.

In addition to the admissibility condition (8.1), there are other properties that may be useful in particular applications. For example, we may want to require that ψ be n times continuously differentiable or infinitely differentiable. If the Haar wavelet is convolved $n+1$ times with the function φ given in Example 8.2.4, then the resulting function $\psi * \varphi * \cdots * \varphi$ is an n times differentiable wavelet. The function in Figure 8.4 is an infinitely differentiable wavelet. The so-called "Mexican hat wavelet" is another example of an infinitely differentiable wavelet.

Example 8.2.10. (Mexican hat wavelet) This wavelet is defined by

$$\psi(t) = (1 - t^2)e^{-t^2/2}$$

and shown in Figure 8.5. □

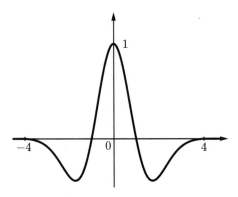

Figure 8.5   Mexican hat wavelet.

Another desirable property of wavelets is the so-called "localization prop-erty." We want ψ to be well localized in both time and frequency. In other words, ψ and its derivatives must decay very rapidly. For frequency localiza-tion, $\widehat{\psi}(\omega)$ must decay sufficiently rapidly as $\omega \to \infty$ and $\widehat{\psi}(\omega)$ should be flat in the neighborhood of $\omega = 0$. The flatness at $\omega = 0$ is associated with the number of vanishing moments of ψ. The kth moment of ψ is defined by

$$m_k = \int_{-\infty}^{\infty} t^k \psi(t)dt.$$

A wavelet is said to have n vanishing moments if

$$\int_{-\infty}^{\infty} t^k \psi(t)dt = 0 \quad \text{for } k = 0, 1, \ldots, n,$$

or, equivalently,

$$\left[\frac{d^k \widehat{\psi}(\omega)}{d\omega^k}\right]_{\omega=0} = 0 \quad \text{for } k = 0, 1, \ldots, n.$$

Wavelets with a larger number of vanishing moments result in more flatness when frequency ω is small.

8.3 The Discrete Wavelet Transform

While the continuous wavelet transform is compared to the Fourier transform, which requires calculating the integral $\int_{-\infty}^{\infty} e^{-i\omega x}f(x)dx$ for all (or almost all) $\omega \in \mathbb{R}$, the discrete wavelet transform can be compared to the Fourier series, which requires calculating the integral $\int_{0}^{2\pi} e^{-inx}f(x)dx$ for integer values of n.

Since the continuous wavelet transform is a two-parameter representation of a function

$$(W_\psi f)(a, b) = |a|^{-1/2} \int_{-\infty}^{\infty} f(t) \overline{\psi\left(\frac{t - b}{a}\right)} dt,$$

we could discretize it by assuming that a and b take only integer values. It turns out that it is better to discretize it in a different way. First we fix two positive constants a_0 and b_0 and then define

$$\psi_{m,n}(x) = a_0^{-m/2} \psi\left(a_0^{-m} x - n b_0\right), \tag{8.15}$$

where m and n range over $\mathbb{Z}$. By the *discrete wavelet coefficients* of $f \in L^2(\mathbb{R})$ we mean the numbers $\langle f, \psi_{m,n} \rangle$, where $m, n \in \mathbb{Z}$. The fundamental question here is whether it is possible to reconstruct f from those coefficients. The weakest interpretation of this problem is whether $\langle f, \psi_{m,n} \rangle = \langle g, \psi_{m,n} \rangle$ for all $m, n \in \mathbb{Z}$ implies $f = g$. In practice we expect much more than that: we want $\langle f, \psi_{m,n} \rangle$ and $\langle g, \psi_{m,n} \rangle$ to be "close" if f and g are "close." This will be guaranteed if there exists a $B > 0$, such that

$$\sum_{m,n=-\infty}^{\infty} \left|\langle f, \psi_{m,n} \rangle\right|^2 \leq B \|f\|^2$$

for all $f \in L^2(\mathbb{R})$. Similarly, we want f and g to be "close" if $\langle f, \psi_{m,n} \rangle$ and $\langle g, \psi_{m,n} \rangle$ are "close." This is important because we want to be sure that when we neglect some small terms in the representation of f in terms of $\langle f, \psi_{m,n} \rangle$, then the reconstructed function will not differ much from f. The representation will have this property if there exists an $A > 0$, such that

$$A \|f\|^2 \leq \sum_{m,n=-\infty}^{\infty} \left|\langle f, \psi_{m,n} \rangle\right|^2$$

for all $f \in L^2(\mathbb{R})$. These two requirements are best studied in terms of the so-called frames.

Definition 8.3.1.   (Frame)
A sequence $(\varphi_1, \varphi_2, \ldots)$ in a Hilbert space H is called a *frame* if there exist $A, B > 0$ such that

$$A \|f\|^2 \leq \sum_{n=1}^{\infty} \left|\langle f, \varphi_n \rangle\right|^2 \leq B \|f\|^2 \tag{8.16}$$

for all $f \in H$. The constants A and B are called *frame bounds*. If $A = B$, then the frame is called *tight*.

If (φ_n) is an orthonormal basis, then it is a tight frame since $\sum_{n=1}^{\infty} |\langle f, \varphi_n \rangle|^2$ $= \|f\|^2$ for all $f \in H$. The vectors $(1, 0)$, $(-\frac{1}{2}, \frac{\sqrt{3}}{2})$, $(-\frac{1}{2}, -\frac{\sqrt{3}}{2})$ form a tight frame in $\mathbb{C}^2$ which is not a basis.

As pointed out above, we want the family of functions $\psi_{m,n}$ to form a frame in $L^2(\mathbb{R})$. (Obviously, the double indexing of the functions is irrelevant.) The following theorem gives fairly general, sufficient conditions for $(\psi_{m,n})$ to constitute a frame in $L^2(\mathbb{R})$.

Theorem 8.3.2. *If ψ and a_0 are such that*

$$\inf_{1 \leq |s| \leq a_0} \sum_{m=-\infty}^{\infty} \left| \widehat{\psi} \left(a_0^m s \right) \right|^2 > 0,$$

$$\sup_{1 \leq |s| \leq a_0} \sum_{m=-\infty}^{\infty} \left| \widehat{\psi} \left(a_0^m s \right) \right|^2 < \infty$$

and

$$\sup_{s \in \mathbb{R}} \sum_{m=-\infty}^{\infty} \left| \widehat{\psi} \left(a_0^m s \right) \right| \left| \widehat{\psi} \left(a_0^m s + \xi \right) \right| \leq C \left(1 + |\xi| \right)^{-(1+\varepsilon)}$$

for some $\varepsilon > 0$ and some constant C, then there exists $\tilde{b}$ such that $(\psi_{m,n})$ is a frame in $L^2(\mathbb{R})$ for any $b_0 \in (0, \tilde{b})$.

Proof: Let $f \in L^2(\mathbb{R})$. Then

$$\sum_{m,n=-\infty}^{\infty} |\langle f, \psi_{m,n} \rangle|^2 = \sum_{m,n=-\infty}^{\infty} \left| \int_{-\infty}^{\infty} f(x) a_0^{-m/2} \overline{\psi(a_0^m x - n b_0)} dx \right|^2$$

$$= \sum_{m,n=-\infty}^{\infty} \left| \int_{-\infty}^{\infty} \hat{f}(s) a_0^{m/2} \overline{\widehat{\psi}(a_0^m s)} e^{i b_0 a_0^m n s} ds \right|^2 = \clubsuit$$

by the general Parseval relation (Theorem 5.11.21), basic properties of the Fourier transform (Theorem 5.11.7), and the fact that we sum over all integers. Since, for any $\omega > 0$, the integral $\int_{-\infty}^{\infty} g(t) dt$ can be written as

$$\sum_{l=-\infty}^{\infty} \int_0^{\omega} g(t + l\omega) dt,$$

by taking $\omega = \frac{2\pi}{b_0 a_0^m}$, we obtain

$$\clubsuit = \sum_{m,n=-\infty}^{\infty} a_0^m \left| \sum_{l=-\infty}^{\infty} \int_0^{\omega} e^{2\pi i n s / \omega} \hat{f}(s + l\omega) \overline{\widehat{\psi}(a_0^m(s + l\omega))} ds \right|^2$$

$$= \sum_{m,n=-\infty}^{\infty} a_0^m \left| \int_0^\omega e^{2\pi i n s/\omega} \left(\sum_{l=-\infty}^{\infty} \hat{f}(s+l\omega) \overline{\hat{\psi}(a_0^m(s+l\omega))} \right) ds \right|^2$$

$$= \sum_{m=-\infty}^{\infty} a_0^m \omega \int_0^\omega \left| \sum_{l=-\infty}^{\infty} \hat{f}(s+l\omega) \overline{\hat{\psi}(a_0^m(s+l\omega))} \right|^2 ds = \diamondsuit,$$

by Parseval's formula for trigonometric Fourier series. Since

$$\left| \sum_{l=-\infty}^{\infty} \hat{f}(s+l\omega) \overline{\hat{\psi}(a_0^m(s+l\omega))} \right|^2$$

$$= \left(\sum_{l=-\infty}^{\infty} \hat{f}(s+l\omega) \overline{\hat{\psi}(a_0^m(s+l\omega))} \right) \left(\sum_{k=-\infty}^{\infty} \overline{\hat{f}(s+k\omega)} \hat{\psi}(a_0^m(s+k\omega)) \right)$$

and

$$F(s) = \sum_{k=-\infty}^{\infty} \overline{\hat{f}(s+k\omega)} \hat{\psi}(a_0^m(s+k\omega))$$

is a periodic function with a period of ω, we have

$$\int_0^\omega \left(\sum_{l=-\infty}^{\infty} \hat{f}(s+l\omega) \overline{\hat{\psi}(a_0^m(s+l\omega))} \right) F(s) ds$$

$$= \int_{-\infty}^{\infty} \hat{f}(s) \overline{\hat{\psi}(a_0^m s)} F(s) ds$$

$$= \sum_{k=-\infty}^{\infty} \int_{-\infty}^{\infty} \hat{f}(s) \overline{\hat{\psi}(a_0^m s)} \overline{\hat{f}(s+k\omega)} \hat{\psi}(a_0^m(s+k\omega)) ds.$$

Consequently,

$$\diamondsuit = \frac{2\pi}{b_0} \sum_{m,k=-\infty}^{\infty} \int_{-\infty}^{\infty} \hat{f}(s) \overline{\hat{f}(s+k\omega)} \, \overline{\hat{\psi}(a_0^m s)} \hat{\psi}(a_0^m(s+k\omega)) ds$$

$$= \frac{2\pi}{b_0} \int_{-\infty}^{\infty} |\hat{f}(s)|^2 \sum_{m=-\infty}^{\infty} |\hat{\psi}(a_0^m s)|^2 ds$$

$$+ \frac{2\pi}{b_0} \sum_{\substack{m,k=-\infty \\ k\neq 0}}^{\infty} \int_{-\infty}^{\infty} \hat{f}(s) \overline{\hat{f}(s+k\omega)} \, \overline{\hat{\psi}(a_0^m s)} \, \hat{\psi}(a_0^m(s+k\omega)) ds.$$

To find a bound on the second summand, we use the Schwarz inequality:

$$
\left| \frac{2\pi}{b_0} \sum_{\substack{m,k=-\infty \\ k\neq 0}} \int_{-\infty}^{\infty} \hat{f}(s)\overline{\hat{f}(s+k\omega)}\,\overline{\widehat{\psi}\left(a_0^m s\right)}\widehat{\psi}\left(a_0^m(s+k\omega)\right)ds \right|
$$

$$
\leq \frac{2\pi}{b_0} \sum_{\substack{m,k=-\infty \\ k\neq 0}} \left(\int_{-\infty}^{\infty} |\hat{f}(s)|^2 \left|\widehat{\psi}\left(a_0^m s\right)\right|\left|\widehat{\psi}\left(a_0^m(s+k\omega)\right)\right| ds \right)^{1/2}
$$

$$
\times \left(\int_{-\infty}^{\infty} |\hat{f}(s+k\omega)|^2 \left|\widehat{\psi}\left(a_0^m s\right)\right|\left|\widehat{\psi}\left(a_0^m(s+k\omega)\right)\right| ds \right)^{1/2} = \heartsuit.
$$

Then, by first changing the variables in the second factor and by using Hölder's inequality (Theorem 1.2.7), we have

$$
\heartsuit = \frac{2\pi}{b_0} \sum_{\substack{m,k=-\infty \\ k\neq 0}} \left(\int_{-\infty}^{\infty} |\hat{f}(s)|^2 \left|\widehat{\psi}\left(a_0^m s\right)\right|\left|\widehat{\psi}\left(a_0^m(s+k\omega)\right)\right| ds \right)^{1/2}
$$

$$
\times \left(\int_{-\infty}^{\infty} |\hat{f}(s)|^2 \left|\widehat{\psi}\left(a_0^m(s-k\omega)\right)\right|\left|\widehat{\psi}\left(a_0^m s\right)\right| ds \right)^{1/2}
$$

$$
\leq \frac{2\pi}{b_0} \sum_{\substack{k=-\infty \\ k\neq 0}} \left(\int_{-\infty}^{\infty} |\hat{f}(s)|^2 \sum_{m=-\infty}^{\infty} \left|\widehat{\psi}\left(a_0^m s\right)\right|\left|\widehat{\psi}\left(a_0^m(s+k\omega)\right)\right| ds \right)^{1/2}
$$

$$
\times \left(\int_{-\infty}^{\infty} |\hat{f}(s)|^2 \sum_{m=-\infty}^{\infty} \left|\widehat{\psi}\left(a_0^m(s-k\omega)\right)\right|\left|\widehat{\psi}\left(a_0^m s\right)\right| ds \right)^{1/2} = \spadesuit.
$$

If we denote

$$
\beta(\xi) = \sup_{s\in\mathbb{R}} \sum_{m=-\infty}^{\infty} \left|\widehat{\psi}\left(a_0^m s\right)\right|\left|\widehat{\psi}\left(a_0^m s + \xi\right)\right|,
$$

then

$$
\spadesuit \leq \frac{2\pi}{b_0} \|f\|^2 \sum_{\substack{k=-\infty \\ k\neq 0}}^{\infty} \left[\beta\left(a_0^m k\omega\right)\beta\left(-a_0^m k\omega\right)\right]^{1/2}
$$

$$
= \frac{2\pi}{b_0} \|f\|^2 \sum_{\substack{k=-\infty \\ k\neq 0}}^{\infty} \left[\beta\left(\frac{2\pi k}{b_0}\right)\beta\left(-\frac{2\pi k}{b_0}\right)\right]^{1/2}.
$$

Consequently, if we denote

$$A = \frac{2\pi}{b_0} \left\{ \sup_{s \in \mathbb{R}} \sum_{m=-\infty}^{\infty} |\widehat{\psi}(a_0^m s)|^2 - \sum_{\substack{k=-\infty \\ k \neq 0}}^{\infty} \left[\beta\left(\frac{2\pi k}{b_0}\right) \beta\left(-\frac{2\pi k}{b_0}\right) \right]^{1/2} \right\}$$

and

$$B = \frac{2\pi}{b_0} \left\{ \inf_{s \in \mathbb{R}} \sum_{m=-\infty}^{\infty} |\widehat{\psi}(a_0^m s)|^2 + \sum_{\substack{k=-\infty \\ k \neq 0}}^{\infty} \left[\beta\left(\frac{2\pi k}{b_0}\right) \beta\left(-\frac{2\pi k}{b_0}\right) \right]^{1/2} \right\},$$

we obtain

$$A\|f\|^2 \leq \sum_{m,n=-\infty}^{\infty} |\langle f, \psi_{m,n} \rangle|^2 \leq B\|f\|^2.$$

Since $\beta(\xi) \leq C(1 + |\xi|)^{-(1+\varepsilon)}$, we have

$$\sum_{\substack{k=-\infty \\ k \neq 0}}^{\infty} \left[\beta\left(\frac{2\pi k}{b_0}\right) \beta\left(-\frac{2\pi k}{b_0}\right) \right]^{1/2} = 2 \sum_{k=1}^{\infty} \left[\beta\left(\frac{2\pi k}{b_0}\right) \beta\left(-\frac{2\pi k}{b_0}\right) \right]^{1/2}$$

$$\leq 2C \sum_{k=1}^{\infty} \left(1 + \frac{2\pi k}{b_0}\right)^{-(1+\varepsilon)}$$

$$\leq 2C \int_0^{\infty} \left(1 + \frac{2\pi t}{b_0}\right)^{-(1+\varepsilon)} dt$$

$$= \frac{Cb_0}{\pi \varepsilon}.$$

Now, since $\frac{Cb_0}{\pi \varepsilon} \to 0$ as $b_0 \to 0$ and $\inf_{1 \leq |s| \leq a_0} \sum_{m=-\infty}^{\infty} |\widehat{\psi}(a_0^m s)|^2 > 0$, there exists a positive number $\tilde{b}$ such that $A > 0$ for any $b_0 \in (0, \tilde{b})$. Finally, since $\sup_{1 \leq |s| \leq a_0} \sum_{m=-\infty}^{\infty} |\widehat{\psi}(a_0^m s)|^2 < \infty$, we also have $B < \infty$ for all $b_0 \in (0, \tilde{b})$. Thus, $\psi_{m,n}$ constitute a frame for all such b_0. $\square$

Now we turn to the problem of reconstruction of f from $\langle f, \psi_{m,n} \rangle$ and representation of f in terms of $\psi_{m,n}$. For a complete orthonormal system (φ_n), both questions are answered by the equality $f = \sum_{n=1}^{\infty} \langle f, \varphi_n \rangle \varphi_n$. As we will see, since we do not have orthogonality, the situation is more complicated for frames.

Definition 8.3.3. (Frame operator)
Let $(\varphi_1, \varphi_2, \ldots)$ be a frame in a Hilbert space H. The operator F from H into l^2

defined by

$$F(f) = (\langle f, \varphi_n \rangle)$$

is called a *frame operator*.

Lemma 8.3.4. *Let F be a frame operator. Then F is a bounded, linear, and invertible operator. Its inverse F^{-1} is also a bounded operator.*

The proof is easy and is left as an exercise.

Consider the adjoint operator F^* of the frame operator F associated with a frame (φ_n). For any $(c_n) \in l^2$, we have

$$\langle F^*(c_n), f \rangle = \langle (c_n), Ff \rangle = \sum_{n=1}^{\infty} c_n \langle \varphi_n, f \rangle = \left\langle \sum_{n=1}^{\infty} c_n \varphi_n, f \right\rangle.$$

Thus, the adjoint operator of a frame operator has the form

$$F^*(c_n) = \sum_{n=1}^{\infty} c_n \varphi_n. \tag{8.17}$$

Note also that since

$$\sum_{n=1}^{\infty} |\langle f, \varphi_n \rangle|^2 = \|Ff\|^2 = \langle F^* Ff, f \rangle,$$

the condition (8.16) can be expressed as

$$A\mathcal{I} \leq F^* F \leq B\mathcal{I},$$

where the inequality $\leq$ is to be understood in the sense defined in Section 4.6.

Theorem 8.3.5. *Let $(\varphi_1, \varphi_2, \ldots)$ be a frame with frame bounds A and B and let F be the associated frame operator. Define*

$$\tilde{\varphi}_n = \left(F^* F \right)^{-1} \varphi_n.$$

Then $(\tilde{\varphi}_n)$ is a frame with frame bounds $\frac{1}{B}$ and $\frac{1}{A}$.

Proof: By Corollary 4.5.6, we have $(F^* F)^{-1} = ((F^* F)^{-1})^*$. Consequently,

$$\langle f, \tilde{\varphi}_n \rangle = \langle f, \left(F^* F \right)^{-1} \varphi_n \rangle = \langle \left(F^* F \right)^{-1} f, \varphi_n \rangle$$

and then

$$\sum_{n=1}^{\infty} |\langle f, (\tilde{\varphi}_n) \rangle|^2 = \sum_{n=1}^{\infty} |\langle (F^* F)^{-1} f, \varphi_n \rangle|^2$$

$$= \left\| F(F^*F)^{-1}f \right\|^2$$
$$= \left\langle F(F^*F)^{-1}f, F(F^*F)^{-1}f \right\rangle$$
$$= \left\langle (F^*F)^{-1}f, f \right\rangle.$$

Now, since $A\mathcal{I} \leq F^*F \leq B\mathcal{I}$, Theorem 4.6.11 implies

$$\frac{1}{B}\mathcal{I} \leq (F^*F)^{-1} \leq \frac{1}{A}\mathcal{I},$$

which gives

$$\frac{1}{B}\|f\|^2 \leq \sum_{n=1}^{\infty} |\langle f, (\tilde{\varphi}_n) \rangle|^2 \leq \frac{1}{A}\|f\|^2. \qquad \Box$$

The sequence $(\tilde{\varphi}_n)$ is called the *dual frame*.

Lemma 8.3.6. *Let F be the frame operator associated with the frame $(\varphi_1, \varphi_2, \ldots)$, and let $\tilde{F}$ be the frame operator associated with the dual frame $(\tilde{\varphi}_1, \tilde{\varphi}_2, \ldots)$. Then*

$$\tilde{F}^*F = \mathcal{I} = F^*\tilde{F}.$$

Proof: Since

$$F(F^*F)^{-1}f = \left(\langle (F^*F)^{-1}f, \varphi_n \rangle \right) = \left(\langle f, \tilde{\varphi}_n \rangle \right) = \tilde{F}f,$$

we have

$$\tilde{F}^*F = \left(F(F^*F)^{-1} \right)^* F = (F^*F)^{-1}F^*F = \mathcal{I}$$

and

$$F^*\tilde{F} = F^*F(F^*F)^{-1} = \mathcal{I}. \qquad \Box$$

Now we are ready to state and prove the theorem which answers the question of reconstructability of f from $(\langle f, \varphi_n \rangle)$.

Theorem 8.3.7. *Let $(\varphi_1, \varphi_2, \ldots)$ be a frame in a Hilbert space H and let $(\tilde{\varphi}_1, \tilde{\varphi}_2, \ldots)$ be the dual frame. Then*

$$f = \sum_{n=1}^{\infty} \langle f, \varphi_n \rangle \tilde{\varphi}_n$$

and

$$f = \sum_{n=1}^{\infty} \langle f, \tilde{\varphi}_n \rangle \varphi_n$$

for any $f \in H$.

Proof: Let F be the frame operator associated with (φ_n) and let $\tilde{F}$ be the frame operator associated with the dual frame $(\tilde{\varphi}_n)$. Since $\mathcal{I} = \tilde{F}^*F$, for any $f \in H$, we have

$$f = \tilde{F}^*Ff = \tilde{F}^*\big(\langle f, \varphi_n\rangle\big) = \sum_{n=1}^{\infty}\langle f, \varphi_n\rangle\tilde{\varphi}_n,$$

by (8.17). The proof of the other equality is similar. $\square$

8.4 Multiresolution Analysis and Orthonormal Bases of Wavelets

Some difficulties in dealing with frames arise from lack of orthogonality. If we have orthogonality, that is, $\langle \psi_{k,\ell}, \psi_{m,n}\rangle = 0$ whenever $(k, \ell) \neq (m, n)$, then the reconstruction of f from $\langle f, \psi_{m,n}\rangle$ is much simpler: $f = \sum_{m,n=-\infty}^{\infty}\langle f, \psi_{m,n}\rangle\psi_{m,n}$. In this section we describe a general method of constructing orthonormal bases of wavelets based on the so-called *multiresolution analysis*. In this construction we take $a_0 = 2$ and $b_0 = 1$. Consequently,

$$\varphi_{m,n}(x) = 2^{-m/2}\varphi\big(2^{-m}x - n\big).$$

Definition 8.4.1.   (Multiresolution analysis)

By a *multiresolution analysis* we mean a sequence $(\ldots, V_{-2}, V_{-1}, V_0, V_1, V_2, \ldots)$ of spaces of functions defined on $\mathbb{R}$ such that the following conditions are satisfied:

(a) V_n is a closed subspace of $L^2(\mathbb{R})$ for every $n \in \mathbb{Z}$,

(b) $V_{n+1} \subset V_n$ for every $n \in \mathbb{Z}$,

(c) $\bigcup_{n=-\infty}^{\infty} V_n$ is dense in $L^2(\mathbb{R})$,

(d) $\bigcap_{n=-\infty}^{\infty} V_n = \{0\}$,

(e) $f \in V_n$ if and only if $f(2^n\cdot) \in V_0$, for all $n \in \mathbb{Z}$,

(f) There exists a $\varphi \in V_0$ such that $\{\varphi_{0,k}\colon k \in \mathbb{Z}\}$ is an orthonormal basis in V_0.

The function φ is called the *scaling function* or *father wavelet*.

The fundamental idea of multiresolution analysis is to represent a function as a limit of successive approximations, each of which is a "smoother" version of the original function. These approximations correspond to different resolutions. Multiresolution analysis is a formal approach to constructing orthogonal wavelet bases using a definite set of rules and procedures. In applications, it is an effective mathematical framework for hierarchical decomposition of a signal or an image into components of different scales. The spaces V_n correspond to those different scales.

Example 8.4.2. Let φ be the characteristic function of the interval $[0, 1]$. Define

$$V_n = \left\{ \sum_{k=-\infty}^{\infty} c_k \varphi_{n,k} : (c_k) \in l^2(\mathbb{Z}) \right\}.$$

Then (V_n) is a multiresolution analysis. This example is discussed in more detail in the next section. $\qquad\square$

The spaces in a multiresolution analysis form a ladder:

$$\cdots \supset V_{-2} \supset V_{-1} \supset V_0 \supset V_1 \supset V_2 \supset \cdots.$$

If any of the spaces V_n is given, then, by condition (e), any other space V_m is known, because

$$V_m = \left\{ f(2^{m-n} \cdot) : f \in V_n \right\}.$$

In most arguments concerning multiresolution analysis, it thus is enough to deal with V_0. Notice that a resolution analysis is completely determined by the scaling function φ. Indeed, for a given φ, we first define

$$V_0 = \left\{ \sum_{k=-\infty}^{\infty} c_k \varphi_{0,k} : (c_k) \in l^2(\mathbb{Z}) \right\}.$$

The remaining spaces V_n are then determined by V_0 as discussed earlier. This does not mean that any choice of φ is going to produce a multiresolution analysis. The first problem we have to deal with is orthogonality of translates of φ. The following theorem can be useful in that matter.

Theorem 8.4.3. *For any $\varphi \in L^2(\mathbb{R})$, the following two conditions are equivalent:*

(a) *The system $\{\varphi_{0,k} : k \in \mathbb{Z}\}$ is orthonormal,*

(b) $\sum_{m=-\infty}^{\infty} |\widehat{\varphi}(\omega + 2m\pi)|^2 = 1/2\pi$ *almost everywhere.*

Proof: Since

$$\widehat{\varphi}_{0,k}(\omega) = e^{-ik\omega} \widehat{\varphi}(\omega),$$

we have

$$\langle \varphi_{0,k}, \varphi_{0,l} \rangle = \langle \varphi_{0,0}, \varphi_{0,l-k} \rangle = \langle \widehat{\varphi}_{0,0}, \widehat{\varphi}_{0,l-k} \rangle$$

$$= \int_{-\infty}^{\infty} e^{-i(l-k)\omega} |\widehat{\varphi}(\omega)|^2 d\omega$$

$$= \sum_{m=-\infty}^{\infty} \int_{2m\pi}^{2(m+1)\pi} e^{-i(l-k)\omega} \left|\widehat{\varphi}(\omega)\right|^2 d\omega$$

$$= \int_0^{2\pi} e^{-i(l-k)\omega} \sum_{m=-\infty}^{\infty} \left|\widehat{\varphi}(\omega + 2m\pi)\right|^2 d\omega.$$

Thus, $\langle \varphi_{0,k}, \varphi_{0,l} \rangle = \delta_{k,l}$ if and only if $\sum_{m=-\infty}^{\infty} |\widehat{\varphi}(\omega + 2m\pi)|^2 = 1/2\pi$ almost everywhere. □

Now we return to the problem of constructing an orthonormal basis of wavelets from a given multiresolution analysis. Let (V_n) be a multiresolution analysis. For every $n \in \mathbb{Z}$, define W_n to be the orthogonal complement of V_n in V_{n-1}, so that we have

$$V_{n-1} = V_n \oplus W_n \quad \text{and} \quad W_n \perp W_m \quad \text{if } m \neq n.$$

Since,

$$V_n = V_{n+1} \oplus W_{n+1}$$

$$= V_{n+2} \oplus W_{n+2} \oplus W_{n+1}$$

$$= V_{n+3} \oplus W_{n+3} \oplus W_{n+2} \oplus W_{n+1}$$

$$= \cdots,$$

for every $N > n$, we have

$$V_n = V_N \oplus \left(\bigoplus_{k=0}^{N-n-1} W_{N-k} \right).$$

This, in view of (c) and (d) of Definition 8.4.1, leads to

$$L^2(\mathbb{R}) = \bigoplus_{n=-\infty}^{\infty} W_n. \tag{8.18}$$

We have thus constructed a decomposition of $L^2(\mathbb{R})$ into mutually orthogonal subspaces. Moreover, for all $n \in \mathbb{Z}$,

$$f \in W_n \quad \text{if and only if} \quad f(2^n \cdot) \in W_0.$$

This is a very desirable property because if $\{\theta_k : k \in \mathbb{Z}\}$ is an orthonormal basis for W_0, then the scaled version $\{\theta_k(2^n \cdot) : k \in \mathbb{Z}\}$ is an orthonormal basis for W_n, for every $n \in \mathbb{Z}$. This fact, together with (8.18), implies that the union of all these bases, that is $\{\theta_k(2^n \cdot) : k, n \in \mathbb{Z}\}$, is an orthonormal basis for $L^2(\mathbb{R})$. Note that the spaces V_n do not have this property. This is the reason for defining the spaces W_n.

The property just described reduces the problem of finding an orthonormal basis for $L^2(\mathbb{R})$, to the problem of finding an orthonormal basis for W_0. We want to go one step further: we want to find a single function ψ such that the translates $\{\psi_{0,n}: n \in \mathbb{Z}\}$ constitute an orthonormal basis in W_0. If we can find a function like that, then the system $\{\psi_{m,n}: m, n \in \mathbb{Z}\}$, will be an orthonormal basis for $L^2(\mathbb{R})$. It will be an orthonormal basis of wavelets. It turns out that such a ψ exists for every multiresolution analysis. The proof of this fact not only guarantees existence of such a function, but actually gives a method of constructing it. The function ψ is often called the *mother wavelet*.

We start the construction by introducing an auxiliary function $\widehat{m}$, which plays an important role in the argument. Since $\varphi \in V_{-1}$ and $(\varphi_{-1,n})$ is an orthonormal basis for V_{-1}, we can express φ in the form

$$\varphi = \sum_{n=-\infty}^{\infty} h_n \varphi_{-1,n} \quad \text{or} \quad \varphi(x) = \sqrt{2} \sum_{n=-\infty}^{\infty} h_n \varphi(2x - n) \tag{8.19}$$

where

$$h_n = \langle \varphi, \varphi_{-1,n} \rangle \quad \text{and} \quad \sum_{n=-\infty}^{\infty} |h_n|^2 = 1. \tag{8.20}$$

By applying the Fourier transform to (8.19), we obtain

$$\widehat{\varphi}(s) = \frac{1}{\sqrt{2}} \sum_{n=-\infty}^{\infty} h_n e^{-ins/2} \widehat{\varphi}(s/2). \tag{8.21}$$

If we define

$$\widehat{m}(s) = \frac{1}{\sqrt{2}} \sum_{n=-\infty}^{\infty} h_n e^{-ins}, \tag{8.22}$$

then (8.21) can be written as

$$\widehat{\varphi}(s) = \widehat{m}(s/2)\widehat{\varphi}(s/2). \tag{8.23}$$

By Theorem 8.4.3, we have

$$\sum_{k=-\infty}^{\infty} |\widehat{\varphi}(s + 2k\pi)|^2 = \frac{1}{2\pi} \tag{8.24}$$

which, together with (8.23), gives

$$\sum_{k=-\infty}^{\infty} |\widehat{m}(s + k\pi)|^2 |\widehat{\varphi}(s + k\pi)|^2 = \frac{1}{2\pi} \quad \text{a.e.} \tag{8.25}$$

By splitting the sum into even and odd k and then using 2π-periodicity of $\widehat{m}$, we obtain

$$\sum_{k=-\infty}^{\infty} |\widehat{m}(s+k\pi)|^2 |\widehat{\varphi}(s+k\pi)|^2 \tag{8.26}$$

$$= \sum_{k=-\infty}^{\infty} |\widehat{m}(s+2k\pi)|^2 |\widehat{\varphi}(s+2k\pi)|^2 \tag{8.27}$$

$$+ \sum_{k=-\infty}^{\infty} |\widehat{m}(s+(2k+1)\pi)|^2 |\widehat{\varphi}(s+(2k+1)\pi)|^2 \tag{8.28}$$

$$= \sum_{k=-\infty}^{\infty} |\widehat{m}(s)|^2 |\widehat{\varphi}(s+2k\pi)|^2 + \sum_{k=-\infty}^{\infty} |\widehat{m}(s+\pi)|^2 |\widehat{\varphi}(s+\pi+2k\pi)|^2 \tag{8.29}$$

$$= \left(|\widehat{m}(s)|^2 + |\widehat{m}(s+\pi)|^2\right)\frac{1}{2\pi}, \tag{8.30}$$

by (8.24) used in its original form and with s replaced by $s+\pi$. Now (8.25) implies

$$|\widehat{m}(s)|^2 + |\widehat{m}(s+\pi)|^2 = 1 \quad \text{a.e.} \tag{8.31}$$

The Fourier transform $\widehat{\varphi}$ of the scaling function φ satisfies the functional equation (8.23). The function $\widehat{m}$ is called the *generating function* of the multiresolution analysis. This function is often called the *discrete Fourier transform* of the sequence (h_n). In signal processing, $\widehat{m}(\omega)$ is called the transfer function of a discrete filter with impulse response (h_n) or the *low-pass filter* associated with the scaling function φ.

Since $|\widehat{\varphi}(0)| = 1 \neq 0$, we have $\widehat{m}(0) = 1$ and $\widehat{m}(\pi) = 0$. Thus $\widehat{m}$ can be considered as a low-pass filter because the transfer function passes the frequencies near $s = 0$ and cuts off frequencies near $\omega = \pi$.

Theorem 8.4.4. *The function $\widehat{\varphi}$ can be represented by the infinite product*

$$\widehat{\varphi}(s) = \prod_{k=1}^{\infty} \widehat{m}\left(\frac{s}{2^k}\right). \tag{8.32}$$

Proof: A simple iteration of (8.23) gives

$$\widehat{\varphi}(s) = \widehat{m}\left(\frac{s}{2}\right)\widehat{\varphi}\left(\frac{s}{2}\right) = \widehat{m}\left(\frac{s}{2}\right)\left[\widehat{m}\left(\frac{s}{4}\right)\widehat{\varphi}\left(\frac{s}{4}\right)\right]$$

which gives, by the $(k-1)$th iteration,

$$\widehat{\varphi}(s) = \widehat{m}\left(\frac{s}{2}\right)\widehat{m}\left(\frac{s}{4}\right)\cdots\widehat{m}\left(\frac{s}{2^k}\right)\widehat{\varphi}\left(\frac{s}{2^k}\right)$$

$$= \prod_{k=1}^{k}\widehat{m}\left(\frac{s}{2^k}\right)\widehat{\varphi}\left(\frac{s}{2^k}\right). \tag{8.33}$$

Since $\widehat{\varphi}(0) = 1$ and $\widehat{\varphi}(s)$ is continuous, we obtain

$$\lim_{k\to\infty}\widehat{\varphi}\left(\frac{s}{2^k}\right) = \widehat{\varphi}(0) = 1.$$

Thus, the limit of (8.33) as $k \to \infty$ gives (8.32). $\qquad\square$

Now we prove a technical lemma.

Lemma 8.4.5. *For every* $f \in W_0$ *there exists a* 2π*-periodic function* v *such that*

$$\widehat{f}(s) = v(s)e^{is/2}\overline{\widehat{m}(s/2+\pi)}\,\widehat{\varphi}(s/2). \tag{8.34}$$

Note that $e^{is/2}\overline{\widehat{m}(s/2+\pi)}\,\widehat{\varphi}(s/2)$ in (8.34) is independent of f.

Proof: Let $f \in W_0$. Since $W_0 \subset V_{-1}$, we can write

$$f = \sum_{n=-\infty}^{\infty} c_n\varphi_{-1,n},$$

where $c_n = \langle f, \varphi_{-1,n}\rangle$. Proceeding as before, we get

$$\widehat{f}(s) = \frac{1}{\sqrt{2}}\sum_{n=-\infty}^{\infty} c_n e^{-ins/2}\widehat{\varphi}(s/2) = \widehat{m}_f(s/2)\widehat{\varphi}(s/2), \tag{8.35}$$

where

$$\widehat{m}_f(s) = \frac{1}{\sqrt{2}}\sum_{n=-\infty}^{\infty} c_n e^{-ins}. \tag{8.36}$$

Since $f \perp V_0$,

$$\int_{-\infty}^{\infty} \widehat{f}(s)\overline{\widehat{\varphi}(s)}e^{ins}\,ds = 0,$$

and hence

$$\int_0^{2\pi} e^{ins}\sum_{k=-\infty}^{\infty} \widehat{f}(s+2k\pi)\overline{\widehat{\varphi}(s+2k\pi)}\,ds = 0,$$

for every $n \in \mathbb{Z}$. Consequently,

$$\sum_{k=-\infty}^{\infty} \hat{f}(s + 2k\pi)\overline{\hat{\varphi}(s + 2k\pi)} = 0. \tag{8.37}$$

Now we substitute (8.23) and (8.35) in (8.37), then split the sum into even and odd k's, and finally use 2π-periodicity of $\hat{m}$ and $\hat{m}_f$:

$$0 = \sum_{k=-\infty}^{\infty} \hat{f}(s + 2k\pi)\overline{\hat{\varphi}(s + 2k\pi)}$$

$$= \sum_{k=-\infty}^{\infty} \hat{m}_f(s/2 + k\pi)\overline{\hat{m}(s/2 + k\pi)}\left|\hat{\varphi}(s/2 + k\pi)\right|^2$$

$$= \sum_{k=-\infty}^{\infty} \hat{m}_f(s/2 + 2k\pi)\overline{\hat{m}(s/2 + 2k\pi)}\left|\hat{\varphi}(s/2 + 2k\pi)\right|^2$$

$$+ \sum_{k=-\infty}^{\infty} \hat{m}_f(s/2 + \pi + 2k\pi)\overline{\hat{m}(s/2 + \pi + 2k\pi)}\left|\hat{\varphi}(s/2 + \pi + 2k\pi)\right|^2$$

$$= \sum_{k=-\infty}^{\infty} \hat{m}_f(s/2)\overline{\hat{m}(s/2)}\left|\hat{\varphi}(s/2 + 2k\pi)\right|^2$$

$$+ \sum_{k=-\infty}^{\infty} \hat{m}_f(s/2 + \pi)\overline{\hat{m}(s/2 + \pi)}\left|\hat{\varphi}(s/2 + \pi + 2k\pi)\right|^2$$

$$= \hat{m}_f(s/2)\overline{\hat{m}(s/2)} \sum_{k=-\infty}^{\infty} \left|\hat{\varphi}(s/2 + 2k\pi)\right|^2$$

$$+ \hat{m}_f(s/2 + \pi)\overline{\hat{m}(s/2 + \pi)} \sum_{k=-\infty}^{\infty} \left|\hat{\varphi}(s/2 + \pi + 2k\pi)\right|^2$$

$$= \left(\hat{m}_f(s/2)\overline{\hat{m}(s/2)} + \hat{m}_f(s/2 + \pi)\overline{\hat{m}(s/2 + \pi)}\right)\frac{1}{2\pi}.$$

The last equality follows from the fact that the system $\{\varphi_{0,k}: k \in \mathbb{Z}\}$ is orthonormal and from Theorem 8.4.3. The obtained result and (8.37) imply

$$\hat{m}_f(s)\overline{\hat{m}(s)} + \hat{m}_f(s + \pi)\overline{\hat{m}(s + \pi)} = 0 \quad \text{a.e.}$$

Thus,

$$\begin{vmatrix} \hat{m}_f(s) & \overline{\hat{m}(s + \pi)} \\ -\hat{m}_f(s + \pi) & \overline{\hat{m}(s)} \end{vmatrix} = 0,$$

which can be interpreted as linear dependence of the vectors $(\widehat{m}_f(s),$ $-\widehat{m}_f(s+\pi))$, and $(\overline{\widehat{m}(s+\pi)}, \overline{\widehat{m}(s)})$. Consequently, there exists a function λ such that

$$\widehat{m}_f(s) = \lambda(s)\overline{\widehat{m}(s+\pi)} \quad \text{a.e.} \tag{8.38}$$

Since both $\widehat{m}$ and $\widehat{m}_f$ are 2π-periodic, so is λ. Moreover,

$$\lambda(s) + \lambda(s+\pi) = 0 \quad \text{a.e.} \tag{8.39}$$

There exists a 2π-periodic function ν defined by

$$\lambda(s) = e^{is}\nu(2s). \tag{8.40}$$

Combining (8.35), (8.38), and (8.40), we obtain the desired representation (8.34). □

Now we return to our main problem. Suppose that we found a function ψ such that $\{\psi_{0,n}: n \in \mathbb{Z}\}$ is a basis for W_0. Then every function $f \in W_0$ has a representation

$$f = \sum_{n=-\infty}^{\infty} \gamma_n \psi_{0,n} = \sum_{n=-\infty}^{\infty} \gamma_n \psi(x-n), \tag{8.41}$$

where $\sum_{n=-\infty}^{\infty} |\gamma_n|^2 < \infty$. When we apply the Fourier transform to (8.41), we obtain

$$\hat{f}(s) = \left(\sum_{n=-\infty}^{\infty} \gamma_n e^{-ins}\right)\widehat{\psi}(s) = \hat{\gamma}(s)\widehat{\psi}(s).$$

The function $\hat{\gamma}(s) = \sum_{n=-\infty}^{\infty} \gamma_n e^{-ins}$ is a 2π-periodic and square integrable on $[0, 2\pi]$. When this is compared with (8.34), then we see that we should take

$$\widehat{\psi}(s) = e^{is/2}\overline{\widehat{m}(s/2+\pi)}\widehat{\varphi}(s/2) = \widehat{m}_1\left(\frac{s}{2}\right)\widehat{\varphi}\left(\frac{s}{2}\right), \tag{8.42}$$

where the function $\widehat{m}_1$ is given by

$$\widehat{m}_1(s) = e^{is}\overline{\widehat{m}(s+\pi)}. \tag{8.43}$$

This function $\widehat{m}_1$ is called the *filter conjugate* to $\widehat{m}$ and hence, in signal processing, $\widehat{m}$ and $\widehat{m}_1$ are called *conjugate quadratic filters* (CQF).

Since

$$\widehat{\psi}(s) = e^{is/2}\overline{\widehat{m}(s/2+\pi)}\widehat{\varphi}(s/2)$$

$$= e^{is/2}\frac{1}{\sqrt{2}}\sum_{n=-\infty}^{\infty} h_n e^{in(s/2+\pi)}\widehat{\varphi}(s/2)$$

$$= \frac{1}{\sqrt{2}} \sum_{n=-\infty}^{\infty} h_n e^{in\pi} e^{i(n+1)(s/2)} \widehat{\varphi}(s/2),$$

by letting $n = -(k+1)$, we obtain

$$\widehat{\psi}(s) = \frac{1}{\sqrt{2}} \sum_{n=-\infty}^{\infty} h_{-k-1}(-1)^{k-1} \exp\left(-\frac{iks}{2}\right) \widehat{\varphi}\left(\frac{s}{2}\right). \tag{8.44}$$

Invoking the inverse Fourier transform to (8.44), with k replaced by n, gives the mother wavelet

$$\psi(x) = \sqrt{2} \sum_{n=-\infty}^{\infty} (-1)^{n-1} h_{-n-1} \varphi(2x - n), \tag{8.45}$$

$$= \sqrt{2} \sum_{n=-\infty}^{\infty} d_n \varphi(2x - n), \tag{8.46}$$

where the coefficients d_n are given by

$$d_n = (-1)^{n-1} h_{-n-1}. \tag{8.47}$$

Thus, the representation (8.46) of a mother wavelet has the same structure as that of the father wavelet φ given by (8.19).

The mother wavelet ψ associated with a given multiresolution analysis is not unique because the coefficients

$$d_n = (-1)^{n-1} h_{2N-1-n} \tag{8.48}$$

define the same mother wavelet (8.45) with suitably selected $n \in \mathbb{Z}$. The wavelet with coefficients d_n given by (8.48) has the Fourier transform

$$\widehat{\psi}(s) = \exp\left\{(2N - 1)\frac{is}{2}\right\} \overline{\widetilde{m}}\left(\frac{s}{2} + \pi\right) \widehat{\varphi}\left(\frac{s}{2}\right). \tag{8.49}$$

Thus, the nonuniqueness property of d_n allows us to define a form of ψ different from (8.45) by

$$\psi(x) = \sqrt{2} \sum_{n=-\infty}^{\infty} d_n \varphi(2x - n), \tag{8.50}$$

with a slightly modified d_n given by

$$d_n = (-1)^n h_{1-n}. \tag{8.51}$$

In practice, any one of the preceding formulas for d_n can be used to find a mother wavelet.

The orthogonality condition (8.31), together with (8.23) and (8.42), implies

$$\left|\widehat{\varphi}(s)\right|^2 + \left|\widehat{\psi}(s)\right|^2 = \left|\widehat{\varphi}\left(\frac{s}{2}\right)\right|^2. \tag{8.52}$$

Or, equivalently,

$$\left|\widehat{\varphi}\left(2^m s\right)\right|^2 + \left|\widehat{\psi}\left(2^m s\right)\right|^2 = \left|\widehat{\varphi}(s)\right|^2. \tag{8.53}$$

Summing both sides of (8.53) from $m = 1$ to infinity leads to the result

$$\left|\widehat{\varphi}(s)\right|^2 = \sum_{m=1}^{\infty} \left|\widehat{\psi}\left(2^m s\right)\right|^2. \tag{8.54}$$

If φ has a compact support, the series (8.46) for the mother wavelet ψ terminates and, consequently, ψ can be represented by a finite linear combination of a translated version of $\varphi(2x)$.

Theorem 8.4.6. *Let* (V_n), $n \in \mathbb{Z}$ *be a multiresolution analysis with the scaling function* φ. *Then the function*

$$\psi(x) = \sqrt{2} \sum_{n=-\infty}^{\infty} (-1)^{n-1} h_{-n-1} \varphi(2x - n),$$

where

$$h_n = \sqrt{2} \int_{-\infty}^{\infty} \varphi(x) \overline{\widehat{\varphi}(2x - n)} dx,$$

is a mother wavelet, that is, the system $\{\psi_{m,n} \colon m, n \in \mathbb{Z}\}$ *is an orthonormal basis for* $L^2(\mathbb{R})$.

Proof: First we need to verify that $\{\psi_{0,n} \colon m, n \in \mathbb{Z}\}$ is an orthonormal set. Indeed, we have

$$\int_{-\infty}^{\infty} \psi(x - k) \overline{\widehat{\psi}(x - l)} dx = \int_{-\infty}^{\infty} e^{-is(k-l)} \left|\widehat{\psi}(s)\right|^2 ds$$

$$= \int_{0}^{2\pi} e^{-is(k-l)} \sum_{n=-\infty}^{\infty} \left|\widehat{\psi}(s + 2n\pi)\right|^2 ds,$$

and

$$\sum_{n=-\infty}^{\infty} \left|\widehat{\psi}(s + 2n\pi)\right|^2 = \sum_{n=-\infty}^{\infty} \left|\widehat{m}(s/2 + (n+1)\pi)\right|^2 \left|\widehat{\varphi}(s/2 + n\pi)\right|^2$$

$$= \sum_{n=-\infty}^{\infty} \left|\widehat{m}(s/2 + (2n+1)\pi)\right|^2 \left|\widehat{\varphi}(s/2 + 2n\pi)\right|^2$$

$$+ \sum_{n=-\infty}^{\infty} \left| \widehat{m}(s/2 + (2n+2)\pi) \right|^2 \left| \widehat{\phi}(s/2 + (2n+1)\pi) \right|^2$$

$$= \left| \widehat{m}(s/2 + \pi) \right|^2 \sum_{n=-\infty}^{\infty} \left| \widehat{\varphi}(s/2 + 2n\pi) \right|^2$$

$$+ \left| \widehat{m}(s/2) \right|^2 \sum_{n=-\infty}^{\infty} \left| \widehat{\varphi}(s/2 + (2n+1)\pi) \right|^2$$

$$= \left(\left| \widehat{m}(s/2 + \pi) \right|^2 + \left| \widehat{m}(s/2) \right|^2 \right) \frac{1}{2\pi} = \frac{1}{2\pi}.$$

Thus,

$$\int_{-\infty}^{\infty} \psi(x - k) \overline{\psi(x - l)} \, dx = \delta_{k,l},$$

which proves that $\{\psi_{m,n}: m, n \in \mathbb{Z}\}$ is an orthonormal system. In view of Lemma 8.4.5 and the discussion following the lemma, to prove that it is a basis it suffices to show that the function v in (8.40) is square integrable over $[0, 2\pi]$. In fact,

$$\int_0^{2\pi} |v(s)|^2 \, ds = 2 \int_0^{\pi} |\lambda(s)|^2 \, ds$$

$$= 2 \int_0^{\pi} |\lambda(s)|^2 \left(\left| \widehat{m}(s + \pi) \right|^2 + \left| \widehat{m}(s) \right|^2 \right) ds$$

$$= 2 \int_0^{2\pi} |\lambda(s)|^2 \left| \widehat{m}(s + \pi) \right|^2 ds$$

$$= 2 \int_0^{2\pi} \left| \widehat{m}_f(s) \right|^2 ds$$

$$= 2\pi \sum_{n=-\infty}^{\infty} |c_n|^2 \quad (\text{where } c_n = \langle f, \varphi_{-1,n} \rangle)$$

$$= 2\pi \|f\|^2 < \infty. \qquad \square$$

8.5 Examples of Orthonormal Wavelets

Since the discovery of wavelets, orthonormal wavelets play an important role in the wavelet theory and have a variety of applications. In this section, we discuss several examples of orthonormal wavelets.

Definition 8.5.1. (Orthonormal wavelet)

A wavelet $\psi \in L^2(\mathbb{R})$ is called *orthonormal* if the family of functions $\psi_{m,n}$ generated from ψ by

$$\psi_{m,n}(x) = 2^{m/2} \psi\left(2^m\left(x - \frac{n}{2^m}\right)\right) = 2^{m/2} \psi\left(2^m x - n\right), \quad m, n \in \mathbb{Z}, \quad (8.55)$$

is orthonormal, that is,

$$\langle \psi_{m,n}, \psi_{k,\ell} \rangle = \int_{-\infty}^{\infty} \psi_{m,n}(x) \psi_{k,\ell}(x) dx = \delta_{m,k} \delta_{n,\ell}, \quad (8.56)$$

for all $m, n, k, \ell \in \mathbb{Z}$.

The following lemma is often useful when dealing with orthogonality of wavelets.

Lemma 8.5.2. *If* $\psi, \varphi \in L^2(\mathbb{R})$, *then*

$$\langle \psi_{m,k}, \varphi_{m,\ell} \rangle = \langle \psi_{n,k}, \varphi_{n,\ell} \rangle, \quad (8.57)$$

for all $m, n, k, \ell \in \mathbb{Z}$.

Proof: We have

$$\langle \psi_{m,k}, \varphi_{m,\ell} \rangle = \int_{-\infty}^{\infty} 2^m \psi\left(2^m x - k\right) \varphi\left(2^m x - \ell\right) dx,$$

which is, by letting $2^m x = 2^n t$,

$$\langle \psi_{m,k}, \varphi_{m,\ell} \rangle = \int_{-\infty}^{\infty} 2^n \psi\left(2^n t - k\right) \varphi\left(2^n t - \ell\right) dx = \langle \psi_{n,k}, \varphi_{n,\ell} \rangle. \qquad \square$$

Example 8.5.3. (The Haar wavelet) The simplest example of an orthonormal wavelet is the classic Haar wavelet. Consider the scaling function $\varphi = \chi_{[0,1)}$. The function φ satisfies the dilation equation

$$\varphi(x) = \sqrt{2} \sum_{n=-\infty}^{\infty} c_n \varphi(2x - n), \quad (8.58)$$

where the coefficients c_n are given by

$$c_n = \sqrt{2} \int_{-\infty}^{\infty} \varphi(x) \varphi(2x - n) dx. \quad (8.59)$$

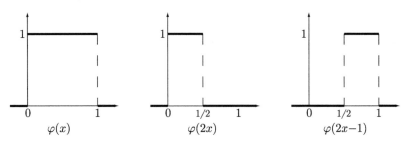

Figure 8.6   Two-scale relation of $\varphi(x) = \varphi(2x) + \varphi(2x - 1)$.

Evaluating this integral with $\varphi = \chi_{[0,1)}$ gives c_n as follows:

$$c_0 = c_1 = \frac{1}{\sqrt{2}} \quad \text{and} \quad c_n = 0 \quad \text{for } n > 1.$$

Consequently, the dilation equation becomes

$$\varphi(x) = \varphi(2x) + \varphi(2x - 1). \tag{8.60}$$

This means that $\varphi(x)$ is a linear combination of the even and odd translates of $\varphi(2x)$ and satisfies a very simple two-scale relation (8.60), as shown in Figure 8.6.

In view of (8.51), we obtain

$$d_0 = c_1 = \frac{1}{\sqrt{2}} \quad \text{and} \quad d_1 = -c_0 = -\frac{1}{\sqrt{2}}.$$

Thus, the Haar mother wavelet is obtained from (8.50) as a simple two-scale relation

$$\psi(x) = \varphi(2x) - \varphi(2x - 1) \tag{8.61}$$

$$= \chi_{[0,1/2]}(x) - \chi_{[1/2,1]}(x)$$

$$= \begin{cases} 1 & \text{if } 0 \le x < \frac{1}{2}, \\ -1 & \text{if } \frac{1}{2} \le x < 1, \\ 0 & \text{otherwise.} \end{cases} \tag{8.62}$$

This two-scale relation (8.61) of ψ is represented in Figure 8.7.

For any $m, n \in \mathbb{Z}$, we have

$$\psi_{m,n}(t) = 2^{-m/2}\psi\left(2^{-m}t - n\right) = \begin{cases} 2^{-m/2}, & 2^m n \le t < 2^m n + 2^{m-1}, \\ -2^{-m/2}, & 2^m n + 2^{m-1} \le t < 2^m n + 2^m, \\ 0, & \text{otherwise.} \end{cases}$$

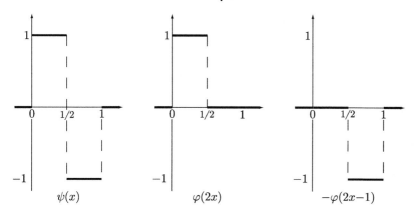

Figure 8.7 Two-scale relation of $\psi(x) = \varphi(2x) - \varphi(2x-1)$.

Clearly, $\|\psi_{m,n}\|_2 = \|\psi\|_2 = 1$, for all $m, n \in \mathbb{Z}$. To verify that $\{\psi_{m,n}\}$ is an orthonormal system, note that

$$\langle \psi_{m,n}, \psi_{k,\ell} \rangle = \int_{-\infty}^{\infty} 2^{m/2} \psi\left(2^m x - n\right) 2^{k/2} \psi\left(2^k x - \ell\right) dx,$$

which gives, by the change of variables $2^m x - n = t$,

$$\langle \psi_{m,n}, \psi_{k,\ell} \rangle = 2^{k/2} 2^{-m/2} \int_{-\infty}^{\infty} \psi(t) \psi\left(2^{k-m}(t+n) - \ell\right) dt. \qquad (8.63)$$

For $m = k$, we obtain

$$\langle \psi_{m,n}, \psi_{m,\ell} \rangle = \int_{-\infty}^{\infty} \psi(t) \psi(t+n-\ell) dt = \delta_{0,n-\ell} = \delta_{n,\ell}, \qquad (8.64)$$

where $\psi(t) \neq 0$ in $0 \le t < 1$ and $\psi(t - \overline{\ell - n}) \neq 0$ in $\ell - n \le t < 1 + \ell - n$, and these intervals are disjoint from each other unless $n = \ell$.

We now consider the case $m \neq k$. In view of symmetry, it suffices to consider the case $m > k$. Putting $r = m - k > 0$ in (8.63), we obtain, for $m > k$,

$$\langle \psi_{m,n}, \psi_{m,\ell} \rangle = 2^{r/2} \int_{-\infty}^{\infty} \psi(t) \psi\left(2^r t + s\right) dt, \qquad (8.65)$$

where $s = 2^r n - \ell$. Thus, it suffices to show that

$$\int_0^{1/2} \psi\left(2^r t + s\right) dt - \int_{1/2}^1 \psi\left(2^r t + s\right) dt = 0.$$

Invoking a simple change of variables $2^r t + s = x$, we find

$$\int_0^{1/2} \psi(2^r t + s)dt - \int_{1/2}^1 \psi(2^r t + s)dt = \int_s^a \psi(x)dx - \int_a^b \psi(x)dx, \quad (8.66)$$

where $a = s + 2^{r-1}$ and $b = s + 2^r$. Since the interval $[s, a]$ contains the support $[0, 1]$ of ψ, the first integral in (8.66) is zero. Similarly, the second integral is also zero. $\qquad \square$

Example 8.5.4. (The Shannon wavelet) The function ψ whose Fourier transform satisfies

$$\widehat{\psi}(\omega) = \chi_I(\omega), \quad (8.67)$$

where $I = [-2\pi, -\pi] \cup [\pi, 2\pi]$, is called the *Shannon wavelet*. The function ψ can be directly obtained from the inverse Fourier transform of $\widehat{\psi}$ so that

$$\psi(t) = \frac{1}{2\pi} \int_{-\infty}^{\infty} e^{i\omega t} \widehat{\psi}(\omega) d\omega$$

$$= \frac{1}{2\pi} \left[\int_{-2\pi}^{-\pi} e^{i\omega t} d\omega + \int_{\pi}^{2\pi} e^{i\omega t} d\omega \right]$$

$$= \frac{1}{\pi t} (\sin 2\pi t - \sin \pi t) = \frac{\sin(\frac{\pi t}{2})}{(\frac{\pi t}{2})} \cos \left(\frac{3\pi t}{2} \right). \quad (8.68)$$

This function is orthonormal to its translates by integers. Indeed, by Parseval's relation,

$$\langle \psi(t), \psi(t-n) \rangle = \frac{1}{2\pi} \langle \widehat{\psi}, e^{in\omega} \widehat{\psi} \rangle$$

$$= \frac{1}{2\pi} \int_{-\infty}^{\infty} \widehat{\psi}(\omega) e^{in\omega} \overline{\widehat{\psi}(\omega)} d\omega$$

$$= \frac{1}{2\pi} \int_{-2\pi}^{2\pi} e^{in\omega} d\omega = \delta_{0,n}.$$

The wavelet basis is now given by

$$\psi_{m,n}(t) = 2^{-m/2} \psi \left(2^{-m} t - n - \frac{1}{2} \right), \quad m, n \in \mathbb{Z}$$

or

$$\psi_{m,n}(t) = 2^{-m/2} \frac{\sin\{\frac{\pi}{2}(2^{-m}t - n)\}}{\frac{\pi}{2}(2^{-m}t - n)} \cos\left\{ \frac{3\pi}{2}(2^{-m}t - n) \right\}. \quad (8.69)$$

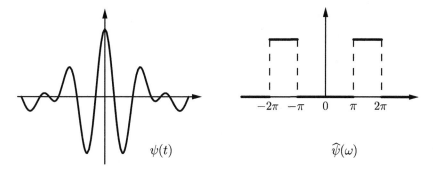

Figure 8.8   The Shannon wavelet and its Fourier transform.

For any fixed $n \in \mathbb{Z}$, the functions $\psi_{m,n}(t)$ form a basis for the space of functions supported on the interval

$$\left[-2^{-m+1}\pi, -2^{-m}\pi\right] \cup \left[2^{-m}\pi, 2^{-m+1}\pi\right].$$

The system $\{\psi_{m,n}(t)\}$, $m, n \in \mathbb{Z}$, is an orthonormal basis for $L^2(\mathbb{R})$. Both $\psi(t)$ and $\widehat{\psi}(\omega)$ are shown in Figure 8.8.

The Fourier transform of $\psi_{m,n}$ is

$$\widehat{\psi}_{m,n}(\omega) = \begin{cases} 2^{m/2} \exp(-i\omega n 2^m) & \text{if } 2^{-m}\pi < |\omega| < 2^{-m+1}\pi, \\ 0 & \text{otherwise.} \end{cases} \tag{8.70}$$

Evidently, $\widehat{\psi}_{m,n}$ and $\widehat{\psi}_{k,\ell}$ do not overlap for $m \neq k$. Hence, by Theorem 5.11.21, it turns out that, for $m \neq k$,

$$\langle \psi_{m,n}, \psi_{k,\ell} \rangle = \frac{1}{2\pi} \langle \widehat{\psi}_{m,n}, \widehat{\psi}_{k,\ell} \rangle = 0. \tag{8.71}$$

For $m = k$, we have

$$\langle \psi_{m,n}, \psi_{k,\ell} \rangle = \frac{1}{2\pi} \langle \widehat{\psi}_{m,n}, \widehat{\psi}_{m,\ell} \rangle$$

$$= \frac{1}{2\pi} 2^{-m} \int_{-\infty}^{\infty} \exp\{-i\omega 2^{-m}(n - \ell)\} \left| \widehat{\psi} \left(2^{-m}\omega \right) \right|^2 d\omega$$

$$= \frac{1}{2\pi} \int_{-\infty}^{\infty} \exp\{-i\sigma (n - \ell)\} d\sigma = \delta_{n,\ell}. \tag{8.72}$$

This shows that $\{\psi_{m,n}(t)\}$ is an orthonormal system.   □

Example 8.5.5. (The Daubechies wavelets and algorithms)   Daubechies (1988, 1992) first developed the theory and construction of continuous orthonormal wavelets with compact support. Wavelets with compact support have many interesting properties. They can be constructed to have a given number of derivatives and to have a given number of vanishing moments.

We assume that the scaling function φ satisfies the dilation equation

$$\varphi(x) = \sqrt{2} \sum_{n=-\infty}^{\infty} c_n \varphi(2x - n), \tag{8.73}$$

where $c_n = \langle \varphi, \varphi_{1,n} \rangle$ and $\sum_{n=-\infty}^{\infty} |c_n|^2 \leq \infty$.

If the scaling function φ has compact support, then only a finite number of c_n have nonzero values. The associated generating function $\widehat{m}$,

$$\widehat{m}(\omega) = \frac{1}{\sqrt{2}} \sum_{n=-\infty}^{\infty} c_n e^{-i\omega n} \tag{8.74}$$

is a trigonometric polynomial and it satisfies the identity (8.31) with special values $\widehat{m}(0) = 1$ and $\widehat{m}(\pi) = 0$. If coefficients c_n are real, then the corresponding scaling function, as well as the mother wavelet ψ, will also be real-valued. The mother wavelet ψ corresponding to φ is given by the formula (8.42) with $|\widehat{\varphi}(0)| = 1$. The Fourier transform $\widehat{\psi}(\omega)$ is m-times continuously differentiable and it satisfies the moment condition

$$\widehat{\psi}^{(k)}(0) = 0 \quad \text{for } k = 0, 1, \ldots, m. \tag{8.75}$$

It follows that $\psi \in C^m(\mathbb{R})$ implies that $\widehat{m}_0$ has a zero at $\omega = \pi$ of order $(m+1)$. In other words,

$$\widehat{m}_0(\omega) = \left(\frac{1 + e^{-i\omega}}{2} \right)^{m+1} \widehat{L}(\omega), \tag{8.76}$$

where $\widehat{L}$ is a trigonometric polynomial.

In addition to the orthogonality condition (8.31), we assume

$$\widehat{m}_0(\omega) = \left(\frac{1 + e^{-i\omega}}{2} \right)^{N} \widehat{L}(\omega), \tag{8.77}$$

where $\widehat{L}(\omega)$ is 2π-periodic and $\widehat{L} \in C^{N-1}(\mathbb{R})$. Evidently,

$$\left| \widehat{m}_0(\omega) \right|^2 = \widehat{m}_0(\omega) \widehat{m}_0(-\omega) \tag{8.78}$$

$$= \left(\frac{1 + e^{-i\omega}}{2} \right)^{N} \left(\frac{1 + e^{i\omega}}{2} \right)^{N} \widehat{L}(\omega) \widehat{L}(-\omega)$$

$$= \left(\cos^2 \frac{\omega}{2} \right)^{N} |\widehat{L}(\omega)|^2, \tag{8.79}$$

where $|\widehat{L}(\omega)|^2$ is a polynomial in $\cos\omega$, that is,

$$|\widehat{L}(\omega)|^2 = Q(\cos\omega).$$

Since $\cos\omega = 1 - 2\sin^2(\frac{\omega}{2})$, it is convenient to introduce $x = \sin^2(\frac{\omega}{2})$ so that (8.79) reduces to the form

$$|\widehat{m}_0(\omega)|^2 = \left(\cos^2\frac{\omega}{2}\right)^N Q(1 - 2x) = (1 - x)^N P(x), \qquad (8.80)$$

where $P(x)$ is a polynomial in x.

We next use the fact that

$$\cos^2\left(\frac{\omega + \pi}{2}\right) = \sin^2\left(\frac{\omega}{2}\right) = x$$

and

$$\begin{aligned} |\widehat{L}(\omega + \pi)|^2 &= Q(-\cos\omega) = Q(2x - 1), \\ &= Q\big(1 - 2(1 - x)\big) = P(1 - x) \end{aligned} \qquad (8.81)$$

to express the identity (8.31) in terms of x so that (8.31) becomes

$$(1 - x)^N P(x) + x^N P(1 - x) = 1. \qquad (8.82)$$

Since $(1 - x)^N$ and x^N are two polynomials of degree N, which are relatively prime, then, by Bezout's theorem (see Daubechies, 1992), there exists a unique polynomial P_N of degree $\leq N - 1$ such that (8.82) holds. An explicit solution for $P_N(x)$ is given by

$$P_N(x) = \sum_{k=0}^{N-1} \binom{N + k - 1}{k} x^k, \qquad (8.83)$$

which is positive for $0 < x < 1$ so that $P_N(x)$ is at least a possible candidate for $|\widehat{L}(\omega)|^2$. There also exist higher degree polynomial solutions $P_N(x)$ of (8.82) which can be written as

$$P_N(x) = \sum_{k=0}^{N-1} \binom{N + k - 1}{k} x^k + x^N R\left(x - \frac{1}{2}\right), \qquad (8.84)$$

where R is an odd polynomial.

Since $P_N(x)$ is a possible candidate for $|\widehat{L}(\omega)|^2$ and

$$\widehat{L}(\omega)\widehat{L}(-\omega) = |\widehat{L}(\omega)|^2 = Q(\cos\omega) = Q(1 - 2x) = P_N(x), \qquad (8.85)$$

the next problem is how to find $\widehat{L}(\omega)$. This can be done by the following lemma.

Lemma 8.5.6. (Riesz's lemma for spectral factorization)   *If*

$$\widehat{A}(\omega) = \sum_{k=0}^{n} a_k \cos^k \omega, \tag{8.86}$$

where $a_k \in \mathbb{R}$ and $a_n \neq 0$, and if $\widehat{A}(\omega) \geq 0$ for all $\omega \in \mathbb{R}$ with $\widehat{A}(0) = 1$, then there exists a trigonometric polynomial

$$\widehat{L}(\omega) = \sum_{k=0}^{n} b_k e^{-ik\omega} \tag{8.87}$$

with real coefficients b_k such that $\widehat{L}(0) = 1$ and

$$\widehat{A}(\omega) = \widehat{L}(\omega)\widehat{L}(-\omega) = \left|\widehat{L}(\omega)\right|^2 \tag{8.88}$$

for all $\omega \in \mathbb{R}$.

We refer to Daubechies (1992) for a proof of the Riesz lemma 8.5.6. We also point out that the factorization of $\widehat{A}(\omega)$ given in (8.88) is not unique.

For a given N, if we select $P = P_N$, then $\widehat{A}(\omega)$ becomes a polynomial of degree $N - 1$ in $\cos \omega$ and $\widehat{L}(\omega)$ is a polynomial of degree $(N - 1)$ in $\exp(-i\omega)$. Therefore, the generating function $\widehat{m}_0$ given by (8.77) is of degree $(2N - 1)$ in $\exp(-i\omega)$. The interval $[0, 2N - 1]$ becomes the support of the corresponding scaling function $_N\varphi$. The mother wavelet $_N\psi$ obtained from $_N\varphi$ is called the *Daubechies wavelet.*

For $N = 2$, it follows from (8.83) that

$$P_2(x) = \sum_{k=0}^{1} \binom{k+1}{k} x^k = 1 + 2x$$

and hence (8.85) gives

$$\left|\widehat{L}^2(\omega)\right|^2 = P_2(x) = P_2\left(\sin^2 \frac{\omega}{2}\right) = 1 + 2\sin^2 \frac{\omega}{2} = 2 - \cos \omega.$$

Using (8.87) in Lemma 8.5.6, we obtain that $\widehat{L}(\omega)$ is a polynomial of degree $N - 1 = 1$ and

$$\widehat{L}(\omega)\widehat{L}(-\omega) = 2 - \frac{1}{2}\left(e^{i\omega} + e^{-i\omega}\right).$$

It follows from (8.87) that

$$\left(b_0 + b_1 e^{-i\omega}\right)\left(b_0 + b_1 e^{i\omega}\right) = 2 - \frac{1}{2}\left(e^{i\omega} + e^{-i\omega}\right). \tag{8.89}$$

Equating the coefficients in this identity gives

$$b_0^2 + b_1^2 = 1 \quad \text{and} \quad 2b_0 b_1 = -1. \tag{8.90}$$

These equations admit solutions

$$b_0 = \frac{1}{2}(1 + \sqrt{3}) \quad \text{and} \quad b_1 = \frac{1}{2}(1 - \sqrt{3}). \tag{8.91}$$

Thus, the generating function (8.77) takes the form

$$\widehat{m}_0(\omega) = \left(\frac{1 + e^{-i\omega}}{2} \right)^2 (b_0 + b_1 e^{-i\omega})$$

$$= \frac{1}{8} \left[(1 + \sqrt{3}) + (3 + \sqrt{3})e^{-i\omega} + (3 - \sqrt{3})e^{-2i\omega} + (1 - \sqrt{3})e^{-3i\omega} \right]$$

$$\tag{8.92}$$

with $\widehat{m}_0(0) = 1$. Comparing coefficients of (8.92) with (8.22) gives $h_n = c_n$ as

$$c_0 = \frac{1}{4\sqrt{2}}(1 + \sqrt{3}), \qquad c_1 = \frac{1}{4\sqrt{2}}(3 + \sqrt{3}),$$

$$c_2 = \frac{1}{4\sqrt{2}}(3 - \sqrt{3}), \qquad c_3 = \frac{1}{4\sqrt{2}}(1 - \sqrt{3}). \tag{8.93}$$

Consequently, the Daubechies scaling function $_2\varphi(x)$ takes the form, dropping the subscript,

$$\varphi(x) = \sqrt{2} \left[c_0 \varphi(2x) + c_1 \varphi(2x - 1) + c_2 \varphi(2x - 2) + c_3 \varphi(2x - 3) \right]. \tag{8.94}$$

Using (8.48) with $N = 2$, we obtain the Daubechies wavelet $_2\psi(x)$, dropping the subscript,

$$\psi(x) = \sqrt{2} \left[d_0 \varphi(2x) + d_1 \varphi(2x - 1) + d_2 \varphi(2x - 2) + d_3 \varphi(2x - 3) \right]$$

$$= \sqrt{2} \left[-c_3 \varphi(2x) + c_2 \varphi(2x - 1) - c_1 \varphi(2x - 2) + c_0 \varphi(2x - 3) \right], \tag{8.95}$$

where the coefficients in (8.95) are the same as for the scaling function $\varphi(x)$, but in reverse order and with alternate terms having their signs changed from plus to minus.

On the other hand, the use of (8.46) with (8.51) also gives the Daubechies wavelet $_2\psi(x)$ in the form

$$_2\psi(x) = \sqrt{2} \left[-c_0 \varphi(2x - 1) + c_1 \varphi(2x) - c_2 \varphi(2x + 1) + c_3 \varphi(2x + 2) \right].$$

The wavelet has the same coefficients as ψ given in (8.95) except that the wavelet is reversed in sign and runs from $x = -1$ to 2 instead of starting from

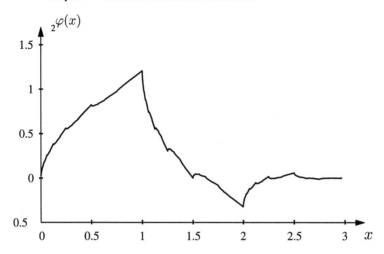

Figure 8.9   The Daubechies scaling function $_2\varphi(x)$.

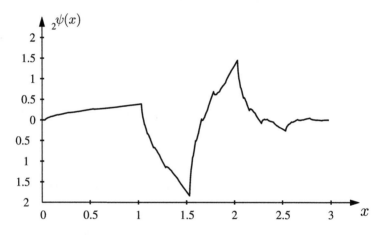

Figure 8.10   The Daubechies wavelet $_2\psi(x)$.

$x = 0$. It is often referred to as the *Daubechies D4 wavelet* since it is generated by four coefficients.

Both Daubechies' scaling function $_2\varphi$ and Daubechies' wavelet $_2\psi$ are shown in Figures 8.9 and 8.10 respectively.   □

For $N = 1$, it follows from (8.83) that $P_1(x) \equiv 1$, and this in turn leads to the fact that $Q(\cos \omega) = 1$, $\widehat{L}(\omega) = 1$ so that the generating function is

$$\widehat{m}_0(\omega) = \frac{1}{2}\left(1 + e^{-i\omega}\right). \tag{8.96}$$

This corresponds to the generating function for the Haar wavelet.

8.6 Exercises

1. If f is a homogeneous function of degree n, show that

$$(W_\psi f)(\lambda a, \lambda b) = \lambda^{n+1/2}(W_\psi f)(a, b).$$

2. In the proof of Theorem 8.2.7 we do not address the difficulty that arises from the fact that a takes both positive and negative values. Provide a more detailed proof that removes the difficulty.

3. Prove Theorem 8.2.9.

4. Prove that the vectors $(1, 0), (-\frac{1}{2}, \frac{\sqrt{3}}{2}), (-\frac{1}{2}, -\frac{\sqrt{3}}{2})$ form a tight frame in $\mathbb{C}$.

5. Prove that if (φ_n) is a tight frame in a Hilbert space H with frame bound A, then

$$A\langle f, g \rangle = \sum_{n=1}^{\infty} \langle f, \varphi_n \rangle \langle \varphi_n, g \rangle$$

for all $f, g \in H$.

6. Prove that if (φ_n) is a tight frame in a Hilbert space H with frame bound 1, then (φ_n) is an orthonormal basis in H.

7. Show that

$$\int_{-\infty}^{\infty} \frac{\sin \pi x}{\pi x} \frac{\sin \pi (2x - n)}{\pi (2x - n)} dx = \frac{1}{2\pi n} \sin \frac{\pi n}{2}.$$

8. Let $\psi_{m,n}(t) = 2^{-m/2} \psi(2^{-m}t - n)$, where $\psi(t)$ is the Haar wavelet, that is,

$$\psi_{m,n}(t) = \begin{cases} 2^{-m/2} & \text{if } 2^m n < t < 2^m n + 2^{m-1}, \\ -2^{-m/2} & \text{if } 2^m n + 2^{m-1} < t < 2^m n + 2^m, \\ 0 & \text{otherwise} \end{cases}$$

and

$$f(t) = \begin{cases} a & \text{if } 0 < t < \frac{1}{2}, \\ b & \text{if } \frac{1}{2} < t < 1, \\ 0 & \text{otherwise.} \end{cases}$$

(a) Find $\langle f, \psi_{m,0} \rangle$.

(b) Show that $\sum_{m=0}^{\infty} \langle f, \psi_{m,0} \rangle \psi_{m,0}(t) = \begin{cases} a & \text{if } 0 < t < \frac{1}{2}, \\ b & \text{if } \frac{1}{2} < t < 1. \end{cases}$

9. For the Shannon wavelet

$$\psi(t) = \frac{\sin(\frac{\pi t}{2})}{\frac{\pi t}{2}} \cos\left(\frac{3\pi t}{2}\right),$$

$$\psi_{m,n}(t) = 2^{-\frac{m}{2}} \frac{\sin(\frac{\pi}{2}(2^{-m}t - n))}{\frac{\pi}{2}(2^{-m}t - n)} \cos\left(\frac{3\pi}{2}(2^{-m}t - n)\right),$$

show that $\psi_{m,n}(t)$ are orthonormal wavelets.

10. The cardinal B-splines $B_n(x)$ of order n are defined by the following convolution product

$$B_n(x) = B_1(x) * B_1(x) * \cdots * B_1(x) = B_1(x) * B_{n-1}(x) \quad (n \geq 2),$$

where n factors are involved in the convolution product and $B_1(x) = \chi_{[0,1]}(x)$. Show the following:

(a) $B_n(x) = \int_{x-1}^{x} B_{n-1}(t)dt.$

(b) $B_2(x) = x\chi_{[0,1]}(x) + (2 - x)\chi_{[1,2]}(x).$

(c) $B_3(x) = \frac{x^2}{2}\chi_{[0,1]} + \frac{1}{2}(6x - 2x^2 - 3)\chi_{[1,2]} + \frac{1}{2}(x - 3)^2\chi_{[2,3]}.$

(d) $B_4(x) = \frac{1}{6}x^3\chi_{[0,1]} + \frac{1}{3}(2 - 6x + 6x^2 - x^3)\chi_{[1,2]} + \frac{1}{2}(x^3 - 2x^2 + 20x - 13)\chi_{[2,3]}.$

11. Use Exercise 10 to show that the two-scale relation for the B-splines of order n is

$$B_n(x) = \sum_{k=0}^{n} 2^{1-n} \binom{n}{k} B_n(2x - k).$$

12. Use the Fourier transform $\widehat{B}_1(\omega)$ of $B_1(x)$ to prove

$$\sum_{k=-\infty}^{\infty} |\widehat{B}_n(2\omega + 2\pi k)|^2 = -\frac{\sin^{2n}(\omega)}{(2n-1)!} \frac{d^{2n-1}}{d\omega^{2n-1}}(\cot \omega).$$

13. The *Franklin wavelet* is generated by the second-order ($n = 2$) splines. Show that the Fourier transform $\widehat{\varphi}(\omega)$ of this wavelet is

$$\widehat{\varphi}(\omega) = \frac{\sin^{2n}\frac{\omega}{2}}{(\frac{\omega}{2})^2}\left(1 - \frac{2}{3}\sin^2\frac{\omega}{2}\right)^{-1/2}.$$

Hence, calculate the Fourier transform of the Franklin wavelet φ.

14. The harmonic wavelet $\psi(t)$ due to Newland (1993, 1994) is defined by its Fourier transform $\widehat{\psi}(\omega) = \widehat{\psi}_e(\omega) + i\widehat{\psi}_0(\omega)$, where

$$\widehat{\psi}_e(\omega) = \begin{cases} \frac{1}{4\pi} & \text{if } -4\pi \le \omega < -2\pi \text{ and } 2\pi \le \omega < 4\pi, \\ 0 & \text{otherwise,} \end{cases}$$

and

$$\widehat{\psi}_0(\omega) = \begin{cases} \frac{i}{4\pi} & \text{if } -4\pi \le \omega < -2\pi, \\ -\frac{i}{4\pi} & \text{if } 2\pi \le \omega < 4\pi, \\ 0 & \text{otherwise.} \end{cases}$$

Show that

$$\widehat{\psi}(t) = \psi_e(t) + i\psi_0(t) = \frac{1}{2\pi it}\left[\exp(4\pi it) - \exp(2\pi it)\right],$$

where ψ_e and ψ_0 are real even and odd functions respectively and the Fourier transform $\widehat{\psi}(\omega)$ is defined by using the factor $\frac{1}{2\pi}$.

15. (a) Show that the family of harmonic wavelets $\psi(2^m t - k)$ forms an orthogonal set.

(b) Show that

$$\int_{-\infty}^{\infty} \left|\psi(2^m t - k)\right|^2 dt = 2^{-m}.$$

Optimization Problems and Other Miscellaneous Applications

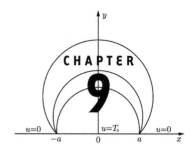

9.1 Introduction

The fundamental problems in classical functional analysis include the following:

(a) The existence and uniqueness of solutions for operator equations in a Hilbert space ($Tx = y$, where T is a bounded linear operator from a Hilbert space H_1 into a Hilbert space H_2 and y is a given element of H_2).

(b) The minimization of the norm in a Hilbert space, the so-called minimum norm optimization problems.

(c) The maximization or minimization of a functional on a Hilbert space, the so-called *variational problems*.

477

In previous chapters, the first problem has been discussed in some detail as an application of the theory of operators on Hilbert spaces. This chapter is essentially concerned with the optimization and variational problems and their applications. To discuss these problems, it is necessary to introduce some ideas of the calculus of operators in Banach spaces. The rest of the chapter deals with applications to miscellaneous problems, including minimization of a quadratic functional, variational inequalities, and optimal control problems for dynamical systems. Also included are approximation theory, the Shannon sampling theorem, linear and nonlinear stability problems, and bifurcation theory.

9.2 The Gateaux and Fréchet Differentials

Throughout this section, B_1 and B_2 are Banach spaces over a field $\mathbb{F}$, which may be either $\mathbb{R}$, the real numbers, or $\mathbb{C}$, the complex numbers. Assume that $T : B_1 \to B_2$ is an operator (not necessarily linear) with the domain $\mathcal{D}(T) = B_1$.

Definition 9.2.1.   (Gateaux differential)
Suppose x is a fixed element of B_1. The operator $T : B_1 \to B_2$ is said to be *Gateaux differentiable* at x if there exists a continuous linear operator A such that

$$\lim_{t \to 0} \left\| \frac{T(x + th) - T(x)}{t} - A(h) \right\| = 0 \qquad (9.1)$$

for every $h \in B_1$, where $t \to 0$ in $\mathbb{F}$. The operator A is called the *Gateaux differential* of T at x, and its value at h is denoted by

$$A(h) = dT(x, h).$$

According to the definition, the Gateaux differential of an operator from B_1 into B_2 at $x \in B_1$ is a linear operator from B_1 into B_2. Note that if T is a linear operator, then $dT(x, h) = T(h)$, that is, $dT(x) = T$ for all $x \in B_1$.

If f is a functional on B_1, that is, $f : B_1 \to \mathbb{F}$, and f is Gateaux differentiable at some $x \in B_1$, then

$$df(x, h) = \left[\frac{d}{dt} f(x + th) \right]_{t=0}$$

and, for each fixed $x \in B_1$, $df(x, h)$ is a linear functional of $h \in B_1$.

Theorem 9.2.2.   *If the Gateaux differential exists, it is unique.*

Proof: Suppose that two operators A_1 and A_2 satisfy (9.1). Then, for every $h \in B_1$ and every $t > 0$, we have

$$\left\| A_1(h) - A_2(h) \right\|$$

$$= \left\| \left(\frac{T(x+th) - T(x)}{t} - A_2(h) \right) - \left(\frac{T(x+th) - T(x)}{t} - A_1(h) \right) \right\|$$

$$\leq \left\| \frac{T(x+th) - T(x)}{t} - A_2(h) \right\| + \left\| \frac{T(x+th) - T(x)}{t} - A_1(h) \right\| \to 0,$$

as $t \to 0$. Therefore, $\|A_1(h) - A_2(h)\| = 0$ for all $h \in B_1$, proving the theorem. $\qquad\square$

Definition 9.2.3. (Gradient of a functional)
The mapping $x \to df(x, \cdot)$ is called the *gradient* of f, and is usually denoted by ∇f. Consequently, ∇ is a mapping from B_1 into the dual space B_1'.

Example 9.2.4. Let $B_1 = \mathbb{R}^N$ and let f be a functional on B_1, that is, $f : B_1 \to \mathbb{R}$. Let $x = (x_1, \ldots, x_N) \in B_1$ and $h = (h_1, \ldots, h_N) \in B_1$. If f has continuous partial derivatives of order one, then the Gateaux differential of f is

$$df(x, h) = \sum_{k=1}^{N} \frac{\partial f(x)}{\partial x_k} h_k. \tag{9.2}$$

For a fixed $x_0 \in B_1$, the Gateaux differential at x_0,

$$df(x_0, h) = \left[\sum_{k=1}^{N} \frac{\partial f(x)}{\partial x_k} h_k \right]_{x=x_0}, \tag{9.3}$$

is a bounded linear operator from $\mathbb{R}^N$ into $\mathbb{R}^N$. (Note that, in this example, $B_1' = B_1$.) We can also write

$$df(x_0, h) = \left\langle \left(\frac{\partial f(x_0)}{\partial x_1}, \ldots, \frac{\partial f(x_0)}{\partial x_N} \right), h \right\rangle. \qquad\square$$

Example 9.2.5. Let $B_1 = \mathbb{R}^N$ and $B_2 = \mathbb{R}^M$. Let

$$f = (f_1, \ldots, f_M) : B_1 \to B_2$$

be Gateaux differentiable at some $x \in \mathbb{R}^N$. The Gateaux differential A can be identified with an $M \times N$ matrix (a_{ij}). If h is the jth coordinate vector, $h = e_j =$

$(0, \ldots, 1, \ldots, 0)$, then

$$\lim_{t \to 0} \left\| \frac{f(x + th) - f(x)}{t} - A(h) \right\| = 0$$

implies

$$\lim_{t \to 0} \left| \frac{f_i(x + te_j) - f_i(x)}{t} - a_{ij} \right| = 0,$$

for every $i = 1, \ldots, M$ and $j = 1, \ldots, N$. This shows that f_i's have partial derivatives at x and

$$\frac{\partial f_i(x)}{\partial x_j} = a_{ij},$$

for every $i = 1, \ldots, M$ and $j = 1, \ldots, N$. The Gateaux differential of f at x has the matrix representation

$$\begin{bmatrix} \dfrac{\partial f_1(x)}{\partial x_1} & \cdots & \dfrac{\partial f_1(x)}{\partial x_N} \\ \vdots & & \vdots \\ \dfrac{\partial f_M(x)}{\partial x_1} & \cdots & \dfrac{\partial f_M(x)}{\partial x_N} \end{bmatrix} = (a_{ij}). \tag{9.4}$$

This is called the *Jacobian matrix* of f at x. Note that if $M = 1$ then the matrix reduces to a row vector, which is the case discussed in Example 9.2.4. □

Example 9.2.6. Let $B = \mathcal{C}([a, b])$ be the normed space of real-valued continuous functions on $[a, b]$ with the norm defined by

$$\|x\| = \sup_{t \in [a,b]} |x(t)|.$$

Let $K(s, t)$ be a continuous real valued function defined on $[a, b] \times [a, b]$, and let $g(t, x)$ be a continuous real valued function on $[a, b] \times \mathbb{R}$ with continuous partial derivative $\partial g / \partial x$ on $[a, b] \times \mathbb{R}$. Define a mapping $f : B \to B$ by

$$f(x)(s) = \int_a^b K(s, t) g(t, x(t)) dt. \tag{9.5}$$

Then

$$df(x, h) = \left[\frac{d}{d\alpha} \int_a^b K(s, t) g(t, x(t) + \alpha h(t)) dt \right]_{\alpha = 0}.$$

Interchange of the order of differentiation and integration is permissible under the given assumption on g, and hence it follows that

$$df(x, h) = \int_a^b K(s, t) \left[\frac{\partial}{\partial x} g(t, x(t)) \right] h(t) dt. \tag{9.6}$$

Thus, the Gateaux differential of the integral operator (9.5) is the linear integral operator (9.6) and its kernel is $K(s, t)g_x(t, x)$.   □

Note that the Gateaux differential is a generalization of the idea of the directional derivative familiar in finite dimensional spaces.

Theorem 9.2.7. (Mean value theorem) *Suppose the functional f has a Gateaux differential $df(x, h)$ at every point $x \in B$. Then, for any two points $x, x + h \in B$, there exists a $\xi \in (0, 1)$ such that*

$$f(x + h) - f(x) = df(x + \xi h, h). \tag{9.7}$$

Proof: Put $\Phi(t) = f(x + th)$. Then

$$\Phi'(t) = \lim_{s \to 0}\left[\frac{\Phi(t + s) - \Phi(t)}{s}\right] = \lim_{s \to 0}\left[\frac{f(x + th + sh) - f(x + th)}{s}\right]$$
$$= df(x + th, h).$$

Application of the mean value theorem for functions of one variable to Φ yields

$$\Phi(1) - \Phi(0) = \Phi'(\xi) \quad \text{for some } \xi \in (0, 1).$$

Consequently,

$$f(x + h) - f(x) = df(x + \xi h, h). \qquad \qquad □$$

The derivative of a function f of a real variable is defined by

$$f'(x) = \lim_{h \to 0}\frac{f(x + h) - f(x)}{h}, \tag{9.8}$$

provided the limit exists. This definition cannot be used in the case of mappings defined on a Banach space because h is then a vector, and division by a vector is meaningless. On the other hand, division by a vector can be easily avoided by rewriting (9.8) as

$$f(x + h) = f(x) + f'(x)h + hw(h), \tag{9.9}$$

where w is a function (which depends on h) such that $w(h) \to 0$ as $h \to 0$. Equivalently, we can now say that $f'(x)$ is the derivative of f at x if

$$f(x + h) - f(x) = f'(x)h + \Phi(h), \tag{9.10}$$

where $\Phi(h) = hw(h)$, and thus $\Phi(h)/h \to 0$ as $h \to 0$, usually denoted by $\Phi(h) = o(h)$. The definition based on (9.10) can be generalized to include mappings from a Banach space into a Banach space. This leads to the concept of the Fréchet differentiability and Fréchet derivative (Maurice René Fréchet (1878–1973)).

Definition 9.2.8.   (Fréchet differential)

Let x be a fixed point in a Banach space B_1. A continuous linear operator $A : B_1 \rightarrow B_2$ is called the *Fréchet differential* of the operator $T : B_1 \rightarrow B_2$ at x if

$$T(x + h) - T(x) = Ah + \Phi(x, h) \tag{9.11}$$

and

$$\lim_{\|h\| \to 0} \frac{\|\Phi(x, h)\|}{\|h\|} = 0 \tag{9.12}$$

or, equivalently,

$$\lim_{\|h\| \to 0} \frac{\|T(x + h) - T(x) - Ah\|}{\|h\|} = 0. \tag{9.13}$$

The Fréchet differential at x will be denoted by $T'(x)$ or $dT(x)$.

In the case of a real-valued function $f : \mathbb{R} \rightarrow \mathbb{R}$, the ordinary derivative at x is a number representing the slope of the graph of the function at x. The Fréchet differential of f is not a number, but a linear operator from $\mathbb{R}$ into $\mathbb{R}$. The existence of the ordinary derivative $f'(x)$ implies the existence of the Fréchet differential at x, and the comparison of (9.9) and (9.11) shows that A is the operator, which multiplies every $h \in \mathbb{R}$ by the number $f'(x)$.

In elementary calculus, the tangent to a curve is the straight line giving the best approximation of the curve in the neighborhood of the point of tangency. Similarly, the Fréchet differential of an operator f can be interpreted as the best local linear approximation. We consider the change in f when its argument changes from x to $x + h$, and then we approximate this change by a linear operator A so that

$$f(x + h) = f(x) + Ah + \varepsilon, \tag{9.14}$$

where ε is the error in the linear approximation. Thus, ε has the same order of magnitude as h, except for the case when A is equal to the Fréchet derivative of f. In such a case, $\varepsilon = o(h)$, so that ε is much smaller than h as $h \to 0$. In this sense, the Fréchet differential gives the best linear approximation of f near x. Finally, if A is a linear operator, then the derivative of A is A itself and the best linear approximation of A is A itself.

Theorem 9.2.9. *If a mapping has the Fréchet differential at a point, then it has the Gateaux differential at that point and both differentials are equal.*

Proof: Let $T : B_1 \rightarrow B_2$, and let $x \in B_1$. If T has the Fréchet differential at x, then

$$\lim_{\|h\| \to 0} \frac{\|T(x + h) - T(x) - Ah\|}{\|h\|} = 0$$

for some continuous linear operator $A: B_1 \to B_2$. In particular, for any fixed nonzero $h \in B_1$, we have

$$\lim_{t \to 0} \left\| \frac{T(x + th) - T(x)}{t} - Ah \right\| = \lim_{t \to 0} \left\| \frac{T(x + th) - T(x) - A(th)}{\|th\|} \right\| \|h\| = 0.$$

Thus, A is the Gateaux differential of T at x. □

Corollary 9.2.10. *If the Fréchet differential exists, it is unique.*

Proof: Suppose A_1 and A_2 are Fréchet differentials of T at some $x \in B_1$. Then A_1 and A_2 are the Gateaux differentials of T at x. Thus, $A_1 = A_2$, by Theorem 9.2.2. □

Example 9.2.11. Let $f: \mathbb{R}^2 \to \mathbb{R}$ be defined by

$$f(x, y) = \begin{cases} \dfrac{x^3 y}{x^4 + y^2} & \text{if } x \neq 0 \text{ and } y \neq 0, \\ 0 & \text{if } x = y = 0. \end{cases}$$

It is easy to check that f is Gateaux differentiable at 0, and the Gateaux differential at that point is 0. On the other hand, since

$$\frac{|f(x, x^2)|}{\|(x, x^2)\|} = \frac{|x^3 x^2|}{(x^4 + x^4)} \frac{1}{\sqrt{x^2 + x^4}} = \frac{1}{2\sqrt{1 + x^2}} \to \frac{1}{2} \quad \text{as } x \to 0,$$

f is not Fréchet differentiable at (0,0). □

Example 9.2.12. Suppose $f: H \to \mathbb{F}$ is a functional on a Hilbert space H, where $\mathbb{F} = \mathbb{R}$ or $\mathbb{C}$. If f is Fréchet differentiable at some $x \in H$, then that differential must be a bounded linear functional on H, that is, $f'(x) \in H'$. By the Riesz Representation Theorem 3.7.7, there exists an element $y \in H$ such that $f(x)(h) = \langle h, y \rangle$ for every $h \in H$. The differential $f'(x)$ can thus be identified with the vector y, which is called the *gradient of* f, ∇f. □

Example 9.2.13. Let T be a bounded linear operator on a real Hilbert space H, and let f be a functional on H defined by $f(x) = \langle x, Tx \rangle$. We will show that the Fréchet differential of f is $f'(x)(h) = \langle h, (T + T^*)x \rangle$.

Let h be an arbitrary nonzero element of H. Then

$$f(x + h) - f(x) - \langle h, (T + T^*)x \rangle$$
$$= \langle x + h, T(x + h) \rangle - \langle x, Tx \rangle - \langle h, Tx \rangle - \langle h, T^*x \rangle$$
$$= \langle x, Tx \rangle + \langle x, Th \rangle + \langle h, Tx \rangle + \langle h, Th \rangle - \langle x, Tx \rangle - \langle h, Tx \rangle - \langle h, T^*x \rangle$$
$$= \langle h, Th \rangle.$$

Consequently,

$$\lim_{\|h\| \to 0} \frac{\|f(x+h) - f(x) - \langle h, T + T^* \rangle x\|}{\|h\|} = \lim_{\|h\| \to 0} \frac{|\langle h, Th \rangle|}{\|h\|} = 0.$$

So the result follows. □

Example 9.2.14. By the *Hammerstein operator* $T : \mathcal{C}([a,b]) \to \mathcal{C}([a,b])$, we mean an operator defined by

$$(Tu)(x) = \int_a^b K(x,t) f\big(t, u(t)\big) dt,$$

where $K : [a,b] \times [a,b] \to \mathbb{R}$ and $f : [a,b] \times \mathbb{R} \to \mathbb{R}$ are given functions. If f is sufficiently smooth, then

$$T(u+h)(x) = \int_a^b K(x,t) \left[f(t,u) + h f_u(t,u) + \frac{1}{2} h^2 f_{uu}(t,u) + \cdots \right] dt$$

$$= (Tu)(x) + Ah + o(h),$$

where the Fréchet differential $A = T'(u)$ is

$$T'(u)(h) = \int_a^b K(x,t) f_u\big(t, u(t)\big) h(t) dt.$$

Thus, the Fréchet derivative of T at u is the linear integral operator with the kernel $K(x,t) f_u(t, u(t))$. □

Theorem 9.2.15. *If an operator defined on an open subset of a Banach space is Fréchet differentiable at a point, then it is continuous at that point.*

Proof: Let Ω be an open set in a Banach space B_1, and let T be an operator from Ω into a Banach space B_2. Let $x \in \Omega$ and let $\varepsilon > 0$ be such that $x + h \in \Omega$ whenever $\|h\| < \varepsilon$. Then

$$\big\| T(x+h) - T(x) \big\| = \big\| Ah + \Phi(x,h) \big\| \to 0$$

as $\|h\| \to 0$. This proves that T is continuous at x. □

Much of the theory, results, and methods of ordinary calculus can be easily generalized to Fréchet derivatives. For example, the usual rules for differentiation of the sum and product (in the case of functionals) of two or more functions apply to Fréchet differentials. The mean value theorem, the implicit function theorem, and Taylor series have satisfactory extensions. The interested reader is referred to Liusternik and Sobolev (1974). In the next theorem, we prove the chain rule for Fréchet differentials.

Theorem 9.2.16. (Chain rule) *Let B_1, B_2, and B_3 be real Banach spaces. If $g : B_1 \to B_2$ is Fréchet differentiable at some $x \in B_1$ and $f : B_2 \to B_3$ is Fréchet differentiable at $y = g(x) \in B_2$, then $\Phi = f \circ g$ is Fréchet differentiable at x and*

$$\Phi'(x) = f'\big(g(x)\big)g'(x).$$

Proof: For $x, h \in B_1$, we have

$$\begin{aligned}
\Phi(x+h) - \Phi(x) &= f\big(g(x+h)\big) - f\big(g(x)\big) \\
&= f\big(g(x+h) - g(x) + g(x)\big) - f(y) \\
&= f(d+y) - f(y),
\end{aligned}$$

where $d = g(x+h) - g(x)$. Thus,

$$\big\| \Phi(x+h) - \Phi(x) - f'(y)d \big\| = o\big(\|d\| \big).$$

In view of $\|d - g'(x)h\| = o(\|h\|)$, we obtain

$$\big\| \Phi(x+h) - \Phi(x) - f'(y)g'(x)h \big\| = o\big(\|h\| \big) + o\big(\|d\| \big).$$

Since g is continuous at x, by Theorem 9.2.15, we have $\|d\| = o(\|h\|)$ and thus

$$\Phi'(x)h = f'\big(g(x)\big)g'(x)h. \qquad \square$$

Theorem 9.2.17. *A linear operator T from a Banach space into a Banach space is Fréchet differentiable if and only if T is bounded. In that case, $T' = T$.*

Proof: If T is a linear operator and Fréchet differentiable at a point, then T is continuous (and hence bounded), by Theorem 9.2.15. Conversely, if T is a bounded linear operator, then

$$\big\| T(x+h) - Tx - Th \big\| = 0,$$

proving that T is Fréchet differentiable and $T' = T$. $\qquad \square$

Definition 9.2.18.   (Second Fréchet differential)
If $T : B_1 \to B_2$ is Fréchet differentiable on an open set $\Omega \subset B_1$ and T' is Fréchet differentiable at $x \in \Omega$, then T is called *twice Fréchet differentiable* at x. The Fréchet differential of T' at x is called the *second Fréchet differential* of T and is denoted by $T''(x)$.

Note that if $T : B_1 \to B_2$ is Fréchet differentiable on an open set $\Omega \subset B_1$, then T' is a mapping from B_1 into $\mathcal{B}(B_1, B_2)$ (recall that $\mathcal{B}(B_1, B_2)$ denotes the space of all bounded linear mappings from B_1 into B_2). Consequently, if $T''(x)$ exists, it is a bounded linear mapping from B_1 into $\mathcal{B}(B_1, B_2)$. If T'' exists at every point of Ω, then $T'' : B_1 \to \mathcal{B}(B_1, \mathcal{B}(B_1, B_2))$.

In ordinary calculus, the analogue of the formula (9.9) for the second derivative can be written as

$$f(x + h) = f(x) + f'(x)h + \frac{1}{2} f''(x)h^2 + o(h^2), \qquad (9.15)$$

where $o(h^2)$ stands for a function $\varphi(h) = w(h)h^2$ such that $\lim_{h \to 0} w(h) = 0$. Clearly, (9.15) is the truncated Taylor expansion of $f(x + h)$ in powers of h. It says that $f'(x)$ is the coefficient of the linear term h and $\frac{1}{2} f''(x)$ is the coefficient of the quadratic term h^2. Thus, (9.15) can be used to find f'' in the same way that (9.9) is used to calculate f'.

To extend (9.15) to the case of the second Fréchet differential of a scalar valued function on a Banach space, it seems reasonable to interpret $f''(x)h^2$ as the result of application of the linear operator $f''(x)$ to the pair (h, h), and $o(h^2)$ as a function φ such that $\|\varphi(h)\|/\|h\|^2 \to 0$ as $h \to 0$.

Example 9.2.19. To calculate $f''(x)$ for a differentiable real-valued function $f : \mathbb{R}^N \to \mathbb{R}$, we use Taylor series expansion

$$f(\mathbf{x} + \mathbf{h}) = f(\mathbf{x}) + \sum_i \frac{\partial f}{\partial x_i}(\mathbf{x})h_i + \frac{1}{2} \sum_{i,j} \frac{\partial^2}{\partial x_i \partial x_j}(\mathbf{x})h_i h_j + \cdots. \qquad (9.16)$$

Thus, $f'(\mathbf{x})$ represents the gradient operator defined by

$$f'(\mathbf{x})\mathbf{h} = \sum_i \frac{\partial f}{\partial x_i}(\mathbf{x})h_i.$$

The second Fréchet differential $f''(\mathbf{x})$ is defined by

$$f''(\mathbf{x})(h, h) = \sum_{i,j} \frac{\partial^2}{\partial x_i \partial x_j}(\mathbf{x})h_i h_j = \sum_{i,j} f_{ij}(\mathbf{x})h_i h_j, \qquad (9.17)$$

where f_{ij} denotes the second partial derivative. In other words, $f''(x, x)$ is the bilinear operator given by the matrix of the second partial derivative of f. For a twice differentiable function $f : \mathbb{R}^N \to \mathbb{R}$, we have $f_{ij} = f_{ji}$, so the matrix is symmetric. This leads to the following result for a general pair (h, k):

$$f''(\mathbf{x})(h, k) = \sum_{i,j} f_{ij}(\mathbf{x})h_i k_j. \qquad (9.18)$$

In general, we can define the second Fréchet differential $f''(\mathbf{x})$ as a symmetric bilinear operator. For a symmetric bilinear mapping φ, $\varphi(h, k)$ is uniquely determined by its values for $\varphi(h, h)$. In fact, we have

$$\varphi(h + k, h + k) = \varphi(h, h) + 2\varphi(h, k) + \varphi(k, k), \tag{9.19}$$

and hence,

$$\varphi(h, k) = \frac{1}{2}\left[\varphi(h + k, h + k) - \varphi(h, h) - \varphi(k, k)\right]. \tag{9.20}$$

This formula can be used to obtain an expression for $f''(\mathbf{x})(h, k)$ in terms of $f''(\mathbf{x})(h, h)$. □

Example 9.2.20. Consider a nonlinear integral operator $T:\mathcal{C}([a, b]) \rightarrow \mathcal{C}([a, b])$ defined by

$$(Tu)(t) = \int_a^b K\big(t, s, u(s)\big)ds, \tag{9.21}$$

where K is a continuous function of three variables $K(t, s, u)$, twice differentiable with respect to u. Then

$$T(x + h)(t) = \int_a^b \left[K\big(t, s, x(s)\big) + K_u\big(t, s, x(s)\big)h(s)\right.$$

$$\left. + \frac{1}{2}K_{uu}\big(t, s, x(s)\big)h^2(s) + \cdots\right]ds$$

$$= T(x) + T'(x)h + \frac{1}{2}T''(x)(h, h) + \cdots, \tag{9.22}$$

where $K_u = \partial K/\partial u$,

$$T'(x)h = \int_a^b K_u\big(t, s, x(s)\big)h(s)ds \tag{9.23}$$

and

$$T''(x)(h, h)(t) = \int_a^b K_{uu}\big(t, s, x(s)\big)h^2(s)ds. \tag{9.24}$$

The left-hand side of (9.24) represents the value at t of the function $T''(x)(h, h)$ obtained by application of the bilinear operator $T''(x)$ to the pair of functions (h, h). Clearly, $T''(x)$ applied to a general pair of functions (h, k) yields

$$T''(x)(h, k)(t) = \int_a^b K_{uu}\big(t, s, x(s)\big)h(s)k(s)ds. \tag{9.25}$$

This can be obtained from (9.24) using (9.20). □

Example 9.2.21. Consider the real Hilbert space $L^2([a, b])$. Define a functional $f : L^2([a, b]) \to \mathbb{R}$ by the double integral

$$f(x) = \int_a^b \int_a^b x(s)K(s, t)x(t)\,dt\,ds, \tag{9.26}$$

where K is a continuous function. If we define a linear operator T on $L^2([a, b])$ by

$$(Tx)(t) = \int_a^b K(t, s)x(s)\,ds, \tag{9.27}$$

we can rewrite (9.26) in a neat form:

$$f(x) = \langle x, Tx \rangle. \tag{9.28}$$

Hence,

$$f(x + h) = \langle x + h, T(x + h) \rangle = \langle x + h, Tx + Th \rangle$$
$$= f(x) + \langle h, Tx \rangle + \langle x, Th \rangle + \langle h, Th \rangle. \tag{9.29}$$

The second Fréchet differential f'' is given by the quadratic term in (9.29), so that

$$f''(x)(h, h) = \langle h, Th \rangle. \tag{9.30}$$

Application of the formula (9.20) gives

$$f''(x)(h, k) = \frac{1}{2}[\langle h, Tk \rangle + \langle k, Th \rangle] = \left\langle h, \frac{1}{2}(T + T^*)k \right\rangle, \tag{9.31}$$

where T^* is the adjoint of T. If $K(s, t) = K(t, s)$, then T is self-adjoint, and (9.31) becomes

$$f''(x)(h, k) = \langle h, Tk \rangle. \tag{9.32}$$

This expression is symmetric in h and k.

This result can be generalized to any functional f on a real Hilbert space. □

Theorem 9.2.22. *Suppose $f : H \to \mathbb{R}$ is a twice differentiable functional on a real Hilbert space H. Then, for each $x \in H$, there exists a self-adjoint bounded linear operator $T : H \to H$ (depending on x) such that*

$$f''(x)(h, k) = \langle h, Tk \rangle$$

for all $h, k \in H$.

Proof: For any fixed $k \in H$, $f''(x)(h, k)$ is a continuous linear functional. Continuity follows from Fréchet differentiability. According to the

Riesz Representation Theorem 3.7.7, there exists an $x_0 \in H$ such that $f''(x)(h, k) = \langle h, x_0 \rangle$. Thus, there exists a map T which to each $k \in H$ assigns $x_0 \in H$, that is, $x_0 = Ak$. Since $f''(x)(h, k)$ is a linear in k for every fixed h, it follows that T is a linear operator. Consequently,

$$f''(x)(h, k) = \langle h, x_0 \rangle = \langle h, Tk \rangle.$$

Moreover, since $f''(x)$ is continuous and symmetric, T is bounded and self-adjoint. $\square$

Definition 9.2.23. (Convex function)
A function $f : E \to \mathbb{R}$, where E is a vector space, is called *convex* if

$$f\big(tx + (1 - t)y\big) \le tf(x) + (1 - t)f(y) \quad \text{for all } x, y \in E, \text{ and } t \in (0, 1). \quad (9.33)$$

It is called *strictly convex* if

$$f\big(tx + (1 - t)y\big) < tf(x) + (1 - t)f(y) \quad \text{for all } x, y \in E, \text{ and } t \in (0, 1). \quad (9.34)$$

Example 9.2.24. One of the most important convex functions is the norm in a normed space:

$$f(x) = \|x\|. \qquad \square$$

Example 9.2.25. Suppose S is a convex subset of a Hilbert space H. Define the distance function

$$d(x, S) = \inf_{y \in S} \|x - y\|.$$

It follows from the triangle inequality that $d(x, S)$ is a convex function. $\square$

Example 9.2.26. Let T be a self-adjoint bounded operator on a Hilbert space H. Consider the quadratic functional Φ defined by

$$\Phi(x) = \frac{1}{2} \langle Tx, x \rangle.$$

Φ is differentiable and $\nabla \Phi = T$. Moreover, Φ is convex if T is strictly positive. $\square$

Theorem 9.2.27. *If a convex function $f : B \to \mathbb{R}$ is Gateaux differentiable at some $x_0 \in B$, then*

$$f(x) \ge f(x_0) + df(x_0)(x - x_0) \quad \text{for all } x \in B.$$

Proof: Since f is convex, for every $t \in (0, 1)$, we have

$$f(x_0 + t(x - x_0)) - f(x_0) \leq tf(x) + (1 - t)f(x_0) - f(x_0) = t(f(x) - f(x_0)),$$

and thus,

$$f(x) - f(x_0) \geq \frac{f(x_0 + t(x - x_0)) - f(x_0)}{t}.$$

Now, by letting $t \to 0$, we obtain

$$f(x) - f(x_0) \geq \lim_{t \to 0+} \frac{f(x_0 + t(x - x_0)) - f(x_0)}{t} = df(x_0)(x - x_0). \qquad \square$$

9.3 Optimization Problems and the Euler–Lagrange Equations

In ordinary calculus, maximum and minimum problems are concerned with those values of the independent variables for which a given function attains its maximum or minimum. If a differentiable function has a maximum or a minimum at a point, then its derivative vanishes at that point. It turns out that this property can be generalized to the case of a maximum or a minimum of a functional on a normed space. We will show that if a real-valued functional defined on a subset of a normed space has a maximum or a minimum at a point, then its Gateaux (or Fréchet) differential at that point is zero. In the subsequent discussion, we use the term *extremum* to refer to either maximum or minimum.

Suppose f is a real-valued function defined on a subset Ω of a normed space. A very general formulation of an optimization problem is

$$\text{Find } x_0 \in \Omega \text{ such that } f(x_0) = \min_{x \in \Omega} f(x). \tag{9.35}$$

This and the following section are concerned with some aspects of such a problem.

Definition 9.3.1. (Relative extremum)
A real-valued functional f defined on a subset Ω of a normed space E is said to have a *relative minimum* (or *relative maximum*) at a point $x_0 \in \Omega$ if there is an open ball $B(x_0, r) \subset E$ such that $f(x_0) \leq f(x)$ (or $f(x_0) \geq f(x)$) holds for all $x \in B(x_0, r) \cap \Omega$. If f has either a relative minimum or relative maximum at x_0, then f is said to have a *relative extremum* at x_0.

A relative extremum is often called a *local extremum*. The set Ω on which an extremum problem is defined is sometimes called the *admissible set*.

Theorem 9.3.2. *If a functional $f : E \to \mathbb{R}$ is Gateaux differentiable at $x_0 \in E$ and has a relative extremum at x_0, then $df(x_0, h) = 0$ for all $h \in E$.*

Proof: For every $h \in E$, the function $f(x_0 + th)$ (of the real variable t) has a relative extremum at $t = 0$. Since it is differentiable at 0, it follows from ordinary calculus that

$$\left[\frac{d}{dt} f(x_0 + th) \right]_{t=0} = 0.$$

This means that $df(x_0, h) = 0$ for all $h \in E$, proving the theorem. $\square$

Corollary 9.3.3. *If a functional $f : E \to \mathbb{R}$ is Fréchet differentiable at $x_0 \in E$ and has a relative extremum at x_0, then $f'(x_0) = 0$.* $\square$

Definition 9.3.4. (Stationary point)
A point x at which $df(x, h) = 0$ for all $h \in E$ (or $f'(x) = 0$) is called a *stationary point*.

The preceding theorem (or corollary) states that relative extrema of a Gateaux (or Fréchet) differentiable function occur at stationary points.

Example 9.3.5. Consider the real-valued functional f defined on a real Hilbert space H_1 by

$$f(x) = \|u - Tx\|^2,$$

where T is a bounded linear operator from H_1 into a Hilbert space H_2, and $u \in H_2$. The Fréchet differential is given by

$$f'(x) = -2T^* u + 2T^* Tx.$$

Therefore, a necessary condition for f to have an extremum at x_0 is that $f'(x_0) = 0$, that is,

$$T^* Tx_0 = T^* u. \qquad \square$$

Theorem 9.3.6. *Suppose f is a real-valued functional on a vector space E and x_0 minimizes f on a convex set $\Omega \subset E$. If f is a Gateaux differentiable at x_0, then*

$$df(x_0, x - x_0) \geq 0$$

for all $x \in \Omega$.

Proof: Since Ω is a convex set, $x_0 + t(x - x_0) \in \Omega$ for all $t \in (0, 1)$ and $x \in \Omega$. Hence,

$$\left[\frac{d}{dt} f\big(x_0 + t(x - x_0)\big) \right]_{t=0} \geq 0. \qquad \square$$

Example 9.3.7. Consider a functional $I : \mathcal{C}([0, 1]) \to \mathbb{R}$ defined by

$$I(x) = \int_0^1 a(t)x^2(t)dt,$$

where a is a given function. Then

$$I(x + h) - I(x) = 2\int_0^1 a(t)x(t)h(t)dt + \int_0^1 a(t)h^2(t)dt.$$

Hence,

$$I'(x)h = 2\int_0^1 a(t)x(t)h(t)dt.$$

Consequently, $I'(x) = 0$ if $x = 0$. $\qquad \square$

Example 9.3.8. Consider the functional

$$I(u) = \langle Au, u \rangle - 2\langle u, f \rangle,$$

where f belongs to a real Hilbert space H, and A is a linear self-adjoint operator on H. Clearly,

$$I(u + h) - I(u) = 2\langle Au, h \rangle - 2\langle f, h \rangle + \langle Ah, h \rangle.$$

Setting $\langle I'(u), h \rangle = 2\langle Au, h \rangle - 2\langle f, h \rangle = 2\langle (Au - f), h \rangle$, we obtain

$$\big\| I(u + h) - I(u) - \langle I'(u), h \rangle \big\| = \big\| \langle Ah, h \rangle \big\| \leq K\|h\|^2,$$

where K is a constant. Thus, it follows that the Fréchet differential of $I(u)$ is

$$I'(u) = 2(Au - f).$$

Furthermore, it follows that

$$\langle I'(u + h), k \rangle - \langle I'(u), k \rangle = \langle 2Ah, k \rangle.$$

Thus, the second Fréchet differential is

$$I''(u)(h, k) = \langle 2Ah, k \rangle$$

which is independent of $u \in H$.

It turns out that if A is a positive operator, then $I(u)$ has a local minimum when u satisfies the operator equation $Au = f$. $\qquad \square$

The most remarkable classical Euler–Lagrange variational problem is to determine a function $u(x)$ on the interval $[a, b]$ satisfying the boundary conditions $u(a) = \alpha$ and $u(b) = \beta$, and extremizing the functional

$$I(u) = \int_a^b F(x, u, u')dx \quad \left(u'(x) = \frac{du}{dx}\right), \tag{9.36}$$

where u is a twice continuously differentiable function on the interval $[a, b]$ ($u \in C^2([a, b])$), F is continuous in x, u, and u', and has continuous partial derivatives with respect to u and u'.

We assume that $I(u)$ has an extremum at some $u \in C^2([a, b])$. Then we consider the set of all variations $u + tv$, for an arbitrary fixed $v \in C^2([a, b])$, such that $v(a) = v(b) = 0$. Then

$$I(u + tv) - I(u) = \int_a^b \left[F(x, u + tv, u' + tv') - F(x, u, u')\right]dx. \tag{9.37}$$

Using the Taylor series expansion

$$F(x, u + tv, u' + tv') = F(x, u, u') + t\left(v\frac{\partial F}{\partial u} + v'\frac{\partial F}{\partial u'}\right)$$
$$+ \frac{t^2}{2!}\left(v\frac{\partial F}{\partial u} + v'\frac{\partial F}{\partial u'}\right)^2 + \cdots,$$

it follows from (9.37) that

$$I(u + tv) = I(u) + tdI(u, v) + \frac{t^2}{2!}d^2I(u, v) + \cdots, \tag{9.38}$$

where the first and the second Fréchet differentials are given by

$$dI(u, v) = \int_a^b \left(v\frac{\partial F}{\partial u} + v'\frac{\partial F}{\partial u'}\right)dx, \tag{9.39}$$

$$d^2I(u, v) = \int_a^b \left(v\frac{\partial F}{\partial u} + v'\frac{\partial F}{\partial u'}\right)^2 dx. \tag{9.40}$$

The necessary condition for the functional I to have an extremum at u is that $dI(u, v) = 0$ for all $v \in C^2([a, b])$ such that $v(a) = v(b) = 0$, that is,

$$0 = dI(u, v) = \int_a^b \left(v\frac{\partial F}{\partial u} + v'\frac{\partial F}{\partial u'}\right)dx. \tag{9.41}$$

Integrating the second term in the integrand in (9.41) by parts, we obtain

$$\int_a^b \left[\frac{\partial F}{\partial u} - \frac{d}{dx}\left(\frac{\partial F}{\partial u'}\right)\right]vdx + \left[v\frac{\partial F}{\partial u'}\right]_a^b = 0. \tag{9.42}$$

Since $v(a) = v(b) = 0$, the boundary terms vanish and the necessary condition becomes

$$\int_a^b \left[\frac{\partial F}{\partial u} - \frac{d}{dx}\left(\frac{\partial F}{\partial u'} \right) \right] v\, dx = 0 \tag{9.43}$$

for all functions $v \in C^2([a, b])$ vanishing at a and b. This is possible only if

$$\frac{\partial F}{\partial u} - \frac{d}{dx}\left(\frac{\partial F}{\partial u'} \right) = 0. \tag{9.44}$$

This is called the *Euler–Lagrange equation*. We therefore have the following:

Theorem 9.3.9. (The Euler–Lagrange Variational Principle) *A necessary condition for the functional $I(u)$ to be stationary at u is that u must satisfy the Euler–Lagrange equation (9.44) in $a \leq x \leq b$ with the boundary conditions $u(a) = \alpha$ and $u(b) = \beta$.*

After we have determined the solution of (9.44), which makes $I(u)$ stationary, the question arises on whether $I(u)$ has a minimum, a maximum, or a saddle point there. To answer this question, we look at the second derivative involved in (9.38). If terms of $o(t^2)$ can be neglected in (9.38), or if they vanish for the case of quadratic F, it follows that a necessary condition for the functional $I(u)$ to have a minimum at u_0 is that $d^2 I(u, v) \geq 0$ for all v. Similarly, a necessary condition for the functional $I(u)$ to have a maximum at u_0 is that $d^2 I(u, v) \leq 0$ for all v. These results enable us to determine the upper and lower bounds for the stationary value $I(u_0)$ of the functional.

Example 9.3.10. (Minimum arc length) Determine the form of the curve in a plane, which will make the distance between two points in the plane minimum.

Suppose the plane curve $y = y(x)$ passes through the points (x_1, y_1) and (x_2, y_2). The length of such a curve is given by the functional

$$I(y) = \int_{x_1}^{x_2} \sqrt{1 + (y')^2}\, dx. \tag{9.45}$$

Thus, the problem is to determine the curve for which the functional $I(y)$ is minimum. Since $F = \sqrt{1 + (y')^2}$ depends on y' only, the Euler–Lagrange equation becomes

$$\frac{d}{dx}\left(\frac{\partial F}{\partial y'} \right) = 0.$$

Hence,

$$y'' = 0. \tag{9.46}$$

This means that the curve extremizing $I(y)$ is a straight line:

$$y = \frac{y_2 - y_1}{x_2 - x_1}(x - x_1) + y_1. \qquad \Box$$

Example 9.3.11. Determine the meridian curve joining two points in a plane which, when revolved about the x-axis, gives the surface of revolution with minimum area.

This is a problem of minimum surface of revolution generated by the rotation of the curve $y = y(x)$ about the x-axis. In this case, the area is given by

$$S = 2\pi \int_{x_1}^{x_2} y(x)\sqrt{1 + (y')^2}\, dx,$$

so that the functional to be minimized is

$$I(y) = \int_{x_1}^{x_2} y(x)\sqrt{1 + (y')^2}\, dx,$$

subject to the conditions

$$y_1 = y(x_1) \quad \text{and} \quad y_2 = y(x_2). \qquad (9.47)$$

This corresponds to

$$F(x, y, y') = y\sqrt{1 + (y')^2},$$

which does not depend on x explicitly. The Euler–Lagrange equation is

$$yy'' - (y')^2 - 1 = 0. \qquad (9.48)$$

Writing p for y', we have $y'' = dp/dx = p\,dp/dy$, and (9.48) becomes

$$py\frac{dp}{dy} = p^2 + 1.$$

Separating the variables and integrating, we obtain

$$y = a\sqrt{1 + p^2},$$

and hence

$$\frac{dy}{dx} = \sqrt{\frac{y^2}{a^2} - 1}.$$

Integrating again, we find

$$y = a\cosh\left(\frac{x - b}{a}\right), \qquad (9.49)$$

where a and b are constants of integration, which can be determined from conditions (9.47). The curve defined by (9.49) is called the *catenary*, and the resulting surface is called a *catenoid* of revolution. □

Example 9.3.12. Consider the functional

$$I(u) = \int_a^b \left[\frac{1}{2} p(x)(u')^2 - \frac{1}{2} q(x)u^2 + f(x)u \right] dx, \qquad (9.50)$$

where p, q, and f are given functions and u belongs to an admissible set Ω of I. Clearly, the Euler–Lagrange equation associated with the functional $I(u)$ is

$$\frac{\partial F}{\partial u} - \frac{d}{dx} \left(\frac{\partial F}{\partial u'} \right) = 0, \qquad (9.51)$$

where

$$F(x, u, u') = \frac{1}{2} p(x)(u')^2 - \frac{1}{2} q(x)u^2 + f(x)u.$$

Consequently, (9.51) becomes

$$(pu')' + qu = f.$$

This is a nonhomogeneous ordinary differential equation of the Sturm–Liouville type. □

Example 9.3.13. (Hamilton's principle)   According to Hamilton's principle, a particle moves on a path that makes the time integral

$$I = \int_{t_1}^{t_2} L(q_i, \dot{q}_i, t) dt \qquad (9.52)$$

stationary, where the *Lagrangian* $L = T - V$ is the difference between the kinetic energy T and the potential energy V. In coordinate space, there are numerous possible paths joining any two positions. From all these paths, which start at a point A at time t_1 and end at another point B at time t_2, nature selects the path $q_i = q_i(t)$ for which $dI = 0$. Consequently, the Euler–Lagrange equation assumes the form

$$\frac{\partial L}{\partial q_i} - \frac{d}{dt} \left(\frac{\partial L}{\partial \dot{q}_i} \right) = 0, \quad i = 1, 2, \ldots, n. \qquad (9.53)$$

In classical mechanics, these are simply called the *Lagrange equations of motion*.

The *Hamilton's function* (or *Hamiltonian*) H is defined in terms of the generalized coordinates q_i, generalized momentum $p_i = \partial L/\partial \dot{q}_i$, and L as

$$H = \sum_{i=1}^n p_i \dot{q}_i - L = \sum_{i=1}^n \dot{q}_i \frac{\partial L}{\partial \dot{q}_i} - L(q_i, \dot{q}_i). \qquad (9.54)$$

It follows that

$$\frac{dH}{dt} = \frac{d}{dt}\left(\sum_{i=1}^{n} \dot{q}_i \frac{\partial L}{\partial \dot{q}_i} - L\right) = \sum_{i=1}^{n} \dot{q}_i \left(\frac{d}{dt}\frac{\partial L}{\partial \dot{q}_i} - \frac{\partial L}{\partial q_i}\right) = 0.$$

Hence, the Hamiltonian H is the constant of motion. $\square$

Example 9.3.14. (Fermat's principle in optics) This principle states that, in an optically homogeneous isotropic medium, light travels from one point (x_1, y_1) to another point (x_2, y_2) along a path $y = y(x)$ for which the travel time is minimum. Since the velocity v is constant in such a medium, the time is minimum along the shortest path. In other words, the path $y = y(x)$ minimizes the integral

$$I = \int_{x_1}^{x_2} \frac{\sqrt{1 + (y')^2}}{v}\,dx = \int_{x_1}^{x_2} F(y, y')dx, \tag{9.55}$$

with $y(x_1) = y_1$ and $y(x_2) = y_2$. The Euler–Lagrange equation is given by

$$\frac{d}{dx}\left(F - y'\frac{\partial F}{\partial y'}\right) = 0.$$

Hence,

$$F - y'\frac{\partial F}{\partial y'} = \text{constant}$$

or

$$\frac{1}{v}\frac{1}{\sqrt{1 + (y')^2}} = \text{constant}. \tag{9.56}$$

To give a physical interpretation, we rewrite (9.56) in terms of the angle φ between the tangent to the minimum path and the vertical y-axis, so that

$$\sin\varphi = \frac{1}{\sqrt{1 + (y')^2}}.$$

Hence,

$$\frac{\sin\varphi}{v} = \text{constant} \tag{9.57}$$

for all points on the minimum path. For a ray of light $1/v$ must be proportional to the refractive index n of the medium through which light is traveling. Equation (9.57) is known as the *Snell law of refraction of light*. This law is often stated as

$$n\sin\varphi = \text{constant}. \tag{9.58}$$

$\square$

Example 9.3.15. (Abel's problem of tautochronous motion)   The problem is to determine the plane curve $y = y(x)$ for which the time of descent of a particle sliding freely along the curve that passes through the origin and the point (x_1, y_1) is minimum (see Example 5.7.1).

The velocity of the particle at the intermediate point (x, y) is found from the energy equation

$$\frac{1}{2}m\left(\frac{ds}{dt}\right)^2 = mg(y - 0),$$

so that the velocity is

$$\frac{ds}{dt} = \sqrt{2gy},$$

or

$$\sqrt{1 + (y')^2}\,dx = \sqrt{2gy}\,dt,$$

which gives the time required for the particle to descend from the origin to the point (x_1, y_1) on a frictionless curve path in a plane as

$$T(y) = \int_0^{x_1} \sqrt{\frac{1 + (y')^2}{2gy}}\,dx.$$

The problem is to minimize this functional subject to the conditions $y(0) = 0$, $y(x_1) = y_1$.

This case corresponds to

$$F(x, y, y') = F(y, y') = \sqrt{\frac{1 + (y')^2}{2gy}}.$$

Thus, the Euler–Lagrange equation

$$\frac{\partial F}{\partial y} - \frac{d}{dx}\left(\frac{\partial F}{\partial y'}\right) = 0$$

can be written as

$$0 = y'\left(\frac{\partial F}{\partial y} - \frac{d}{dx}\frac{\partial F}{\partial y'}\right) = \frac{d}{dx}\left(F - y'\frac{\partial F}{\partial y'}\right),$$

so that

$$F - y'\frac{\partial F}{\partial y'} = c,$$

where c is a constant. More explicitly,

$$c\sqrt{2gy(1 + (y')^2)} = 1,$$

or

$$y' = \pm\sqrt{\frac{a-y}{y}},$$

where $a^{-1} = 2gc^2$. This can be integrated to obtain

$$x = \int_0^y \frac{y\,dy}{\sqrt{ay - y^2}},$$

where the positive sign is appropriate. We rewrite this integral in the form

$$x = -\frac{1}{2}\int_0^y \frac{(a - 2y)dy}{\sqrt{ay - y^2}} + \frac{a}{2}\int_0^y \frac{dy}{\sqrt{ay - y^2}},$$

so that the first integral can be evaluated at once, and the second one can be evaluated by making the substitution $(a/2) - y = (a/2)\cos\theta$. The final result is

$$x = -\sqrt{ay - y^2} + \frac{a}{2}\cos^{-1}\left(\frac{a - 2y}{a}\right).$$

This is the equation of the curve of minimum time of descent, where the constant a is to be determined so that the curve passes through the point (x_1, y_1). It is convenient to write the equation in a parametric form by letting $(a/2) - y = (a/2)\cos\theta$. Then

$$x = \frac{a}{2}(\theta - \sin\theta), \qquad y = \frac{a}{2}(1 - \cos\theta).$$

They represent a cycloid.   □

Example 9.3.16. A uniform elastic beam of length l is fixed at each end. The beam of line density ρ, cross-sectional moment of inertia I, and modulus of elasticity E, perform small transverse oscillations in the horizontal x-y-plane. Derive the equation of the motion of the beam.

The potential energy of the beam is

$$V = \frac{1}{2}\int_0^l \frac{M^2}{EI}dx = \frac{1}{2}\int_0^l EI(y'')^2 dx, \qquad (9.59)$$

where the bending moment M is proportional to the curvature so that

$$M = EI\frac{y''}{\sqrt{1 + (y')^2}} \sim EIy'' \quad \text{for small } y'.$$

The kinetic energy is

$$T = \frac{1}{2}\int_0^l \rho\dot{y}^2 dx. \qquad (9.60)$$

The variational principle gives

$$d \int_{t_1}^{t_2} L dt = d \int_{t_1}^{t_2} (T - V) dt = d \int_{t_1}^{t_2} F(\dot{y}, y'') dt = 0, \tag{9.61}$$

where

$$F(\dot{y}, y'') = \frac{1}{2} \int_0^l \left(\rho \dot{y}^2 - EI(y'')^2 \right) dx. \tag{9.62}$$

The associated Euler–Lagrange equation is

$$- \int_0^l \left(\rho \ddot{y} + EI y^{(iv)} \right) dx = 0$$

or

$$\rho \ddot{y} + EI y^{(iv)} = 0. \tag{9.63}$$

This represents the equation of the motion for the transverse vibration of the beam.   □

Example 9.3.17. Find $u(x, y, z)$ which minimizes the functional

$$I[u(x, y, z)] = \iiint_\Omega \left(u_x^2 + u_y^2 + u_z^2 \right) dx \, dy \, dz,$$

where $\Omega \subset \mathbb{R}^3$.

The Euler–Lagrange equation is

$$\frac{\partial}{\partial x} \left(\frac{\partial F}{\partial u_x} \right) + \frac{\partial}{\partial y} \left(\frac{\partial F}{\partial u_y} \right) + \frac{\partial}{\partial z} \left(\frac{\partial F}{\partial u_z} \right) = 0$$

or

$$u_{xx} + u_{yy} + u_{zz} = 0,$$

which is the three-dimensional Laplace equation.   □

We next consider optimization problems with a given auxiliary condition. It is often required to determine the function $y = y(x)$ which minimizes the functional

$$I(y) = \int_{x_1}^{x_2} F(x, y, y') dx \tag{9.64}$$

subject to auxiliary condition

$$J(y) = \int_{x_1}^{x_2} G(x, y, y') dx = C. \tag{9.65}$$

It can easily be shown that this problem is equivalent to the problem already discussed earlier, namely, that of determining the function $y = y(x)$, which minimizes the functional

$$I_1(y) = I(y) + \lambda J(y) = \int_{x_1}^{x_2} \left[F(x, y, y') + \lambda G(x, y, y') \right] dx. \tag{9.66}$$

The constant λ involved in (9.66) must be determined from the auxiliary condition (9.65). The resulting Euler–Lagrange equation is

$$\frac{\partial}{\partial y}(F + \lambda G) - \frac{d}{dx}\left[\frac{\partial}{\partial y'}(F + \lambda G) \right] = 0, \tag{9.67}$$

which must be solved with the auxiliary condition (9.65). However, this method breaks down in the case where

$$\frac{d}{dx}\left(\frac{\partial G}{\partial y'} \right) - \frac{\partial G}{\partial y} = 0.$$

Example 9.3.18. The problem is to find the curve $y = y(x)$ of the shortest length between two points (x_1, y_1) and (x_2, y_2) such that the area under the curve is A.

The length of the curve is given by the functional

$$I(y) = \int_{x_1}^{x_2} \sqrt{1 + (y')^2} \, dx \tag{9.68}$$

and the area under the curve is

$$\int_{x_1}^{x_2} y(x) dx = A. \tag{9.69}$$

This is a constrained optimization problem which reduces to that of finding the extremum of

$$I_1(y) = \int_{x_1}^{x_2} \left(\sqrt{1 + (y')^2} + \lambda y \right) dx. \tag{9.70}$$

The associated Euler–Lagrange equation is

$$\frac{\partial}{\partial y}\left(\sqrt{1 + (y')^2} + \lambda y \right) - \frac{d}{dx}\left[\frac{\partial}{\partial y'}\left(\sqrt{1 + (y')^2} + \lambda y \right) \right] = 0$$

or

$$\frac{d}{dx} \frac{y'}{\sqrt{1 + (y')^2}} = \lambda. \tag{9.71}$$

This differential equation can be integrated twice to obtain the equation for y:

$$\left(x - \frac{\alpha}{\lambda}\right)^2 + (y - \beta)^2 = \frac{1}{\lambda^2}. \tag{9.72}$$

Thus, the curve of shortest length is an arc of a circle, where the constants of integration α and β together with the constant λ can be determined from the condition that the curve passes through the points (x_1, y_1) and (x_2, y_2) and the given constrained condition (9.69). □

Example 9.3.19. (Geodesic) A *geodesic* is a curve of minimum length between two points on a smooth surface $G(x, y, z) = 0$ when the whole curve is confined to the surface.

The geodesic of the surface $G = 0$ is obtained by minimizing the functional

$$I(x, y, z) = \int_{t_1}^{t_2} \sqrt{\dot{x}^2 + \dot{y}^2 + \dot{z}^2} \, dt, \tag{9.73}$$

subject to the auxiliary condition

$$G(x, y, z) = 0.$$

This extremum problem is equivalent to finding the minimum of the functional

$$I_1(x, y, z) = \int_{t_1}^{t_2} \left[\sqrt{\dot{x}^2 + \dot{y}^2 + \dot{z}^2} + \lambda(t) G(x, y, z)\right] dt. \tag{9.74}$$

A necessary condition for this minimum problem leads to the Euler–Lagrange equations:

$$\frac{d}{dt} \frac{\dot{x}}{\sqrt{\dot{x}^2 + \dot{y}^2 + \dot{z}^2}} + \lambda(t) G_x = 0, \tag{9.75a}$$

$$\frac{d}{dt} \frac{\dot{y}}{\sqrt{\dot{x}^2 + \dot{y}^2 + \dot{z}^2}} + \lambda(t) G_y = 0, \tag{9.75b}$$

$$\frac{d}{dt} \frac{\dot{z}}{\sqrt{\dot{x}^2 + \dot{y}^2 + \dot{z}^2}} + \lambda(t) G_z = 0. \tag{9.75c}$$

These equations, combined with the constraint condition $G(x, y, z) = 0$, can be solved for the desired geodesic $x(t)$, $y(t)$, $z(t)$, and $\lambda(t)$.

In particular, the geodesic on a sphere $G(x, y, z) = x^2 + y^2 + z^2 - r^2 = 0$ can be determined. In this case, $G_x = 2x$, $G_y = 2y$, and $G_z = 2z$. Let

$$ax + by + cz = 0$$

be an equation of the plane which contains two given points on the sphere and the origin. Let the curve of minimum length be given by

$$p(t) = ax(t) + by(t) + cz(t).$$

Clearly, $p(t)$ satisfies the equation

$$\frac{d}{dt}\frac{\dot{p}(t)}{\sqrt{\dot{x}^2 + \dot{y}^2 + \dot{z}^2}} + 2\lambda(t)p(t) = 0,$$

and $p(t_1) = p(t_2) = 0$. It follows from the uniqueness of solutions of differential equations that $p(t) = 0$. Therefore, the geodesic lies on the plane and is a segment of a great circle of the sphere.   □

The variational problem can readily be extended for functionals depending on functions of several variables. Many physical problems require us to determine a function of several variables which will give rise to an extremum of such functionals. The following two-variable case will serve as an illustration of such a problem. The associated functional has the form

$$I[u(x, y)] = \iint_\Omega F(x, y, u, u_x, u_y)dx\,dy, \tag{9.76}$$

where u is a function of two real variables. We want to find the function u that extremizes I. The values of u are prescribed on the boundary $\partial\Omega$ of Ω. The function F is defined on the domain Ω of x and y, and is assumed to have continuous second-order partial derivatives.

In accordance with the preceding case of functionals depending on a function of one independent variable, the first variation δI of I is defined by

$$\delta I(u, \varepsilon) = I(u + \varepsilon) - I(u).$$

By Taylor's expansion theorem, this reduces to

$$\delta I = \iint_\Omega (\varepsilon F_u + \varepsilon_x F_p + \varepsilon_y F_q)dx\,dy, \tag{9.77}$$

where $\varepsilon = \varepsilon(x, y)$ is small, $p = u_x$, and $q = u_y$.

A necessary condition for the functional I to have an extremum is that the first variation of I vanishes, that is,

$$0 = \delta I = \iint_\Omega (\varepsilon F_u + \varepsilon_x F_p + \varepsilon_y F_q)dx\,dy$$

$$= \iint_\Omega \varepsilon\left(F_u - \frac{\partial}{\partial x}F_p - \frac{\partial}{\partial y}F_q\right)dx\,dy$$

$$+ \iint_{\Omega} \left[\varepsilon \left(\frac{\partial}{\partial x} F_p + \frac{\partial}{\partial y} F_q \right) + (\varepsilon_x F_p + \varepsilon_y F_q) \right] dx\, dy$$

$$= \iint_{\Omega} \varepsilon \left(F_u - \frac{\partial}{\partial x} F_p - \frac{\partial}{\partial y} F_q \right) dx\, dy$$

$$+ \iint_{\Omega} \left[\frac{\partial}{\partial x} (\varepsilon F_p) + \frac{\partial}{\partial y} (\varepsilon F_q) \right] dx\, dy. \tag{9.78}$$

We assume that the boundary curve $\partial \Omega$ has a piecewise continuously turning tangent so that Green's theorem can be applied to the second double integral in (9.78). Consequently, (9.78) becomes

$$0 = \delta I = \iint_{\Omega} \varepsilon \left(F_u - \frac{\partial}{\partial x} F_p - \frac{\partial}{\partial y} F_q \right) dx\, dy + \int_{\partial \Omega} \varepsilon (F_p\, dy - F_q\, dx). \tag{9.79}$$

Since ε is assumed to vanish on the curve $\partial \Omega$, the second integral in (9.79) equals 0. Moreover, since ε is otherwise arbitrary, it follows that the integrand of the first integral in (9.79) must vanish identically. Consequently, the extremizing function $u(x, y)$ satisfies

$$F_u - \frac{\partial}{\partial x} F_p - \frac{\partial}{\partial y} F_q = 0. \tag{9.80}$$

This is the Euler–Lagrange partial differential equation for the variational problem involving two independent variables. A necessary condition for an extremum of $I[u(x, y)]$ is that u satisfies this equation. If the values of u are prescribed on the boundary $\partial \Omega$, the solution of the Euler–Lagrange equation which satisfies the given boundary conditions must be found.

Example 9.3.20. (Vibrating string)   We derive the equation of motion for free vibration of an elastic string of length l and line density ρ.

We assume that, initially, the string is stretched along the x-axis from $x = 0$ to $x = l$. The string will be given a small lateral displacement, which at each point along the x-axis is denoted by $u(x, t)$. The kinetic energy of the string is given by

$$T = \frac{1}{2} \int_0^l \rho u_t^2 \, dx, \tag{9.81}$$

provided longitudinal motion of the string is neglected.

If the displacement of the string is small, the change in length of the string caused by the displacement is given by

$$\int_0^l \sqrt{1 + u_x^2} \, dx - l \sim \int_0^l \left(1 + \frac{1}{2} u_x^2 \right) dx - l = \frac{1}{2} \int_0^l u_x^2 \, dx.$$

Since the potential energy is proportional to the elongation, we have

$$V = \frac{T_1}{2} \int_0^l u_x^2 dx, \tag{9.82}$$

where T_1 is the constant tension of the string.

According to the Hamilton principle,

$$0 = d \int_{t_1}^{t_2} L \, dt = d \int_{t_1}^{t_2} (T - V) \, dt = d \int_{t_1}^{t_2} \frac{1}{2} \int_0^l \left(\rho u_t^2 - T_1 u_x^2 \right) dx \, dt. \tag{9.83}$$

In this case, $L = \frac{1}{2}(\rho u_t^2 - T_1 u_x^2)$, which does not depend explicitly on x, t, or u, and hence, the above result has the form

$$d \int_{t_1}^{t_2} \int_0^l L(u_t, u_x) \, dx \, dt = 0.$$

Thus, the Euler–Lagrange equation is

$$\frac{\partial}{\partial t}(\rho u_t) - \frac{\partial}{\partial x}(T_1 u_x) = 0$$

or

$$u_{tt} - c^2 u_{xx} = 0, \tag{9.84}$$

where $c^2 = T_1/\rho$. This is the familiar partial differential equation for the vibrating string. □

9.4 Minimization of Quadratic Functionals

Suppose A is a real symmetric positive definite operator defined on a Hilbert space H. We study the variational formulation of the operator equation

$$Au = f, \tag{9.85}$$

where f is a given element of H. The solution of (9.85), if it exists, is unique due to the fact that the operator A is positive definite. Moreover, the solution of (9.85) can be shown to be the function that minimizes the quadratic functional

$$I(u) = \langle Au, u \rangle - 2\langle f, u \rangle. \tag{9.86}$$

Conversely, if we can find a solution u that minimizes $I(u)$ on H, then u is the desired solution of (9.85). This fundamental result is stated in the following theorem.

Theorem 9.4.1. *Suppose $A:H \to H$ is a linear symmetric positive definite operator on a real Hilbert space H and f is a given element of H. Then the quadratic functional $I(u) = \langle Au, u \rangle - 2\langle f, u \rangle$ attains its minimum value for some $u_0 \in H$ if and only if u_0 is the solution of the equation $Au = f$.*

Proof: Suppose u_0 is the solution of (9.85). Let u be any element of H. Then

$$I(u) - I(u_0) = \langle Au, u \rangle - 2\langle f, u \rangle - \langle Au_0, u_0 \rangle + 2\langle f, u_0 \rangle$$

$$= \langle Au, u \rangle - 2\langle Au_0, u \rangle + \langle Au_0, u_0 \rangle$$

$$= \langle A(u - u_0), u - u_0 \rangle.$$

Since A is a positive definite operator, $\langle A(u - u_0), u - u_0 \rangle \geq 0$, where the equality holds if and only if $u = u_0$. Hence,

$$I(u) \geq I(u_0).$$

This shows that $I(u)$ attains its minimum at the solution u_0 of (9.85).

Conversely, suppose u_0 is an element of H which minimizes $I(u)$, that is, $I(u) \geq I(u_0)$ for all $u \in H$. In particular,

$$I(u_0 + tv) \geq I(u_0)$$

for any real number t and any $v \in H$. More explicitly,

$$I(u_0 + tv) = \langle A(u_0 + tv), u_0 + tv \rangle - 2\langle f, u_0 + tv \rangle$$

$$= \langle Au_0, u_0 \rangle + 2t\langle Au_0, v \rangle + t^2 \langle Av, v \rangle - 2\langle f, u_0 \rangle - 2t\langle f, v \rangle,$$

or

$$\frac{I(u_0 + tv) - I(u_0)}{t} = 2\langle Au_0 - f, v \rangle + t\langle Av, v \rangle.$$

In the limit $t \to 0$, this expression leads to the Gateaux differential

$$dI(u_0, v) = 2\langle Au_0 - f, v \rangle$$

for any $v \in H$. Since $I(u)$ has a local minimum at $u = u_0$, $dI(u_0, v) = 0$ for all $v \in H$. This implies that $Au_0 - f = 0$, which proves that u_0 is a solution of (9.85). □

This minimization problem can be interpreted as a maximization problem for $-I(u)$. Theorem 9.4.1 can be generalized to a symmetric positive definite operator A (not necessarily real) on a complex Hilbert space H. The functional I is then modified to

$$I(u) = \langle Au, u \rangle - \langle f, u \rangle - \langle u, f \rangle.$$

Note that Theorem 9.4.1 can be interpreted physically as a minimum energy principle.

Example 9.4.2. Suppose $A:H \to K$ is a symmetric bounded linear operator, where H and K are real Hilbert spaces. We want to minimize

$$I(u) = \|Au - b\|^2,$$

where $u \in H$ and $b \in K$.

We have

$$I(u) = \langle Au - b, Au - b \rangle = \langle Au, Au \rangle - 2\langle b, Au \rangle + \langle b, b \rangle$$
$$= \langle A^*Au, u \rangle - 2\langle A^*b, u \rangle + \langle b, b \rangle,$$

and thus,

$$I(u + h) - I(u) = \langle 2A^*Au - 2A^*b, h \rangle + \langle A^*Ah, h \rangle.$$

Clearly, $I(u)$ has the first differential given by

$$I'(u) = 2A^*Au - 2A^*b.$$

Thus,

$$\langle I'(u + h), k \rangle - \langle I'(u), k \rangle = \langle 2A^*Ah, k \rangle.$$

Hence, the second differential is

$$I''(u)(h, k) = \langle 2A^*Ah, k \rangle,$$

which is independent of u.

So I has an extremum at $u = u_0$ if $I'(u_0) = 0$, that is, $A^*Au_0 = A^*b$. In particular, if A^*A is a positive operator, then $I''(u) \geq 0$, and u_0 given by the above equation is a local minimum of $I(u)$. $\qquad\square$

9.5 Variational Inequalities

Most partial differential equations in physical and engineering sciences arise from a variational principle. Usually, there is a class of admissible solutions and an energy functional associated with these admissible functions. We seek to minimize the energy to determine the solution of the problem which satisfies the Euler–Lagrange equation. Many physical problems can be expressed in terms of an unknown function u, for example, displacement of a mechanical system, satisfying an inequality

$$a(u, v - u) \geq F(v - u) \quad \text{for all } v \in S, \tag{9.87}$$

where the set S of admissible functions is a closed convex subset of a Hilbert space H, $a(\cdot, \cdot)$ is a bilinear form, and F is a bounded linear functional on S. Such inequalities are called *variational inequalities*. So the existence and uniqueness of the solutions of this type of inequality are of special interest. Under suitable conditions, the problem of the variational inequality is found to be equivalent to the minimization problem. Find an element u such that

$$u \in S \quad \text{and} \quad I(u) = \inf_{v \in S} I(v),$$

where the functional $I : S \to \mathbb{R}$ is defined by

$$I(v) = \frac{1}{2} a(v, v) - F(v). \tag{9.88}$$

The problem of finding $u \in S$ is called an *abstract minimization problem*.

The problem of finding $u \in S$ such that

$$a(u, v - u) \geq F(v - u) \tag{9.89}$$

for all $v \in S$ is called a *variational inequality problem*, and u is called its *solution*.

The following theorem gives the relation between the abstract minimization problem and the variational inequality problem.

Theorem 9.5.1. *Suppose $a(\cdot, \cdot)$ is a continuous, symmetric, and elliptic bilinear form on a Hilbert space H, $f \in H$, and $S \subset H$ is a closed convex subset. Then there exists a unique $u \in S$ such that*

$$a(u, v - u) \geq \langle f, v - u \rangle \quad \text{for each } v \in S. \tag{9.90}$$

Further, u can be characterized by

$$u \in S, \tag{9.91a}$$

$$I(u) = \min_{v \in S} I(v), \tag{9.91b}$$

where

$$I(v) = \frac{1}{2} a(v, v) - \langle f, v \rangle. \tag{9.92}$$

Proof: Since $a(u, v)$ is a symmetric and elliptic bilinear form on H, it defines an inner product $\langle \cdot, \cdot \rangle_a$ and a norm $\| \cdot \|_a$ on H:

$$\langle u, v \rangle_a = a(u, v), \tag{9.93a}$$

$$\|u\|_a = \sqrt{a(u, u)}. \tag{9.93b}$$

By the continuity and ellipticity of $a(u, v)$, we have

$$K \|u\|^2 \leq a(u, v) \leq M \|u\|^2, \tag{9.94}$$

and so the norm $\| \cdot \|_a$ is equivalent to the original norm $\| \cdot \|$ on H. Consequently, H is a Hilbert space with respect to the new inner product. Now, by the Riesz Representation Theorem 3.7.7, there exists $\hat{f} \in H$ such that, for every $v \in H$,

$$a(\hat{f}, v) = \langle f, v \rangle. \tag{9.95}$$

Now,

$$\frac{1}{2} \| v - \hat{f} \|_a^2 = \frac{1}{2} a(v - \hat{f}, v - \hat{f})$$

$$= \frac{1}{2} a(v, v) - a(v, \hat{f}) + \frac{1}{2} a(\hat{f}, \hat{f})$$

$$= \frac{1}{2} a(v, v) - \langle f, v \rangle + \frac{1}{2} \| \hat{f} \|_a^2 \quad \text{(by (9.95))}$$

$$= I(v) + \frac{1}{2} \| \hat{f} \|_a^2.$$

Since $\| \hat{f} \|_a^2$ is a constant, minimizing $I(v)$ over S is equivalent to minimizing $\| v - \hat{f} \|_a^2$ over S. So, by Theorem 3.6.5, there exists a unique $u \in S$ such that

$$\left\langle \hat{f} - u, v - u \right\rangle_a \leq 0 \quad \text{for every } v \in S,$$

or

$$\left\langle \hat{f}, v - u \right\rangle_a \leq \langle u, u - v \rangle_a \quad \text{for every } v \in S,$$

or, finally,

$$a(u, u - v) \geq \langle f, v - u \rangle \quad \text{for every } v \in S. \qquad \square$$

In the following theorem, we prove existence and uniqueness of the solution of the variational inequality (9.90) without assuming that the form $a(\cdot, \cdot)$ is symmetric.

Theorem 9.5.2. (Lions–Stampacchia) *Suppose H is a Hilbert space, $a(\cdot, \cdot)$ is a continuous and elliptic bilinear form on H, and S is a closed convex subset of H. Then the variational inequality (9.90) has a unique solution.*

Proof: For a fixed $u \in H$, we consider a map $v \to a(u, v)$. By the continuity of $a(\cdot, \cdot)$, this is a continuous linear functional. Hence, there exists $Au \in H$ such that

$$\langle Au, v \rangle = a(u, v) \quad \text{for every } v \in H. \tag{9.96}$$

Since $a(u, v)$ is bilinear, the map $u \to Au$ is linear. Further, it follows from continuity of $a(u, v)$ that

$$\|Au\| \leq M \|u\|$$

for some $M > 0$. Hence A is continuous. Moreover, since $a(u, v)$ is elliptic, we have

$$\langle Au, u \rangle \geq K \|u\|^2$$

for some $K > 0$.

With this new notation, we seek $u \in S$ such that

$$\langle Au, v - u \rangle \geq \langle f, v - u \rangle \quad \text{for every } v \subset S. \tag{9.97}$$

Let $\alpha > 0$ be a positive constant to be chosen shortly. Then (9.97) is equivalent to

$$\langle \alpha f - \alpha Au + u - u, v - u \rangle \leq 0 \quad \text{for every } v \in S.$$

In other words, by Theorem 3.6.5, we seek $u \in S$ such that

$$u = P(\alpha f - \alpha Au + u),$$

where P is the projection on S. Hence, we look for a fixed point of the continuous map $F : H \to H$, whose range lies in S, defined by

$$F(v) = P(\alpha f - \alpha Av + v).$$

Now, if $v_1, v_2 \in H$, we have

$$\left\| F(v_1) - F(v_2) \right\|^2 = \left\| P(\alpha f - \alpha Av_1 + v_1) - P(\alpha f - \alpha Av_2 + v_2) \right\|^2$$
$$\leq \left\| (v_1 - v_2) - \alpha A(v_1 - v_2) \right\|^2.$$

Thus,

$$\left\| F(v_1) - F(v_2) \right\|^2 \leq \|v_1 - v_2\|^2 - 2\alpha \langle a(v_1 - v_2), (v_1 - v_2) \rangle$$
$$+ \alpha^2 \left\| A(v_1 - v_2) \right\|^2$$
$$\leq \left(1 - 2\alpha K + \alpha^2 M^2 \right) \|v_1 - v_2\|^2.$$

We next choose α such that $0 < \alpha < 2K/M^2$; then $\beta^2 = 1 - 2\alpha K + \alpha^2 M^2 < 1$. Clearly,

$$\left\| F(v_1) - F(v_2) \right\| \leq \beta \|v_1 - v_2\|$$

with $0 < \beta < 1$, and so F is a contraction. By the Banach Fixed Point Theorem (Theorem 1.6.4), F has a unique fixed point u which must belong to S. $\qquad \square$

9.6 Optimal Control Problems for Dynamical Systems

The theory of optimal control is one of the major areas of applications of Hilbert space methods. A large class of optimal control problems can be formulated as variational or optimizational problems. Indeed, optimal control problems are a special kind of optimization problem where the constraints specify the dynamics of the system described by differential equations.

A control problem involves, first of all, a dynamical system—an ordinary differential equation with time as the independent variable:

$$\frac{dx}{dt} = f(t, x(t), u(t)), \quad 0 \le t \le T, \qquad x(0) = x_0, \qquad (9.98)$$

where $x(t)$, a function with a range in a Euclidean space, is called the *state function*, and $u(t)$ is referred to as the *control function* and is assumed to be in some specified set of admissible controls. The control problem is to find a control function $u(t)$ which minimizes an "index of performance" or "cost functional"

$$I(u) = \int_0^T l(x, u, t)dt + m(x(T)), \qquad (9.99)$$

subject to constraints on the control such as $u(t)$ belongs to a convex set Ω, and $x(T)$ is in a given set Γ, where l and m are given functions.

The control problem is called *time invariant* if f and l are independent of t. The control problem is called *linear-quadratic* if the cost functional $I(u)$ is quadratic, that is, l and m are quadratic in x and u, and the differential equation (9.98) is linear.

A control that attains the minimum of performance index in the admissible class of controls is known as *optimal control*, and the corresponding $x(t)$ as an *optimal trajectory*. The upshot of the theory is that under certain conditions an optimal control must satisfy the maximum principle of Pontrjagin.

Suppose $I = (t_1, t_2) \subset \mathbb{R}$ is an interval, $A = [\alpha_{ij}]$ is a constant $N \times N$ matrix, and $\mathbf{A}(t) = [a_{ij}(t)]$ is an $N \times N$ matrix with elements $a_{ij}(t)$, which are continuous functions on the interval I. Let $\mathbf{x}^T = (x_1, \ldots, x_N)$ be an N dimensional vector, and let $\mathbf{v}^T(t) = (v_1(t), \ldots, v_N(t))$ be an N dimensional vector with components $v_k(t)$, which are piecewise continuous on I. We put

$$D = \{(t, \mathbf{x}) : t \in I, \ \mathbf{x} \in \mathbb{R}^N \ (\text{or } \mathbb{C}^N)\}.$$

With this notation, we consider linear systems of first order ordinary differential equations

$$\frac{d\mathbf{x}}{dt} = \mathbf{A}(t)\mathbf{x} + \mathbf{v}(t), \qquad (9.100)$$

$$\frac{d\mathbf{x}}{dt} = \mathbf{A}(t)\mathbf{x}, \qquad (9.101)$$

and

$$\frac{d\mathbf{x}}{dt} = A\mathbf{x}. \qquad (9.102)$$

According to the general theory of ordinary differential equations, systems (9.100), (9.101), and (9.102) possess unique solutions for every $(t, \mathbf{x}) \in D$ that exist over the entire interval I and that depend continuously on the initial data.

A set on N linearly independent solutions of (9.101) on I is called the *fundamental set of solutions*. An $N \times N$ matrix $\mathbf{\Phi}$ whose N columns are linearly independent solutions of (9.101) on I is called a *fundamental matrix*.

If $\{\mathbf{\Phi}_1, \ldots, \mathbf{\Phi}_N\}$ is a set of N linearly independent solutions of (9.101) and if $(\mathbf{\Phi}_k)^T = \{\mathbf{\Phi}_{1k}, \ldots, \mathbf{\Phi}_{Nk}\}$, then

$$\mathbf{\Phi} = \begin{bmatrix} \mathbf{\Phi}_{11} & \mathbf{\Phi}_{12} & \cdots & \mathbf{\Phi}_{1N} \\ \mathbf{\Phi}_{21} & \mathbf{\Phi}_{22} & \cdots & \mathbf{\Phi}_{2N} \\ \vdots & \vdots & \vdots & \vdots \\ \mathbf{\Phi}_{N1} & \mathbf{\Phi}_{N2} & \cdots & \mathbf{\Phi}_{NN} \end{bmatrix}$$

is a fundamental matrix.

We use the natural basis for the $\mathbf{x}$-space:

$$[u_1, \ldots, u_N] = \begin{bmatrix} 1 & 0 & \cdots & 0 \\ 0 & 1 & \cdots & 0 \\ \vdots & \vdots & \vdots & \vdots \\ 0 & 0 & \cdots & 1 \end{bmatrix}.$$

A fundamental matrix $\mathbf{\Phi}$ for equation (9.101), whose columns are determined by the linearly independent solutions $\mathbf{\Phi}(\tau) = u_k$, $k = 1, \ldots, N$, $\tau \in I$, is called the *state transition matrix* $\mathbf{\Phi}$ of equation (9.101).

We now formulate a linear-quadratic optimal problem described by the dynamical equation for the state as

$$\frac{d\mathbf{x}}{dt} = \mathbf{A}\mathbf{x}(t) + \mathbf{B}\mathbf{u}(t), \qquad \mathbf{x}(0) = \mathbf{x}_0, \tag{9.103}$$

where $\mathbf{x}(t) \in \mathbb{R}^N$ is the state vector, $\mathbf{u}(t) \in \mathbb{R}^M$ is the control vector, $t \in [0, T]$, and $\mathbf{B}$ is an $N \times M$ matrix.

According to the theory of ordinary differential equations, if every element of the vector $\mathbf{u}(t)$ is a continuous function of t, then the unique solution of (9.103) is given by

$$\mathbf{x}(t) = \mathbf{\Phi}(t, 0)\mathbf{x}(0) + \int_0^t \mathbf{\Phi}(t, \tau)\mathbf{B}\mathbf{u}(\tau)d\tau, \tag{9.104}$$

where $\mathbf{\Phi}(t, \tau)$ is the state transition matrix for the dynamical system (9.103).

We next define the class of vector valued functions

$$L_M^2([0, T]) = \left\{ \mathbf{u} \colon \mathbf{u}^T = (u_1, \ldots, u_M), \ u_k \in L^2([0, T]), \ k = 1, \ldots, M \right\}.$$

Clearly, $L^2_M([0, T])$ is a vector space. We introduce an inner product in $L^2_M([0, T])$ by

$$\langle \mathbf{u}, \mathbf{v} \rangle = \int_0^T \mathbf{u}^T(t)\mathbf{v}(t)dt. \tag{9.105}$$

Since $L^2_M([0, T])$ is complete, it is a Hilbert space (see Exercise 8). Next we define a linear operator $L : L^2_M([0, T]) \to L^2_N([0, T])$ by

$$[L\mathbf{u}](t) = \int_0^t \mathbf{\Phi}(t, \tau)\mathbf{B}\mathbf{u}(\tau)d\tau. \tag{9.106}$$

Since the elements of $\mathbf{\Phi}(t, \tau)$ are continuous functions on $[0, T] \times [0, T]$, L is a compact operator.

The linear-quadratic control problem is to find $\mathbf{u}(t) \in L^2_M([0, T])$, which minimizes the cost functional

$$I(\mathbf{u}) = \int_0^T \mathbf{x}^T(t)\mathbf{x}(t)dt + \alpha \int_0^T \mathbf{u}^T(t)\mathbf{u}(t)dt, \tag{9.107}$$

where α is a positive constant which, without loss of generality, can be taken to be 1. If we replace $-\mathbf{\Phi}(t, 0)\mathbf{x}_0$ by $\mathbf{v}(t)$ in (9.104) and rewrite the resulting equation in terms of the operator L, we obtain $\mathbf{x}(t) = L\mathbf{u} - \mathbf{v}$. Thus, the control problem reduces to a minimization problem in a Hilbert space with the cost functional

$$I(\mathbf{u}) = \langle L\mathbf{u} - \mathbf{v}, L\mathbf{u} - \mathbf{v} \rangle + \alpha \langle \mathbf{u}, \mathbf{u} \rangle. \tag{9.108}$$

The required minimizing $\mathbf{u}(t)$ can be determined from the following general result.

Theorem 9.6.1. *Suppose H and K are real Hilbert spaces and $L : H \to K$ is a compact operator with L^* as its adjoint. Let v be a given fixed element in K and let $\alpha \in \mathbb{R}$. Define a functional $I : H \to \mathbb{R}$ by*

$$I(u) = \|Lu - v\|^2 + \alpha\|u\|^2, \tag{9.109}$$

where $\| \cdot \|$ is the norm in K. If $\alpha > 0$, then there exists a unique $u_0 \in H$ such that $I(u_0) \leq I(u)$ for all $u \in H$. Moreover, u_0 is a solution of the equation

$$L^*Lu_0 + \alpha u_0 = L^*v. \tag{9.110}$$

Proof: Let $A = L^*L$. Then A is a positive compact operator. Therefore, $-\alpha$ cannot be an eigenvalue of L^*L. It can be shown that the equation $Ax - \lambda x = y$ has a unique solution provided α is not an eigenvalue of A and $\alpha \neq 0$. The solution is given by

$$x = \frac{P_0 y}{-\lambda} + \sum_{n=1}^{\infty} \frac{P_n y}{\lambda_n - \lambda}, \tag{9.111}$$

where $\lambda_1, \lambda_2, \ldots$ are nonzero distinct eigenvalues of A, P_n is the projection onto the null space of $A - \lambda_n \mathcal{I}$, and P_0 is the projection onto the null space of A.

Suppose $u_0 \in H$ is the unique solution of (9.110). It follows from (9.109) that, for an arbitrary $h \in H$, we have

$$
\begin{aligned}
I(u_0 + h) &= \langle Lu_0 + Lh - v, Lu_0 + Lh - v \rangle + \alpha \langle u_0 + h, u_0 + h \rangle \\
&= \langle Lu_0 - v, Lu_0 - v \rangle + 2\langle Lh, Lh_0 - v \rangle \\
&\quad + \langle v, v \rangle + \alpha \langle u_0, u_0 \rangle + 2\alpha \langle u_0, h \rangle + \alpha \langle h, h \rangle \\
&= \langle Lu_0 - v, Lu_0 - v \rangle + \langle v, v \rangle + \alpha \langle u_0, u_0 \rangle \\
&\quad + 2\alpha \langle h, L^* Lu_0 + \alpha u_0 - L^* v \rangle + \alpha \langle h, h \rangle \\
&= \| Lu_0 - v \|^2 + \| v \|^2 + \alpha \| u_0 \|^2 + \alpha \| h \|^2. \qquad (9.112)
\end{aligned}
$$

Hence, $I(u_0 + h)$ is minimum if and only if $h = 0$. $\qquad\square$

The solution of (9.110) can be obtained from (9.111). However, a more convenient technique is available for determination of the solution when L is given by (9.106). The method is formulated in the following theorem.

Theorem 9.6.2. *Suppose* $I(\mathbf{u})$ *is given by*

$$
I(\mathbf{u}) = \int_0^T \mathbf{x}^T(t)\mathbf{x}(t)dt + \alpha \int_0^T \mathbf{u}^T(t)\mathbf{u}(t)dt,
$$

where $\alpha > 0$. *Suppose* $\mathbf{x}(t)$ *satisfies the dynamical system* (9.103):

$$
\frac{d\mathbf{x}}{dt} = \mathbf{A}\mathbf{x}(t) + \mathbf{B}\mathbf{u}(t), \qquad \mathbf{x}(0) = \mathbf{x}_0.
$$

If the control vector

$$
\mathbf{u}(t) = -\frac{1}{\alpha} \mathbf{B}^T \mathbf{P}(t)\mathbf{x}(t), \qquad (9.113)
$$

for all $t \in [0, T]$, *and* $\mathbf{P}(t)$ *is the solution of the matrix Riccati equation*

$$
\dot{\mathbf{P}}(t) = \mathbf{A}^T \mathbf{P}(t) + \mathbf{P}(t)\mathbf{A} - \frac{1}{\alpha} \mathbf{P}(t)\mathbf{B}\mathbf{B}^T \mathbf{P}(t) + I = 0 \qquad (9.114)
$$

with $\mathbf{P}(T) = 0$, *then* $\mathbf{u}(t)$ *minimizes* $I(\mathbf{u})$.

Proof: We will show that $\mathbf{u}(t)$ satisfies (9.110), where $L\mathbf{u}$ is given by (9.106). If $\mathbf{u}(t)$ satisfies (9.110), then

$$
\mathbf{u} = -\frac{1}{\alpha} L^*(L\mathbf{u} - \mathbf{v}) = -\frac{1}{\alpha} L^* \mathbf{x}.
$$

We next find a formula for evaluating $L^*\mathbf{w}$ for an arbitrary $\mathbf{w} \in L^2_M([0, T])$. To do this, we calculate

$$\langle L\mathbf{u}, \mathbf{w} \rangle = \int_0^T \left[\int_0^s \mathbf{\Phi}(s, t)\mathbf{B}\mathbf{u}(t)dt \right]^T \mathbf{w}(s)ds$$

$$= \int_0^T \int_0^s \mathbf{u}^T(t)\mathbf{B}^T\mathbf{\Phi}^T(s, t)\mathbf{w}(s)dtds$$

$$= \int_0^T \mathbf{u}^T(t) \left[\int_t^T \mathbf{B}^T\mathbf{\Phi}^T(s, t)\mathbf{w}(s)ds \right]dt = \langle \mathbf{u}, L^*\mathbf{w} \rangle. \quad (9.115)$$

Consequently,

$$[L^*\mathbf{w}](t) = \int_t^T \mathbf{B}^t\mathbf{\Phi}^t(s, t)\mathbf{w}(s)ds$$

for all $t \in [0, T]$.

We next assume that there exists a matrix $\mathbf{P}(t)$ such that

$$\mathbf{P}(t)\mathbf{x}(t) = \int_t^T \mathbf{\Phi}^T(s, t)\mathbf{x}(s)ds \quad (9.116)$$

with $\mathbf{P}(T) = 0$.

It is necessary to find conditions for existence of such a matrix $\mathbf{P}(t)$. Differentiating (9.116) with respect to t and using the equality

$$\dot{\mathbf{P}}^T(s, t) = -\mathbf{A}^T\mathbf{\Phi}(s, t),$$

we obtain

$$\dot{\mathbf{P}}(t)\mathbf{x}(t) + \mathbf{P}(t)\dot{\mathbf{x}}(t) = -\mathbf{x}(t) - \mathbf{A}^T \int_t^T \mathbf{\Phi}^T(s, t)\mathbf{x}(s)ds$$

$$= -\mathbf{x}(t) - \mathbf{A}^T\mathbf{P}(t)\mathbf{x}(t). \quad (9.117)$$

In view of (9.103), this equation becomes

$$\dot{\mathbf{P}}(t)\mathbf{x}(t) + \mathbf{P}(t)\left[\mathbf{A}\mathbf{x}(t) + \mathbf{B}\mathbf{u}(t)\right] = -\mathbf{x}(t) - \mathbf{A}^t\mathbf{P}(t)\mathbf{x}(t).$$

We next use (9.113) to replace $\mathbf{u}(t)$, so that the above equation becomes

$$\dot{\mathbf{P}}(t)\mathbf{x}(t) + \mathbf{P}(t)\mathbf{A}\mathbf{x}(t) - \frac{1}{\alpha}\mathbf{P}(t)\mathbf{B}\mathbf{B}^T\mathbf{P}(t)\mathbf{x}(t) + \mathbf{x}(t)I + \mathbf{A}^T\mathbf{P}(t)\mathbf{x}(t) = 0.$$

Clearly, $\mathbf{P}(t)$ satisfies (9.114) with $\mathbf{P}(T) = 0$.

If

$$\mathbf{u}(t) = -\frac{1}{\alpha}\mathbf{B}^T\mathbf{P}(t)\mathbf{x}(t),$$

it turns out that $\mathbf{u}(t)$ satisfies

$$L^*L\mathbf{u} + \alpha\mathbf{u} = L^*\mathbf{v},$$

where $\mathbf{v} = -\mathbf{\Phi}(t, 0)\mathbf{x}_0$, and hence by Theorem 9.6.1, $\mathbf{u}(t)$ minimizes the cost functional $I(\mathbf{u})$.   □

The matrix Riccati equation (9.114) is often called the *state equation* of the linear-quadratic control problem, and it can be shown that this equation has a unique solution for all $t < T$.

Example 9.6.3. Find the control u which minimizes the cost functional

$$I(u) = \int_0^T \left(x^2 + u^2\right)dt,$$

where $x(t)$ satisfies $\dot{x}(t) = u(t)$ with $x(0) = x_0$. In this problem, $A = 0$, $B = 1$, and $\alpha = 1$. Hence, (9.114) becomes

$$\dot{P}(t) - P^2(t) + 1 = 0, \qquad P(T) = 0.$$

The solution of this equation is

$$P(t) = \left(1 - e^{2(t-T)}\right)/\left(1 + e^{2(t-T)}\right).$$

Thus, this optimal control $u(t)$ is obtained from (9.113) in the form $u(t) = -P(t)x(t)$, where $x(t)$ can be solved from the given equation, $\dot{x}(t) = u(t)$.   □

Example 9.6.4. (Control of chemical reaction with nonlinear cost)   Consider a chemical mixture which is added to a tank at a constant rate for a fixed time interval $[0, T]$. Suppose that the pH value x at which the reaction occurs determines the quality of the final product and that this pH value can be controlled by the strength $u(t)$ of some component of the mixture. Suppose the chemical reaction takes place so that the rate of change of pH value $x(t)$ is proportional to the sum of the current pH value and the strength $u(t)$ of the controlling component, that is,

$$\dot{x} = ax + bu,$$

where a and b are known positive constants. Assume that the decrease in yield due to variation of the pH value is $\int_0^T x^2 dt$ and the rate of cost of maintaining the strength $u(t)$ is proportional to u^2. Then the total cost associated with the control function $u(t)$ in $[0, T]$ is

$$I(u) = \int_0^T \left(cx^2 + u^2\right)dt,$$

where $c > 0$ is an appropriate constant. Suppose the initial pH is given, that is, $x(0) = x_0$. We seek a control function $u^*(t)$ on $[0, T]$, which determines a response $x^*(t)$ so that the cost functional is minimum. This is an example of a linear-quadratic optimal control problem. $\square$

9.7 Approximation Theory

Typical problems in approximation theory deal with the determination of an element from a given set of elements that is closest to a prescribed element not in the given set. Such a closest point may or may not exist in general. More precisely, given an element x and a set S in a normed linear space E, an element y of S is said to be a *best approximation* to x from S if, for all $z \in S$,

$$\|x - y\| \leq \|x - z\|. \tag{9.118}$$

The problem of determining such an element y is usually called a *best approximation problem*. One of the most common and remarkable examples of such a problem is the approximation of continuous functions by polynomials.

In the language of the optimization theory, the closest element y is called an *optimal solution* of the problem:

$$\text{Minimize } \|x - z\|,$$

where $x \in E$ and $z \in S \subset E$. The element $x - y$ is called an *optimal error*.

The theory of approximation is essentially concerned with the following four basic problems:

(a) Existence of best approximations.

(b) Uniqueness of best approximations.

(c) Characterization of best approximations.

(d) Methods for determining the best approximation.

A basic existence Theorem 3.6.4 was proved in Section 3.6. Of primary interest here are the uniqueness and characterization of best approximations.

Theorem 9.7.1. (Uniqueness of best approximation) *Let S be a subspace of a Hilbert space H, and $x \in H$. If a best approximation of x from S exists, then it is unique. A necessary and sufficient condition that $y \in S$ be the unique best approximation to x from S is that the optimal error $x - y$ be orthogonal to S.*

Proof: Let $y \in S$ be a best approximation to x from S. To prove $(x - y) \perp S$, we consider $z \in S$ with $\|z\| = 1$. Then $w = y + \langle x - y, z \rangle z \in S$. We have

$$\|x - y\|^2 \leq \|x - w\|^2 \tag{9.119}$$

$$= \langle x - w, x - w \rangle$$

$$= \langle (x - y) - \langle x - y, z \rangle z, (x - y) - \langle x - y, z \rangle z \rangle$$
$$= \langle x - y, x - y \rangle - |\langle x - y, z \rangle|^2$$
$$= \|x - y\|^2 - |\langle x - y, z \rangle|^2. \tag{9.120}$$

Consequently,

$$|\langle x - y, z \rangle|^2 = 0,$$

that is, $\langle x - y, z \rangle = 0$ or $(x - y) \perp z$.

Conversely, let $y \in S$ and $(x - y) \perp S$. Then, for any $z \in S$, we have $y = z \in S$ so that $(x - y) \perp (y - z)$. Hence, by Pythagorean Formula (3.8),

$$\|x - z\|^2 = \|x - y + y - z\|^2 = \|x - y\|^2 + \|y - z\|^2.$$

Thus,

$$\|x - z\| > \|x - y\| \quad \text{if } y \neq z.$$

This shows that y is the unique best approximation to x from S. $\qquad \square$

Note that the element y is the projection of x onto S. If S is a plane through the origin in $\mathbb{R}^3$, the previous theorem states that the optimal error is obtained by dropping a perpendicular from x onto S.

One of the most important problems in approximation theory is the least-squares polynomial approximation in an interval. We first give the characterization of least-squares approximation and then demonstrate the unique importance of orthogonal polynomials in approximation theory. Let Π_n denote the space of all polynomials of a single variable of degree at most n.

It is a consequence of the basic existence theorem (Theorem 3.6.4) that, given a continuous function $f(x)$ on the interval $I = [-1, 1]$ and an integrable *weight function* $w(x)$, which is positive on I, except possibly at a finite number of points of I at which $w(x) = 0$, there exists a unique polynomial $Q_n^*(x) \in \Pi_n$ such that

$$\|f - Q_n^*\|_2 = \left[\int_{-1}^1 \left[f(x) - Q_n^*(x) \right]^2 w(x) dx \right]^{1/2} < \|f - P\|_2 \tag{9.121}$$

for any $P \in \Pi_n$, $P \neq Q_n^*$. This Q_n^* is called the *least-squares approximation* to f out of Π_n, with respect to the weight function $w(x)$.

The following theorem characterizes the least-squares approximation.

Theorem 9.7.2. *If $f \in C(I)$, and w is a weight function in I, then Q_n^* is the least-squares approximation to f out of Π_n if and only if*

$$\int_{-1}^1 \left[f(x) - Q_n^*(x) \right] P(x) w(x) dx = 0 \tag{9.122}$$

for every $P \in \Pi_n$.

Proof: It suffices to use Theorem 9.7.1 with $S = \Pi_n$ and H defined as the space of functions such that

$$\int_{-1}^{1} |f(x)|^2 w(x)dx < \infty$$

with the inner product

$$\langle f, g \rangle = \int_{-1}^{1} f(x)\overline{g(x)}w(x)dx. \tag{9.123}$$

$\square$

To find Q_n^* explicitly, we put $P(x) = x^k$, $k = 0, 1, 2, \ldots, n$, successively in (9.122) to obtain

$$\int_{-1}^{1} x^k Q_n^*(x)w(x)dx = \int_{-1}^{1} x^k f(x)w(x)dx. \tag{9.124}$$

This is a system of $(n+1)$ linear equations for the $(n+1)$ unknown coefficients $Q_n^*(x)$. Suppose we write

$$Q_n^*(x) = a_0 + a_1 x + \cdots + a_n x^n; \tag{9.125}$$

then the system (9.124) may be expressed as

$$\sum_{m=1}^{n} c_{km} a_m = b_k, \quad k = 0, 1, 2, \ldots, n, \tag{9.126}$$

where

$$c_{km} = \int_{-1}^{1} x^{k+m} w(x)dx, \tag{9.127}$$

$$b_k = \int_{-1}^{1} x^k f(x)w(x)dx. \tag{9.128}$$

In principle, the system of normal equations (9.126) can be solved to determine $a_0, a_1, \ldots, a_n$, and hence Q_n^* is determined explicitly. However, when n is large, say $n \geq 6$, there are formidable numerical difficulties in solving the normal system even in the simple case $w(x) = 1$. These computational difficulties can be avoided by noting that $\{1, x, x^2, \ldots, x^n\}$ is not the only set of polynomials that spans Π_n. If $P_0, P_1, \ldots, P_n \in \Pi_n$ are linearly independent, then every $P \in \Pi_n$ has a unique representation of the form

$$P(x) = C_0 P_0 + C_1 P_1 + \cdots + C_n P_n, \tag{9.129}$$

where the set $\{P_0, P_1, \ldots, P_n\}$ can be determined by assuming that this set is orthogonal in $-1 \le x \le 1$ with respect to the given weight function $w(x)$. In other words, the condition of orthogonality with respect to the inner product (9.123) requires

$$\int_{-1}^{1} P_m(x)P_s(x)w(x)dx = \delta_{ms}, \tag{9.130}$$

where $m, s = 0, 1, 2, \ldots, n$.

It is important to point out that $\{P_0, P_1, \ldots, P_n\}$ is a set of *orthonormal* polynomials with respect to $w(x)$, and such an orthonormal set can always be constructed by the Gram–Schmidt orthonormalization process. One of the remarkable features of the orthonormal set is the ability to simplify the least-squares approximation problem. This can be shown as follows:

Suppose the set $P_0, P_1, \ldots, P_n \in \Pi_n$ satisfies (9.130). Then the set is linearly independent. We next suppose

$$Q_n^*(x) = \sum_{k=0}^{n} \lambda_k P_k. \tag{9.131}$$

We substitute $P = P_m$, $m = 0, 1, \ldots, n$, successively in (9.122) to obtain

$$\int_{-1}^{1} P_m(x)Q_n^*(x)w(x)dx = \int_{-1}^{1} P_m(x)f(x)w(x)dx.$$

In view of (9.131), this result reduces to

$$\sum_{k=0}^{n} \lambda_k \int_{-1}^{1} P_m(x)P_k(x)w(x)dx = \int_{-1}^{1} P_m(x)f(x)w(x)dx \tag{9.132}$$

which, together with (9.130), determines λ_m's:

$$\lambda_m = \int_{-1}^{1} P_m(x)f(x)w(x)dx, \quad m = 0, 1, \ldots, n. \tag{9.133}$$

The system of normal equations is completely uncoupled, and the computational difficulties in solving the normal system (9.126) disappear as we choose an orthogonal basis for Π_n.

To obtain a set of polynomials orthonormal in $-1 \le x \le 1$ with respect to $w(x)$, we define

$$\tilde{P}_0(x) = 1, \qquad \tilde{P}_1(x) = x - \frac{\langle 1, x \rangle}{\langle 1, 1 \rangle}. \tag{9.134}$$

Note that $\langle 1, 1 \rangle > 0$ and $\langle \tilde{P}_0, \tilde{P}_1 \rangle = 0$. We then construct an orthogonal set of polynomials $\{\tilde{P}_0, \tilde{P}_1, \ldots, \tilde{P}_k\}$ by mathematical induction. Suppose $\{\tilde{P}_0, \tilde{P}_1, \ldots, \tilde{P}_k\}$ form an orthogonal set with $\tilde{P}_m \in \Pi_m$, $\tilde{P}_m \ne 0$, $m = 0$,

$1, 2, \ldots, k$. We determine a_k and b_k so that

$$\tilde{P}_{k+1}(x) = (x - a_k)\tilde{P}_k(x) - b_k\tilde{P}_{k-1}(x) \tag{9.135}$$

is orthogonal to $\tilde{P}_1, \tilde{P}_2, \ldots, \tilde{P}_2$. Then (9.135) implies that

$$\langle \tilde{P}_{k+1}, \tilde{P}_m \rangle = \langle x\tilde{P}_k, \tilde{P}_m \rangle - a_k\langle \tilde{P}_k, P_m \rangle - b_k\langle \tilde{P}_{k-1}, \tilde{P}_m \rangle.$$

If $m < k - 1$, the inductive hypothesis implies that $\langle \tilde{P}_{k-1}, \tilde{P}_m \rangle = \langle \tilde{P}_k, \tilde{P}_m \rangle = 0$. Furthermore, $\langle x\tilde{P}_k, \tilde{P}_m \rangle = \langle \tilde{P}_k, x\tilde{P}_m \rangle$. In view of the fact that $x\tilde{P}_m \in \Pi_{k-1}$, we can write

$$x\tilde{P}_m = C_0\tilde{P}_0 + C_1\tilde{P}_1 + \cdots + C_{k-1}\tilde{P}_{k-1},$$

and this leads to $\langle \tilde{P}_k, x\tilde{P}_m \rangle = 0$ and hence $\langle \tilde{P}_{k+1}, \tilde{P}_m \rangle = 0$, $m = 0, 1, \ldots, k - 2$. Also,

$$\langle \tilde{P}_{k+1}, \tilde{P}_k \rangle = \langle x\tilde{P}_k, \tilde{P}_k \rangle - a_k\langle \tilde{P}_k, \tilde{P}_k \rangle.$$

If we choose

$$a_k = \frac{\langle x\tilde{P}_k, \tilde{P}_k \rangle}{\langle \tilde{P}_k, \tilde{P}_k \rangle}, \tag{9.136}$$

then $\langle \tilde{P}_{k+1}, \tilde{P}_k \rangle = 0$.

On the other hand,

$$\langle \tilde{P}_{k+1}, \tilde{P}_{k-1} \rangle = \langle x\tilde{P}_k, P_{k-1} \rangle - b_k\langle \tilde{P}_{k-1}, \tilde{P}_{k-1} \rangle.$$

If we choose

$$b_k = \frac{\langle x\tilde{P}_k, \tilde{P}_{k-1} \rangle}{\langle \tilde{P}_{k-1}, \tilde{P}_{k-1} \rangle}, \tag{9.137}$$

then $\langle \tilde{P}_{k+1}, \tilde{P}_{k-1} \rangle = 0$. Since $\tilde{P}_{k+1} \in \Pi_{k+1}$, $\{\tilde{P}_0, \tilde{P}_1, \ldots, \tilde{P}_{k+1}\}$ is an orthogonal set. It follows from the mathematical induction that $\{\tilde{P}_0, \tilde{P}_1, \ldots, \tilde{P}_n\}$, $\tilde{P}_m \in \Pi_m$, is a set of orthogonal polynomials obtained from

$$\tilde{P}_0 = 1 \quad \text{and} \quad \tilde{P}_1 = (x - a_0)$$

by the three-term recurrence relation (9.135) for $k = 0, 1, \ldots, n - 1$, where a_k is given by (9.136) for $k = 0, 1, 2, \ldots, n - 1$, and b_k by (9.137) for $k = 1, 2, 3, \ldots, n - 1$.

We next introduce

$$P_k = \frac{\tilde{P}_k}{\|\tilde{P}_k\|}, \quad k = 0, 1, \ldots, n, \tag{9.138}$$

so that $\{P_0, P_1, \ldots, P_n\}$ forms an orthonormal set with respect to the weight function $w(x)$. Also, $P_k \in \Pi_k$ and the leading coefficient of P_k is positive for

$k = 0, 1, \ldots, n$. Thus, the set is unique. Finally, the least-square polynomial approximation of degree n of f is then determined by (9.131), where P_k is given by (9.138) and

$$\lambda_k = \langle f, P_k \rangle, \quad k = 0, 1, \ldots, n. \tag{9.139}$$

To find $Q_{n+1}^*(x)$, we need only find $P_{n+1}(x)$ by (9.135) and (9.138) and then calculate λ_{n+1} by (9.139). The final result is

$$Q_{n+1}^* = Q_n^* + \lambda_{n+1} P_{n+1}. \tag{9.140}$$

As a concluding remark, we must mention that a lot of progress has been made on the theory of least-squares approximations by using various orthogonal polynomials in $-1 \le x \le 1$. These include Jacobi polynomials $P_n^{(\alpha, \beta)}(x)$, $(w(x) = (1 - x)^\alpha (1 + x)^\beta, \alpha, \beta > -1)$; Legendre polynomials $P_n(x)$ ($\alpha = \beta = 0$, $w(x) = 1$); and Chebyshev polynomials $T_n(x)$ ($\alpha = \beta = -\frac{1}{2}$, $w(x) = (1 - x^2)^{-1/2}$).

Other orthogonal polynomials are also of interest and can be obtained from the Chebyshev polynomials $T_n(x)$, which satisfy the recurrence relation

$$T_n(x) = 2x T_{n-1}(x) - T_{n-2}(x), \quad n \ge 2,$$

with $T_0(x) = 1$ and $T_1(x) = x$. It follows from $T_n(x) = \cos n\theta$ ($n = 0, 1, 2, \ldots$), where $x = \cos\theta$, $0 \le \theta \le \pi$, that $T_n'(x) = n \sin n\theta / \sin\theta$.

We then define the new polynomials $U_n(x)$ of degree at most n by

$$U_n(x) = \frac{\sin(n+1)\theta}{\sin\theta}, \quad n = 0, 1, 2, \ldots, \tag{9.141}$$

where $x = \cos\theta$. These are called the *Chebyshev polynomials of the second kind*. It is easy to check that polynomials (9.141) are orthogonal with respect to $w(x) = (1 - x^2)^{1/2}$ and hence are constant multiples of Jacobi's polynomials $P_n^{(1/2, 1/2)}(x)$. Using L'Hôpital's rule, it follows that

$$U_n(1) = (n + 1),$$

and then

$$P_n^{(1/2, 1/2)}(1) = \frac{1 \cdot 3 \cdot 5 \cdots (2n + 1)}{2^n (n + 1)!} U_n(1).$$

There are many identities connecting $T_n(x)$ and $U_n(x)$. Some of them are given as exercises.

9.8 The Shannon Sampling Theorem

An analog signal $f(t)$ is a continuous function of time t defined in $-\infty < t < \infty$ with the exception of perhaps a countable number of jump discontinuities. Almost all analog signals $f(t)$ of interest in engineering have finite energy. By this we mean that $f \in L^2(-\infty, \infty)$. The norm of f,

$$\|f\| = \left[\int_{-\infty}^{\infty} |f(t)|^2 dt \right]^{1/2},$$

represents the square root of the total energy content of the signal $f(t)$. The *spectrum* of a signal $f(t)$ is represented by its Fourier transform $F(\omega)$, defined by (5.106), where ω is called the *frequency*. The frequency is measured by $\nu = \frac{\omega}{2\pi}$ in terms of Hz.

A signal $f(t)$ is called *band-limited* if its Fourier transform $F(\omega)$ has a compact support, that is

$$F(\omega) = 0 \quad \text{for } |\omega| > \omega_0 \tag{9.142}$$

for some $\omega_0 > 0$. If ω_0 is the smallest value for which (9.142) holds, then it is called the *bandwidth* of the signal. Even if an analog signal $f(t)$ is not band-limited, we can reduce it to a band-limited signal by what is called an *ideal low-pass filtering*. To reduce $f(t)$ to a band-limited signal $f_{\omega_0}(t)$ with bandwidth less than or equal to ω_0, we consider

$$F_{\omega_0}(\omega) = \begin{cases} F(\omega) & \text{for } |\omega| \leq \omega_0, \\ 0 & \text{for } |\omega| > \omega_0, \end{cases} \tag{9.143}$$

and find $f_{\omega_0}(t)$, called the *low-pass filter function*, by the inverse Fourier transform

$$f_{\omega_0}(t) = \frac{1}{2\pi} \int_{-\infty}^{\infty} e^{i\omega t} F_{\omega_0}(\omega) d\omega = \frac{1}{2\pi} \int_{-\omega_0}^{\omega_0} e^{i\omega t} F_{\omega_0}(\omega) d\omega.$$

In particular, if

$$F_{\omega_0}(\omega) = \begin{cases} 1 & \text{for } |\omega| \leq \omega_0, \\ 0 & \text{for } |\omega| > \omega_0, \end{cases} \tag{9.144}$$

then $F_{\omega_0}(\omega)$ is called the *gate function* and its associated signal $f_{\omega_0}(t)$ is given by

$$f_{\omega_0}(t) = \frac{1}{2\pi} \int_{-\omega_0}^{\omega_0} e^{i\omega t} d\omega = \frac{\sin \omega_0 t}{\pi t}. \tag{9.145}$$

This function is called the *Shannon sampling function*. When $\omega_0 = \pi$, $f_\pi(t)$ is called the *Shannon scaling function*. Both $f_{\omega_0}(t)$ and $F_{\omega_0}(\omega)$ are shown in Figure 9.1.

In engineering, linear analog filtering is defined by the *time-domain convolution*. If $\varphi(t)$ is the filter function, then the input-output relation of this filter is given by

$$g(t) = (\varphi * f)(t) = \int_{-\infty}^{\infty} \varphi(\tau) f(t - \tau) d\tau. \tag{9.146}$$

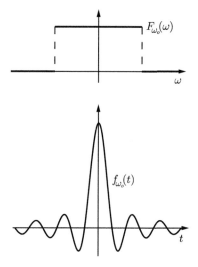

Figure 9.1   The low-pass filter function and the low-pass filter characteristic.

In the frequency domain, the filtering process is represented by pointwise multiplication as

$$G(\omega) = \Phi(\omega)F(\omega), \tag{9.147}$$

where $G(\omega)$ is the Fourier transform of $g(t)$ and $\Phi(\omega)$ is the Fourier transform of $\varphi(t)$, usually called the *transfer function of the filter*.

Consider the limit, as $\omega_0 \to \infty$, of the Fourier integral

$$
\begin{aligned}
1 &= \lim_{\omega_0 \to \infty} F_{\omega_0}(\omega) = \lim_{\omega_0 \to \infty} \int_{-\infty}^{\infty} e^{-i\omega t} f_{\omega_0}(t)\,dt \\
&= \lim_{\omega_0 \to \infty} \int_{-\infty}^{\infty} e^{-i\omega t} \frac{\sin \omega_0 t}{\pi t}\,dt = \int_{-\infty}^{\infty} e^{-i\omega t} \lim_{\omega_0 \to \infty} \frac{\sin \omega_0 t}{\pi t}\,dt \\
&= \int_{-\infty}^{\infty} e^{-i\omega t} \delta(t)\,dt. \tag{9.148}
\end{aligned}
$$

Clearly, the delta function can be thought of as the limit of the sequence of signal functions $f_{\omega_0}(t)$. More explicitly,

$$\delta(t) = \lim_{\omega_0 \to \infty} \frac{\sin \omega_0 t}{\pi t}.$$

The band-limited signal $f_{\omega_0}(t)$ has the representation

$$f_{\omega_0}(t) = \frac{1}{2\pi} \int_{-\omega_0}^{\omega} F(\omega)e^{i\omega t}\,d\omega = \frac{1}{2\pi} \int_{-\infty}^{\infty} F(\omega)F_{\omega_0}(\omega)e^{i\omega t}\,d\omega, \tag{9.149}$$

which gives, by the Convolution Theorem 5.11.11,

$$f_{\omega_0}(t) = \int_{-\infty}^{\infty} f(\tau) f_{\omega_0}(t-\tau) d\tau = \int_{-\infty}^{\infty} \frac{\sin \omega_0(t-\tau)}{\pi(t-\tau)} f(\tau) d\tau. \qquad (9.150)$$

This gives the *sampling integral representation* of a band-limited signal $f_{\omega_0}(t)$.

Next, we consider the Fourier series expansion of the Fourier transform $F_{\omega_0}(\omega)$ of a band-limited signal $f_{\omega_0}(t)$ on the interval $-\omega_0 < \omega < \omega_0$ in terms of the orthogonal set of functions $(\exp(-\frac{in\pi\omega}{\omega_0}))$ in the form

$$F_{\omega_0}(\omega) = \sum_{n=-\infty}^{\infty} a_n \exp\left(-\frac{in\pi\omega}{\omega_0}\right), \qquad (9.151)$$

where the Fourier coefficients a_n are given by

$$a_n = \frac{1}{2\pi} \int_{-\omega_0}^{\omega_0} F_{\omega_0}(\omega) \exp\left(-\frac{in\pi\omega}{\omega_0}\right) d\omega = \frac{1}{2\pi} f_{\omega_0}\left(\frac{n\pi}{\omega_0}\right).$$

Thus, the Fourier series expansion (9.151) reduces to the form

$$F_{\omega_0}(\omega) = \frac{1}{2\omega_0} \sum_{n=-\infty}^{\infty} f_{\omega_0}\left(\frac{n\pi}{\omega_0}\right) \exp\left(-\frac{in\pi\omega}{\omega_0}\right). \qquad (9.152)$$

Multiplying (9.152) by $e^{i\omega t}$ and integrating over $(-\omega_0, \omega_0)$ leads to the reconstruction of the signal function $f_{\omega_0}(t)$ in the form

$$\begin{aligned}
f_{\omega_0}(t) &= \int_{-\omega_0}^{\omega_0} F_{\omega_0}(\omega) e^{i\omega t} d\omega \\
&= \frac{1}{2\pi} \int_{-\omega_0}^{\omega_0} \left[\sum_{n=-\infty}^{\infty} f_{\omega_0}\left(\frac{n\pi}{\omega_0}\right) \exp\left(-\frac{in\pi\omega}{\omega_0}\right) \right] d\omega \\
&= \frac{1}{2\pi} \sum_{n=-\infty}^{\infty} f_{\omega_0}\left(\frac{n\pi}{\omega_0}\right) \int_{-\omega_0}^{\omega_0} \exp\left[i\omega\left(t - \frac{n\pi}{\omega_0}\right) \right] d\omega \\
&= \sum_{n=-\infty}^{\infty} f_{\omega_0}\left(\frac{n\pi}{\omega_0}\right) \frac{\sin \omega_0(t - \frac{n\pi}{\omega_0})}{\omega_0(t - \frac{n\pi}{\omega_0})}. \qquad (9.153)
\end{aligned}$$

This formula is referred to as the *Shannon sampling theorem*. It represents an expansion of a band-limited signal $f_{\omega_0}(t)$ in terms of its discrete values $f_{\omega_0}\left(\frac{n\pi}{\omega_0}\right)$. This is very important in practice, because most systems receive discrete samples as an input.

9.9 Linear and Nonlinear Stability

We consider linear and nonlinear problems of stability and instability for differential systems. In dynamical systems, the state at any time t can be represented by an element of a Banach (or Hilbert) space E. Suppose that the dynamics of a physical system are governed by the evolution equation

$$\frac{du}{dt} = F(\lambda, u, t), \tag{9.154}$$

where $\lambda \in \Lambda$ is a parameter, Λ is a set of parameters (for instance $\Lambda = \mathbb{R}$), u is a function of a real variable t with values in E, and F is a mapping from $\Lambda \times E \times \mathbb{R}$ into E.

Definition 9.9.1.   (Autonomous dynamical system)
The dynamical system governed by (9.154) is called *autonomous* if the function F does not depend explicitly on t.

For autonomous systems, (9.154) can be written in the form $du/dt = F(\lambda, u)$.

Definition 9.9.2.   (Equilibrium solution)
If $F(\lambda_0, u_0) = 0$ for some $\lambda = \lambda_0$ and $u = u_0$, then u_0 is called an *equilibrium solution*.

Definition 9.9.3.   (Stable, unstable, and asymptotically stable solutions)
Let u_0 be an equilibrium solution of equation (9.154).

(a) u_0 is called *stable* if for every $\varepsilon > 0$ there exists a $\delta > 0$ such that

$$\left\| u(t) - u_0 \right\| < \varepsilon$$

for all solutions $u(t)$ of (9.154) such that $\| u(0) - u_0 \| < \delta$.

(b) u_0 is called *unstable* if it is not stable.

(c) u_0 is called *asymptotically stable* if it is stable and $\| u(t) - u_0 \| \to 0$ as $t \to \infty$.

Example 9.9.4. Consider the scalar equation $\dot{x} = 0$. Every solution of this equation has the form $x = c$, where c is a constant. Thus, every solution is stable but not asymptotically stable. $\square$

Example 9.9.5. Consider the system $du/dt = \lambda u$, $u(0) = u_0$, where $u(t)$ is real for each t and $\lambda \in \mathbb{R}$. This equation has the equilibrium solution $u_0(t) = 0$. The general solution is

$$u(t) = u_0 e^{\lambda t}.$$

If $\lambda \leq 0$, then the zero solution is stable. If $\lambda > 0$, the solution is unstable because $u(t) \to \infty$, no matter how small u_0 is. □

Example 9.9.6. Consider the equation $\dot{x} = x^2$ with $x(0) = x_0$. The solution of this equation is obtained by separating the variables and has the form

$$x(t) = \frac{x_0}{1 - x_0 t}.$$

The solution is not defined for $t = 1/x_0$. Thus, $x(t) \equiv 0$ is a solution which is unstable. □

Example 9.9.7. Consider a linear autonomous system

$$\dot{u} = Lu + v, \tag{9.155}$$

where $u(t) \in E$ for each t, $L : E \to E$ is a linear operator which does not depend on t, and v is a given element of E. Clearly, $u_0 \in E$ is an equilibrium solution of (9.155) if $Lu_0 = -v$. We suppose the solution of (9.155) is of the form $u(t) = u_0 + e^{\lambda t} w$, where λ is a constant and $w \in E$. Clearly, $u(t)$ satisfies (9.155) provided

$$Le^{\lambda t} w = \lambda e^{\lambda t} w.$$

This means that λ is an eigenvalue of L with eigenvector w. If the eigenvalue λ has a positive real part and w is a normalized eigenvector, then, for any $\varepsilon > 0$, the function $u(t) = u_0 + \varepsilon w e^{\lambda t}$ is a solution of (9.155) such that $\|u(0) - u_0\| = \varepsilon$ and $\|u(t) - u_0\| \to \infty$ as $t \to \infty$. This shows that the equilibrium solution u_0 is unstable provided there is an eigenvalue with a positive real part. □

This example leads to the *Principle of Linearized Stability*, which can be described as follows: Consider a system of ordinary differential equation

$$\dot{u} = F(\lambda, u), \tag{9.156}$$

where $u = (u_1, u_2, \ldots, u_n)$, $F = (F_1, F_2, \ldots, F_n)$, and λ is a parameter. Let u_0 be the equilibrium solution with $\lambda = \lambda_0$, so that $F(u_0, \lambda_0) = 0$. Suppose the solution of (9.156) can be written as $u(t) = v(t) + u_0$, where $v(t)$ is the perturbation from equilibrium. It follows from $\dot{u} = F(u, \lambda_0)$ that

$$\dot{v} = \dot{u} = F(v + u_0, \lambda_0) = F(u_0, \lambda_0) + \left[\frac{\partial F_i}{\partial u_j} \right](v) + O(\|v\|^2)$$

or

$$\dot{v} = Av + O(\|v\|^2), \qquad (9.157)$$

where $A = [\partial F_i / \partial u_j]_{(u_0, \lambda_0)}$ and $G(v) = O(\|v\|^2)$ represent a term such that

$$\|G(v)\| \le c\|v\|^2,$$

where c is a constant. Neglecting the second term in (9.157), we obtain the linear equation

$$\dot{v} = Av. \qquad (9.158)$$

The solution of this equation is

$$v(t) = e^{tA} u_0. \qquad (9.159)$$

Clearly, all solutions of this equation decay if the spectrum of A lies in the left half-plane. Some solutions of (9.159) may grow exponentially provided A has eigenvalues in the right half-plane. In general, the second order term is negligible when the perturbations are small. This heuristic argument can be justified by Lyapunov's theorem (Aleksander Mikhailovich Lyapunov (1857–1918)):

Theorem 9.9.8. (Lyapunov's theorem)   *If all eigenvalues of A have negative real parts, then u_0 is a stable equilibrium solution of (9.156). If some eigenvalues of A have positive real parts, then u_0 is an unstable solution.*

A rigorous proof of this theorem is beyond the scope of this book. However, the reader is referred to Coddington and Levinson (1955).

The following example shows that the weak inequality $\text{Re}(\lambda) \le 0$ for all eigenvalues does not ensure stability.

Example 9.9.9.   Consider the equation $\dot{u} = Au$, where $u(t) \in \mathbb{R}^2$ and A is the matrix operator $\begin{pmatrix} 0 & 1 \\ 0 & 0 \end{pmatrix}$.

If u_0 is an equilibrium solution of this equation, then $Au_0 = 0$. Clearly, $u_0 = (a, 0)$ represents an equilibrium solution for any number a. The only eigenvalue of A is zero. If we write $u = (x, y)$, then the given equation becomes $\dot{x} = y$ and $\dot{y} = 0$. Hence, the general solution is $y = m$, $x = mt + c$, where m and c are constants. For sufficiently small m and c, the solution $u(t) = (mt + c, m)$ can be made sufficiently close to $u_0 = (a, 0)$ at $t = 0$. But $\|u(t) - u_0\| \to \infty$ as $t \to \infty$. This shows that the equilibrium solution is unstable. $\square$

Theorem 9.9.10. (Stability criterion)   *If A is a linear operator on a space E and $A + A^*$ is negative semidefinite, that is $\langle v, (A + A^*)v \rangle \le 0$ for all $v \in E$, then all equilibrium solutions of the equation*

$$\dot{u} = Au + f \qquad (9.160)$$

are stable, where u is an element of a Hilbert space E for each t, $A: E \to E$ is independent of t, and f is a given element of E.

Proof: Suppose u_0 is an equilibrium solution of (9.160), that is $Au_0 = 0$, and $u(t)$ is any other solution. If $v = u - u_0$, then $\dot{v} = Av$. Thus,

$$\frac{d^2}{dt^2}\|v\|^2 = \frac{d^2}{dt^2}\langle v, v \rangle = \langle v, \dot{v} \rangle + \langle \dot{v}, v \rangle$$

$$= \langle v, Av \rangle + \langle Av, v \rangle$$

$$= \langle v, (A + A^*)v \rangle.$$

If $A + A^*$ is negative semidefinite, then

$$\frac{d^2}{dt^2}\|v\|^2 \leq 0.$$

This means that $\|v\|$ is a non-increasing function. Consequently, if $\|u(0) - u_0\| < \varepsilon$, then $\|u(t) - u_0\| < \varepsilon$ for all $t > 0$. This shows that all equilibrium solutions are stable.   □

We next consider the stability of a general nonlinear autonomous equation

$$\dot{u} = Nu. \tag{9.161}$$

The question of stability of an equilibrium solution u_0 of (9.161) is concerned with the effects of small initial displacements of u from u_0, and it only involves values of u in the neighborhood of u_0. If N is Fréchet differentiable, then the operator N can be approximated by the linear operator $N'(u)$ in the neighborhood of u_0, and linear stability theory can be used. Hence,

$$Nu = Nu_0 + N'(u_0)(u - u_0) + o(u - u_0), \tag{9.162}$$

where $Nu_0 = 0$.

Neglecting the term $o(u - u_0)$, equation (9.162) is approximately equal to

$$\dot{u} = N'(u_0)(u - u_0). \tag{9.163}$$

This equation may be called the linearized approximation of the nonlinear equation (9.161). Its stability can be determined by stability criteria discussed earlier. When u is near u_0, (9.163) is the linearized approximation to (9.161), so it is naturally assumed that the stability of the linearized equations determines that for the nonlinear equations. This principle is generally accepted as valid in the applied literature, and stability is determined formally by solving the associated linear eigenvalue problem. However, this general principle is not necessarily true as shown by a counterexample.

Example 9.9.11. Consider the nonlinear equation $\dot{u} = u^3$, where $u(t) \in \mathbb{R}$ for each t. The equilibrium solution is $u_0 = 0$. It can be explicitly solved by using the initial condition $u(0) = u_0$, and the solution is

$$u^2 = \frac{u_0^2}{1 - 2u_0^2 t},$$

which is not defined for $t = \frac{1}{2u_0^2}$. Thus, u_0 is an unstable equilibrium. However, the linearized equation $\dot{u} = 0$ admits a stable solution. Thus, the stability of the linearized equation does not imply stability of the nonlinear equation.

The difficulty associated with this example is that the linearized equation has eigenvalue $\lambda = 0$ (critical case when $\mathrm{Re}\,\lambda = 0$). In other words, the linearized system is only marginally stable. This means that an arbitrarily small perturbation can push the eigenvalue into the right half-plane, and make the system unstable. The eigenvalue zero corresponds to a constant solution of the linearized equation, and an arbitrarily small perturbation can change this constant solution and thus lead to instability. However, if all the eigenvalues of a linearized problem are negative, then its solutions tend to u_0 exponentially. The small perturbations involved in going from the linearized to the nonlinear problem cannot change exponential decay of $u - u_0$ into growth, so in this case the nonlinear problem will be stable. $\qquad\square$

9.10 Bifurcation Theory

Bifurcation is a phenomenon involved in nonlinear problems and is closely associated with the loss of stability. Section 9.9 showed that the stability of a dynamical system depends on whether the eigenvalues of the linearized operator are positive or negative. These eigenvalues correspond to bifurcation points.

We shall discuss bifurcation theory in terms of operator equations in a real Banach (or Hilbert) space. By a nonlinear eigenvalue problem, we usually mean the problem of determining appropriate solutions of a nonlinear equation of the form

$$F(\lambda, u) = 0, \tag{9.164}$$

where $F : \mathbb{R} \times E \to B$ is a nonlinear operator, depending on the parameter λ, which operates on the unknown function or vector u, and E and B are real Banach (or Hilbert) spaces.

Bifurcation theory deals with the existence and behavior of solutions $u(\lambda)$ of equation (9.164) as a function of the parameter λ. Of particular interest is the process of bifurcation (or branching) where a given solution of (9.164) splits into two or more solutions as λ passes through a critical value λ_0, called a *bifurcation point*.

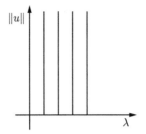

Figure 9.2   Bifurcation diagram.

Definition 9.10.1.   (Bifurcation point)
The solution of (9.164) is said to bifurcate from the solution $u_0(\lambda_0)$ at the value $\lambda = \lambda_0$ if the equation has at least two distinct solutions $u_1(\lambda)$ and $u_2(\lambda)$ such that they tend to $u_0 \equiv u_0(\lambda_0)$ as $\lambda \to \lambda_0$. The points (λ_0, u_0) satisfying equation (9.164) are referred to as *bifurcation* (or *branch*) points if, in every neighborhood of (λ_0, u_0), there exists a solution (λ, u) different from (λ_0, u_0).

The first problem of bifurcation theory is to determine the solution u_0 and the parameter λ_0 at which bifurcation occurs. The second problem is to find the number of solutions which bifurcate from $u_0(\lambda_0)$. The third problem is to study the behavior of these solutions for λ near λ_0.

To illustrate bifurcation, we consider the linear eigenvalue problem

$$Lu = \lambda u, \tag{9.165}$$

where L is a linear operator acting on a function or a vector u in some Banach space and $\lambda \in \mathbb{R}$. For every value of λ, (9.165) has a trivial solution $u = 0$ with the norm $\|u\| = 0$. Suppose there is a sequence of eigenvalues $\lambda_1 < \lambda_2 < \lambda_3 < \cdots$, and the corresponding normalized eigenfunctions $u_1, u_2, u_3, \ldots$ such that

$$Lu_k = \lambda_k u_k, \quad \|u_k\| = 1, \quad k = 1, 2, 3, \ldots. \tag{9.166}$$

Then, for any real number a, nontrivial solutions are $u = au_k$, $k = 1, 2, 3, \ldots$, with the norm $\|u\| = a$. The norms of both trivial and nontrivial solutions are shown graphically by Figure 9.2.

Many examples of bifurcation phenomena occur in both differential and integral equations. One such example is as follows:

Example 9.10.2.   Consider a thin elastic rod with pinned ends lying in the x–z plane. The shape of the rod is described by two functions $u(x)$ and $w(x)$, which are the dimensionless displacement functions in the x and z directions. The x-displacements of its end points are prescribed. The displacement functions $u(x)$ and $w(x)$ satisfy the following differential equations and boundary conditions:

$$\frac{d^2w}{dx^2} + \lambda w(x) = 0, \quad 0 \le x \le 1, \tag{9.167}$$

$$\left(\frac{du}{dx}\right) + \frac{1}{2}\left(\frac{dw}{dx}\right)^2 = -\mu\lambda, \quad 0 \le x \le 1, \tag{9.168}$$

$$w(0) = w(1) = 0, \qquad u(0) = -u(1) = a > 0, \tag{9.169}$$

where the parameter λ is proportional to the axial stress in the rod, the constant a in (9.169) is proportional to the prescribed end displacement and is referred to as the *end-shortening*, and μ is a positive physical constant.

Consider the linearized problem where the nonlinear term w_x^2 is absent. The solution of the linearized equation (9.168) is

$$u(x) = a(1 - 2x), \tag{9.170}$$

where $a = \lambda\mu/2$.

The solution of (9.167) and (9.169) is $w(x) \equiv 0$ unless λ is an eigenvalue λ_n given by

$$\lambda = \lambda_n = n^2\pi^2, \quad n = 1, 2, 3, \ldots. \tag{9.171}$$

In this case, w is a multiple of the eigenfunctions w_n given by

$$w(x) = A_n w_n(x) = A_n \sin n\pi x, \quad n = 1, 2, 3, \ldots, \tag{9.172}$$

where the A_n are constants.

From $a = \frac{1}{2}\lambda\mu$ and $\lambda = \lambda_n = n^2\pi^2$, we conclude that if $a = a_n = \frac{1}{2}\mu\lambda_n$, then the rod buckles into a shape given by (9.170) and (9.172) with an undetermined amplitude A_n. The numbers a_n are called the *critical end-shortenings*. For $a \ne a_n$, $n = 1, 2, \ldots$, the rod remains straight because the solution of (9.167) and (9.169) is

$$w(x) \equiv 0. \tag{9.173}$$

We now consider the nonlinear problem (9.167) to (9.169). The solution of the problem is still given by (9.172) when $\lambda = \lambda_n$ and by (9.173) when $\lambda \ne \lambda_n$. To find $u(x)$ when $\lambda = \lambda_n$, we put (9.172) into (9.168) and integrate using (9.169) at $x = 0$ to obtain

$$u(x) = u_n(x) = a - \mu\lambda_n\left(1 + \frac{A_n^2}{4\mu}\right)x + \frac{1}{8}n\pi A_n^2 \sin 2n\pi x. \tag{9.174}$$

In view of the boundary condition $u(1) = -a$, we obtain

$$a = a_n\left(1 + \frac{A_n^2}{4\mu}\right). \tag{9.175}$$

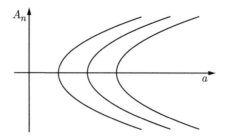

Figure 9.3   Bifurcation diagram for the thin rod.

This is a relation between the end-shortening and the amplitude. The bifurcation diagrams for the thin rod are given in Figure 9.3.

The diagram shows that, for $a < a_1$, the only solution is the trivial solution $w \equiv 0$. At $a = a_1$, the nontrivial solution $w_1 = A_1 \sin \pi x$ bifurcates from the trivial solution and continues to exist for all $a > a_1$. The point $a = a_1$ is called the first bifurcation point, and the nontrivial solution is called the first bifurcation solution.

For each n, nontrivial solutions of (9.175) for A_n are possible if and only if $a \geq a_n$. The solutions bifurcate from the trivial (unbuckled) state $A_n = 0$ at $a = a_n$. Thus, the solution of the linearized problem determines the bifurcation points of the nonlinear problem. For any a in $a_n \leq a \leq a_{n+1}$, there are $2n + 1$ solutions. For $a < a_1$, no buckling is possible. We also note from (9.175) that $da/dA_n = a_n A_n / 2\mu$. Hence, for a fixed amplitude A, the parabola in Figure 9.3 bifurcating from a_n has a steeper slope than that bifurcating from a_m if $m < n$. Clearly, these parabolas do not intersect.

For any fixed value of a, the bifurcation solutions can be classified by the values of the potential energy associated with them. We also observe that the potential energy is equal to the internal energy, since the displacements are specified at the ends of the rod. Consequently, the potential energy is proportional to the functional V defined by

$$V(w) = \frac{1}{2} \int_0^1 \left[w_{xx}^2 + \frac{1}{\mu} \left(u_x + \frac{1}{2} w_x^2 \right)^2 \right] dx. \tag{9.176}$$

In the unbuckled state, equations (9.170) and (9.173) hold with $a = \lambda \mu / 2$, and the corresponding potential energy is

$$V_\infty = \frac{1}{2} \mu \left(\frac{2a}{\mu} \right)^2. \tag{9.177}$$

The potential energy V_n of the buckled state given by (9.172) is obtained by substituting (9.171), (9.172), and (9.174) into (9.176) in the form

$$V_n = \lambda_n(2a - a_n). \qquad (9.178)$$

Hence,

$$V_n - V_\infty = -\frac{2}{\mu}(a - a_n)^2 \leq 0, \qquad a \geq a_n, \qquad (9.179)$$

$$(V_n - V_m) = \frac{2}{\mu}(a_n - a_m)\big[(a - a_n) + (a - a_m)\big] \geq 0, \qquad a \geq a_n \geq a_m. \qquad (9.180)$$

It follows from (9.179) and (9.180) that, for fixed $a > a_1$, the straight state has the largest energy, and the branch originating from a_1 has the smallest energy. For fixed a in the interval $a_n \leq a \leq a_{n+1}$, the energies of the branches are ordered as $V_\infty > V_n > V_{n-1} > \cdots > V_1$. For the state of smallest energy, the displacement function of this state is

$$w = A_1 w_1 = \pm 2\sqrt{\mu}\left(\frac{a}{a_1} - 1\right) \sin \pi x \quad \text{for all } a > a_n. \qquad (9.181)$$

$\square$

Suppose the solutions of (9.164) represent equilibrium solutions for a dynamical system which evolves according to the time-dependent equations

$$u_t = F(\lambda, u), \qquad (9.182)$$

where $u : \mathbb{R} \to E$ and E is a Banach (or Hilbert) space. An equilibrium solution u_0 is stable if small perturbations from it remain close to u_0 as $t \to \infty$; u_0 is asymptotically stable if small perturbations tend to zero as $t \to \infty$ (see Section 9.9). When the parameter λ changes, one solution may persist but become unstable as λ passes a critical value λ_0, and it is at such a transition point that new solutions may bifurcate from the known solution.

One of the simple nonlinear partial differential equations which exhibits the transition phenomena shown in Figure 9.3 is

$$u_t = \nabla^2 u + \lambda u + u^3 \quad \text{in } D, \qquad (9.183)$$

$$u = 0 \quad \text{on } \partial D, \qquad (9.184)$$

where D is a smooth bounded domain in $\mathbb{R}^N$. The equilibrium states of (9.183) are given by solutions of the time-independent equation ($u_t \equiv 0$). One solution is obviously $u = 0$, which is valid for all λ; this solution becomes unstable at $\lambda = \lambda_1$, the first eigenvalue of the Laplacian: $\nabla^2 u_1 + \lambda_1 u_1 = 0$ on D, $u_1 = 0$ on ∂D. For $\lambda > \lambda_1$, there are at least three solutions of the nonlinear equilibrium equation. The nature of the solution set in the neighborhood of $(\lambda_1, 0)$

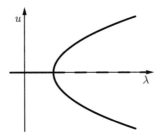

Figure 9.4   Bifurcation diagram where unstable solutions are represented by dashed lines.

is given in Figure 9.4; the new bifurcating solutions are stable. The Laplacian has a set of eigenvalues $\lambda_1 < \lambda_2 < \lambda_3 < \cdots$, which tend to infinity, and all of these eigenvalues are potential bifurcation points.

In the theory of calculus in Banach spaces, the following version of the Implicit Function Theorem is concerned with the existence, uniqueness, and smoothness properties of the solution of the equation (9.164).

Theorem 9.10.3. (Implicit Function Theorem) *Suppose Λ, E, B are real Banach spaces and F is a Fréchet differentiable mapping from a domain $D \subset \Lambda \times E$ to B. Assume $F(\lambda_0, u_0) = 0$ and the Fréchet differential $F'(\lambda_0, u_0)$ is an isomorphism from E to B. Then locally, for $\|\lambda - \lambda_0\|$ sufficiently small, there is a differentiable mapping $u(\lambda)$ from Λ to E, with $(\lambda, u(\lambda)) \in D$, such that $F(\lambda, u(\lambda)) = 0$. Moreover, $(\lambda, u(\lambda))$ is the only solution of $F = 0$ in a sufficiently small neighborhood $D' \subset D$. If F is C^n then u is C^n. If λ, E, and B are complex Banach spaces and F is Fréchet differentiable, then F is analytic and u is analytic in λ.*

The proof of the theorem is beyond the scope of this book. However, the theorem can be proved by using a contraction mapping argument and is adequate for most physical applications. The reader is referred to Sattinger (1973) or Dieudonné (1969) for a detailed discussion of proofs.

Bifurcation phenomena typically accompany the transition to instability when a characteristic parameter crosses a critical value, and hence they play an important role in applications to mechanics. Indeed, the area of mechanics is a rich source of bifurcation and instability phenomena, and the subject has always stimulated the rapid development of functional analysis. For more on bifurcation theory, see Arrowsmith and Place (1992) and Guckenheimer and Holmes (1983).

9.11 Exercises

1. Show that the norm in an inner product space is strictly convex.

2. Let H_1 and H_2 be real Hilbert spaces. Show that if T is a bounded linear operator from H_1 into H_2, and f is a real functional on H_1 defined by

$$f(x) = \|u - Tx\|^2,$$

where u is a fixed vector in H_2, then f has a Fréchet derivative at every point given by

$$f'(x) = -2T^*u + 2T^*Tx,$$

where T^* is the adjoint of T.

3. Suppose $T : B_1 \to B_2$ is Fréchet differentiable on an open set $\Omega \subset B_1$. Show that if $x \in \Omega$ and $h \in B_1$ are such that $x + th \in \Omega$ for every $t \in [0, 1]$, then

$$\left\| T(x+h) - T(x) \right\| \le \|h\| \sup_{0 < \alpha < 1} \left\| T'(x + \alpha h) \right\|.$$

4. Suppose T is a twice Fréchet differentiable on an open domain Ω in a Banach space B_1. Let $x \in \Omega$ and $x + \alpha h \in \Omega$ for every $\alpha \in [0, 1]$. Prove that

$$\left\| T(x+h) - T(x) - T'(x)h \right\| \le \frac{1}{2} \|h\|^2 \sup_{0 < \alpha < 1} \left\| T''(x + \alpha h) \right\|.$$

5. Find the extrema of the following functionals:

(a) $I(y) = \displaystyle\int_1^2 \left((y')^2 - 2xy \right) dx,$ $y(1) = 0,$ $y(2) = -1.$

(b) $I(y) = \displaystyle\int_0^1 e^{2x} \left((y')^2 - y^2 - y \right) dx,$ $y(0) = 0,$ $y(1) = \dfrac{1}{e}.$

(c) $I(y) = \displaystyle\int_0^{2\pi} \left((y')^2 - y^2 \right) dx,$ $y(0) = 1,$ $y(2\pi) = 1.$

(d) $I(y) = \displaystyle\int_0^a \left((y')^2 - y^2 \right) dx,$ $y(0) = 1,$ $y(a) = 1,$

$$a \ne \pi k.$$

6. Find the extrema of the following functional:

$$I\big(u(x), v(y)\big) = \int_1^2 \left((u')^2 + v^2 + (v')^2 \right) dx,$$

$$u(1) = 1, \qquad u(2) = 2, \qquad v(1) = 0, \qquad v(2) = 1.$$

7. Determine the extrema of the functional

$$I(y) = \int_a^b \frac{1}{y}(1 + (y')^2)^{1/2} dx,$$

which passes through the given points (a, y_1) and (b, y_2) in the upper half-plane.

8. Find the shortest distance from the point $(1, 1, 1)$ to the sphere

$$x^2 + y^2 + z^2 = 1.$$

9. Solve the following isoperimetric problem: Minimize the functional

$$I(y) = \int_0^1 ((y')^2 - y^2) dx,$$

subject to $\int_0^1 \sqrt{1 + (y')^2}\, dx = \sqrt{2}$, $y(0) = 0$, $y(1) = 1$.

10. Find the minimum of the functional

$$I(y) = \int_0^\pi (y')^2(x) dx,$$

subject to the condition $\int_0^\pi y^2(x)dx = 1$, $y(0) = 0$, $y(\pi) = 0$.

11. Derive Newton's Second Law of Motion from Hamilton's principle.

12. Derive the equation of a simple harmonic oscillator in a nonresisting medium from Hamilton's principle.

13. Find the curve $y = y(x)$, $0 \leq x \leq 1$, with $y(0) = 0$, $y(1) = 0$, and fixed arc length L that has maximal area

$$A = \int_0^1 y(x)dx.$$

14. Show that the potential function $u(x, y)$, which minimizes the functional

$$I(u) = \iint_\Omega \frac{1}{2}(u_x^2 + u_y^2)dx\, dy, \quad \Omega \subset \mathbb{R}^2,$$

satisfies the Laplace equation $u_{xx} + u_{yy} = 0$.

15. The distance between two points on a sphere is given by

$$ds = \sqrt{a^2 d\theta^2 + a^2 \sin^2 \theta d\varphi^2}.$$

Show that the shortest curve between the two points on a sphere lies on a great circle.

16. The kinetic and potential energies of a spherical pendulum are given by

$$T = \frac{1}{2}(l^2\dot{\theta}^2 + l^2\sin^2\theta\dot{\varphi}^2), \qquad V = mgl(1 - \cos\theta).$$

Find the equation of motion.

17. Complete the proof of Theorem 9.6.1 by showing that the equation $Ax - \lambda x = y$ has a solution given by (9.111) only if α is not an eigenvalue of A and $\alpha \neq 0$.

18. Show that the Chebyshev polynomial T_n is even for even n and odd for odd n.

19. Show that $T_{n+1}(x) = 2xT_n(x) - T_{n-1}(x)$ for $n = 2, 3, \ldots$.

20. Show that

$$T_0(x) = 1,$$
$$T_1(x) = x,$$
$$T_2(x) = 2x^2 - 1,$$
$$T_3(x) = 4x^3 - 3x,$$
$$T_4(x) = 8x^4 - 8x^2 + 1,$$
$$T_5(x) = 16x^5 - 20x^3 + 5x.$$

21. Find the best approximation of $f(x) = x^{n+2}$ out of Π_n.

22. Prove that, for every $n \in \mathbb{N}$, T_n satisfies the differential equation

$$(1 - x^2)y''(x) - xy'(x) + n^2 y(x) = 0.$$

23. Show that every $P(x) \in \Pi_n$ has a unique representation of the form

$$P(x) = \sum_{k=0}^{n} a_k T_k(x).$$

24. Show that

$$T_n(x) = x^n - \binom{n}{2}x^{n-2}(1 - x^2) + \binom{n}{4}x^{n-4}(1 - x^2)^2 - \cdots.$$

25. (a) Let $L_n f$ denote the polynomial of degree less than or equal to n, which agrees with a given function $f \in C([a, b])$ at the fixed nodes $x_0, x_1, \ldots, x_n \in [a, b]$. Show that L_n is a linear operator on $C([a, b])$.

(b) Show that $L_n P = P$ if and only if P is a polynomial of degree less than or equal to n.

(c) Show that the error in the Lagrange interpolation is

$$(L_n f - f)(x) = \sum_{k=0}^{n} [f(x_k) - f(x)] l_k(x),$$

where

$$l_k(x) = \prod_{\substack{k=0 \\ k \neq j}} [(x - x_k)/(x_j - x_k)].$$

26. Prove that the Lagrange interpolating polynomial for nodes defined as zeros of $T_n(x)$ is

$$P(x) = \frac{1}{n} \sum_{k=0}^{n} f(x_k) \frac{T_n(x)(-1)^{k-1} \sin \theta_k}{x - x_k},$$

where $x_k = \cos \theta_k$ and $\theta_k = (2k - 1)\pi/2n$.

27. Show that the least-squares approximation of degree $n - 1$ to $f(x) = x^n$ is $Q_{n-1}^* = x^n - \tilde{P}_n(x)$, where $\tilde{P}_n$ is the orthogonal polynomial of degree n determined by $w(x)$ and (9.136).

28. Show that another form of (9.137) is

$$b_k = \frac{\langle \tilde{P}_k, \tilde{P}_k \rangle}{\langle \tilde{P}_{k-1}, \tilde{P}_{k-1} \rangle}.$$

29. Prove that, if $w(x) = 1$, the orthogonal polynomials defined by (9.135) satisfy

$$\langle \tilde{P}_n, \tilde{P}_n \rangle = \frac{1}{2n + 1} \tilde{P}_n^2(1), \quad n = 0, 1, 2, \ldots.$$

30. Prove Rodrigues' formula:

$$P_n(x) = \frac{1}{2^n n!} D^n [(x^2 - 1)^n],$$

where P_n is the nth degree Legendre polynomial.

31. Show that the Legendre polynomials can be represented in the following form:

$$P_n(x) = \frac{1}{2^n} \sum_{k=0}^{[n/2]} (-1)^k \binom{n}{k} \binom{2n - 2k}{n} x^{n-2k}.$$

Use this result to show that

$$P_0(x) = 1,$$
$$P_1(x) = x,$$
$$P_2(x) = \frac{3}{2}x^2 - \frac{1}{2},$$
$$P_3(x) = \frac{5}{2}x^3 - \frac{3}{2}x,$$
$$P_4(x) = \frac{35}{8}x^4 - \frac{15}{4}x^2 + \frac{3}{8}.$$

Sketch graphs of these polynomials.

32. Prove the following recurrence relations for Legendre polynomials:

(a) $(n+1)P_{n+1}(x) = (2n+1)xP_n(x) - nP_{n-1}(x)$.

(b) $P_n'(x) = xP_{n-1}'(x) + nP_{n-1}(x)$.

33. Show that $\langle P_n, P_n \rangle = 2/(2n+1)$, $n = 0, 1, 2, \ldots$.

34. Show the following:

(a) $P_{n+1}'(x) - P_{n-1}'(x) = (2n+1)P_n(x)$.

(b) $xP_n'(x) - P_{n-1}'(x) = nP_n(x)$.

35. Show the following:

(a) $(x^2 - 1)P_n'(x) = nxP_n(x) - nP_{n-1}(x)$.

(b) $\dfrac{1-x^2}{n^2}\left[P_n'(x)\right]^2 + \left[P_n(x)\right]^2 = \dfrac{1-x^2}{n^2}\left[P_{n-1}'(x)\right]^2 + \left[P_{n-1}(x)\right]^2$.

36. Show the following:

(a) $T_n(x) = U_n(x) - xU_{n-1}(x)$.

(b) $(1 - x^2)U_{n-1}(x) = xT_n(x) - T_{n+1}(x)$.

37. Show that $U_n(x)$ is generated by the following three-term recurrence relation:

$$U_n(x) = 2xU_{n-1}(x) - U_{n-2}(x), \quad n \geq 2,$$
$$U_0(x) = 1, \qquad U_1(x) = 2x.$$

38. Find the general solution of the system

$$\dot{x} = y + x\left(1 - x^2 - y^2\right), \qquad \dot{y} = -x + y\left(1 - x^2 - y^2\right).$$

Show that the unit circle is the orbit of a solution of the system.

39. Show that the equilibrium points of the Volterra system

$$\dot{x} = ax - Ax^2 - bxy, \qquad \dot{y} = -cy + dxy$$

are

$$(0,0), \qquad \left(\frac{a}{A}, 0\right), \qquad \left(\frac{c}{d}, \frac{a}{b}\left(1 - \frac{Ab}{ad}\right)\right).$$

40. Consider a diffusion–reaction system

$$u_t = Lu, \qquad u(0, t) = u(1, t) = 0,$$

in a Hilbert space $L^2([0, 1])$ where $Lu = u_{xx} + a(x)u$. If a is a constant, show that the eigenvalues of L are $a - n^2\pi^2$, $n = 1, 2, 3, \ldots$. Hence, show that the equilibrium solution is stable or unstable according to if $a < \pi^2$ or $a > \pi^2$. If a is not a constant, discuss the stability of the system.

41. Find the nontrivial solutions of the linear eigenvalue problem

$$w'' + \lambda w = 0, \qquad 0 \le x \le \pi,$$
$$w(0) = w(\pi) = 0.$$

Draw the bifurcation diagram.

42. Find small nontrivial solutions of the nonlinear eigenvalue problems

(a) $\quad w'' + \left[\lambda - \frac{2}{\pi}\int_0^\pi w^2(x)dx\right]w(x) = 0, \qquad 0 \le x \le \pi;$

(b) $\quad \left[\frac{2}{\pi}\int_0^\pi w^2(x)dx\right]w'' + \lambda w = 0, \qquad 0 \le x \le \pi;$

with the boundary conditions $w(0) = w(\pi) = 0$. Draw the bifurcation diagrams in each case.

43. Find small nontrivial solutions of the nonlinear eigenvalue problem

$$u'' + \lambda u - u^2 = 0, \qquad 0 \le x \le 1,$$
$$u(0) = u(1) = 0.$$

Discuss their behavior as a function of λ.

44. The Euler equation for the displacement of a thin elastic rod with end-shortening proportional to λ is

$$w'' + \left(\lambda - \frac{1}{2}\int_0^1 (w')^2 ds\right)w = 0, \qquad 0 \le x \le 1,$$

$$w(0) = w(1) = 0.$$

Describe the behavior of the solution as a function of λ.

45. Consider the nonlinear boundary value problem for a pinned inextensi-
ble rod subject to prescribed axial thrust. The shape of the rod is determined by $\theta(x)$, the angle between the centerline of the deformed rod and
the x-axis, and the displacements $u(x)$ and $w(x)$ parallel and normal to
the x-axis, respectively. The governing equations of the problem for the
elastica are

$$\theta'' + \lambda \sin \theta = 0, \quad 0 \le x \le 1,$$

$$\theta'(0) = \theta'(1) = 0,$$

$$u' = \cos \theta - 1, \qquad w' = \sin \theta, \quad 0 \le x \le 1,$$

$$u(0) = w(0) = w(1) = 0,$$

where the constant λ is proportional to the applied thrust.

Find the eigenvalues and eigenfunctions of the linearized problem. Show
that the linearized eigenvalue problem yields the points of bifurcation for
the nonlinear problem.

46. The nonlinear integro-differential system

$$u'' + \left[\lambda - 2\int_0^1 u^2(x)dx\right]u = 0, \qquad u(0) = u(1) = 0$$

can be solved exactly. Describe its solutions as a function of λ.

47. Show that the solution of nonlinear integral equation

$$u = 1 + \int_0^1 u^2 dx$$

is

$$u = \frac{1}{2\lambda}\left[1 \pm \sqrt{1 - 4\lambda}\right].$$

Draw the graph of u as a function of λ.

48. Show that the following boundary value problem has no bifurcation
from any eigenvalue of the linearized problem:

$$\ddot{u} + \lambda\left[u + v(u^2 + v^2)\right] = 0, \qquad \ddot{v} + \lambda\left[v - u(u^2 + v^2)\right] = 0,$$

$$u(0) = u(a) = v(0) = v(a) = 0.$$

49. Consider the nonlinear system

$$\dot{x} = y - \frac{1}{2}x(x^2 + y^2), \qquad \dot{y} = -x - \frac{1}{2}y(x^2 + y^2).$$

Show that the linearized system has the periodic solution $(x, y) = (a\cos t, a\sin t)$, but the nonlinear system has no nontrivial periodic solution.

50. Discuss the stability and bifurcation phenomena for the following differential equations:

    (a) $\dot{x} = ax \pm x^2$.

    (b) $\dot{x} = ax \pm x^3$.

51. Solve the equation

$$\dot{x} = 1 - x^2, \qquad x(t_0) = x_0 \in (-1, 1).$$

Examine the stability of the solution $x(t) = -1$ for all t.

52. Solve the isoperimetric problem that is to maximize the area under a curve

$$\int_{x_1}^{x_2} y(x)\,dx$$

subject to the fixed arc-length

$$\int_{x_1}^{x_2} \sqrt{1 + (y'(x))^2}\,dx = l.$$

53. Show that the geodesic on a cylinder is a spiral curve.

54. Find the Euler–Lagrange equation of

$$I(\theta, \varphi) = \int_0^T \frac{1}{2}\left[ml^2(\dot{\theta}^2 + \sin^2\theta\,\dot{\varphi}^2) + mgl\cos\theta\right]dt,$$

where the dot represents the time derivative. Hence, determine Hamilton's equations of motion for the pendulum in space.

55. Consider the following nonlinear systems:

    (a) $\dot{x} = -\varepsilon + \lambda x - x^2$.

    (b) $\dot{x} = \lambda x + 2\varepsilon x^2 - x^3$.

    When $\varepsilon = 0$, (a) represents the normal form of a transcritical bifurcation. Draw the bifurcation diagram for $\varepsilon < 0$ and $\varepsilon > 0$.

When $\varepsilon = 0$, (b) represents the normal form of a supercritical bifurcation. Draw the bifurcation diagram for a small nonzero ε.

56. Obtain the equilibrium points of the following systems:

   (a) $\dot{x} = \lambda x \pm x^3$.

   (b) $\dot{x} = \lambda x - x^2$.

   Draw the bifurcation diagram for both (a) and (b).

57. Draw the bifurcation diagram for the following nonlinear systems:

   (a) $\dot{x} = \lambda - x^2$,

   (b) $\dot{x} = \lambda - 2\lambda x - x^2$,

   where λ is a parameter. Draw the graph of $\dot{x}$ versus x for $\lambda > 0$ and $\lambda < 0$.

58. Consider the following nonlinear systems:

   (a) $\dot{x} = -x - 2y^2, \dot{y} = xy - y^3$.

   (b) $\dot{x} = y, \dot{y} = -(x + x^2 + y)$.

   Show that for (a) the equilibrium point $(0, 0)$ is both Lyapunov and asymptotically stable. In the case of (b), show that there are two equilibrium points $(-1, 0)$ and $(0, 0)$. Construct a Lyapunov function to show that the origin is Lyapunov stable. Draw the phase portrait for cases (a) and (b).

59. Determine the equilibrium points for the nonlinear systems:

   (a) $\dot{x} = x^2 - y^2, \dot{y} = -2xy$.

   (b) $\dot{x} = 2x - x^2, \dot{y} = x - y$.

   Show that the origin is unstable for (a).

60. Draw the various phase portraits when $\lambda > 0$, $\lambda = 0$ and $\lambda < 0$ for the second-order nonlinear systems:

   (a) $\dot{x} = \lambda - x^2, \dot{y} = y$.

   (b) $\dot{x} = \lambda x - x^2, \dot{y} = -y$.

   (c) $\dot{x} = \lambda x \pm x^3, \dot{y} = -y$.

   Show that system (a) represents a saddle-node bifurcation and that systems (b) and (c) describe supercritical and subcritical bifurcations.

61. The Lorenz equations (1963) represent a system of three autonomous ordinary differential equations

$$\dot{x} = -\frac{8}{3}x + yz, \qquad \dot{y} = -10(y - z), \qquad \dot{z} = -xy + 28y - z.$$

This system describes as the leading order approximation to the behavior of an ideal model of the Earth's atmosphere. Find the equilibrium points of the system.

Hints and Answers to Selected Exercises

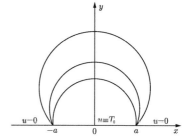

1.7 Exercises

8. Consider the function $f(x) = \frac{1}{p}x + \frac{1}{q} - x^{1/p}$. Show that f is decreasing on $[0, 1]$ and that $f(1) = 0$.

9. If $|x| < 1$, then $|x|^q \le |x|^p$. Consider the sequence $(n^{-1/p})$.

10. The system

$$\lambda_1 = x_1$$
$$\lambda_1 + \lambda_2 = x_2$$
$$\lambda_1 + \lambda_2 + \lambda_3 = x_3$$

has a solution for any $(x_1, x_2, x_3) \in \mathbb{R}^3$.

12. Prove that if $\lambda_0 + \lambda_1 x + \cdots + \lambda_n x^n = 0$ for all x, then $\lambda_0 = \lambda_1 = \cdots = \lambda_n = 0$.

14. See Exercise 12 or 13.

15. $(1, 0, 0, \ldots), (0, 1, 0, 0, \ldots), (0, 0, 1, 0, 0, \ldots), \ldots$.

16. If $x = \alpha_1 e_1 + \cdots + \alpha_n e_n$ and $x = \beta_1 e_1 + \cdots + \beta_n e_n$, then $(\alpha_1 - \beta_1)e_1 + \cdots + (\alpha_n - \beta_n)e_n = 0$.

17. $d_1 + d_2 + \cdots + d_n$.

18. Use the triangle inequality $\|a + b\| \le \|a\| + \|b\|$ with $a = x - y$ and $b = y$.

26. No. If $f_n(x) = \max\{0, 1 - |x - n|\}$, then $\alpha_n f_n \to 0$ as $n \to \infty$ for any sequence of scalars α_n.

28. Consider $x_n = (1, \frac{1}{2}, \frac{1}{3}, \ldots, \frac{1}{n}, 0, 0, \ldots)$. If $1 \leq p < q$, the convergence in l^p implies convergence in l^q, but not conversely.

29.

$$\max\{|z_1|, \ldots, |z_N|\} \leq \sqrt{\sum_{n=1}^{N} |z_n|^2} \leq \sum_{n=1}^{N} |z_n| \leq N \max\{|z_1|, \ldots, |z_N|\}.$$

30.

$$\|f\|_1 = \max_{x \in [0,1]} |f(x)|, \qquad \|f\|_2 = \int_0^1 |f(x)| dx.$$

38. No.

41. Modify the proof in Example 1.4.6.

42. Consider the space of polynomials on $[0, 1]$ with the norm

$$\|f\| = \max_{x \in [0,1]} |f(x)|.$$

43. Every sequence (x_n) convergent to 0 has a subsequence (x_{p_n}) such that $\|x_{p_n}\| \leq 1/2^n$.

44. No.

45.

$$\max_{x \in [0,1]} \left| \int_0^x f(t) dt \right| \leq \max_{x \in [0,1]} |f(x)|.$$

46. See Exercise 52.

52. No.

53. $(0, 0), (1, 1), (-1, -1)$.

54. No.

55.

$$\max_{1 \leq k \leq N} \sum_{n=1}^{N} |a_{kn}| < 1.$$

2.16 Exercises

3. Show that if $\lambda_1 f_1 + \cdots + \lambda_n f_n = 0$, then $\lambda_1 \int f_1 + \cdots + \lambda_n \int f_n = 0$.

5. If g_k denotes the characteristic function of $[a_k, b_k)$, then

$$-M(g_1 + \cdots + g_n) \leq f \leq M(g_1 + \cdots + g_n).$$

8. The proof is easy if f is the characteristic function of an interval $[a, b)$. Then generalize for step functions.

10. Expand f into a series of step functions, and then replace every step function in that series by its basic representation.

13. No.

16. First prove it for a positive function. Then use the representation $f = f^+ - f^-$, where f^+ and f^- are as defined in Exercise 15.

17. Expand the characteristic function of an interval $[a, b)$ into a series of tent functions, and conversely.

22. See Exercise 17.

24.

$$f(x) = \sum_{n=1}^{\infty} n e^{-n^6 (x-n)^2}.$$

27. If $f \simeq f_1 + f_2 + \cdots$ and $f_1, f_2, \ldots$ are step functions, then $f_1 g, f_2 g, \ldots$ are integrable and $fg \simeq f_1 g + f_2 g + \cdots$.

29. Use the Lebesgue dominated convergence theorem.

31. Redefine f_n's at those points x where $|f_n(x)| > h(x)$.

36. Show that for

$$f_n(x) = \begin{cases} f(x) & \text{if } x \in A_n, \\ 0 & \text{if } x \notin A_n, \end{cases}$$

we have $f_n \to 0$ a.e. Then use the Lebesgue dominated convergence theorem.

46. Note that

$$\frac{(f * g)(x + h) - (f * g)(x)}{h} = \int f(t) \frac{g(x + h - t) - g(x - t)}{h} dt,$$

and then prove that

$$\lim_{h \to 0} \int f(t) \frac{g(x + h - t) - g(x - t)}{h} dt$$
$$= \int f(t) \lim_{h \to 0} \frac{g(x + h - t) - g(x - t)}{h} dt.$$

47. Compare with Exercise 46.

3.8 Exercises

4. (a) No. (b) It is an inner product space, but it is not a Hilbert space.

5. It is an inner product space, but it is not a Hilbert space.

7. Prove, using Schwarz's inequality, that $\|x + y\| = \|x\| + \|y\|$ if and only if x and y are linearly independent.

9. Use

$$\langle x \pm \alpha y, x \pm \alpha y \rangle = \|x\|^2 \pm \overline{\alpha}\langle x, y \rangle \pm \alpha \langle y, x \rangle + |\alpha|^2 \|y\|^2.$$

11. Use Schwarz's inequality and the fact that $\|x\| = \langle x, \|x\|/x \rangle$ if $x \neq 0$.

12. See, for example, Friedman (1982), Theorem 6.1.5.

13. No. The space does not satisfy the parallelogram law.

15. Use the parallelogram law or verify by direct calculations.

18. Yes. Use continuity of the inner product.

19. For (c) and (d), use an orthonormal sequence.

34. $a = c = 0$, $b = \frac{\sqrt{6}}{5}$.

36. $x = \sum_{n=1}^{\infty} \alpha_n a_n$.

46.

$$\frac{1}{\sqrt{b-a}},$$

$$\sqrt{\frac{2}{b-a}} \cos \frac{n\pi(2x-b-a)}{b-a}, \quad n = 1, 2, \ldots,$$

$$\sqrt{\frac{2}{b-a}} \sin \frac{n\pi(2x-b-a)}{b-a}, \quad n = 1, 2, \ldots.$$

47. The set of all functions in $L^2(\mathbb{R})$ such that $f(-x) = -f(x)$ almost everywhere.

48. The set of all functions in $L^2([-\pi, \pi])$ such that $f(-x) = f(x)$ almost everywhere.

49. Use Theorem 3.6.2.

52. Consider, for example, the set $S = \{(x, 0): x \in \mathbb{R}\}$ and the point $x_0 = (0, 1)$.

53. $y = \sum_{n=1}^{\infty} \langle x, e_n \rangle e_n$.

55. Consider, for example, $x_n(t) = ne^{-n^6(t-t_0)^2}$.

58. The sequence

$$(1, 0, 0, \ldots), (0, 1, 0, 0, \ldots), (0, 0, 1, 0, 0, \ldots), \ldots$$

is a complete orthonormal system in l^2.

4.12 Exercises

2. Define $\varphi(x, y) = \langle Ax, y \rangle$ and use the polarization identity.

5. Compare with the proof of Schwarz's inequality.

7. $\lambda_1 = 4, x_1 = (2, 3), \lambda_2 = -1, x_2 = (1, -1)$.

8. $T^*(x, y) = (x + 2y, 3x + y)$.

28. No. Consider, for example, $T : \mathbb{R}^2 \to \mathbb{R}^2$ defined by

$$T(x, y) = \left(\frac{x}{2} + \frac{y}{3}, \frac{\sqrt{3}x}{2} - \frac{y}{3\sqrt{3}} \right).$$

32. See, for example, Friedman (1982), Lemma 6.6.6.

34. $\sqrt{\mathcal{I}} = \mathcal{I}$.

37. Use Theorem 4.6.14.

39. No. Consider, for example, $P = -\mathcal{I}$.

41. Find an example in $\mathbb{R}^2$.

50. Use Corollaries 4.8.13 and 4.10.2.

55. See Example 4.5.18.

5.13 Exercises

5. Show that $(Tf)(x) = \int_0^1 e^{-sx} \cos(\alpha f(s)) ds$ is a contraction in $\mathcal{C}([0, 1])$.

7. Set $c_2 = 1$. Then $c_1 = \sqrt{3}, \lambda_1 = \frac{\sqrt{3}}{2}, f_1(x) = \frac{\sqrt{3}}{2}(\sqrt{3}x + 1); c_1 = -\sqrt{3}$, $\lambda_2 = \frac{\sqrt{3}}{2}, f_1(x) = \frac{\sqrt{3}}{2}(\sqrt{3}x - 1)$.

8. $\lambda_1 = \frac{\pi}{2}$,

$$f_1(x) = \frac{\pi c}{2} \left[\frac{1}{2} \pi^2 x^2 - \left(1 + \frac{\pi}{4} \right) \pi x \right],$$

where c is an arbitrary constant. Similarly, f_2 can be found for $\lambda_2 = -\frac{\pi}{2}$.

12. Apply the method of separable kernel. $D(\lambda) = \frac{1}{12}(12 - 12\lambda - \lambda^2)$. For $D(\lambda) \neq 0$,

$$c_1 = \frac{b_1 - \lambda\left(\frac{b_1}{2} - b_2\right)}{D(\lambda)}, \qquad c_2 = \frac{b_2 + \lambda\left(\frac{b_1}{3} - \frac{b_2}{2}\right)}{D(\lambda)}.$$

The resolvent kernel is

$$\Gamma(x, t, \lambda) = \frac{D(x, t, \lambda)}{D(\lambda)},$$

where

$$D(x, t, \lambda) = \begin{bmatrix} 0 & M_1(x) & M_2(x) \\ N_1(t) & 1 - \lambda a_{11} & -\lambda a_{12} \\ N_2(t) & -\lambda a_{21} & 1 - \lambda a_{22} \end{bmatrix}$$

$$= -\left((x + t) - \frac{\lambda}{2}(x + t) + \lambda tx + \frac{\lambda}{3}\right).$$

Answer: $f(x) = \varphi(x) + \lambda \int_0^1 \Gamma(x, t, \lambda)dt.$

14. (a) $\lambda_1 = \lambda_2 = \frac{1}{\pi}, f(x) = c_1 \cos x + c_2 \sin x,$

   (b) $\lambda_1 = \frac{\sqrt{3}}{2}i, f_1(x) = 1 - i\sqrt{3}x, \lambda_2 = -\frac{\sqrt{3}}{2}i, f_2(x) = 1 + i\sqrt{3}x.$

15. (d) $f(x) = -2.$

17. (a) $f(x) = \frac{1}{2}(3x + 1)$, (b) $f(x) = \sin x$, (c) $f(x) = \sinh x$, (d) $f(x) = e^{-x^2}.$

20.

$$\langle Lu, v \rangle = \int_0^1 v(x)\left(e^x \frac{d^2}{dx^2} + e^x \frac{d}{dx}\right)u(x)dx$$

$$= \int_0^1 v(x)\left(e^x u'(x)\right)' dx$$

$$= \int_0^1 uLvdx = \langle u, Lv \rangle.$$

22.

$$\lambda_n = \frac{(2n - 1)^2}{4}, \qquad u_n(x) = B_n \sin\left(\frac{2n - 1}{2}\right)x, \qquad n = 1, 2, \ldots.$$

23. $\lambda_n = n^2\pi^2, u_n(x) = B_n \sin(n\pi \ln x), n = 1, 2, \ldots.$

25. $\langle Tu, u \rangle = \langle (DpD + q)u, u \rangle = \langle DpDu, u \rangle + q\langle u, u \rangle = -p\|Du\|^2 + q\|u\|^2.$

26.

$$\langle u, v \rangle = \int_a^b u(x)\overline{v(x)}r(x)dx.$$

29. (a) $\sqrt{\frac{2}{\pi}}\frac{\sin a\omega}{\omega}$, (b) $\sqrt{\frac{2}{\pi}}(\frac{\sin\omega}{\omega})^2$.

31. (a) π, (b) $\frac{3\pi}{4}$, (c) $\frac{2\pi}{3}$.

36.

$$I(t) = \left(\frac{\omega E_0}{AL} + I_0\right)e^{-Rt/L} + \frac{E_0}{AL}\left(\frac{R}{L} - \alpha\right)e^{-\alpha t}\sin\omega t - \frac{\omega E_0}{AL}e^{-\alpha t}\cos\omega t,$$

where $A = (\alpha - R/L)^2 + \omega^2$.

37. $\hat{I}(\omega) = \frac{C}{\sqrt{2\pi}}(1 + ikRC - LCk^2)^{-1}$.

41. (b) $\mathcal{F}\{Tu\} = \frac{2}{1+k^2}\mathcal{F}\{u\}$. Since $\mathcal{F}$ is unitary, we have

$$\|\mathcal{F}\{Tu\}\| = \|Tu\| \le 2\|u\|.$$

6.7 Exercises

1. If f is a continuous function that is not identically zero, then there exists a $\varphi \in C^\infty(\mathbb{R}^N)$ with compact support such that $\int_{\mathbb{R}^N} f(x)\varphi(x)dx \neq 0$.

5. Use the function

$$f(t) = \begin{cases} 0 & \text{if } t \le 0, \\ e^{-1/t^2} & \text{if } t > 0. \end{cases}$$

6. (a), (b), (e), (i) Yes. (c), (f), (g), (h), (j) No. (d) No, if (x_n) has a convergent subsequence.

10. $f^{(n)} = 2\delta^{(n-1)}$.

11. Use the Riemann–Lebesgue lemma (Theorem 5.11.6).

13. Note that, for every $\varepsilon > 0$,

$$\lim_{n\to\infty}\int_{|x|\ge\varepsilon} e^{-n^2x^2}dx = 0.$$

14. Use the Riemann–Lebesgue lemma (Theorem 5.11.6) to show that

$$\lim_{n\to\infty}\int_{-\infty}^{\infty}\sin nx\frac{\varphi(x) - \varphi(0)}{x}dx = 0.$$

15. $K = \int_{-\infty}^{\infty} \varphi(x)dx$, $\varphi_1 = \varphi - K\varphi_0$.

18. Use the Laplace transform and the double Fourier transforms of $u(x, y, t)$ to obtain

$$\tilde{u}(k, l, s) = \left[s + K\left(k^2 + l^2\right)\right]\tilde{f}(k, l).$$

Application of the inverse transforms combined with the convolution theorem gives the solution.

19. The Green's function associated with the given equation satisfies the equation

$$G_{tt} - c^2 \nabla^2 G = \delta(x)\delta(y)\delta(z)$$

with the same boundary and initial data. Application of the joint Laplace and the three-dimensional Fourier transform gives

$$\overline{\tilde{G}} = (2\pi)^{-3/2}\left(s^2 + c^2\kappa^2\right), \quad \kappa^2 = k^2 + l^2 + m^2.$$

The joint inverse transforms give

$$G(\mathbf{x}, t) = \frac{1}{(2\pi)^{3/2}} \int_{-\infty}^{\infty}\int_{-\infty}^{\infty}\int_{-\infty}^{\infty} \kappa^{-1} \sin(ckt) \exp(i\underline{\kappa} \cdot \mathbf{x})d\underline{\kappa},$$

where $\underline{\kappa} = (k, l, m)$. The use of spherical polar coordinates with the polar axis along the vector $\mathbf{x}$, so that $\underline{\kappa} \cdot \mathbf{x} = \kappa r \cos\theta$, $r = |\mathbf{x}|$, θ is the polar angle, and $d\underline{\kappa} = \kappa^2 d\kappa \sin\theta \, d\theta \, d\phi$, gives the Green's function.

20. Apply the joint Laplace and triple Fourier transform to find

$$\overline{\tilde{G}}(k, s) = \frac{1}{(2\pi)^{3/2}} \frac{1}{s^2 + c^2\kappa^2 + d^2}, \quad \kappa^2 = k^2 + l^2 + m^2.$$

The joint inverse transform gives the solution

$$G(\mathbf{x}, t) = \frac{1}{(2\pi)^{3/2}} \int_{-\infty}^{\infty}\int_{-\infty}^{\infty}\int_{-\infty}^{\infty} e^{i(\underline{\kappa}\cdot\mathbf{x})} \frac{\sin\alpha t}{\alpha}d\underline{\kappa},$$

where $\underline{\kappa} = (k, l, m)$, and $\alpha = (c^2\kappa^2 + d^2)^{1/2}$. In terms of the spherical polar coordinates with the vector $\mathbf{x}$ as the polar axis, so that $\underline{\kappa} \cdot \mathbf{x} = \kappa r \cos\theta$, $r = |\mathbf{x}|$, θ is the polar angle, and $d\underline{\kappa} = \kappa^2 d\kappa \sin\theta \, d\theta \, d\phi$, we obtain

$$G(\mathbf{x}, t) = \frac{1}{(2\pi)^3} \int_0^{2\pi} d\phi \int_0^{\infty} \frac{\sin\alpha t}{\alpha}\kappa^2 d\kappa \int_0^{\pi} \exp(i\kappa r \cos\theta) \sin\theta \, d\theta$$

$$= -\frac{1}{2\pi^2 r} \int_0^{\infty} \frac{\sin\alpha t}{\alpha} \frac{\partial}{\partial r}(\cos\kappa r)d\kappa$$

$$= -\frac{1}{4\pi rc} \frac{\partial}{\partial r} J_0\left[\frac{d}{c}\sqrt{c^2 t^2 - r^2}\right]H\left(c^2 t^2 - r^2\right)$$

$$= \frac{1}{2\pi c}\left[\delta(c^2 t^2 - r^2) J_0\left(\frac{d}{c}\sqrt{c^2 t^2 - r^2}\right)\right.$$

$$\left. - \frac{d}{2c}\frac{J_1(\frac{d}{c}\sqrt{c^2 t^2 - r^2})}{(c^2 t^2 - r^2)^{1/2}} H(c^2 t^2 - r^2)\right].$$

21. Apply the joint Laplace and Fourier transform to obtain

$$\tilde{\bar{u}}(k, s) = \left(s^2 + c^2 k^2\right)^{-1}\left[s\tilde{f}(k) + \tilde{g}(k) + \tilde{\bar{q}}(k, s)\right].$$

The inverse Laplace transform combined with the convolution theorem yields

$$\tilde{u}(k, t) = \tilde{f}(k)\cos(ckt) + \frac{\tilde{g}(k)}{ck}\sin(ckt) + \frac{1}{ck}\int_0^t \sin ck(t - \tau)\tilde{q}(k, \tau)d\tau.$$

The use of the inverse Fourier transform leads to the final answer:

$$u(x, t) = \frac{1}{2}\left[f(x - ct) + f(x + ct)\right] + \frac{1}{2c}\int_{x-ct}^{x+ct} g(\xi)d\xi$$

$$+ \frac{1}{2c}\int_0^t d\tau \int_{x-c(t-\tau)}^{x+c(t-\tau)} p(\xi, \tau)d\xi.$$

22. The use of the joint Laplace and Fourier transform gives

$$\widehat{\bar{G}}(k, s) = \frac{1}{\sqrt{2\pi}}\frac{1}{s^2 + c^2 k^2}.$$

The inverse Fourier transform yields

$$\bar{G}(x, s) = \frac{1}{2cs}\exp\left(-\frac{s}{c}|x|\right)$$

and hence the inverse Laplace transform gives the answer:

$$G(x, t) = \frac{1}{2c}H(t)\left[H(x + ct) - H(x - ct)\right] = \frac{1}{2c}H(ct - |x|).$$

23. (b) $u(x, t) = \int_{-\infty}^{\infty} G(x, \xi, t)g(\xi)d\xi$, where $G(x, \xi, t)$ is obtained in part (a). Since $G(x, \xi, t) = \frac{1}{2c}$ if $x - ct < \xi < x + ct$ and 0 elsewhere, we have $u(x, t) = \frac{1}{2c}\int_{x-ct}^{x+ct} g(\xi)d\xi$.

29. Define a new function $v(x, y) = u_y(x, y)$ so that $u(x, y) = \int^y v(x, \eta)d\eta$. Then $v(x, y)$ satisfies the Laplace equation $v_{xx} + v_{yy} = 0$ with

$v(x, 0) = g(x)$. Then use Example 6.6.3 to find $v(x, y)$ so that

$$u(x, y) = \int^y v(x, \eta) d\eta = \frac{1}{2\pi} \int_{-\infty}^{\infty} g(\xi) \log[(x - \xi)^2 + y^2] d\xi.$$

30.

$$u(x, y) = \frac{1}{2a} \sin\left(\frac{\pi y}{a}\right) \int_{-\infty}^{\infty} \frac{f(t) dt}{\cosh\left(\frac{\pi}{a}(x - t)\right) - \cos\frac{\pi y}{a}}$$

$$+ \frac{1}{2a} \sin\left(\frac{\pi y}{a}\right) \int_{-\infty}^{\infty} \frac{g(t) dt}{\cosh\left(\frac{\pi}{a}(x - t)\right) + \cos\frac{\pi y}{a}}.$$

31.

$$u(x, y) = \sin\left(\frac{\pi y}{2a}\right) \int_{-\infty}^{\infty} \frac{f(t) \cosh\left(\frac{\pi}{2a}(x - t)\right) dt}{\cosh\left(\frac{\pi}{a}(x - t)\right) - \cos\frac{\pi y}{a}}.$$

33.

$$\varphi(x, y) = \frac{1}{2a} \int_{-\infty}^{\infty} \frac{J_0(ak)}{k} e^{ikx - |k|y} dk.$$

36. (a) Matching conditions at $x = \pm a$ are $\psi_2(a) = \psi_3(a)$ and $\psi_1(-a) = \psi_2(-a)$ that give $B \sin ak + C \cos ak = De^{-ak}$ and $-B \sin ak + C \cos ak = Ae^{-ak}$. Matching the derivatives at $x = \pm a$ are $\psi_2'(a) = \psi_3'(a)$ and $\psi_1'(-a) = \psi_2'(-a)$ which yield $Bk \cos ak - Ck \sin ak = -Dke^{-ak}$ and $Bk \cos ak + Ck \sin ak = Ake^{-ak}$. Adding and subtracting two sets of matching conditions and putting $A + D = A_1$ and $A - D = -A_2$ give $2B \sin ak - A_2 e^{-ak} = 0$ and $2Bk \cos ak + A_2 \kappa e^{-ak} = 0$, $2C \cos ak - A_1 e^{-ak} = 0$ and $2Ck \sin ak - A_1 \kappa e^{-ak} = 0$. For nontrivial solutions, we obtain

$$\begin{vmatrix} 2 \sin ak & -e^{-ak} \\ 2k \cos ak & \kappa e^{-ak} \end{vmatrix} = 0 \quad \text{and} \quad \begin{vmatrix} 2 \cos ak & -e^{-ak} \\ 2k \sin ak & -\kappa e^{-ak} \end{vmatrix} = 0.$$

These give the desired solutions.

37. Use the Fourier transform of $g_t(x)$ and the Convolution theorem of the Fourier transform.

7.12 Exercises

1. (a) $\frac{d}{dt} \frac{\partial L}{\partial \dot{x}_i} - \frac{\partial}{\partial x_i} = 0$ implies $m\ddot{x}_i - kx_i = 0$. Multiply this equation by $\dot{x}_i$ and integrate to obtain $\frac{1}{2} m\dot{x}_i^2 + \frac{1}{2} kx_i^2 = $ constant.

(b) Use (7.259abc) and L in 1(a) and show that it becomes the expression for L in 1(b).

2.

$$p_r = \frac{\partial L}{\partial \dot{r}} = m\dot{r}, \qquad p_\theta = \frac{\partial L}{\partial \dot{\theta}} = mr^2\dot{\theta},$$

$$\text{where } L = T - V = \frac{1}{2}m(\dot{r}^2 + r^2\dot{\theta}^2) - V(r);$$

$$H = T + V = \frac{1}{2}m(\dot{r}^2 + r^2\dot{\theta}^2) + V(r) = \frac{1}{2}\frac{p_r^2}{m} + \frac{p_\theta^2}{2mr^2} + V(r),$$

$$\dot{r} = \frac{\partial H}{\partial p_r} = \frac{p_r}{m}, \qquad \dot{\theta} = \frac{\partial H}{\partial p_\theta} = \frac{p_\theta}{mr^2}.$$

Then use $\dot{p}_r = -\frac{\partial H}{\partial r}$ and $\dot{p}_\theta = -\frac{\partial H}{\partial \theta}$.

4. (a)

$$p = \frac{\partial T}{\partial \dot{x}} = m\dot{x}, \qquad H = \frac{1}{2}m\left(\frac{p}{m}\right)^2 + \frac{1}{2}kx^2 = \frac{1}{2}\left(\frac{p^2}{m} + kx^2\right),$$

$$\frac{\partial H}{\partial p} = \frac{p}{m} = \dot{x}, \qquad \frac{\partial H}{\partial x} = kx = -\dot{p} \quad \text{implies} \quad (\ddot{x}, \ddot{p}) = -\frac{k}{m}(x, p).$$

(b)

$$p_r = \frac{\partial T}{\partial \dot{r}} = m\dot{r}, \qquad p_\theta = \frac{\partial T}{\partial \dot{\theta}} = mr^2\dot{\theta},$$

$$H = \frac{1}{2m}\left(p_r^2 + \frac{p_\theta^2}{r^2}\right) + m\mu\left(\frac{1}{2a} - \frac{1}{r}\right), \qquad \frac{\partial H}{\partial p_r} = \frac{p_r}{m} = \dot{r},$$

$$\frac{\partial H}{\partial r} = -\frac{p_\theta^2}{mr^3} + \frac{m\mu}{r^2} = -\dot{p}_r, \qquad \frac{\partial H}{\partial p_\theta} = \frac{p_\theta}{mr^2} = \dot{\theta},$$

$$\frac{\partial H}{\partial \theta} = 0 = -\dot{p}_\theta.$$

Thus, $\ddot{r} - r\dot{\theta}^2 = -\frac{\mu}{r^2}$ and $\frac{d}{dt}(r^2\dot{\theta}) = 0$.

8. (c)

$$[\hat{x}^2, \hat{p}_x^2] = \hat{x}^2\hat{p}_x^2 - \hat{p}_x^2\hat{x}^2$$

$$= \hat{x}[\hat{x}, \hat{p}_x]\hat{x} + \hat{x}\hat{p}_x[\hat{x}, \hat{p}_x] + \hat{p}_x[\hat{x}, \hat{p}_x]\hat{x} + [\hat{x}, \hat{p}_x] = 4i\hbar\hat{x}\hat{p}_x + 2\hbar^2.$$

(d)

$$[\hat{p}_x, \hat{x}^2] = \hat{x}[\hat{p}_x, \hat{x}] + [\hat{p}_x, \hat{x}]\hat{x} = -\hat{x}[\hat{x}, \hat{p}_x] - [\hat{x}, \hat{p}_x]\hat{x} = -2i\hbar\hat{x}.$$

10. $\left[\hat{L}^2, \hat{L}_x\right] = \left[\hat{L}_x^2 + \hat{L}_y^2 + \hat{L}_z^2, \hat{L}_x\right].$

11. Use (7.267ab).

12. Use (7.306).

15.

$$\left[\hat{L}_x, \hat{x}\right] = \left[\hat{y}\hat{p}_z - \hat{z}\hat{p}_y, \hat{x}\right] = \left[\hat{y}\hat{p}_z, \hat{x}\right] - \left[\hat{z}\hat{p}_y, \hat{x}\right],$$

$$\left[\hat{L}_x, \hat{L}_y\right] = \left[\hat{y}\hat{p}_z - \hat{z}\hat{p}_y, \hat{z}\hat{p}_x - \hat{x}\hat{p}_z\right]$$

$$= \hat{y}\hat{p}_x\left[\hat{p}_z, \hat{z}\right] + \hat{x}\hat{p}_y\left[\hat{z}, \hat{p}_z\right] = i\hbar(\hat{x}\hat{p}_y - \hat{y}\hat{p}_x) = i\hbar\hat{L}_z.$$

19.

$$E_N = E_{n_1} + E_{n_2} + E_{n_3} = \left(n_1 + \frac{1}{2}\right)\hbar\omega + \left(n_2 + \frac{1}{2}\right)\hbar\omega + \left(n_3 + \frac{1}{2}\right)\hbar\omega.$$

21.

$$\langle\psi|\psi\rangle = \left\langle \sum_{n=1}^{\infty}\langle\psi_n|\psi\rangle\psi_n \left| \sum_{k=1}^{\infty}\langle\psi_k|\psi\rangle\psi_k \right. \right\rangle$$

$$= \sum_{n=1}^{\infty}\sum_{k=1}^{\infty}\langle\psi_n|\psi\rangle^*\langle\psi_k|\psi\rangle\langle\psi_n|\psi_k\rangle.$$

22. (a) Use the fact that $\hat{A}'$ is the difference of two Hermitian operators, and then show that $\langle\psi|\hat{A}'\varphi\rangle = \langle\hat{A}'\psi|\varphi\rangle$ for any ψ and ψ.

(b) Use the fact that $\hat{A}\langle\hat{B}\rangle = \langle\hat{B}\rangle\hat{A}$, since $\hat{A}$ is a linear operator and $\langle\hat{B}\rangle$ is a scalar.

(c) $\langle\hat{A}'\psi|\hat{A}'\psi\rangle = \langle\psi|(\hat{A}')^2\psi\rangle = \langle\psi|[\hat{A} - \langle A\rangle]^2\psi\rangle.$

24. Note that

$$\left[\hat{x}\hat{p}_x, \hat{H}\right] = \left[\hat{x}\hat{p}_x, \frac{\hat{p}_x^2}{2m} + V\right] = \left[\hat{x}\hat{p}_x, \frac{\hat{p}_x^2}{2m}\right] + \left[\hat{x}\hat{p}_x, V\right],$$

and then $\hat{x}\hat{p}_x - \hat{p}_x\hat{x} = i\hbar.$

28. Write $|\varphi\rangle = \hat{a}|\psi\rangle$. Then

$$\langle\varphi|\varphi\rangle = \langle\psi|\hat{a}^*|\hat{a}|\psi\rangle = \langle\psi\left|\hat{H} - \frac{1}{2}\hbar\omega\right|\psi\rangle = \left|E - \frac{1}{2}\hbar\omega\right|$$

if $|\psi\rangle$ is normalized.

By positive definiteness of the inner product, $E \geq \frac{1}{2}\hbar\omega$, and $E = \frac{1}{2}\hbar\omega$ implies $a|\psi\rangle = 0$.

29. Show that $[\hat{H}, \hat{L}_k] = +ci\hbar(\hat{p} \times \hat{a})_k \neq 0, k = 1, 2, 3$.

30.

$$(E - mc^2)a_1 - cp_1a_3 - c(p_1 - ip_2)a_4 = 0,$$
$$(E - mc^2)a_2 - c(p_1 + ip_2)a_3 + cp_3a_4 = 0,$$
$$(E + mc^2)a_3 - cp_3a_1 - c(p_1 - ip_2)a_2 = 0,$$
$$(E + mc^2)a_4 - c(p_1 + ip_2)a_1 + cp_3a_2 = 0.$$

Existence of a nontrivial solution of this system of equations implies that the determinant of the coefficient matrix is zero, which gives $(E^2 - m^2c^4 - c^2\mathbf{p}^2)^2 = 0$.

33. $\psi_3(x) = B\sin kx + C\cos kx$, where B and C are arbitrary constants, is the general solution of $\psi_{xx} + k^2\psi = 0$, $k^2 = \frac{2mE}{\hbar^2}$. For matching the solutions at $x = \pm a$, $\psi_3(a) = B\sin ak + C\cos ak = 0 = \psi_2(a)$ and $\psi_3(-a) = -B\sin ak + C\cos ak = 0 = \psi_1(-a)$. For nontrivial solution of B and C, $\sin ak\cos ak = 0$. For an even solution, $\cos ak = 0$ and hence $ak = (2n+1)\frac{\pi}{2}$, which gives the energy levels, $E_n = \frac{(2n+1)^2\pi^2\hbar^2}{8ma^2}$. Similarly, for an odd solution, $ak = n\pi$ and $E_n = \frac{n^2\pi^2\hbar^2}{2ma^2}$.

34. For the free Dirac equation, $(c\boldsymbol{\alpha} \cdot \hat{\mathbf{p}} + \beta mc^2)\psi = E\psi$. Or $(c\alpha_z p_z + \beta mc^2)\psi = E\psi$. Hence,

$$\begin{pmatrix} mc^2\mathbf{1} & cp_z\sigma_3 \\ cp_z\sigma_3 & -mc^2\mathbf{1} \end{pmatrix}\psi = E\psi,$$

which can be explicitly written in the form

$$\begin{pmatrix} mc^2 - E & 0 & cp_z & 0 \\ 0 & mc^2 - E & 0 & -cp_z \\ cp_z & 0 & -mc^2 - E & 0 \\ 0 & -cp_z & 0 & -mc^2 - E \end{pmatrix}\begin{pmatrix} \psi_1 \\ \psi_2 \\ \psi_3 \\ \psi_4 \end{pmatrix} = 0.$$

This leads to $(E^2 - c^2p_z^2 - m^2c^4)^2 = 0$, and hence $E = E_+, E_+, E_-, E_- = \pm\sqrt{c^2p_z^2 + m^2c^4}$.

35. (a) $(p_\mu p^\mu)\psi = (\frac{E^2}{c^2} - \mathbf{p}^2)\psi = m^2c^2\psi$.

(b)

$$p^\mu \rightarrow i\hbar\partial^\mu = i\hbar\frac{\partial}{\partial x_\mu} = \left(\frac{i\hbar}{c}\frac{\partial}{\partial t}, -i\hbar\nabla\right),$$

$$p_\mu \to i\hbar\partial_\mu = i\hbar\frac{\partial}{\partial x^\mu} = \left(\frac{i\hbar}{c}\frac{\partial}{\partial t}, i\hbar\nabla\right),$$

$$i\hbar\partial_\mu\left(i\hbar\partial^\mu\right)\psi = m^2c^2\psi.$$

Or,

$$-\hbar^2\left(\frac{1}{c^2}\frac{\partial^2}{\partial t^2} - \nabla^2\right)\psi = m^2c^2\psi.$$

This leads to the KG equation.

(c) Plane wave solutions of the form

$$\psi(\mathbf{r}, t) = \exp\left[i(\mathbf{k}\cdot\mathbf{r} - \omega t)\right].$$

They are eigenfunctions of $i\hbar\frac{\partial}{\partial t}$ and $-i\hbar\nabla$ with eigenvalues $E = \hbar\omega$ and $\mathbf{p} = \hbar\mathbf{k}$ respectively. Substituting in the KG equation gives

$$\hbar\omega = \pm\sqrt{(\hbar c k)^2 + m^2c^4} \quad \text{(positive and negative energies).}$$

This is a consequence of the fact that the KG equation, unlike the Schrödinger equation, is a second-order equation in the time derivative.

(d) Both ψ and $\overline{\psi}$ satisfy the KG equation

$$\left(\nabla^2 - \frac{1}{c^2}\frac{\partial^2}{\partial t^2}\right)\psi = \frac{m^2c^2}{\hbar^2}\psi.$$

Multiplying ψ equation by $\overline{\psi}$ and $\overline{\psi}$ equation by ψ, and subtracting these results gives the desired answer. Substituting the plane wave solutions in equation (d) reveals that $\rho(\mathbf{r}, t) = (\hbar\omega/mc^2)$ which is not positive definite since $(\hbar\omega)$ can be positive as well as negative. Unlike the case of the Schrödinger equation, $\rho(\mathbf{r}, t)$ in the present case cannot be thought of as a probability density, which has to be strictly nonnegative. For this reason, the Klein–Gordon equation was abandoned for a long time as inadequate in quantum mechanics.

8.6 Exercises

2. Consider $\int_{-\infty}^{0}\ldots da$ and $\int_{0}^{\infty}\ldots da$ separately.

5. Use the polarization identity.

6. Use Theorem 3.4.14.

7. Use Theorem 5.11.21.

10. See Debnath (2002).

11. See Debnath (2002).

12. See Debnath (2002).

13. See Debnath (2002).

14. See Newland (1993, 1994) or Debnath (2002).

15. See Newland (1993, 1994) or Debnath (2002).

9.11 Exercises

2. Use the fact that

$$\langle u - Tx, u - Tx \rangle = \langle u, u \rangle - 2\langle u, Tx \rangle + \langle Tx, Tx \rangle$$
$$= \langle u, u \rangle - 2\langle T^*u, x \rangle + \langle x, T^*Tx \rangle,$$

and then use $F(x) = -2T^*u + 2T^*Tx$.

5. (b) $y(x) = \frac{1}{2}e^{-x} + \frac{1}{2}(1 + e)xe^{-x} - \frac{1}{2}$.

   (d) $y_0(x) = \frac{\sin x}{\sin a}$.

8. $I(y, z) = \int_{x_1}^{1} \sqrt{1 + (y')^2 + (z')^2}\,dx$, where $Q(x_1, x_2, x_3)$ lies on the sphere.

13. Maximize A subject to the condition $\int_0^1 \sqrt{1 + (y')^2}\,dx = L$. In other words, maximize the functional $I_1(y) = \int_0^1 \left[y(x) + \sqrt{1 + (y')^2} - L \right] dx$. Answer: $(x - \alpha)^2 + (y - \beta)^2 = \lambda^2$ where α, β and λ are constants.

24. This polynomial is the real part of the binomial expansion of $(\cos\theta + i\sin\theta)^n$, where $x = \cos\theta$.

30. Use repeated integration by parts to show that

$$\int_{-1}^{1} D^n\left[(x^2 - 1)^n \right] x^m\,dx = 0, \quad m = 0, 1, \ldots, n - 1,$$

and then find the leading coefficient of $D^n\left[(x^2 - 1)^n \right]$.

31. Use Rodrigues' formula and the binomial expansion of $(x^2 - 1)^n$.

32. (b)

$$D^{n+1}\left[(x^2 - 1)^n \right] = D^n\{ D[(x^2 - 1)^{n-1}(x^2 - 1)] \}$$
$$= 2nD^n\left[x(x^2 - 1)^{n-1} \right]$$
$$= 2n\{ xD^n\left[(x^2 - 1)^{n-1} \right] + nD^{n-1}\left[(x^2 - 1)^{n-1} \right] \}$$

and then use Rodrigues' formula.

34. (a) Use the recurrence relation 32(a) and 32(b). (b) Use 32(b) and 34(a).

35. (a) Multiply the equality in 34(a) by x and subtract from the equality in 32(b). (b) Square and add the equalities in 32(b) and 34(b).

38. Transform into polar coordinates.

41. $w = 0$ unless $\lambda = n^2$. If $\lambda = n^2$, $w_n = A_n \sin nx$.

42. (a) $w = A \sin nx$, $A[-n^2 - (\lambda - A^2)] = 0$.

   (b) $w = A \sin nx$, $A(\lambda - A^2 n^2) = 0$.

44. The term within the first bracket of the equation can be replaced by a constant μ. Answer: $\mu = n^2 \pi^2$, $w = A_n \sin n\pi x$, $A_n^2 = 4\left(\frac{\lambda}{n^2 \pi^2} - 1\right)$.

46. Note that the equality in the square bracket is a constant and can be replaced by a constant a. Then $a = a_n = n^2 \pi^2$, $u = A \sin n\pi x$, $\lambda - |A|^2 = n^2 \pi^2$. Draw the bifurcation diagram.

47. Square both sides of the equation and integrate from 0 to 1.

52. Maximize the functional

$$I(y) = \int_{x_1}^{x_2} \left[y(x) + \lambda \{ y' - \sqrt{1 + (y')^2} \} \right] dx.$$

Answer: $(y(x) - b)^2 + (x + a)^2 = \lambda^2$, where a, b, and λ are constants.

53. In terms of cylindrical coordinates $(x, y, z) = (r \cos \theta, r \sin \theta, z)$, points lying on $f(x, y, z,) = x^2 + y^2 - R^2 = 0$. On the surface $r = R$ and $\dot{x}(t) = -R \sin \theta \dot{\theta}$ and $\dot{y} = R \cos \theta \dot{\theta}$. The problem is to minimize the functional

$$I(x, y, z) = \int_{t_1}^{t_2} \left[R^2 \left(\sin^2 \theta + \cos^2 \theta \right) \dot{\theta}^2 + \dot{z}^2 \right]^{1/2} dt$$

$$= \int_{t_1}^{t_2} \sqrt{R^2 \dot{\theta}^2 + \dot{z}^2} \, dt.$$

The Euler–Lagrange equations are

$$\frac{d}{dt} \left[\frac{\dot{\theta}}{\sqrt{R^2 \dot{\theta}^2 + \dot{z}^2}} \right] = 0 \quad \text{and} \quad \frac{d}{dt} \left[\frac{\dot{z}}{\sqrt{R^2 \dot{\theta}^2 + \dot{z}^2}} \right] = 0.$$

These give $\frac{dz}{d\theta} = \frac{\dot{z}}{\dot{\theta}} = \text{constant} = c$. Consequently, $z = a\theta + b$, which is a spiral.

55. (a) In one case, there are two saddle-node bifurcation points, and in the other case, there are no bifurcation points.

56. (a) The normal form $\dot{x} = \lambda x - x^3$ has a single equilibrium at $x = 0$ for $\lambda < 0$, and three equilibrium points at $x = 0, \pm \sqrt{\lambda}$ when $\lambda > 0$. The

bifurcation at $x = 0$ and $\lambda = 0$ is called a *supercritical bifurcation*. The normal form $\dot{x} = \lambda x - x^3$ produces two equilibrium points at the bifurcation point which are unstable. This is called *subcritical bifurcation*.

(b) This normal system has equilibrium points at $x = 0$ and $x = \lambda$. When $\lambda = 0$, $\dot{x} = -x^2$ gives the nonhyperbolic equilibrium point. When $\lambda \neq 0$, there are two equilibrium points and hence,

$$\frac{d\dot{x}}{dx} = \lambda - 2x = \begin{cases} -\lambda & \text{at } x = \lambda \\ \lambda & \text{at } x = 0. \end{cases}$$

Consequently, $x = \lambda$ is stable for $\lambda > 0$, and unstable for $\lambda < 0$; and $x = -\lambda$ is stable for $\lambda < 0$ and unstable for $\lambda > 0$. This is called a *transcritical bifurcation*. It follows from the bifurcation diagram that, at the bifurcation point, the two equilibrium solutions pass through each other and exchange stabilities.

57. (a) For $\lambda < 0$, there are no equilibrium points, while for $\lambda > 0$, there are two hyperbolic equilibrium points at $x = \pm\sqrt{\lambda}$. When $\lambda = 0$, $X(x) = -x^2$ and $x = 0$ is a nonhyperbolic equilibrium point. Draw a bifurcation diagram for equilibrium solutions as a function of λ and find stable and unstable equilibria.

(b) Equilibrium points at $x = -\lambda \pm \sqrt{\lambda + \lambda^2}$, and hence no real equilibrium points for $-1 < \lambda < 0$ and two equilibrium points otherwise. There are saddle-node bifurcation at $\lambda = 0, x = 0$ and $\lambda = -1, x = 1$. At equilibrium points

$$\frac{d\dot{x}}{dx} = -2\lambda - 2x = \mp\sqrt{\lambda + \lambda^2}.$$

For large x, equilibrium point is stable while the other is unstable. For $x \ll 1$ and $\lambda \ll 1$, $\dot{x} = \lambda - 2\lambda x - x^2 \approx \lambda - x^2$, which is the same as 57 (a). Without any assumption about how large λ is compared to x^2, we examine the neighborhood of the other bifurcation point. Shift the origin using $\lambda = -1 + a\mu$, $x = 1 + by$ where a and b are constants to be determined later. In terms of y and μ, $\dot{y} = -\frac{a}{b}\mu - 2a\mu y - by^2 \approx -\frac{a}{b}\mu - by^2$ for $\mu \ll 1$ and $y \ll 1$. With $a = -1$ and $b = 1$, this gives the normal form of the saddle-node bifurcation. Thus, $\lambda = -(1 + \mu)$ indicates that the sense of the bifurcation is reversed with respect to λ.

58. (a) The origin is the only equilibrium point and the linearized system is $\dot{x} = -x$, $\dot{y} = 0$. The eigenvalues are 0 and -1 and so this is a nonhyperbolic equilibrium point. We construct a Lyapunov function by trying $V = x^2 + ay^2$ which is positive definite for $a > 0$ and $V(0, 0) = 0$. Further, $\frac{dV}{dt} = -2x^2 + 2(a - 2)xy^2 - 2ay^4$. With $a = 2$, $\dot{V} < 0$ for all x and y except $(x, y) = (0, 0)$. This leads to the solution.

(b) $\ddot{x} + \dot{x} + x + x^2 = 0$. Consider $V = \frac{1}{2}(x^2 + y^2) + \frac{1}{3}x^3$. $V(0,0) = 0$ and V is positive in the region $D = \{(x,y)|y^2 > -(x^2 + \frac{2}{3}x^3)\}$. Consider the curve $V = \frac{1}{6}$ which passes through the saddle point $(-1, 0)$ and the point $(\frac{1}{2}, 0)$. If $V = V_0 < \frac{1}{6}$, there is a curve that encloses the origin, but not the saddle point. For the domain of attraction of the equilibrium point at $(0, 0)$, consider a domain D_0 which is a subset of D given by $V < V_0 < \frac{1}{6}$ with V_0 close to $\frac{1}{6}$. Since $\frac{dV}{dt} = -y^2 \le 0$, it follows that the origin is Lyapunov stable.

59. (a) $(0, 0)$ is an equilibrium point of this nonlinear system. At $(0, 0)$, the linearized system is $\dot{x} = \dot{y} = 0$. Both eigenvalues are zero. Consider a Lyapunov function in the form $V = axy^2 - x^3$, where $V(0, 0) = 0$. Since $\frac{dV}{dt} = 3(1 - a)x^2y^2 - ay^4 - 3x^4$. With $a = 1$, $\dot{V} = -y^4 - 3x^4 < 0$, $V = x(y^2 - x^2) = 0$ when $x = 0$ or $y = \pm x$ so that V changes sign six times on any circle with center at the origin. In every neighborhood of $(0, 0)$, there is at least one point where V and $\dot{V}$ have the same sign. Thus, the origin is unstable.

(b) The Jacobian at the equilibrium point, $(0, 0)$ is

$$J = \begin{pmatrix} 2 & 0 \\ 1 & -1 \end{pmatrix},$$

which has eigenvalues 2 and -1 with the corresponding eigenvectors $(3, 1)^T$ and $(0, 1)^T$. Hence, the local stable manifold is the y-axis and the local unstable manifold is the line $3y = x$. The solution is

$$x(t) = \frac{2a}{a + 2e^{-t}}, \qquad y(t) = e^{-t}\left[b + 2e^{-t} - \frac{4}{a}\log(2 + ae^t)\right],$$

where $a \ne 0$ and b are constants. There is also the obvious solution $x = 0$, $y = be^{-t}$, y-axis, which yields the local stable manifold and the global stable manifold.

61. There are three equilibrium points, one at $x = y = z = 0$ and two at $x = 27$, $y = z = \pm 6\sqrt{2}$, each of which is unstable. The two equilibrium points away from the origin each have a two-dimensional unstable manifold, associated with complex eigenvalues, and thus oscillatory nature, and a one-dimensional stable manifold. The Lorenz system has very interesting dynamics. Solutions bounce back and forth between the equilibrium points away from the origin, continually being attracted toward an equilibrium point along a trajectory close to the stable manifold and then spiraling away close to the unstable manifold for the solution with $x(0) = y(0) = z(0) = 1$.

Bibliography

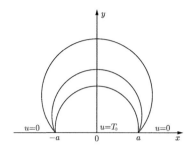

The following bibliography is not, by any means, a complete one for the subject. For the most part, it consists of books and papers to which reference is made in the text. Many other selected books and papers related to material in this book have been included so that they may serve to stimulate new interest in future study and research.

Adams, R.A., and Fournier, P., *Sobolev Spaces*, Pure and Applied Mathematics, Vol. 65 (Second Edition), Academic Press, 2003.

Antosik, P., and Swartz, C., *Matrix Methods in Analysis*, Springer-Verlag, Berlin, 1985.

Arrowsmith, D.K., and Place, C.M., *An Introduction to Dynamical Systems*, Cambridge University Press, Cambridge, 1992.

Balakrishnan, A.V., *Applied Functional Analysis*, Springer-Verlag, New York, 1976.

Balakrishnan, A.V., *Introduction to Optimization Theory in a Hilbert Space*, Springer-Verlag, New York, 1971.

Banach, S., *Théorie des opérations linéaires*, Chelsea, New York, 1955.

Battle, G., A block spin construction of ondelettes, Part I: Lemarié functions, *Commun. Math. Phys.* **110** (1987), 601–615.

Berkovitz, L., *Optimal Control Theory*, Springer-Verlag, New York, 1975.

Beylkin, G., On the representation of operators in bases of compactly supported wavelets, *SIAM J. Numer. Anal.* **29** (1992), 1716–1740.

Beylkin, G., Coifman, R., and Rokhlin, V., Fast wavelet transforms and numerical algorithms, *Comm. Pure Appl. Math.* **44** (1991), 141–183.

Cheney, E.W., *Introduction to Approximation Theory*, McGraw-Hill, New York, 1966.

Chui, C.K., *An Introduction to Wavelets*, Academic Press, Boston, 1992.

Chui, C.K., and Wang, J.Z., A cardinal spline approach to wavelets, *Proc. Amer. Math. Soc.* **113** (1991), 785–793.

Chui, C.K., and Wang, J.Z., On compactly supported spline wavelets and a duality principle, *Trans. Amer. Math. Soc.* **330** (1992), 903–915.

Coddington, E.A., and Levinson, N., *Theory of Ordinary Differential Equations*, McGraw-Hill, New York, 1955.

Cohen, A., *Biorthogonal Wavelets, in Wavelets — A Tutorial in Theory and Applications*, Academic Press (1992b), 123–152.

Cohen, A., Daubechies, I., and Feauveau, J., Biorthogonal basis of compactly supported wavelets, *Comm. Pure and Appl. Math.* **45** (1992a), 485–560.

Coifman, R.R., Jones, P., and Semmes, S., Two elementary proofs of the L^2 boundedness of Cauchy integrals on Lipschitz curves, *J. Amer. Math. Soc.* **2** (1989), 553–564.

Coifman, R.R., Meyer, Y., and Wickerhauser, M.V., Wavelet analysis and signal processing, in: *Wavelets and Their Applications* (Eds. Ruskai *et al.*) (1992a), 153–178; and *Size Properties of Wavelet Packets* (1992b), 453–470, Jones and Bartlett, Boston.

Curtain, R.F., and Pritchard, A.J., *Functional Analysis in Modern Applied Mathematics*, Academic Press, New York, 1977.

Daubechies, I., *Ten Lectures on Wavelets*, CBMS-NSF Regional Conference Series in Applied Math., Vol. 61, SIAM, Philadelphia, PA, 1992.

Daubechies, I., The wavelet transform, time-frequency localization and signal analysis, *IEEE Trans. Inform. Theory* **36** (1990), 961–1005.

Daubechies, I., Orthogonal bases of compactly supported wavelets, *Commun. Pure Appl. Math.* **41** (1988), 909–996.

Daubechies, I., Grossmann, A., and Meyer, Y., Painless nonorthogonal expansions, *J. Math. Phys.* **27** (1986), 1271–1283.

Debnath, L., *Integral Transforms and Their Applications* (Second Edition), CRC Press, Boca Raton, FL, 2005.

Debnath, L., *Nonlinear Partial Differential Equations for Scientists and Engineers* (Second Edition), Birkhäuser, Boston, 2005.

Debnath, L., *Wavelet Transforms and Their Applications*, Birkhäuser, Boston, 2002.

De Boor, C., *Approximation Theory*, Proceedings of Symposia in Applied Mathematics, Vol. 36, American Mathematical Society, Providence, RI, 1986.

Dieudonné, J., *Foundations of Modern Analysis*, Academic Press, New York, 1969.

Dirac, P.A.M., *The Principles of Quantum Mechanics* (Fourth Edition), Oxford University Press, Oxford, 1958.

Duffin, R.J., and Schaeffer, A.C., A class of nonharmonic Fourier series, *Trans. Amer. Math. Soc.* **72** (1952), 341–366.

Dunford, N., and Schwartz, J.T., *Linear Operators, Part I, General Theory*, Interscience, New York, 1958.

Friedman, A., *Foundations of Modern Analysis*, Dover, New York, 1982.

Gabor, D., Theory of communications, *J. Inst. Electr. Eng. London* **93** (1946), 429–457.

Garabedian, P.R., *Partial Differential Equations*, Wiley, New York, 1964.

Glimm, J., and Jaffe, A., *Quantum Physics* (Second Edition), Springer-Verlag, New York, 1987.

Gould, S.H., *Variational Methods for Eigenvalue Problems*, Toronto University Press, Toronto, 1957.

Grossmann, A., and Morlet, J., Decomposition of Hardy functions into square integrable wavelets of constant shape, *SIAM J. Math. Anal.* **15** (1984), 723–736.

Guckenheimer, J., and Holmes, P.J., *Nonlinear Oscillations, Dynamical Systems and Bifurcation of Vector Fields*, Springer-Verlag, New York, 1983.

Halmos, P.R., *Measure Theory*, Springer-Verlag, New York, 1974.

Hilbert, D., *Grundzüge einer allgemeinen Theorie der linearen Integralgleichungen*, Leipzig, 1912.

Holschneider, M., *Wavelets, An Analysis Tool*, Oxford University Press, Clarendon Press, Oxford, 1995.

Hutson, V., and Pym, J.S., *Applications of Functional Analysis and Operator Theory*, Academic Press, New York, 1980.

Iooss, G., and Joseph, D.D., *Elementary Stability and Bifurcation Theory*, Springer-Verlag, New York, 1981.

Jacuch, J.M., *Foundations of Quantum Mechanics*, Addison-Wesley, Reading, MA, 1968.

Jones, D.S., *Generalized Functions*, McGraw-Hill, New York, 1966.

Kantorovich, L.V., and Akilov, G.P., *Functional Analysis in Normed Spaces*, Pergamon Press, London, 1964.

Kanwal, R.P., *Generalized Functions* (Second Edition), Birkhäuser, 1998.

Keller, J.B., and Antman, S., *Bifurcation Theory and Nonlinear Eigenvalue Problems*, W.A. Benjamin, New York, 1969.

Kolmogorov, A.N., and Fomin, S.V., *Elements of the Theory of Functions and Functional Analysis*, Vol. 1, Graylock Press, Rochester, New York, 1957; Vol. 2, Graylock Press, Albany, New York, 1961.

Kolmogorov, A.N., and Fomin, S.V., *Introductory Real Analysis*, Prentice-Hall, New York, 1970.

Kreyn, S.G., *Functional Analysis*, Foreign Technology Division WP-AFB, Ohio, 1967.

Kreyszig, E., *Introductory Functional Analysis with Applications*, Wiley, New York, 1978.

Landau, L.D., and Lifshitz, E.M., *Quantum Mechanics, Non-relativistic Theory*, Pergamon Press, London, 1959.

Lax, P.D., and Milgram, A.N., Parabolic equations, contribution to the theory of partial differential equations, *Ann. of Math. Stud.*, No. 33, Princeton, 1954, 167–190.

Lemarié, P.G., Bases d'ondelettes sur les groupes de Lie stratifiés, *Bull. Soc. Math. France* **117** (1989), 211–232.

Lemarié, P.G., Ondelettes à localisation exponentielle, *J. Math. Pures Appl.* **67** (1988), 227–236.

Lemarié, P.G., and Meyer, Y., Ondellettes et bases hilbertiennes, *Revista Mat. Iberoamer.* **2** (1986), 1–18.

Lions, J.L., and Stampacchia, G., Variational inequalities, *Comp. Pure Appl. Math.* **20** (1967), 493–519.

Liusternik, L.A., and Sobolev, V.J., *Elements of Functional Analysis* (Third English Edition), Hindustan Publishing Co., New Delhi, 1974.

Lorenz, E.N., Deterministic non-periodic flows, *J. Atmos. Sci.* **20** (1963), 130–141.

Luenberger, D.G., *Optimization by Vector Space Methods*, Wiley, New York, 1969.

Mackey, G.W., *The Mathematical Foundations of Quantum Mechanics*, W.A. Benjamin, New York, 1963.

MacNeille, H.M., A unified theory of integration, *Proc. Nat. Acad. Sci. USA* **27** (1941), 71–76.

Mallat, S., Multiresolution approximations and wavelet orthonormal basis of $L^2(\mathbb{R})$, *Trans. Amer. Math. Soc.* **315** (1989a), 69–88.

Mallat, S., A theory for multiresolution signal decomposition: The wavelet representation, *IEEE Trans. Pattern Anal. Machine Intell.* **11** (1989b), 678–693.

Mallat, S., Multiresolution representation and wavelets, Ph.D. Thesis, University of Pennsylvania, 1988.

Merzbacher, E., *Quantum Mechanics* (Second Edition), Wiley, New York, 1961.

Meyer, Y., *Ondelettes, fonctions splines et analyses graduées*, Lecture Notes, University of Torino, Italy, 1986.

Meyer, Y., *Wavelets and Operators*, Cambridge University Press, Cambridge, 1992.

Meyer, Y., *Wavelets, Algorithms and Applications*, SIAM, Philadelphia, PA, 1993.

Mikusiński, J., A theorem on vector matrices and its applications in measure theory and functional analysis, *Bull. Acad. Polon. Sci. Sér. Sci. Math. Astronom. Phys.* **18** (1970), 193–196.

Mikusiński, J., *Bochner Integral*, Birkhäuser, Basel, 1978.

Mikusiński, P., and Taylor, M.D., *An Introduction to Multivariable Analysis: From Vector to Manifold*, Birkhäuser, 2001.

Myint-U, T., and Debnath, L., *Partial Differential Equations for Scientists and Engineers* (Fourth Edition), Birkhäuser, Boston, 2005.

Neumann, J.V., *Mathematical Foundations of Quantum Mechanics*, Princeton University Press, Princeton, 1955.

Newland, D.E., Harmonic wavelet analysis, *Proc. Roy. Soc. London Ser. A* **443** (1993), 203–225.

Newland, D.E., Harmonic and musical wavelet, *Proc. Roy. Soc. London Ser. A* **444** (1994), 605–620.

Reed, M., and Simon, B., *Methods of Modern Mathematical Physics*, Vol. 1, *Functional Analysis*, Academic Press, New York, 1972.

Riesz, F., and Sz-Nagy, B., *Functional Analysis* (Second Edition), Frederick Ungar, New York, 1955.

Rivlin, T.J., *An Introduction to the Approximation of Functions*, Dover, New York, 1969.

Roach, G.F., *Green's Functions* (Second Edition), Cambridge University Press, Cambridge, 1982.

Sattinger, D.H., *Topics in Stability and Bifurcation Theory*, Lecture Notes in Mathematics, Vol. 309, Springer-Verlag, New York, 1973.

Schechter, M., *Modern Methods in Partial Differential Equations*, McGraw-Hill, New York, 1977.

Schwartz, L., *Théorie des distributions*, Vols. I and II, Herman and Cie, Paris, 1950, 1951.

Shilov, G.E., *Generalized Functions and Partial Differential Equations*, Gordon and Breach, New York, 1968.

Sobolev, S.L., *Partial Differential Equations of Mathematical Physics*, Pergamon Press, London, 1964.

Stakgold, I., *Boundary Value Problems of Mathematical Physics*, Macmillan, New York, 1968.

Swartz, C., *An Introduction to Functional Analysis*, Marcel Dekker, New York, 1992.

Taylor, A.E., *Introduction to Functional Analysis*, Wiley, New York, 1958.

Tricomi, F.G., *Integral Equations*, Interscience, New York, 1957.

Walter, G., *Wavelets and Other Orthogonal Systems with Applications*, CRC Press, Boca Raton, FL, 1994.

Yosida, K., *Functional Analysis* (Fourth Edition), Springer-Verlag, New York, 1974.

Young, L.C., *Calculus of Variations and Optimal Control Theory*, W.B. Saunders Company, Philadelphia, PA, 1969.

Zayed, A.I., *Advances in Shannon's Sampling Theory*, CRC Press, Boca Raton, FL, 1993.

Zemanian, A.H., *Distribution Theory and Transform Analysis*, McGraw-Hill, New York, 1965.

Ziemer, W.P., *Weakly Differentiable Functions*, Springer-Verlag, New York, 1989.

Index

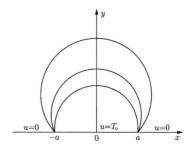

Printed and bound by CPI Group (UK) Ltd, Croydon, CR0 4YY

08/05/2025

01864790-0005